Lecture Notes in Bioinformatics 6293

Edited by S. Istrail, P. Pevzner, and M. Waterman

Subseries of Lecture Notes in Computer Science

Vincent Moulton Mona Singh (Eds.)

Algorithms in Bioinformatics

10th International Workshop, WABI 2010
Liverpool, UK, September 6-8, 2010
Proceedings

Springer

Series Editors

Sorin Istrail, Brown University, Providence, RI, USA
Pavel Pevzner, University of California, San Diego, CA, USA
Michael Waterman, University of Southern California, Los Angeles, CA, USA

Volume Editors

Vincent Moulton
University of East Anglia
School of Computing Sciences
Norwich, NR4 7TJ, UK
E-mail: vincent.moulton@cmp.uea.ac.uk

Mona Singh
Princeton University
Lewis-Sigler Institute for Integrative Genomics
Department of Computer Science
Princeton, NJ 08544, USA
E-mail: mona@cs.princeton.edu

Library of Congress Control Number: 2010932425

CR Subject Classification (1998): J.3, F.1, I.2, F.2.2, H.2.8, E.1

LNCS Sublibrary: SL 8 – Bioinformatics

ISSN	0302-9743
ISBN-10	3-642-15293-7 Springer Berlin Heidelberg New York
ISBN-13	978-3-642-15293-1 Springer Berlin Heidelberg New York

springer.com

Printed in Germany

Typesetting: Camera-ready by author, data conversion by Scientific Publishing Services, Chennai, India
Printed on acid-free paper 06/3180

Preface

We are pleased to present the proceedings of the 10th Workshop on Algorithms in Bioinformatics (WABI 2010) which took place in Liverpool, UK, September 6–8, 2010. The WABI 2010 workshop was part of the four ALGO 2010 conference meetings, which, in addition to WABI, included ESA, ATMOS, and WAOA. WABI 2010 was hosted by the University of Liverpool Department of Computer Science, and sponsored by the European Association for Theoretical Computer Science (EATCS) and the International Society for Computational Biology (ISCB). See `http://algo2010.csc.liv.ac.uk/wabi/` for more details.

The Workshop in Algorithms in Bioinformatics highlights research in algorithmic work for bioinformatics, computational biology and systems biology. The emphasis is mainly on discrete algorithms and machine-learning methods that address important problems in molecular biology, that are founded on sound models, that are computationally efficient, and that have been implemented and tested in simulations and on real datasets. The goal is to present recent research results, including significant work-in-progress, and to identify and explore directions of future research.

Original research papers (including significant work-in-progress) or state-of-the-art surveys were solicited for WABI 2010 in all aspects of algorithms in bioinformatics, computational biology and systems biology. In response to our call, we received 83 submissions for papers and 30 were accepted. In addition, WABI 2010 hosted distinguished lectures by Eran Halperin, of Tel Aviv University and ICSI, Berkeley, and, together with ESA, Paolo Ferragina of University of Pisa. We would like to sincerely thank the authors of all submitted papers and the conference participants. We also thank the Program Committee and their sub-referees for their hard work in reviewing and selecting papers for the workshop.

We would espcially like to thank Bernard Moret and Tandy Warnow for all of their advice and support in carrying out the role of being Co-chairs.

Thanks once again to all who participated in making WABI's 10th anniversary such a success. For us it has been an exciting and rewarding experience.

June 2010

Vincent Moulton
Mona Singh

Organization

Program Committee

Tatsuya Akutsu	Kyoto University, Japan
Bonnie Berger	MIT, USA
Tanya Berger-Wolf	University of Illinois, USA
Mathieu Blanchette	McGill University, Canada
Sebastian Böcker	University of Jena, Germany
Magnus Bordewich	University of Durham, UK
Mike Brudno	University of Toronto, Canada
Philipp Bucher	EPFL, Switzerland
Benny Chor	Tel Aviv University, Israel
Anne Condon	University of British Columbia, Canada
Lenore Cowen	Tufts University, USA
Keith Crandall	Brigham Young University, USA
Bhaskar Das Gupta	University of Illinois, USA
Nadia El-Mabrouk	University of Montreal, Canada
Liliana Florea	University of Maryland, USA
Olivier Gascuel	University of Montpellier, France
Barbara Holland	Massey University, New Zealand
Katharina Huber	University of East Anglia, UK
Daniel Huson	University of Tuebingen, Germany
Lydia Kavraki	Rice University, USA
Junhyong Kim	University Penn, USA
Carl Kingsford	University of Maryland, USA
Mehmet Koyuturk	Case Western, USA
Jens Lagergren	KTH, Sweden
Chris Langmead	CMU, USA
Ryan Lilien	University of Toronto, Canada
Ion Mandiou	University of Connecticut, USA
Joao Meidanis	Campinas University, Brazil
Satoru Miyano	Tokyo University, Japan
Bernard M.E. Moret	Swiss Federal Institute of Technology, Switzerland
Burkhard Morgenstern	University of Göttingen, Germany
Vincent Moulton	University of East Anglia, UK, Co-chair
Gene W. Myers	Janelia Farms, USA
Mihai Pop	University of Maryland, USA
Teresa Przytycka	NIH, USA
Cenk Sahinalp	Simon Fraser University, USA
David Sankoff	University of Montréal, Canada

Table of Contents

Biomolecular Structure: RNA, Protein and Molecular Comparison

Comparative Genomics

Haplotype and Genotype Analysis

Phylogenetics

Sequences, Strings and Motifs

A Worst-Case and Practical Speedup for the RNA Co-folding Problem Using the *Four-Russians* Idea

Yelena Frid and Dan Gusfield

Department of Computer Science, U.C. Davis

Abstract. The computational formulation for finding the optimal simultaneous alignment and fold *(optimal Co-fold)* of RNA sequences was first introduced by Sankoff in 1985. Since then the importance of Co-Folding has grown as conservation of structure and its relationship to function have been widely observed in RNA. For two sequences, the computation time of Sankoff's Algorithm is $\theta(N^6)$. Existing literature on cofolding attempts to improve efficiency through simplifying the original problem formulation.

We present here a practical and worst-case speed up using the Four-Russians method, without placing any added constraints on the types of alignments or folds allowed. Our algorithm, *Fast Cofold*, finds the optimal Co-fold in $O(N^6/\log(N^2))$-time, a speedup which is observed in practice.

Because the solution matrix produced by our algorithm is identical to the one produced by the Sankoff algorithm, the contribution of the algorithm lays not only in its standalone practicality but also in the ability to implement it alongside heuristic speed ups leading to even greater reductions in time.

1 Introduction

The algorithmic goal of finding alignments together with structure prediction is motivated by the understanding that RNA structure helps to determine function. It has been observed particularly that in eukaryotic genomes ncRNA (Non coding RNA) function is seen more clearly from conserved structure then from alignment alone [17,13,16]. In trRNAs, srpRNA and tRNAs there are also observed relationships between structure and function. Alignment methods that take structure into account can also allow biologists to identify non-functional transcripts as well as structure motifs for RNA[4].

Algorithms that produce both folds and alignments can be classified into three groups:(1) folding methods that use aligned sequences as input to find a common structure [11,18,12,15];(2) algorithms that compute structure and then align [8]; and (3) *cofolding* algorithms i.e. those that do alignment and folding simultaneously [14,7,3,9,5,19,1].

Sankoff's Algorithm was the first dynamic programming algorithm to simultaneously find alignment and RNA folding for a set of sequences [14]. For L

V. Moulton and M. Singh (Eds.): WABI 2010, LNBI 6293, pp. 1–12, 2010.

sequences of equal length N the algorithm required O(N^{3L})-time and O(N^{2L})-space[14].

Because of the large run-time and space requirement of Sankoff's solution, his original problem formulation has often been restricted to allow for greater efficiency, but at the cost of not solving the original problem[7,3,9,5]. Problem simplification and restrictions on the recurrences were explored by [7,3,9,5]. FoldAlign[7] removed the possibility of a branch in the recurrences, achieving an $O(N^4)$ time algorithm. Dynalign [9] constrained the distance d between aligned nucleotides, thereby reducing the computation time to $O(N^3 * d^3)$. Eddy et al. used covariance models to achieve $O(N^3 * r)$ run-time where $|r|$ is the number of states in the model [5]. Consan[3] used pairSCFG to constrain the algorithm leading to an asymptotic time between $O(N^6)$ and $O(N^3)$.

Ziv-Ukelson et al. introduced a time reduction algorithm that retained the Sankoff style recurrence. Based on some simple pruning of the branching points, the algorithm was able to achieve practical time reduction and asymptotic bound $O(N^4 * K)$[1] where K is constrained by $N \leq K \leq N^2$ and converges to $O(N)$ when assuming the polymer folding model[19]. An O(N^3+Z)space algorithm was developed based on the pruning formulation by Backofen et al. [2]. While Z ranges from N^2 to N^4, in practice it was seen to be lower then N^4.

Surprisingly, the Four Russians method, which is widely used and known to speed up dynamic programming, has not previously been applied to the cofolding problem. Traditionally the Four-Russians method performs some preprocessing for a subset of all possible inputs and then computes using that preprocessing. We take advantage of the idea discussed in Frid and Gusfield [6] interleaving the computation and preprocessing to create a speedup.

The algorithm as presented by Sankoff does not easily lend itself to subset precomputation, and we reorganized the order of evaluation. We also create a function that choose the optimal subset size that leads to the greatest speedup.

The ***Fast Cofold algorithm*** presented in this paper formulates a Four Russians speed up to the Sankoff's original cofolding problem and reduces the asymptotic computation time to $O(N^6/log(N^2))$.

2 Sankoff Algorithm for Two Sequences

Let $s1$ and $s2$ be two RNA sequences over the four-letter alphabet {A,U,C,G}, where each letter in the alphabet represents an RNA *nucleotide*. We are interested in finding the optimal cofold or optimal alignment and common structure of the two sequences, using a scoring scheme that accounts for alignment, folding, and substitutions that conserve structure. We will make use of a modified version of Sankoff's Algorithm for cofolding as described by Ziv-Ukelson et al. [19] . That version restricts the original algorithm to computing the optimal cofold of two sequences($L = 2$) by maximizing the pair contributions to the fold instead of minimizing energy of a fold.

[1] Still $O(N^6)$ time in terms of the length N.

The basic optimal Alignment problem. Define a scoring scheme α such that $\alpha(x, y)$ is the score for substituting nucleotide x for y where $x, y \in \{$A,U,C,G,$-\}$. Let $s1$ and $s2$ be two RNA sequences of length N over the four-letter alphabet {A,U,C,G}. ***Alignment sequences*** $s1'$ and $s2'$ of $s1, s2$ are created by inserting gaps or '_' into each sequence such that $|s1'| = |s2'|$ and $\{\neg\exists i | s1'[i] = s2'[i] = \text{'\_'}\}$.

Let $AligScore = \sum_{i=0}^{|s1'|} \alpha(s1'[i], s2'[i])$ be the score associated with the alignment sequences $s1'$,$s2'$.

The *optimal alignment problem*: Given $s1$ and $s2$ find alignment sequences $s1'$ and $s2'$ for which $AligScore$ is maximum. Alignments can also be enhanced by creating more complicated scoring schemes, for example adding larger penalties for introducing gaps versus extending gaps.

The basic optimal Folding Problem. We present below the maximum matching folding problem as introduced by Nussinov et al.[10]. However, it is slightly modified to incorporate sequences that have gaps as characters. A nucleotide pair (x, y) is a ***permitted pair*** if (x, y) or $(y, x) \in \{$(A,U), (C,G), (G,U), (_,_)$\}$. For a given sequence *seq* of length N over the alphabet {A,U,C,G,_} we define the ***folding set*** M as a set containing disjoint permitted pairs of sites in sequence *seq*, such that for any i, i', j, j' where $i < i' < j < j'$, M does not contain both pairs (i, j) and (i', j'). Let β be a scoring scheme such that $\beta(x, y)$ returns the contribution of pairing nucleotide at site x with the nucleotide at site y. The basic scoring scheme sets $\beta(x, y)$ equal to one if (x, y) is a permitted pair with $|y - x| > d$ and set $\beta(x, y)$ to zero otherwise. However, richer scoring schemes as those formulated by the cofold problem, allow non-permitted pairs to match. Let $foldScore$ be the score associated with a folding set M where $foldScore = \sum_{(i,j)\in M} \beta(seq[i], seq[j])$.

The *optimal folding problem*: Find the set M for which $foldScore$ is maximum.

The Optimal Cofolding Problem. The cofold of $s1$ and $s2$ consists of alignment sequences $s1'$, $s2'$, and a *folding set M* [19]. Let $cofoldScore = \sum_{i=0}^{|s1'|} \alpha(s1'[i], s2'[i]) + \sum_{(i,j)\in M} \beta(s1'[i], s1'[j]) + \beta(s2'[i], s2'[j]) + \tau(s1'[i], s1'[j], s2'[i], s2'[j])$

where $\tau(s1'[i], s1'[j], s2'[i], s2'[j])$ is a score for aligning s1'[i] for s2'[i] and substituting s1'[j] for s2'[j] taking into account compensator mutations that preserve structure. In general all the scoring schemes α, β and τ can be modified to fit richer biological models.

The *optimal cofold problem*: Find the M and alignment sequences $s1'$ and $s2'$ for which the $cofoldScore$ is maximum.

Recurrences for finding the Optimal Co-fold. Let $\mathbf{S[i,j;k,l]}$ contain the score for the optimal cofold of subsequence $s1(i..j)$[2] and subsequence $s2(k...l)$. S is therefore a four dimensional matrix and optimal cofold score for the entire sequence is equal to $S[1,n;1,n]$.

We make use of the following recurrences derived from Sankoff's algorithm in [19].

$$S[i,j;k,l] = \max \begin{cases} \textbf{Rule a.} & S[i+1,j;k,l] + \alpha(s1[i],'-') \\ \textbf{Rule b.} & S[i,j;k+1,l] + \alpha(s2[k],'-') \\ \textbf{Rule c.} & S[i,j-1;k,l] + \alpha(s1[j],'-') \\ \textbf{Rule d.} & S[i,j;k,l-1] + \alpha(s2[l],'-') \\ \textbf{Rule e.} & S[i+1,j;k+1,l] + \alpha(s1[i],s2[k]) \\ \textbf{Rule f.} & S[i,j-1;k,l-1] + \alpha(s1[j],s2[l]) \\ \textbf{Rule g.} & S[i+1,j-1;k,l] + \beta(s1[i],s1[j]) + \alpha(s1[i],'-') + \alpha(s1[j],'-') \\ \textbf{Rule h.} & S[i,j;k+1,l-1] + \beta(s2[k],s2[l]) + \alpha(s2[k],'-') + \alpha(s2[l],'-') \\ \textbf{Rule i.} & S[i+1,j-1;k+1,l-1] + \beta(s1[i],s1[j]) + \beta(s2[k],s2[l]) + \tau(s1[i],s1[j],s2[k],s2[l]) \\ \textbf{Rule j.} & \max_{i \le m \le j \wedge k \le n \le l}\{S[i,m;k,n] + S[m+1,j;n+1,l]\} \end{cases} \quad (1)$$

Rules a to *d* account for the possibility of placing gaps and not adding any new pair to folding set M. *Rules e* and f account for the possibility of aligning either the right, or left end of the sequences but not adding any pair to the folding set M. Rules g and h account for the possibility of adding a pair to folding set M where the characters of the pair in one sequence are aligned with inserted gaps in the other subsequence [3]. Rule i accounts for adding a pair to folding set M and aligning both ends of the sequences.

Let us call Rule j the Branch Rule. The Branch Rule covers the case where the optimal solution comes from *breaking up* $s1$ at index m, *breaking up* $s2$ at index n and cofolding $s1(i..m)$ with $s2(k..n)$ and $s1(m+1..j)$ with $s2(n+1..l)$ or $S[i,m;k,n] + S[m+1,j;n+1,l]$. The branching rule looks at all combination of m and n and finds the combination which maximizes $S[i,m;k,n] + S[m+1,j;n+1,l]$.

We will call every break up index a *branch point* and call S[i,m;k,n] the ***head*** of the branch and S[m+1,j;n+1,l] the ***tail***. We will call each possible $\{m,n\}$ a ***branch point combination***.

2.1 Cofold Algorithm

The S matrix can be computed by an algorithm that goes through all the possible subsequences of $s1$ and $s2$ and finds the optimal cofold for each pair. There are $O(N^4)$ such pairs of subsequences . For each pair a *branch_function* computes the Branching Rule in an $O(N^2)$ time, searching through the possible branch point combinations.

As shown in *Cofold Algorithm* below, the recurrences are evaluated in increasing order of the right endpoints of $s1$ and $s2$.

[2] Notational note: All subsequences will be represented as seq(a..b) where a is the starting index of the subsequence and b is the index of the of the final character in that subsequence.

[3] Constraint: *Rule g* is applicable only if $(s1(i+1), s1(j-1)) \in$ folding set M. *Rule h* is applicable only if $(s2(k+1), s2(l-1)) \in$ folding set M.

Cofold Algorithm

```
for j=1 to N do
  for l=1 to N do
    for i=j − 1 to 1 do
      for k=l − 1 to 1 do
        const_oper_max=max(Rules a to i)
        branch_max= branch_function(i, j; k, l)
        S[i,j;k,l]=max( const_oper_max, branch_max)
```

branch_function(i,j;k,l)::

```
  for m=j − 1 to i + 1 do
    for n=l − 1 to k + 1 do
      cur_max= max( cur_max,S[i,m;k,n]+S[m+1,j;n+1,l] )
  return cur_max
```

This algorithm is correct based on the following facts.

Fact 1. During computation of $S[i, j; k, l]$ we have already computed the optimal solution for $S[i', j'; k', l']$ where $j' \leq j - 1$, $l' \leq l - 1$ and($i' < j'$ and $k' < l'$.)

Fact 2. For a particular i and k at the time $S[i, j; k, l]$ is computed all $S[i', j; k', l]$ have been computed where $i' > i$ and $k' > k$.

It is clear that the Cofold Algorithm takes $O(N^6)$ time operations to compute the solution matrix S. We present a method that produces the identical solution matrix S as the the above $O(N^6)$ time algorithm. Moreover, we will reduce the asymptotic time to $O(N^6/log_b(N^2))$, by speeding up the *branch_function* through the adaptation of the Four-Russians method.

3 Conceptually Speeding Up the *Branch_function*

For $s1(i..j)$ and $s2(k..l)$ let $\{m^*, n^*\}$ be the branch point combination that maximizes $S[i, m; k, n] + S[m + 1, j; n + 1, l]$ over all possible branch point combinations $\{m, n\}$ where m belongs to the set $\{i + 1, i + 2, i + 3, ...j - 1\}$ and n belongs to the set $\{k + 1, k + 2, ...l - 1\}$.

Overall there are there are $O(N^2)$ branch points combinations to evaluate i.e. $\{i + 1, k + 1\}\{i + 1, k + 2\}...\{i + 2, k + 1\}...\{j - 1, l - 1\}$. Therefore, the time to a compute branch function for a fixed i, j, k, and l is $O(N^2)$.

The *Four Russians* method applied to the branch function lowers the computation to $O(N^2/q^2)$. The value of q will play an important role in the speedup and will be examined in the time analysis section.

We conceptually divide all the possible branch points of $s1$ into sets of size q called **Mgroups**. Let $M_{g=0}$ be the first such group that contains the possible branch points $\{0, 1, ...q - 1\}$, let $M_{g=1}$ contain $\{q, ...2q - 1\}$ and so on ... the last group of which $M_{g=n/q} = \{n - q, ...n - 1\}$. We will also conceptually divide all the possible branch points of $s2$ into sets of size q called **Ngroups** such that $N_{g'=0} = \{0, 1, ..., q - 1\}$ and so on. In general:

$$M_g = \{g * q, g * q + 1, ...g * q + q - 1\}$$
$$N_g = \{g' * q, g' * q + 1, ...g' * q + q - 1\}$$

Let $\{\mathbf{m}^*_{\mathbf{g}}, \mathbf{n}^*_{\mathbf{g}'}\}$ be the branch point combination that maximizes the sum $S[i, m; k, n] + S[m + 1, j, n + 1, L]$ such that m $\in M_g$ and n $\in N_{g'}$.

Let $\{\mathbf{M}^*, \mathbf{N}^*\}$ be equal to the pair $\{m^*_g, n^*_{g'}\}$ where $S[i, m^*_g, k, n^*_{g'}] + S[m^*_g + 1, j, n^*_{g'}+1, l]$ is maximum for all M_g,$N_{g'}$ sets g, g' in $\{\frac{i+1}{q}, ..., \frac{j-1}{q}\}$, $\{\frac{k+1}{q}, ..., \frac{l-1}{q}\}$ respectively.

Fact 3. $\{m^*, n^*\} = \{M^*, N^*\}$.
Based on *Fact 3.* we change the `Branch Rule` from

$$\max_{\{\{m,n\}|m\in\{i+1,...,j-1\}\wedge n\in\{k+1,...,l-1\}\}} \mathtt{S}[\mathtt{i}, \mathtt{m}; \mathtt{k}, \mathtt{n}] + \mathtt{S}[\mathtt{m} + 1, \mathtt{j}, \mathtt{n} + 1, \mathtt{l}]$$

to

$$\max_{\{\{m^*_g,n^*_{g'}\}|g\in\{\frac{i+1}{q},...,\frac{j-1}{q}\}\wedge g'\in\{\frac{k+1}{q},...,\frac{l-1}{q}\}\}} \mathtt{S}[\mathtt{i}, \mathtt{m}^*_{\mathtt{g}}, \mathtt{k}, \mathtt{n}^*_{\mathtt{g}'}] + \mathtt{S}[\mathtt{m}^*_{\mathtt{g}} + 1, \mathtt{j}, \mathtt{n}^*_{\mathtt{g}'} + 1, \mathtt{l}]$$

Now assume there is a precomputed table R^2 that returns $\{m^*_g, n^*_{g'}\}$ in $O(1)$ time for any i, j, k, l. Such a table would reduce computation of the *branch_function* to $O(N^2/q^2)$ time.

3.1 Implementing the Branch Function with Table R^2

Encoding. For a particular j, l, let $V_{g,g'}$ be a q by q matrix that contains the possible tails for the branches in $M_g, N_{g'}$. Where $V_{g,g'}(1, 1) = S[gq+1, j; g'q+1, l]$ ... $V_{g,g'}(1, q) = S[gq + q, j; g'q + q, l]$ and so on.

More precisely $V_{g,g'}(m+1-gq, n+1-g'q) = S[m+1, j; n+1, l]$ (see Figure 1).

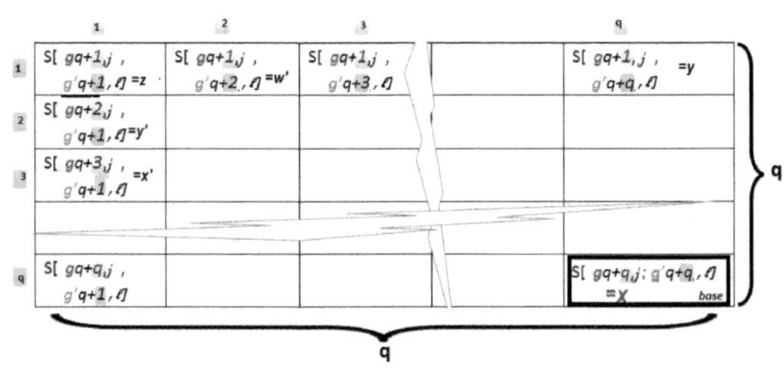

Fig. 1. The example V matrix shown in figure 1 is for Mgroup g and Ngroup g' and some j,l. The integers x,y,z,y', x' are example values in $V_{g,g'}$. These values equal the designated values of the S matrix. The *base*= $S[gq + q, j; g'q + q, l]$=x.

Optimal cofold scores stored in $S[\mathbf{i}, j; k, l]$ and $S[\mathbf{i} + \mathbf{1}, j; k, l]$ can differ by the effect of only one more nucleotide i.e. $s1[i]$. Therefore we can observe that for the scoring scheme and the recurrences of the Colfold Algorithm, $|S[i, j; k, l] - S[i+1, j; k, l]|$ belongs to a finite set of differences D, where D is the set of scores created by combinations of scores from the sets α, β, and τ. The cardinality or size of D is $O(1)$ as a function of N. Clearly, $|S[i, j; k - 1, l] - S[i, j; k, l]|$ also belongs to D.

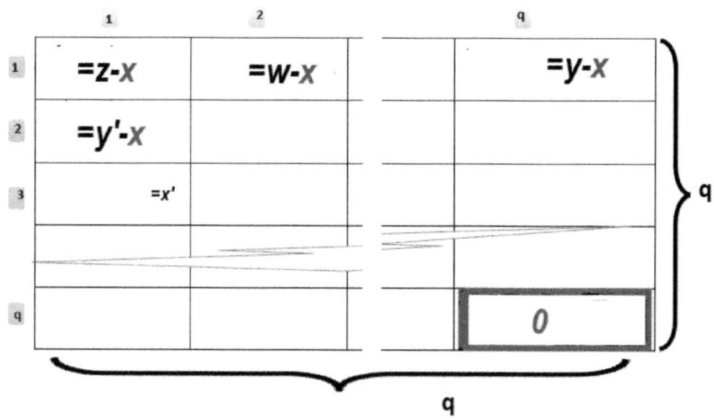

Fig. 2. The corresponding E matrix to $V_{g,g'}$ in figure 1. A few example values are shown i.e. $E_{g,g'}[1,1]$=$V_{g,g'}[1,1]$-base=$z-x \in D$.

Let $S[gq+q,j;g'q+q,k]$ be called the **base** of $V_{g,g'}$ and let E be q by q matrix of differences from the base. We define $\mathbf{E_{g,g'}(x,y)} = (S[a+x,j;b+y,l] - S[a+q,j;b+q,l])$, where S[a+q,j;b+q,l] is the base of $V_{g,g'}$ and $a = gq$ and $b = g'q$.

For a particular j,l we can create and store matrices $E_{g,g'}$ as soon as the corresponding values in the S matrix are computed. Once computed, retrieval of any desired E clearly takes $O(1)$ time. The overall overhead for encoding the S matrix into a set of E matrices for the entire algorithm requires an addition of $O(N^4)$ time.

Theorem 1. *Given $E_{g,g}$ and the base we can reconstruct all the values of $V_{g,g'}$.*

Proof. $V_{g,g'}(e,f)$= $E_{i,k}(e,f)$+*base*

Fact 4. For a specific i,j,k,l and g,g', if $\{m,n\}$ is the branch point combination that leads to the maximum of the sum of $S[i,m;k,n]+E_{g,g'}(m+1-gq,n+1-g'q)$ where $m \in Mg$ and $n \in Ng$ then $\{m^*_g, n^*_{g'}\}$ = {m,n}.

3.2 R^2 Table Integration into Branching

The *fast_branch_function* below calculates the new Branch Rule. Taking advantage of the precomputed R^2 table as well as the precomputed and stored E matrices, the total time for this function will be $O(N^2/q^2)$.

fast_branch_function(i,j;k,l)::
  for g=$\frac{\lfloor j-1 \rfloor}{q}$ to $\frac{\lfloor i+1 \rfloor}{q}$ **do**
    for g'=$\frac{\lfloor l-1 \rfloor}{q}$ to $\frac{\lfloor k+1 \rfloor}{q}$ **do**
      retrieve $E_{g,g'}$
      $\{m^*_g, n^*_{g'}\}$=$\mathbf{R^2}(i,k,g,g',E_{g,g'})$
      cur_max= `max(` *cur_max*, $S[i,m^*_g;k,n^*_{g'}] + S[m^*_g+1,j;n^*_{g'}+1,l]$)
  return *cur_max*

4 Precomputing the R^2 Table

Finally we present how to precompute table R^2 for all possible variations of E. For simplicity of exposition we reorganize the Cofolding Algorithm as follows.

for g=0 to N/q **do**
for g'=0 to N/q **do**
  for j=$g*q$ to $gq+q-1$ **do**
  for l=$g'q$ to $g'q+q-1$ **do**
    for i=$j-1$ to 1 **do**
    for k=$l-1$ to 1 **do**
      const_oper_max=max(Rules a to i)
      branch_max= ***branch_function***$(i,j;k,l)$
      S[i,j;k,l]=max(*const_oper_max, branch_max*)

Note that neither the run-time nor accuracy is affected by this change. Also note that facts 1,and 2, still hold true.

Assume we have completed the iteration of algorithm above where $g = 0$ and $g' = 0$ and have the optimal solutions for all i, k $S[i, m'; k, n']$ where $m' < q$ and $n' < q$. At this point we have computed all the heads for branch points in Mgroup M_0, and Ngroup N_0 ($g = 0$,$g' = 0$). For any matrix E the following algorithm computes $\{m_0^*, n_0^*\}$.

for each matrix v of size q by q such $v[x,y] \in D$ **do**
  compute (E from v) [4]
  for each i such that $i < q-1$ **do**
    for each k such that $k < q-1$ **do**
      $\mathbf{R^2}(i,k,g=0,g'=0,E)$ is the to the branch combination $\{m,n\}$ such that $S[i,m;k,n]+E[m+1,n+1]$ is maximum.

We can generalize this algorithm for any g, g', by creating an ***update_table*** function that is called once any g' iteration is complete.

update_table function(g,g') ::
for each matrix size q by q v such $v[x,y] \in D$ **do**
  compute E from v
  for each i such that $i < gq-1$ **do**
    for each k such that $k < g'q-1$ **do**
      $\mathbf{R^2}(i,k,g,g,E)$ is set to branch combination $\{m,n\}$ such that $S[i,m;k,n]+E(m+1-g*q, n+1-g'*q)$ is maximum.

4.1 Fast Cofold Algorithm

We present the speedup algorithm combining both preprocessing and use of table R^2.

[4] For example compute E from v function sets $E(i,j) = \sum_{x=q-1}^{i} v[x,q] + \sum_{y=q-1}^{j} v[i,y]$.

```
Fast Cofold Algorithm
for g=0 to N/q do
for g'=0 to N/q do
    for j=g*q to gq+q-1 do
    for l=g'q to g'q+q-1 do
        for i=j-1 to 1 do
        for k=l-1 to 1 do
            const_oper_max=max(Rules a to i)
            branch_max= fast_branch_function(i,j;k,l)
            S[i,j;k,l]=max( const_oper_max, branch_max)
    update_table(g,g')
```

Boundary case for branch_function. We define the boundary case of the fast_branch_function(i,j,k,l) the case where $g = j/q$ and/or $g' = l/q$. Because, the update_table function has not yet precomputed $\{m^*_g, n^*_{g'}\}$ in this case, we must explicitly compute $\{m^*_g, n^*_{g'}\}$ comparing all q^2 branch point combinations. The *fast_branch_function* including the Boundary case is shown below.

```
fast_branch_function(i,j;k,l)::
  for g=⌊j-1⌋/q to ⌊i+1⌋/q do
    for g'=⌊l-1⌋/q to ⌊k+1⌋/q do
      if(boundary case) compute {m*_g, n*_g'} directly; continue
      retrieve E_{g,g'}
      {m*_g, n*_g'}=R^2(i,k,g,g',E_{g,g'})
      cur_max=max( cur_max, S[i,m*_g,k;n*_g'] + S[m*_g+1,j;n*_g'+1,l] )
  return cur_max
```

5 Asymptotic Time Analysis

The Fast-Cofold algorithm can be grouped into 3 sections: (1)The computations of Rules a-i, (2) the computation of Branch Rule j or the Fast_Branch_function, (3)the preprocessing done by the update_table function.

The loops g and g' are each called $O(N/q)$ times, loops j and l are each called $O(q)$ times, loops i and k are each called $O(N)$ time. Therefore, the computation time of *Fast Cofold* Algorithm for Rules a-i equals to $O(\frac{N}{q} * \frac{N}{q} * q * q * N * N) = O(N^4)$ and remains unchanged from the Cofolding algorithm.

The *fast_branch_function* is called $O(N^4)$ times. In the branch function, loops for g and g' will reference the R^2 table a total of $O(N^2/q^2)$ times. There are $O(N/q + N/q)$ boundary cases during each call to the *fast_branch_function* that take $O(q^2)$ time to compute. Therefore, each call to the *fast_branch_function* takes $O(N^2/q^2 + (2N/q) * (q^2)) = O(N^2/q^2 + 2qN)$ time.

In the *Fast Cofold* algorithm the Branch Rule j is computed in total $O(N^6/q^2 + 2qN^5)$ time.

The *update_table* function is called for every new Mgroup, Ngroup combination completed, or on every iteration of loop g'. In total there are N^2/q^2

such iterations. With in the *update_table* function we have three loops. The outer loop iterates over every possible E matrix of which there are D^{q^2}. The next two loops then maximize for every i,k taking $O(N^2)$ time to do $O(q^2)$ maximizations. Therefore, the asymptotic time of the update_table function is $O(N^2/q^2 * D^{q^2} * N^2 * q^2) = O(N^4 * D^{q^2})$.

The entire *Fast Cofold algorithm* algorithm has a runtime of $O(N^6/q^2 + 2qN^5)$ $+O(N^4 * D^{q^2}) + O(N^4)$.

Theorem 2. *The Fast Cofold algorithm has an asymptotic time bound of* $O(N^6/log_b(N^2))$ *if* $q = \sqrt{\log_b(N^2)}$ *where the log base b is is constrained by* $D < b < N$.

Proof. If q is set to $\sqrt{\log_b(N^2)}$ then the algorithm takes $O(N^6/\log_b(N^2) + 2\log_b(N^2) * N^5) + O(N^4/\log_b(N^2) * D^{\log_b(N^2)}) + O(N^4)$ time.

So if $O(N^4 * D^{\log_b(M^2)}) = O(N^6/log_b(N^2))$ then *Fast Cofold algorithm* computes in $O(N^6/\log_b(N^2))$ time.

Let $N' = N^2$ and $Q' = q^2$ then the Fast Cofold Algorithm has an asymptotic time of $O(N'^2 * D^{Q'})$. Base on theorem 1. in Frid and Gusfield $O(N'^2 * D^{Q'}) = O(N'^3/\log_b(N'))$ for $D < b < N$ [6]. Therefore, $O(N'^3/\log_b(N')) = O((N^2)^3/\log_b(N^2)) = O(N^6/\log_b(N^2))$ q.e.d.

5.1 Memory

Unchanged from Sankoff's algorithm, the Fast Cofold algorithm will also require $O(N^4)$-space to store matrix S. However, there is an additional memory cost of $O(N^4/q^2 * D^{q^2})$-space for storing Table R^2.

6 Empirical Results

We compare our Fast Cofold Algorithm with the Sankoff Cofold Algorithm for two sequences described in the paper. The purpose of these empirical results is to show that our algorithm not only achieves a theoretical speedup but can also lead to practical improvements. As discussed above we produce the same solution matrix S produced by the $O(N^6)$ algorithm. Therefore, we don't test different values for the scoring schemes α, β, τ, but do test the change in cardinality D. Our algorithm also performs identically for randomly generated and real sequences geneBank sequences of the same length N. In fact the practicality and speed up of the Fast Cofold algorithm is dependent only on the size of the sequence N, the base of the log b, and the cardinality of set D. An optimization function was created that sets base b to the value that would create the greatest speedup. The function calculated the optimal b for a set D, sequence length n and the memory constraints of the computer. We report the results in Table 1 for average times in seconds for 30 random generated and 10 geneBank sequences (standard deviation for all tests is less than .5 seconds).

Table 1. Empirical Results

$\|D\|$	base b	Size (N)	Cofold Algorithm run-time(seconds)	Fast Cofold Algorithm (seconds)	ratio
7	9	150	21307.45	12085.9	1.76
3	4	150	21307.45	7446.04	2.88
2	4	150	21307.45	5866.39	3.66
3	4	100	1770.33	733.502	2.42
2	3	100	1770.33	631.525	2.80
5	6	50	24.41	20.79	1.17
3	4	50	24.41	18.58	1.31
2	3	50	24.41	10.63	2.30

7 Conclusion and Future Work

The Fast Cofold algorithm presented formulates Four Russians speedup for the problem of finding an optimal simultaneous alignment and fold. The algorithm produces the same solution matrix S as the modified Sankoff's Cofolding algorithm but in the reduced time of $O(N^6/\log_b(N^2))$ from $O(N^6)$. This compatibility makes it possible to apply other speed ups and memory reduction algorithms alongside the Four Russians speedup. As discussed in the introduction Ziv-Ukelson et al. [19] and Backofen et al. [2] improved computation time and lowered memory costs by filtering the branch points that the *branch_function* examines. Excluding groups that don't have any members that are co-terminus co-folding from computation in the update table function and branch function would lead to an additional speedup of $O(N^4W)$ where W is the number of group combinations that contain co-terminus co-co folding members. W is constrained by $\frac{N}{log(N)} \leq W \leq min(K, \frac{N^2}{log(N)})^5$. There is also interest in extending the speedup to the algorithms that compute cofolds for more than two sequences, and algorithms that compute local alignments. We also note that based on `Theorem` 2 *Mgroups* and *Ngroups* don't have to be the same size. In fact `Theorem` 2 holds for $Q' = q1 * q2$ where $q1$ is the size of any *Mgroup* and $q2$ of any *Ngroup*. The variation in group sizes can be implemented with a few small changes to the algorithm leading to an even greater speedup.

Acknowledgments

This research was partially supported by grants SEI-BIO 0513910 and IIS-0803564 from the National Science Foundation.

References

1. Backofen, R., Landau, G.M., Möhl, M., Tsur, D., Weimann, O.: Fast RNA structure alignment for crossing input structures. In: Kucherov, G., Ukkonen, E. (eds.) CPM 2009. LNCS, vol. 5577, pp. 236–248. Springer, Heidelberg (2009)

[5] K is constrained by $N \leq K \leq N^2$ [19].

2. Backofen, R., Tsur, D., Zakov, S., Ziv-Ukelson, M.: Sparse RNA folding: Time and space efficient algorithms. In: Kucherov, G., Ukkonen, E. (eds.) CPM 2009. LNCS, vol. 5577, pp. 249–262. Springer, Heidelberg (2009)
3. Dowell, R., Eddy, S.: Efficient pairwise RNA structure prediction and alignment using sequence alignment constraints. BMC Bioinformatics 7(1), 400 (2006)
4. Eddy, S.R.: Computational genomics of noncoding RNA genes. Cell 109(2), 137–140 (2002)
5. Eddy, S.R., Durbin, R.: RNA sequence analysis using covariance models. Nucl. Acids Res. 22(11), 2079–2088 (1994)
6. Frid, Y., Gusfield, D.: A simple, practical and complete $O(n^3/\log(n))$ -time algorithm for RNA folding using the *four russians* speedup. In: Salzberg, S.L., Warnow, T. (eds.) WABI 2009. LNCS, vol. 5724, pp. 97–107. Springer, Heidelberg (2009)
7. Gorodkin, J., Heyer, L.J., Stormo, G.D.: Finding common sequence and structure motifs in a set of RNA sequences. In: ISMB, pp. 120–123 (1997)
8. Hofacker, I.L., Fontana, W., Stadler, P.F., Bonhoeffer, S.L., Tacker, M., Schuster, P.: Fast folding and comparison of RNA secondary structures. Chemical Monthly 125, 167–188 (1994)
9. Mathews, D.H., Turner, D.H.: Dynalign: an algorithm for finding the secondary structure common to two RNA sequences. Journal of Molecular Biology 317(2), 191–203 (2002)
10. Nussinov, R., Pieczenik, G., Griggs, J.R., Kleitman, D.J.: Algorithms for loop matchings. SIAM Journal on Applied Mathematics 35(1), 68–82 (1978)
11. Pedersen, J.S., Bejerano, G., Siepel, A., Rosenbloom, K., Lindblad-Toh, K., Lander, E.S., Kent, J., Miller, W., Haussler, D.: Identification and classification of conserved RNA secondary structures in the human genome. PLoS Comput Biol. 2(4), e33 (2006)
12. Rivas, E., Eddy, S.: Noncoding RNA gene detection using comparative sequence analysis. BMC Bioinformatics 2(1), 8 (2001)
13. Rose, D., Hackermuller, J., Washietl, S., Reiche, K., Hertel, J., FindeiSZ, S., Stadler, P., Prohaska, S.: Computational rnomics of drosophilids. BMC Genomics 8(1), 406 (2007)
14. Sankoff, D.: Simultaneous solution of the RNA folding, alignment and protosequence problems. SIAM Journal on Applied Mathematics 45(5), 810–825 (1985)
15. Seemann, S.E., Gorodkin, J., Backofen, R.: Unifying evolutionary and thermodynamic information for RNA folding of multiple alignments. In: NAR (2008)
16. Torarinsson, E., Yao, Z., Wiklund, E.D., Bramsen, J.B., Hansen, C., Kjems, J., Tommerup, N., Ruzzo, W.L., Gorodkin, J.: Comparative genomics beyond sequence-based alignments: RNA structures in the encode regions. Genome Res. 18(2), 242–251 (2008)
17. Torarinsson, E., Havgaard, J.H., Gorodkin, J.: Multiple structural alignment and clustering of RNA sequences. Bioinformatics 23(8), 926–932 (2007)
18. Washietl, S., Hofacker, I.L.: Consensus folding of aligned sequences as a new measure for the detection of functional RNAs by comparative genomics. Journal of Molecular Biology 342(1), 19–30 (2004)
19. Ziv-Ukelson, M., Gat-Viks, I., Wexler, Y., Shamir, R.: A faster algorithm for RNA co-folding. In: Crandall, K.A., Lagergren, J. (eds.) WABI 2008. LNCS (LNBI), vol. 5251, pp. 174–185. Springer, Heidelberg (2008)

Sparse Estimation for Structural Variability

Raghavendra Hosur[1,3], Rohit Singh[1], and Bonnie Berger[1,2,*]

[1] Computer Science and Artificial Intelligence Laboratory, MIT
Massachusetts Institute of Technology
Cambridge MA 02139
[2] Dept. Of Mathematics, MIT
[3] Dept. Of Materials Science and Eng., MIT
bab@csail.mit.edu

Abstract. Proteins are dynamic molecules that exhibit a wide range of motions; often these conformational changes are important for protein function. Determining biologically relevant conformational changes, or true variability, efficiently is challenging due to the noise present in structure data. In this paper we present a novel approach to elucidate conformational variability in structures solved using X-ray crystallography. We first infer an ensemble to represent the experimental data and then formulate the identification of truly variable members of the ensemble (as opposed to those that vary only due to noise) as a sparse estimation problem. Our results indicate that the algorithm is able to accurately distinguish genuine conformational changes from variability due to noise. We validate our predictions for structures in the Protein Data Bank by comparing with NMR experiments, as well as on synthetic data. In addition to improved performance over existing methods, the algorithm is robust to the levels of noise present in real data. In the case of Ubc9, variability identified by the algorithm corresponds to functionally important residues implicated by mutagenesis experiments. Our algorithm is also general enough to be integrated into state-of-the-art software tools for structure-inference.

1 Introduction

A central tenet of molecular biology is that a protein's three-dimensional (3-D) structure is crucial to its function. Indeed the structural genomics initiative is producing ever increasing number of structures at high resolution, providing accurate coordinates for each atom in the structure [2]. A protein's structure, however, is rarely static. Proteins are dynamic molecules, capable of exhibiting a wide range of motions and conformational variability [11,21]. Such conformational changes are important in biological functions such as enzymatic catalysis, cellular transport, and signaling [27,8]. It has been postulated that even subtle conformational changes may have important functional consequences [16].

A multi-conformer model, or ensemble, attempts to model variability by explaining the data using an ensemble of conformers, rather than just one conformer. Indeed, conformational variability in a protein might be present even

* Corresponding author.

V. Moulton and M. Singh (Eds.): WABI 2010, LNBI 6293, pp. 13–27, 2010.

in a single experiment, where the observed data is an average over multiple conformations [6,9]. Multi-conformer approaches have long been the norm when modeling NMR data. It has been suggested that, for an accurate representation of the physical heterogeneity in a protein, such multiple-conformer models also be used to explain X-ray crystallography data [9,24,14].

An open problem— and the focus of this paper— is understanding the nature of conformational variability implied by experimental data. The key challenge here is to distinguish variability resulting due to noise in experimental data from functionally relevant physical motion [24,20,5]. The problem is particularly difficult to solve with single-conformer approaches, given their limited ability to model the data. Indeed, this issue has been a driving force in the efforts toward ensemble approaches [9]. Even with the current ensemble approaches, it is difficult to disentangle a protein's physical motion (e.g. hinge or loop motions) from other kinds of protein motion (e.g., vibrational motion). The key problem is that limited sampling (i.e. number of conformations) and multiplicity of the problem make for weak statistical estimates [9,14,12]. While a growing number of tools address the problem of using ensembles to implicitly model conformational variability [6,14,5], they generally do not distinguish between variability due to noise vs. physical motion.

There have been some attempts to analyze structural variability, but using pairs of structures rather than ensembles. Conventional parameters such as torsional angle differences, temperature factors and root-mean-squared-distance (RMSD) values have been used to identify flexible regions. But they combine estimation noise and true variability into a single quantity; thus, they are of limited usefulness under noisy data (e.g. for low-to-medium resolution structures) (see Related Work, [20]). More importantly, conformational variability is best described over a population (i.e ensemble) of conformations; pairwise comparison between structures implies such limited sampling of the conformational space that it may be unreliable for all but the least noisy datasets.

In this paper, we take a different approach to analyzing variability. Our approach is inspired by recent developments in regression-based predictive models in machine learning. The basic intuition behind the approach is to construct an ensemble of conformers that explain the experimental data and then use sparse estimation to distinguish between conformers that are just noisy versions of a base conformation (e.g., the PDB structure) and those that capture true conformational variability (relative to the base). Accordingly, structures sampled from a Gaussian distribution about the base structure should be more predictive of the base structure than structures displaying true variability. This allows us to separate out the biologically relevant variability due to physical motion using a feature selection technique, Lasso [25]. Lasso, which stands for "least absolute shrinkage and selection operator", is a regularized regression technique in which only the most significant predictor features are selected [25]. We illustrate the approach on X-ray crystallographic data, as it is the most common source of structural data. Our results demonstrate that the method compares favorably with previous approaches. It is more robust to specific parameter choices and produces fewer false positives and false negatives (see *Comparative Analysis*). In contrast to conventional approaches of pairwise structure comparison, we use Electron Density Maps (EDM) for identification of true variability; this allows us greater power in accurately identifying true structural outliers without the need for any artificial parameters to model noise [13]. Finally, our predictions of true variable regions are in good agreement with the dynamics inferred from solution NMR experiments; the latter are presumably closer to the physical reality.

One of the key contributions of our work is in framing the problem as a sparse estimation problem, in a way that allows a wealth of machine learning knowledge to be applied to it. In particular, the problem of identifying sparse models that can be physically interpreted has recently gained much attention in machine learning, data mining and statistics due to the exponential growth in publicly available data [10]. We show here that identification of true variable regions in an ensemble is naturally formulated as a sparse learning problem via Lasso. This formulation allows us to rigorously deal both with noise in the experimental data and uncertainty associated with the structure-building process. Our approach of using Lasso is quite general, and can be applied to any structural data. Application of our method to proteins of interest may reveal interesting conformational changes that might go unnoticed due to the absence of alternate structural evidence, i.e. independently solved alternate conformations, which are still expensive and cumbersome to obtain.

A key intuition driving our approach is as follows: to identify true variability in a protein fragment, rather than performing a per-atom statistical test, we perform a whole-model statistical test. A per-atom test will essentially ignore correlated motions (even if small) between neighboring atoms; in contrast, a whole-model test will be able to identify even small correlated motions. We formalize this approach using the Lasso-based test. We exploit the idea of borrowing information from all the samples to make a reliable statistical inference on a particular sample. In contrast, a pairwise *t-statistic* approach uses information from only a single sample to make a decision [12].

2 Related Work

Coordinate-based methods using pairwise comparisons have had reasonable success in identifying flexible regions [20]. However these techniques were designed to identify true flexibility in conformations that have been solved independently, where there is already some evidence of variability. Nigham et al. give a statistical test based on pairwise RMSD to identify regions showing true variability in the presence of noise. Key to their method is the assumption of a uniform, normal independent noise (artificially added) at each coordinate. However, this assumption typically does not hold in reality [26].

Related approaches rely on the use of various parameters such as torsion angle differences, temperature factors and RMSD. Torsion angle differences are highly sensitive to noise: small deviations in coordinates might cause significant changes in torsion angles [20]. Temperature factors (B-factors) are parameters used to model uncertainty in atomic positions; the value of the B-factor corresponding to an atom represents the degree of uncertainty in that atom's location in the model. This distribution accounts for small vibrations about an atom's position. However, B-factors tend to encapsulate in one value the conformational variability, as well as ambiguities related to inadequacies in data (e.g. related to crystal imperfections, errors in measurement of intensities). This problem is aggravated at medium-to-low resolutions ($> 1.5\AA$). At such resolutions, B-factors act as "error-sinks", absorbing any errors (not necessarily related to protein motion) in the optimization and model building process [13].

A number of methods have been proposed to model multiple conformations that might give rise to X-ray crystallographic data from a single crystal [6,24,14,5]. Although independently optimized multi-conformer representations prove to be a very attractive solution, interpretation of what the ensemble represents is a gray area [24,5]. Knight et al. (2008) give a simple residue-level heuristic test based on the variance in the ensemble to identify true variability. However, there is no consensus method to identify true structural variability, and the interpretation of such ensembles is still the subject of debate [24,5].

3 Methods

Our method consists of two steps: a) construction of an ensemble representative of the observed data, and b) analysis of the variability in this ensemble using Lasso. The ensemble generation algorithm is independent of the classification of variability; the ensemble can be obtained from any other method. However it is important to ensure that all the structures in the ensemble are of high-quality, and represent the data almost as well as the PDB structure.

3.1 Ensemble Construction

To obtain a diverse, high-quality ensemble representing the X-ray diffraction data, we seed a single-conformer maximum likelihood optimization procedure (e.g. PHENIX) with a diverse set of conformations [1]. We assume that realistic conformations explaining the crystallographic data will be within a limited RMSD distance of the published PDB structure; this follows similar assumptions in previous work [6,14]. However, hinge motion, if present in a single crystal specimen, can also be detected by sampling in a larger conformational space around the PDB structure. Starting from the backbone coordinates in the PDB, we construct alternate backbone conformations within 2Å[1] RMSD using ChainTweak, a state-of-the-art inverse-kinematics based neighborhood-sampling algorithm [22]. ChainTweak can, in principle, exhaustively sample from the neighborhood of a conformation; leading to a highly variable and diverse ensemble. For each backbone, we assign side-chains using RAPPER [6], based on their fit to the

[1] We tried using higher cutoffs, but RAPPER often fails to find a rotamer-assignment compatible with the EDM for conformations greater than 2Å RMSD from the PDB backbone.

Electron Density Map (EDM). The final ensemble is obtained by subsequent optimization using PHENIX and filtering based on fit-to-data, measured using a cross-validation parameter R_{free}[2];lower R_{free} implies better fit-to-data. The final ensemble consists of structures that are of high quality, and collectively represent the data as well as the PDB structure (Fig 1).

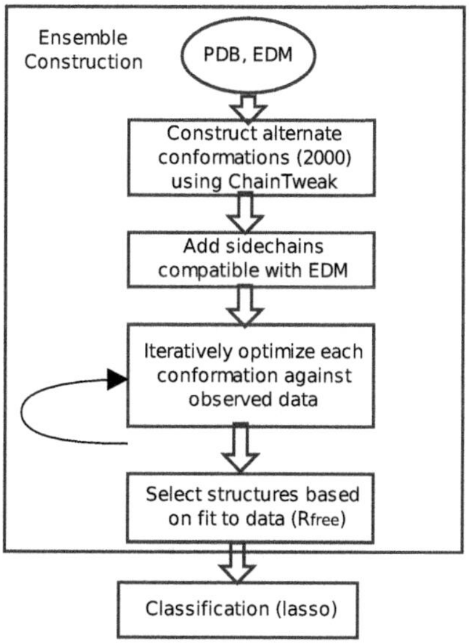

Fig. 1. Overview of the ensemble generation and classification algorithm

3.2 Analysis of Variability Using Lasso

Given an ensemble of conformations, our goal in this section is to identify the subset of conformations whose variation from a given base conformation is most likely due to only noise in the experimental data. The remaining conformations can then be interpreted as demonstrating true variability compared to the base conformation. The choice of a base conformation here is arbitrary; a natural choice for it is the PDB structure, since one is often interested in conformational variability not captured by the published PDB structure. To achieve this goal, we formulate a Lasso regression problem: we express the base conformation as a linear combination of the ensemble members (each such conformation is thus a feature); we use experimental data (i.e. diffraction data) to fit this regression. As part of the Lasso framework for feature selection, we assign (unknown) weights to each feature. The key strength of Lasso is that it is likely to make the weights

[2] R is a measure of agreement between the amplitudes of the structure factors calculated from a structure and those from the original diffraction data. R_{free} is the corresponding cross-validation parameter, calculated on diffraction data not used in the structure optimization process [26].

for irrelevant features exactly zero, clearly identifying them. The intuition here is that structures sampled from a Gaussian distribution (i.e. modeled by B-factors) about the PDB structure should be more predictive of the PDB structure than structures displaying true variability. The former structures will be assigned a non-zero weight during Lasso and can then be classified as not displaying true structural variability, since they are adequately represented by the PDB structure and do not represent biologically relevant long time-scale motion.

Lasso regression is often an effective technique for shrinkage and feature selection in cases where feature selection must be performed with noisy, limited data [25,17,29]. The loss function of Lasso regression is defined as:

$$L = \sum_i (y_i - \sum_p \beta_p x_{ip})^2 + \lambda \sum_p ||\beta_p||_1 \tag{1}$$

where x_{ip} denotes the pth predictor (feature) in the ith data point, y_i denotes the value of the response for this data point, and β_p denotes the regression coefficient of the pth feature. The l_1 regularizer leads to a sparse solution in the feature space, which means that regression coefficients for the most irrelevant and redundant features shrink to zero. Interestingly, recent theoretical work recovers Lasso as a formulation of a linear robust regression problem under feature-wise uncorrelated and norm-bounded noise [29]. The authors suggest that such problems are of interest when values of the features are obtained with noisy pre-processing steps, and the magnitudes of such noises are bounded.

We exploit this parallel in our formulation, where we compute each feature (i.e. each structure in the ensemble) by optimizing against the observed data. The PDB structure is the observed quantity, and the individually optimized structures in the ensemble are our noisy predictor features. A sparse solution in the β space will then represent structures which are variable due to noise ($\beta_p > 0$), thus decomposing the variability observed in the ensemble. To get the regularization penalty λ, we follow suggestions based on other applications of Lasso and use cross-validation [25,19].

3.3 Electron Density Map

Lasso regression can be performed either in the coordinate space or the electron density space (EDM). In contrast to previous approaches, which use coordinate based methods for pairwise structure comparison, we have designed the test using EDMs, since the former cannot distinguish between model errors and genuine structural outliers [13]. EDMs are obtained by taking an inverse-fourier transform of the observed diffraction data, which are appropriately scaled using B-factors [7]. Another advantage of using an EDM is that it directly includes the B-factors of the models, and hence can also inherently deal with isotropic or anisotropic B-factors. This circumvents the problem of estimating actual uncertainty from B-factors; which is often a challenge for coordinate based methods. The simple regression test quantifies the relevance of each structure in the ensemble to the Gaussian distribution around the PDB (as given by the B-factors).

As part of our Lasso formulation, we assume as the observed variable, the EDM computed from the PDB structure. The predictor variables, or features, are EDMs of structures in the ensemble. The electron density at a point 'g' on a grid describing the observed EDM (ρ^g_{PDB}), is then modeled as a linear combination of electron densities at the point 'g' of the predictor EDMs (ρ^g_i). We assume that the observed electron density is noisy with respect to our generative model and model this using a normally distributed noise component ϵ_g. We then minimize the Lasso loss function:

$$\rho^g_{PDB} = \sum_i w_i \rho^g_i + \epsilon_g \tag{2}$$

$$min \sum_{g \epsilon G} (\rho^g_{PDB} - \sum_i w_i \rho^g_i)^2 + \lambda \sum_i w_i \tag{3}$$

Here, w_i are the regression coefficients. The structures for which w_i approaches zero are the ones most irrelevant compared to the PDB, and hence exhibit true variability. To optimize over a fragment (e.g. one residue), G is restricted to the bounding box for the fragment.

All EDMs are constructed using Clipper [3], and are described on the same unit cell with the same symmetry as that of the PDB structure. The optimization was carried out using the non-linear optimization libraries IPOPT. IPOPT uses an interior point method, combined with an efficient line-search procedure to minimize the non-linear objective function [28].

4 Results

4.1 Synthetic Data

Our algorithm successfully models variability in a simulated crystal having two conformations, one the PDB structure (*conformer 1*) and the other constructed computationally (*conformer 2*) (Fig 2A; RMSD = 0.989 Å). The second conformer was constructed using ChainTweak; we randomly selected a conformation from a set of 100. Side chains were built using RAPPER and all atoms were assigned a B-factor of 30 $\AA^2$. Synthetic diffraction data were computed by averaging the simulated structure factors of the two conformers using the experimental resolution cutoffs [5,7].

Starting from an EDM of the simulated crystal, our algorithm generates structures similar to both the original structures (Fig 2A,B). Of the 13 structures output by the algorithm, 4 structures were non-redundant; remaining structures were almost identical to these 4 structures. Lasso regression on these 4 structures show that the ensemble correctly identifies the heterogeneity in the original data, with 2 structures having coefficients $w \approx 0$ with regression done with EDM of *conformer 1* (as per a *t-test*; colored light gray in Fig 2A), corresponding to structures with true variability. Moreover, the same conformations had statistically significant coefficients ($w > 0$) in the regression with EDM of *conformer 2*. Indeed, these conformations are closer to *conformer 2* (RMSDs= 0.298, 0.128 Å) than the conformations classified as non-variable (RMSDs = 0.456, 0.765 Å). The algorithm thus appears to recover the heterogeneity in the data (Fig 2B).

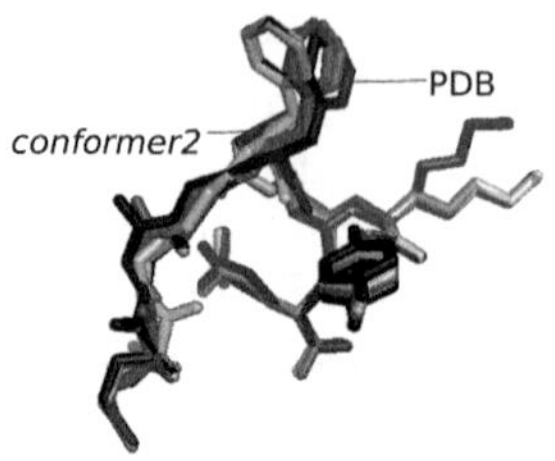

(A)

Num. Initial Conformers	2
Output Ensemble	
Ensemble Size	4
R^2	0.796
RMSDs for $\omega = 0$	0.952, 0.712
RMSDs for $\omega > 0$	0.324, 0.613

(B)

Fig. 2. Example of ensemble construction and classification. A) PDB structure and the second conformer in the synthetic crystal are in black. The two structures classified by Lasso as variable are shown in light gray and the two as variable due to noise, in dark gray. B) Summary of the algorithm output using synthetic data. RMSD is calculated with respect to the PDB structure. Suitability of the linear model and statistical significance of the regression coefficients were evaluated using standard techniques (R^2 and *t-test*).

Performance analysis. Our method is robust and consistent (Fig 3A,B). The consistency and accuracy of our method depends on the extent of correlation between the features. Correlation between structures that are truly variable and ones variable due to noise, will make the regression convoluted; different regularization penalties (λ) will select different structures, leading to highly varying regression weights [19]. Our simulations indicate that the features (i.e. conformations in the ensemble) are uncorrelated to a large extent (Fig 3A), indicated by the overall smooth trends for ω as we increase the regularization penalty λ. Increasing λ shrinks the individual weights of the features towards zero, thereby decreasing the ratio $|\omega|_1/\max|\omega|_1$. We believe the overall smoothness of the regularization path may be due to the efficiency of the sampling algorithm–ChainTweak, which constructs highly diverse and uncorrelated conformations. In our simulations we find that, of the four structures in the ensemble, only one structure (dashed line) is dominant for all regularization penalties. A second structure (dashed line) is selected only at low λ's (< 50; Fig3A).

We find that our overall classifications are quite robust to the size of optimized grid region 'G'. The average weight of a structure, calculated by averaging over all fragments, is consistent across varying fragment and window sizes; structures represented by dashed lines do indeed have the highest average weights and those by solid lines, negligible average weights (Fig 3B). One could vary G in two ways: by splitting the chain into separate fragments and carrying out Lasso on each one, or by sliding a window centered around each residue and optimizing over each window. Our results on the fragment-based approach are identical to Fig 3B; we used fragment sizes of 1,2,4 and 8 (data not shown). For the second approach, we use sliding windows of sizes 3 and 5 centered on each residue (Fig 3B), and optimize over the bounding box enclosing the residues in the window.

Comparative analysis. Lasso compares favorably to other methods in identifying true flexibility. The pairwise comparison method of Nigham et al. (*Pflex*)

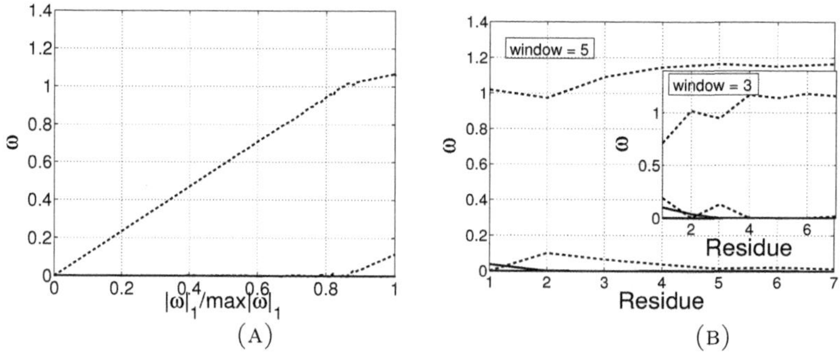

Fig. 3. Performance analysis. A) Regularization path for the ensemble ($|\omega|_1 \rightarrow 0$ as $\lambda \rightarrow \infty$ towards left). B) Residue-level lasso with varying window sizes centered on each residue ($\lambda = 10$). Dashed lines are for structures classified as variable due to noise, solid lines are for those classified as truly variable.

is sensitive to the standard deviation of added noise (σ). *Pflex* computes a flexibility measure, 'f', for each residue based on RMSD, σ and a threshold p-value. A lower f implies higher flexibility. We used the values suggested by Nigham et al. for σ ($0.1 <= \sigma <= 0.2$) and the threshold p-value (= 0.0001). *Pflex* tends to easily classify structures as variable at low levels of added noise ($\sigma = 0.1$, Fig 4A); three of the four structures in the ensemble are classified as variable. At higher noise levels it fails to classify any structure as truly variable, leading to false negatives; f remains at 8 for all residues for all structures in the ensemble ($\sigma = 0.2$, Fig 4A Inset). While B-factors can correctly identify the regions of high variability, they fail to distinguish between noise and true variability, as evidenced by the similar profiles (Fig 4B). RMSD (best-fit) provides some indication of the true variability, but the interpretation may be sensitive to noise levels. The extent of the initial variability in the crystal, represented by each structure in the ensemble can be analyzed by looking at normalized RMSD: RMSD from the PDB structure normalized by the RMSD of *conformer 2* (from the PDB structure). A higher normalized RMSD implies the structure is closer to *conformer 2*, and a lower score implies it is closer to the PDB structure (Fig 4C). However, it is not clear what RMSD cutoff one should use in the presence of noise to robustly classify a structure as variable.

4.2 Real Data

Our algorithm performs well on experimental diffraction data from 5 crystal structures across a range of resolutions (Table 1). We evaluated our models by comparing them with the best available single-conformer model (i.e. PDB). Analysis of data fits and variability amongst the models emphasizes the advantages of representing the data using multiple conformers. Even when our ensemble contains models differing by 1 Å, we get an equivalent/improved fit to data: R^{ens}_{free} is lower than or equal to the PDB R_{free}. Our average improvements in

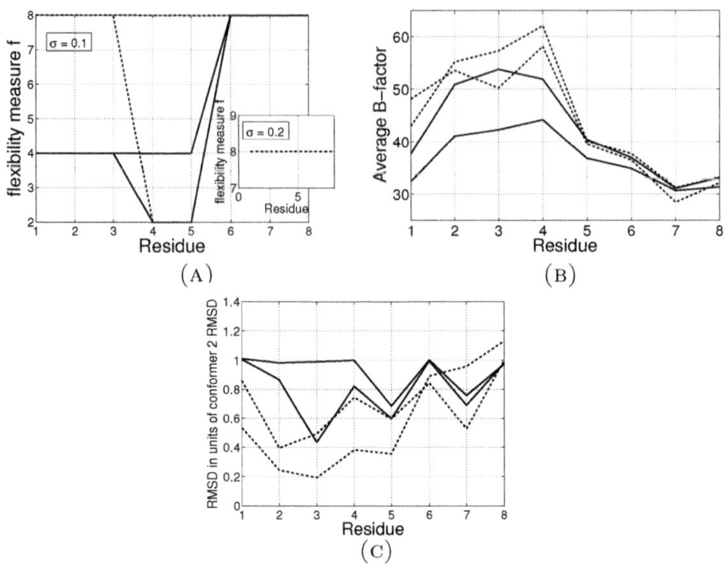

Fig. 4. Comparative analysis. A) *Pflex* is sensitive to the parameter σ, producing false-positives at low values and false-negatives at higher values (inset). B) Average B-factors correctly identify the regions of variability, but cannot distinguish between true variability and variability due to noise. C) Choosing a RMSD cutoff for classification is difficult with noisy coordinates. The line types are the ones used in Figure 3.

R_{free} are competitive with other approaches that construct multiple-conformer representations [24,5,6].

Tests on real data show that multi-conformer models add the most value at low resolutions; at high resolutions ($< 1.5\AA$) the ensemble is not able to significantly improve upon the fit-to-data (Table 1, PDB $R_{free} <= R_{free}^{ens}$)[3]. It is possible that the truly variable conformers themselves cluster into a small number of sets. This may be especially true for structures 3di9 and 1ew4, where the larger number of observations might have a bearing on the larger size of the ensemble. Moreover, for low resolutions, it is interesting to note that most of the variability observed is due to noise– less than 8 alternate conformers are truly variable in most cases. This re-confirms the importance of analyzing the basis of variability, particularly in multi-conformer representations of low resolution data. Our method is suited for this analysis as the structures are selected robustly and the resulting sparsity can be physically interpreted.

We observe that Lasso can classify variability effectively for most cases; structures classified as variable appear to differ more than those classified as non-variable (Fig 5A, B). Since we use an iterative method to solve the regression problem, interpretation of variability in the ensemble can be further analyzed by

[3] R_{free}^{ens} is calculated in the same way as R_{free}, except, amplitudes of structure factors averaged over the ensemble are used to calculate the residual [6].

Table 1. Summary of the models obtained using real diffraction data. PDB R and R_{free} are calculated after 6 iterations of optimization in PHENIX. These may differ from published values. "No. of reflections" gives the total number of experimental observations (i.e. intensity measurements). R^{ens}_{free} measures the collective ability of the ensemble to represent the data. Small RMSD ranges observed for the ensemble highlight the challenges of identifying true variability in real data.

PDB id	**1ew4**	**1q4r**	**3di9**	**9ilb**	**1a3s**
Resolution($\AA$)	1.4	1.9	2.0	2.3	2.8
No. Of reflections	22183	7578	22017	9535	5605
PDB R	0.206	0.187	0.244	0.156	0.176
PDB R_{free}	0.229	0.245	0.264	0.193	0.236
R^{ens}_{free}	0.228	0.216	0.237	0.193	0.240
Ensemble Size	77	4	40	5	11
RMSD ($\AA$)	0.792-1.13	0.678-0.859	0.728-1.085	0.805-1.413	0.826-1.238

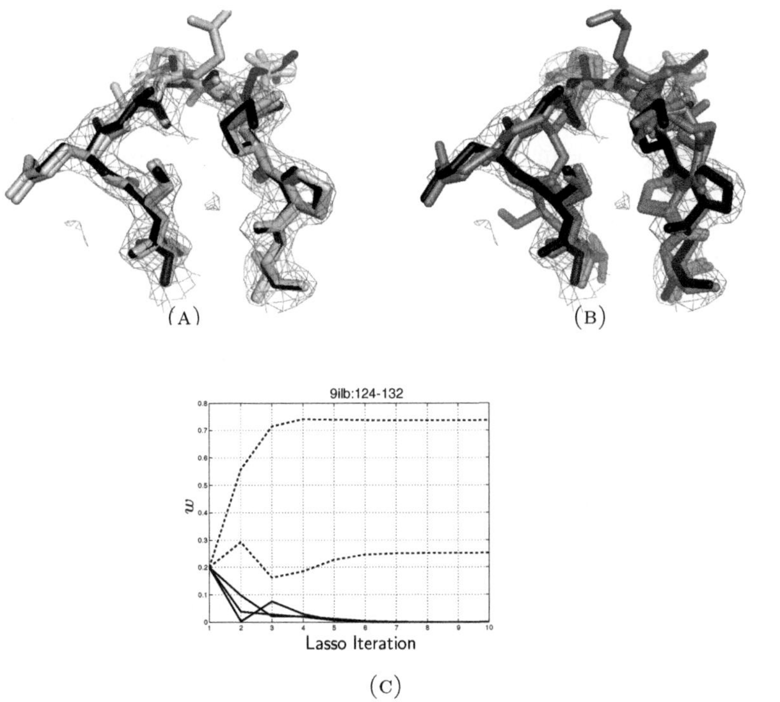

Fig. 5. Interpretation of ensembles on real data. A) Lasso tests on 9ilb:124-132 classifies 2 structures as non-variable (light gray). B) For the same loop, structures classified as truly variable (dark gray) deviate more from the PDB structure (black). C) Trajectory of the solution can give a qualitative knowledge of the landscape in the vicinity of the native structure. Line types are as in Figure 3 and 4. All density maps are contoured at 1.5σ for clarity. Figures were generated using PyMol [4].

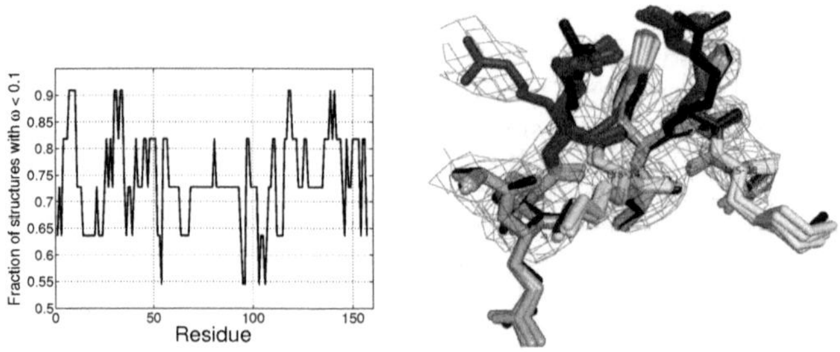

(A) (B)

Fig. 6. Flexibility analysis of the 1a3s ensemble. A) Residue level Lasso with a window size of 5 reveals variable regions. B) The N-terminal region (12-20) of 1a3s, with multiple rotamers of R13 (left, dark gray). The black structure represents PDB coordinates.

looking at the solution trajectory for Lasso (ω vs Lasso iteration; Fig 5C). The trajectory can help give a qualitative picture of the landscape near the native conformation: structures whose coefficients go to zero faster are farther away from the native structure.

We then asked the question "is there is any biological insight from the ensemble that can help us in understanding protein function ?" To this end, our results on the crystal structure of the human ubiquitin-conjugating enzyme (Ubc9, PDB: 1a3s) give some interesting anecdotal evidence. Using a window-size of 5 centered on each residue we used Lasso to identify the most variable regions for 1a3s (11 structures; Fig 6A). Four fragments turn out to be highly variable: the N-terminal helix (6-20), 30-40, 115-120 and C-terminus residues 135-145. This is in good agreement with NMR experiments, which reveal that Leu6, Ala10, Arg13, Arg17, Leu38, Leu119, Ala129, Glu132, Ile136 and Asn140 are amongst the most flexible residues in an otherwise rigid structure [18]. These residues overlap with our predictions of the true variable regions (Fig 6A). Our method is thus able to identify physically relevant variabilities.

Additionally, it is known that the N-terminus is important for Ubc9s specificity for SUMO rather than ubiquitin [18]. However, the molecular mechanisms responsible for substrate identification and interaction are not well understood [23]. Tatham et al. (2003) conducted site-directed mutagenesis experiments on Ubc9 to discover that mutations R13A/K14A and R17A/K18A disrupted Ubc9's interaction with SUMO-1. More recently, through a crystal structure of Ubc9-SUMO-1 complex, R13 and R17 have been observed to be involved in key non-covalent interactions with SUMO-1 [15]. A closer look at the heterogeneity modeled by our method suggests two possible conformational states for R13 (Fig 6B). Proximity of the two arginines at positions 13 and 17 indicate that such conformational changes might influence the binding interface with an E1-ubiquitin conjugate [18,15]. Further detailed analysis in light of our results could give some insight into the molecular mechanisms underlying such specific interactions.

5 Conclusions

We have introduced a novel technique for analyzing conformational changes that may be present in a real protein crystal. Our method first constructs a high-quality, diverse ensemble of structures respresentative of the crystallographic data. We then use a sparse estimation algorithm (Lasso) to distinguish structures that are genuinely variable from those that appear variable due to noise.

Unlike previous approaches, our method involves the estimation of variability by operating in the EDM space rather than in the 3-D coordinate space. This allows us to avoid the errors that are implicitly introduced in inferring the 3-D coordinates from the EDM. In particular, our method is able to effectively deal with correlated motions, without assuming i.i.d noise - a key assumption in earlier approaches [20]. Tests on real data show that the algorithm is able to capture physically relevant conformational changes, even for low resolution structures where the amount of noise is significant. Another advantage of operating in EDM-space is that our current technique is independent of any structure inference packages, and can be integrated to improve structure inference at an earlier stage in the structure-building process (e.g. from an initial experimental EDM). We believe that this approach is particularly useful in inferring/analyzing low-resolution structures. A common criticism of ensemble modeling approaches at low-resolutions is that they over-fit the data [24,5]. In contrast, our use of Lasso enables us to identify and discard structures that are variable only due to noise, permitting simultaneous optimization of the ensemble against the data without significant over-fitting risk. This, in turn, should improve automated structure determination at low resolutions where ambiguous EDMs often lead to error-prone single conformer models [24,5].

A key contribution of this paper is the Lasso-based statistical test to distinguish variability due to noise from that due to true heterogeneity. We believe that the general approach we have introduced – to evaluate noise using the entire ensemble, rather than on a per-atom pairwise basis– may be of value in other ensemble based analyses also. Lasso's performance as a statistical test here could be further improved by using kernel-based algorithms that can effectively deal with correlations and non-linear generative models [17].

Acknowledgements

The authors thank Dr C. Stultz and Dr M.V. Hosur for insightful discussions.

References

1. Adams, P., Grosse-Kunstleve, R., Hung, L., Loerger, T., McCoy, A., Moriarty, N., Read, R., Sacchettini, J., Sauter, N., Terwilliger, T.: Phenix:building new software for automated crystallographic structure determination. Acta Crystallographica (D) 58, 1948–1954 (2002)
2. Bourne, P., Weissig, H.: Structural Bioinformatics. Wiley-Liss, Inc., NJ (2003)

3. Cowtan, K.: Clipper Libraries, http://www.ysbl.york.ac.uk/~cowtan/clipper/clipper.html
4. Delano, W.: The pymol molecular graphics system (2002), http://www.pymol.org
5. Bedem van den, H., Dhanik, A., Latombe, J., Deacon, A.: Modeling discrete heterogeneity in x-ray diffraction data by fitting multi-conformers. Acta Cryst. (D) D65, 1107–1117 (2009)
6. DePristo, M., de Bakker, P., Blundell, T.: Heterogeneity and inaccuracy in protein structures solved by x-ray crystallography. Structure 12, 831–838 (2004)
7. Drenth, J.: Principles of Protein x-ray crystallography. Springer, New York (1999)
8. Eissenmesser, E., Millet, O., Labeikovsky, W., Korzhnev, D., Wolf-Watz, M., Bosco, D., Skalicky, J., Kay, L., Kern, D.: Intrinsic dynamics of an enzyme underlies catalysis. Nature 438, 117–121 (2005)
9. Furnham, N., Blundell, T., DePristo, M., Terwilliger, T.: Is one solution good enough. Nature Struct. and Mol. Biol. 13(3), 184–185 (2006)
10. Hastie, T., Tibshirani, R., Friedman, J.: The Elements of Statistical Learning: Data Mining, Inference and Prediction. Springer, Heidelberg (2009)
11. Jensen, L.: Methods in Enzymology, pp. 353–366 (1997)
12. Ji, H., Liu, S.: Analyzing 'omics data using hierarchical models. Nature Biotechnology 28, 337–340 (2010)
13. Kleywegt, G.: Validation of protein crystal structures. Acta Crystallographica (D) 56, 249–265 (2000)
14. Knight, J., Zhou, Z., Gallichio, E., Himmel, D., Friesner, R., Arnold, E., Levy, R.: Exploring structural variability in x-ray crystallographic models using protein local optimization by torsion angle sampling. Acta Crystallographica (D) 64, 383–396 (2008)
15. Knipscheer, P., van Dijk, W., Olsen, J., Mann, M., Sixma, T.: Noncovalent interaction between ubc9 and sumo promoted sumo chain formation. The EMBO Journal 26, 2797–2807 (2007)
16. Koshland, D.: Conformational changes: How small is big enough? Nature Medicine 4, 1112–1114 (1998)
17. Li, F., Yang, Y., Xing, E.: From lasso regression to feature vector machine. Neural Information Processing Systems (NIPS) 18 (2005)
18. Liu, Q., Yuan, Y., Shen, B., Chen, D., Chen, Y.: Conformational flexibility of a ubiquitin conjugation enzyme (e2). Biochemistry 38, 1415–1425 (1999)
19. Meinshausen, N., Rocha, B., Yu, B.: Discussion: A tale of three cousins: Lasso, l2boosting and dantzig. Annals of Statistics 35, 2373–2384 (2007)
20. Nigham, A., Hsu, D.: Protein conformational flexibility analysis with noisy data. Journal of Computational Biology 15, 813–828 (2008)
21. Ringe, G., Petsko, G.: Study of protein dynamics by x-ray diffraction. Methods in Enzymology 131, 389–433 (1986)
22. Singh, R., Berger, B.: Chaintweak: Sampling from the neighbourhood of a protein conformation. In: Pacific Symposium on Biocomputing, pp. 52–63 (2005)
23. Tatham, M., Kim, S., Yu, B., Jaffray, E., Song, J., Zheng, J., Rodriguez, M., Hay, R., Chen, Y.: Role of n-terminal site of ubc9 in sumo-1,-2, and -3 binding and conjugation. Biochemistry 42, 9959–9969 (2003)
24. Terwilliger, T., Grosse-Kunstleve, R., Afonine, P., Adams, P., Moriarty, N., Zwart, P., Read, R., Turk, D., Hung, L.W.: Interpretation of ensembles created by multiple iterative rebuilding of macromolecular models. Acta Crystallographica (D) 63, 597–610 (2007)

25. Tibshirani, R.: Regression shrinkage and selection via the lasso. Journal of the Royal Stat. Soc. Series B 58, 267–288 (1996)
26. Vitkup, D., Ringe, D., Karplus, M., Petsko, G.: Why proteins r-factors are so large: a self consistent analysis. Proteins 46, 345–354 (2002)
27. Volkman, B., Lipson, D., Wemmer, D., Kern, D.: Two state allosteric behaviour in a single domain signalling protein. Science 291, 2429–2433 (2001)
28. Wachter, A., Biegler, T.: On the implementation of a primal-dual interior point filter line search algorithm for large-scale nonlinear programming. Mathematical Programming 106, 25–57 (2006)
29. Xu, H., Caramanis, C., Mannor, S.: Robust regression and lasso. Neural Information Processing Systems, NIPS (2008)

Data Structures for Accelerating Tanimoto Queries on Real Valued Vectors

Thomas G. Kristensen and Christian N.S. Pedersen

Bioinformatics Research Center
Aarhus University
C. F. Møllers Alle 8, 8000 Aarhus C.,
Denmark
{tgk,cstorm}@birc.au.dk

Abstract. Previous methods for accelerating Tanimoto queries have been based on using bit strings for representing molecules. No work has gone into examining accelerating Tanimoto queries on real valued descriptors, even though these offer a much more fine grained measure of similarity between molecules. This study utilises a recently discovered reduction from Tanimoto queries to distance queries in Euclidean space to accelerate Tanimoto queries using standard metric data structures. The presented experiments show that it is possible to gain a significant speedup and that general metric data structures are better suited than a data structure tailored for Euclidean space on vectors generated from molecular data.

1 Introduction

When developing novel drugs, researchers are faced with the task of selecting a subset of all commercially available molecules for further experiments. There are more than 8 million available molecules [7], and it is therefore not possible to perform computationally expensive calculations on each one. The need therefore arise for fast screening methods for identifying the molecules that are most likely to have an effect on a disease or illness. It is often the case that a molecule with some effect is already known, e.g. from an already existing drug. An obvious initial screening method presents itself, namely to identify the molecules which are similar to this known molecule. To implement this screening method one must decide on a representation of the molecules and a similarity measure between representations of molecules. Several representations and similarity measures have been proposed [5,10,17]. This study focuses on *real valued molecular descriptors*.

Vectors are often used for representing molecular structures when searching for chemical compounds with similar properties. The entries of these vectors can be binary, in which case they are referred to as fingerprints, or real valued, in which case they are referred to as descriptors. A diverse set of similarity measures are available for dealing with these vectors [18]. This study focuses on the Tanimoto coefficient, which is applicable to fingerprints as well as descriptors.

V. Moulton and M. Singh (Eds.): WABI 2010, LNBI 6293, pp. 28–39, 2010.

The *Tanimoto coefficient* $T(A,B)$ between two vectors $A, B \in \mathbb{R}^n$ is calculated as

$$T(A,B) = \frac{AB}{||A||^2 + ||B||^2 - AB}.$$

A *Tanimoto query* consists of a target vector A and a minimum coefficient t. The result of a Tanimoto query are the vectors B in a database for which $T(A,B) \geq t$.

If A and B are binary, that is if their entries take on values either zero or some entry specific value, $T(A,B)$ will lie in the interval $[0,1]$. In that case it has been proven that the Tanimoto distance, defined as $1-T$, is a metric [13,11]. This means that the triangle inequality holds and standard data structures, such as μ-, vp-, M- and GNAT-trees [6,19,4,3], can be used for accelerating Tanimoto queries.

If the entries of the vectors are allowed to take on arbitrary values, [18] states that the codomain of T extends to $[-\frac{1}{3}, 1]$, and $1-T$ ceases to be a metric [11]. The metric data structures for dealing with fingerprints are therefore no longer applicable. However, it is possible to convert the query into that of a distance query in Euclidean space, probably the best known metric. This not only allows the use of data structures based on the triangle inequality, but it also enables the use of data structures tailored for Euclidean space, such as the kd-tree [2], even for binary valued vectors.

Previous studies have focused on decreasing the query time of Tanimoto queries into databases of fingerprint vectors [15,1,8]. This article focusses on speeding up Tanimoto queries into databases of real valued descriptors. These descriptors are able to contain more fine grained information such as molecular weight. Results from fingerprints can not be used on real valued descriptors, as those techniques were tailored for binary vectors.

2 Data Structures

A distance query consists of a query point q and a query radius q_r. When performed on a set of points P, a distance query should return all the points $p \in P$ for which $||q-p|| \leq q_r$. A Tanimoto query can be transformed into a distance query using the result from [9] which states that if $A, B \in \mathbb{R}^n$ and $t \in]0,1]$. Then $T(A,B) \geq t$ if and only if

$$||\frac{t+1}{2t}A - B|| \leq \frac{\sqrt{-4t^2 + (t+1)^2}}{2t}||A||.$$

Furthermore, if $t \in [-\frac{1}{3}; 0[$ then $T(A,B) \geq t$ if and only if

$$||\frac{t+1}{2t}A - B|| \geq \frac{\sqrt{-4t^2 + (t+1)^2}}{2t}||A||.$$

Figure 1 illustrates both relations in the plane.

Distance queries can be accelerated using a variety of data structures. This study examine three of these, namely kd-trees, vp-trees and GNATs.

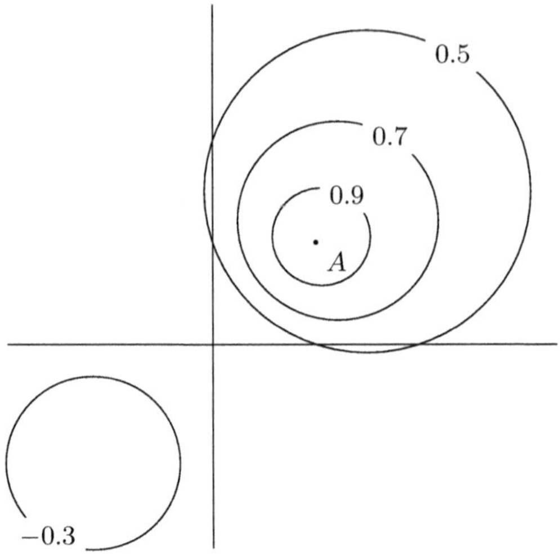

Fig. 1. An example in the plane. All the vectors inside the circle labelled 0.9 have a Tanimoto coefficient larger than 0.9 to A. All vectors outside the circle labelled -0.3 have a Tanimoto coefficient larger than -0.3 to A.

A kd-tree is constructed by recursively dividing a set of points based on one of their entries [2]. On the ith level of the kd-tree the (i mod k)th entry of the points is used to construct a hyper plane dividing the points into two equally sized parts. Figure 2a illustrates this for a set of five points in two dimensional space. kd-trees are used for searching for points, retrieving nearest neighbours or, as in this case, performing distance queries. A distance query can discard parts of the tree if the search radius of the distance query does not overlap with the split plane defined by the nodes in the tree. kd-trees are memory efficient, easy to construct and easy to implement. However, some studies indicate that they have poor performance on high dimensional data [16].

vp-trees are based on *vantage points*, stored in the inner nodes of the trees [19]. Each node has two subtrees: one for the points closest to the nodes vantage point, and one for those further away. Figure 2b illustrates a vp-tree on a set of points, with p_1 as the vantage point. vp-trees are constructed by recursively selecting vantage points and splitting the points into two subsets according to their distance from the vantage points. Each node stores the vantage point p along with the vantage point radius p_r within which all the closest points are located. A distance query in a vp-tree is performed by traversing the tree, pruning away subtrees when it can be proven that the points in the subtree cannot possibly fall within the query radius. This is done by calculating the distance from the vantage point p to the query point q and using this along with the query radius q_r and the vantage point radius p_r. If $||p - q|| + r_q < r_p$ it is the case that the query area falls fully within the closest points and it is therefore possible to exclude the subtree representing points far away from any further

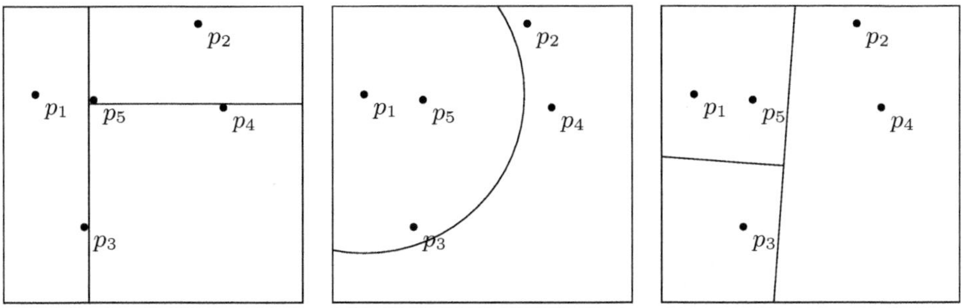

a) kd-tree b) vp-tree c) GNAT

Fig. 2. Illustration of the three different data structures using the same five points. The kd-tree first splits on the x-axis and thereafter on the y-axis. The vp-tree has p_1 as its first vantage point. The GNAT has two reference points, p_1 and p_4.

search. Likewise, if $||p - q|| - q_r > p_r$, the closest points can be skipped. Unlike the kd-tree, vp-trees are designed for general metrics.

A GNAT (Geometric Near-neighbor Access Tree) is similar to a vp-tree in that it is based on assigning points to a closest reference point [3]. GNAT nodes have m subtrees, each with a reference point p. All other points are assigned to the subtree to whose reference point they are closest, as illustrated in Fig. 2c. As in the vp-tree, a radius p_r is stored with each reference point. In the simplest case, m is two and the points are divided into two sets, U and V, according to their distance to the two reference points u and v. Given a query point q and a query radius q_r it is possible to decide if the points in the two sets should be visited using the reference points. When calculating $||u - q||$ it is possible to skip all the points $v_i \in V$ if

$$\min_{v_i \in V}(||u - v_i||) > ||u - q|| + q_r$$

or

$$\max_{v_i \in V}(||u - v_i||) < ||u - q|| - q_r.$$

Therefore, each node stores $\min_{v_i \in V}(||u - v_i||)$ and $\max_{v_i \in V}(||u - v_i||)$ to accelerate queries. These extra calculations require more computations but the gain is more discriminatory power.

Some additional speed can be obtained by collapsing subtrees with under *max-points* points into leafs. For kd and vp-trees, this is the only parameter, whereas the GNAT also has the extra parameter m. Searching for points *outside* the query radius (when t is less than zero) can be done by modifying the pruning technique slightly for all three data structures. This has been done in the experiments; how it is done is trivial and outside the scope of this paper. The three trees are illustrated on a larger data set in Fig. 3.

Fig. 3. The three data structures illustrated on a data set containing 300 points

3 Experimental Setup

To examine the performance of the methods, experiments have been performed on real descriptors, calculated from commercially available molecules from ZINC version 8 [7]. Two different sets of descriptors were generated: one using the Molegro Virtual Docker (MVD) [12] descriptor wizard, and one using the Chemistry Development Kit (CDK) [14]. The MVD descriptors consist of the following nine values: molecular weight; atom count; heavy atom count; number of rotatable bonds; number of rotatable bonds excluding terminal atoms; number of hydrogen donors; number of hydrogen acceptors; number of rings; and the number of aromatic rings. The CDK descriptors consists of the following 14 values: $a \log p$; $a \log_2 p$; AMR; $x \log p$; molecular weight; bond count; aromatic bond count; rotatable bond count; largest chain; largest pi chain; longest aliphatic chain; element count; aromatic atoms count; and Lipinski's rule of five. In a real setting descriptors will often be normalised to inhibit descriptors with large values to dominate the Tanimoto coefficient. Normalised versions of the descriptors were therefore generated in which values were mean centered and given unit variance.

All data structures were implemented and tested in Python 2.6. The methods were tested on thresholds ranging from -0.3 to 1.0 in 0.05 increments. For all data points 100 different molecules were used as queries into the data structures and the average query time and average pruning degree is presented.

4 Results

Initial experiments were focused on parameter tuning of the three data structures. In these experiments the number of descriptors were kept at 100,000 and the Tanimoto threshold was varied from -0.3 to 1.0 in 0.05 increments. *max-points* was varied from two to 20 in increments of two; for the GNAT, ms of 2, 4, 6 and 8 were tested. Judging from the query times (not presented) the best overall *max-points* for both *k*d- and vp-trees was chosen to be 14, while *max-points* for GNAT was chosen to be 18 and m was chosen to be four. Of the

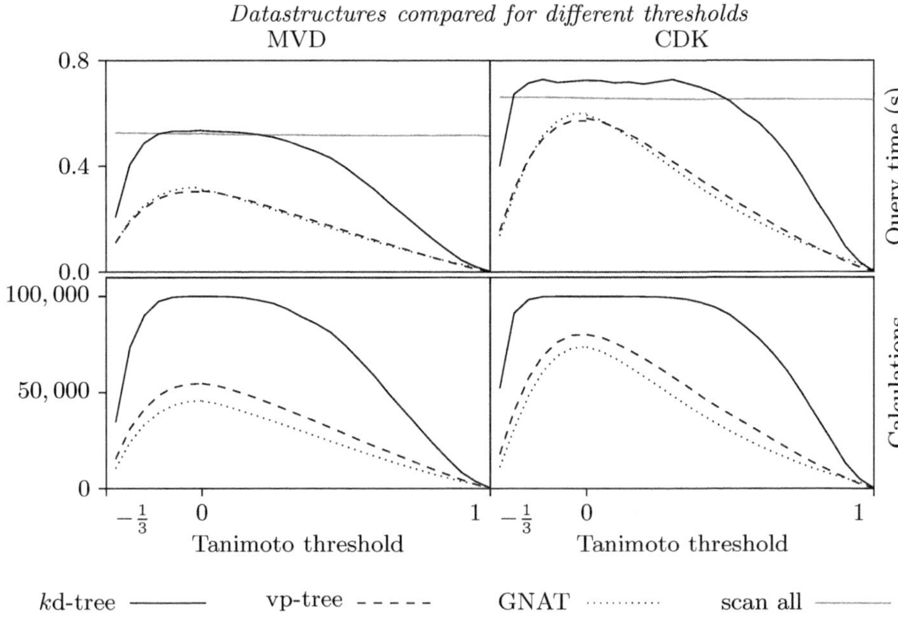

Fig. 4. Comparison of the three data structures on both data sets. The comparison is on both time and number of distance calculations. The threshold is varied on the primary axis while the data base size is kept fixed at 100,000 descriptors.

three data structures, the vp-tree seems to have the most stable query time for different *max-points*.

Figure 4 presents the three methods along with a linear scan measured on query time and the number of distance calculations performed. In the upper part of the graph, where query time is presented, it is clear that the *k*d-tree performs much worse than both the vp-tree and the GNAT. For some thresholds, particularly on the CDK set, it is even worse than a linear scan. The reason for this is easily explained by the bottom part of Fig. 4, in which it is clear that the *k*d-tree performs many more distance calculations than the other two data structures. Not surprisingly, the GNAT performs significantly fewer distance calculations than the vp-tree, due to its tighter bound when pruning subtrees. This does not, however, grant the GNAT faster query times on the tested data, as seen from the top part of Fig. 4.

The data structures were also tested with a fixed threshold of 0.9 and a database size that varied from 100,000 to 1,000,000 in 100,000 increments. The result of this comparison is presented in Fig. 5, from which it is seen that the internal relationship between the three methods seems to repeat itself for larger database sizes, and that all three methods are far superior to a simple linear scan. Interestingly, the vp-tree becomes faster than the GNAT when moving from the MVD to the CDK data base. This is explained by the number of

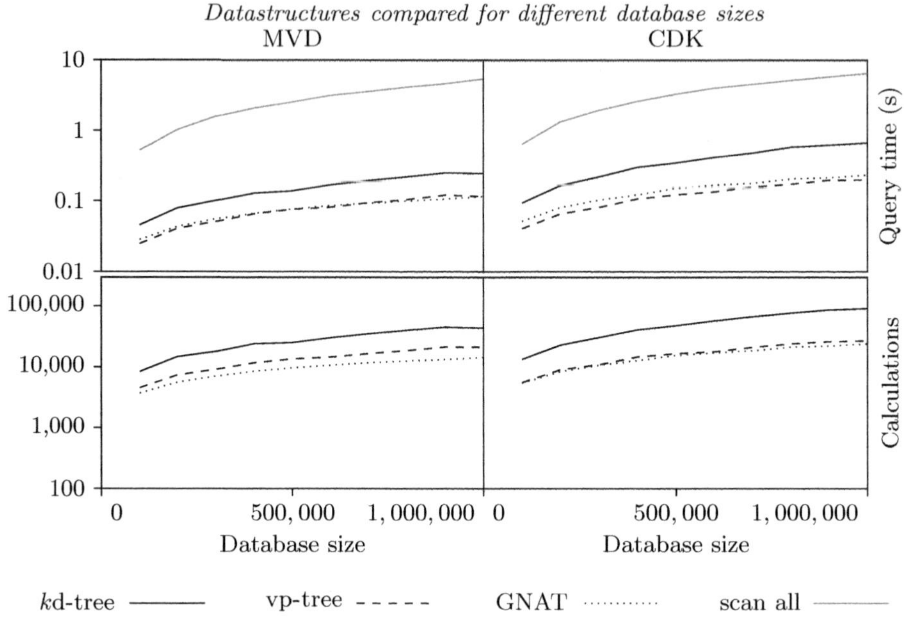

Fig. 5. Comparison of the three data structures on both data sets. The comparison is on both time and number of distance calculations. The data base size is varied on the primary axis while the threshold is kept steady at 0.9. The secondary axis is kept on a log scale to highlight the difference between the data structures.

distance calculations which is practically the same for the two methods on the CDK set, while the vp-tree still has a much smaller overhead. However, this is not to be expected for *all* thresholds.

The data used in the experiments was generated from real descriptors, but an interesting question is if the underlying structure of the data sets matter, or if random entries would give rise to the exact same results. Observing the data structures on random data would also render it possible to examine if the only reason for the difference between the results on MVD and CDK is solely due to dimensionality of descriptors.

As the MVD and CDK data is normalised, random descriptors were drawn from a standard normal distribution. One hundred thousand vectors of length nine and 14 were generated for comparison with MVD and CDK data respectively. The methods were tested with a range of different thresholds, as in the tuning experiments with the real data. Experiments were performed on all three data structures, but for brevity only vp-tree results are presented in Fig. 6; the results for *k*d-trees and GNATs are similar.

From Fig. 6 it is clear that the vp-tree is far slower and prunes far less on the random data than on the original MVD and CDK data. A possible explanation could be that many entries (e.g. number of bonds) in the MVD and CDK sets are

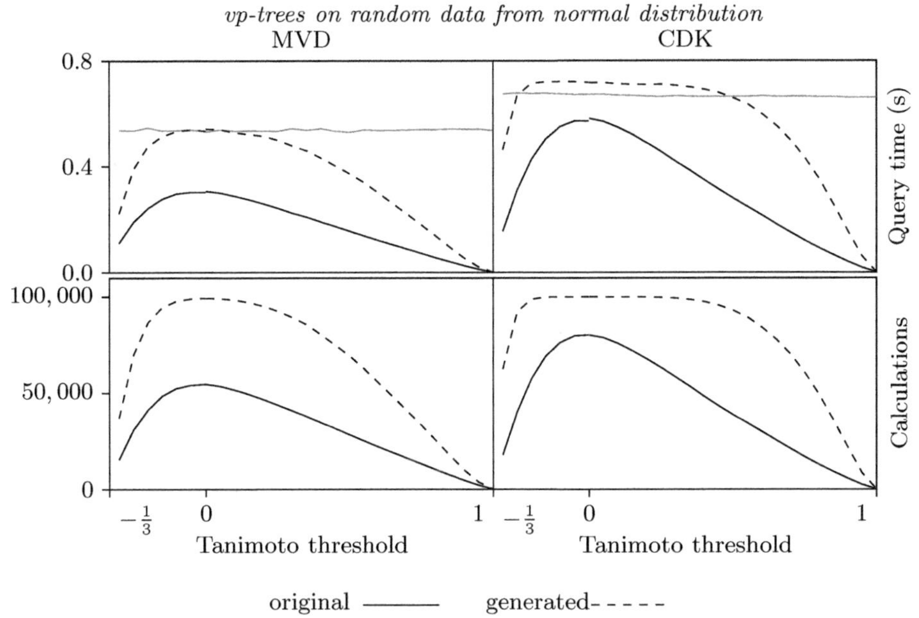

Fig. 6. Comparison of vp-trees performance and pruning degree on real data and data drawn from a normal distribution. The data from the normal distribution has the same dimensionality as the vectors generated by MVD and CDK.

discrete, and therefore data points cluster together rather than being distributed evenly; and therefore they are easy to discriminate against.

To examine if the discrete nature of the data explains the observations, experiments in which descriptors were drawn from a binomial distribution (eight trials, each with success probability of 0.5), were performed. Experiments with other numbers of trials were also performed with similar results. The discrete descriptors were normalised as the original data before the experiments were carried out.

Surprisingly, the queries on discrete random data are just as slow and performs just as many distance calculations as on the data from the normal distribution (Fig. 7). The explanation of these observations might very well be highly correlated entries within the MVD and CDK data. Closer examination of the MVD and CDK data reveals that some entries of the vectors are very highly correlated. For example, the number of aromatic bonds and the number of aromatic atoms of the CDK descriptors have a r^2 of 0.995.

To test if high correlation has an influence on the execution time of queries random data was generated by, for each vector, setting all entries for that vector to the same random number from a standard normal distribution. This entails that all points lie on a line in $\mathbb{R}^9$ for the MVD data and $\mathbb{R}^{14}$ for the CDK data. Running the same experiments show that the vp-tree and the GNAT are very fast on this data; vp-tree results are presented in Fig. 8. The reason why this

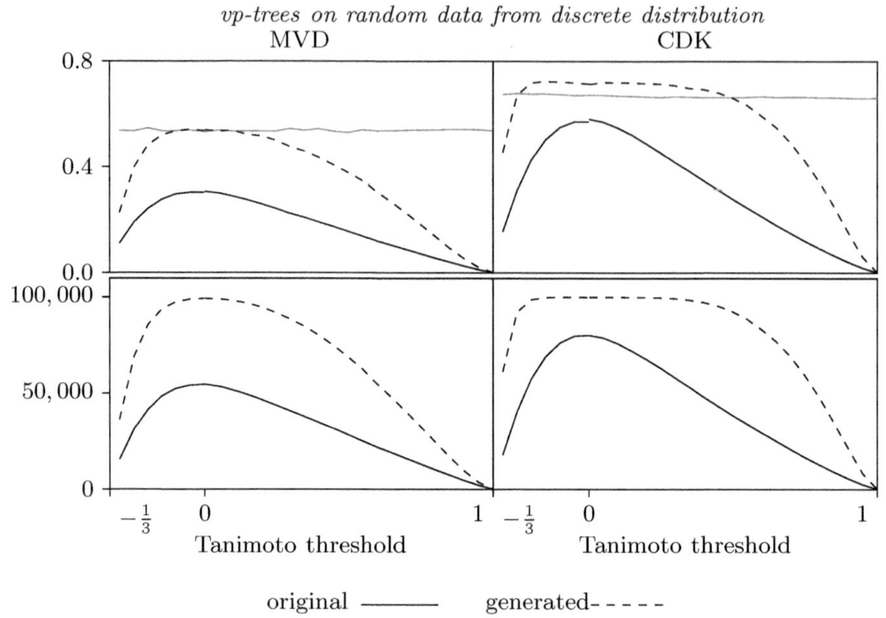

Fig. 7. Comparison of vp-trees performance and pruning degree on real data and data drawn from a binomial distribution. The data has the same dimensionality as the vectors generated by MVD and CDK.

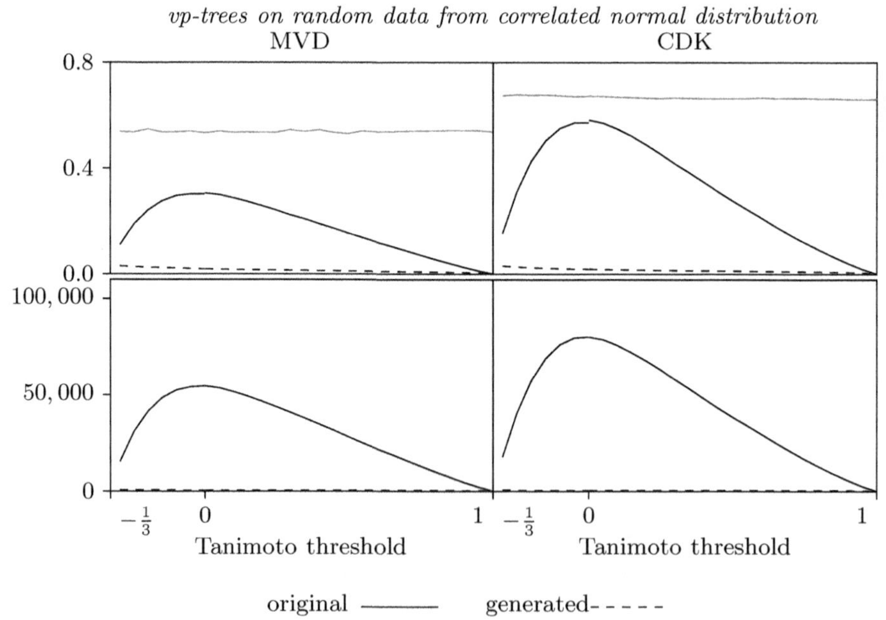

Fig. 8. Comparison of vp-trees performance and pruning degree on very highly correlated data. The data has the same dimensionality as the vectors generated by MVD and CDK.

happens is, that both data structures use a strategy in which the entire subtree is accumulated without performing any distance calculations when it can be reasoned that all points in a subtree lie within a query radius. The *k*d-tree can not use this strategy and its query time is almost identical to that on the real data (Fig. 9).

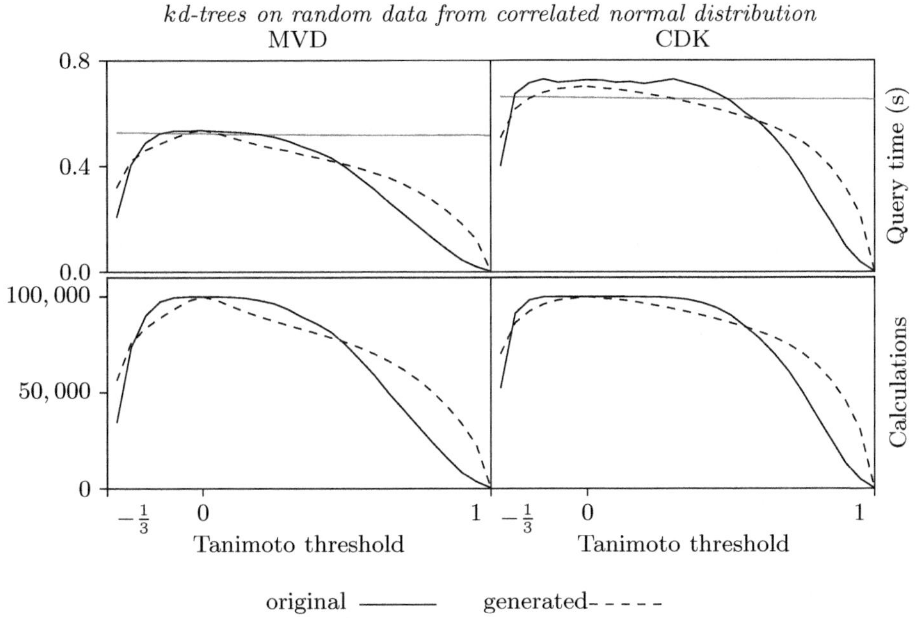

Fig. 9. Comparison of *k*d-trees performance and pruning degree on very highly correlated data. The data has the same dimensionality as the vectors generated by MVD and CDK.

5 Conclusion

The work presented in this paper allows for very fast querying of chemical databases in which molecules are represented as real valued descriptors. The experiments indicate that the vp-tree or the GNAT are the best choice as accelerating data structure, and that *k*d-trees do not work well on chemical data. Furthermore, it seems vp-trees are least affected by changing parameters, and they are therefore recommended as the data structure to use, especially for larger dimensional data where the pruning degree becomes closer to that of GNATs which have a larger overhead. All programs developed and experimental data generated as part of this paper, are available upon request.

Future research could focus on more closely examining the highly correlated data to find out what the underlying dimensionality of the data is, and if special data structures could be created for handling this property. The methods presented here should also work on fingerprints and it would be interesting to see a comparison with data structures tailored to handle these. There are also other metric data structures not covered by this study, and especially IO efficient data structures would be interesting as more data arrives. Parallelising the data structure and the queries would also be a potential area of further investigation.

References

1. Baldi, P., Hirschberg, D.S., Nasr, R.J.: Speeding up chemical database searches using a proximity filter based on the logical exclusive OR. Journal of Chemical Information and Modeling 48(7), 1367–1378 (2008)
2. Bentley, J.L.: Multidimensional binary search trees used for associative searching. Commun. ACM 18(9), 509–517 (1975)
3. Brin, S.: Near neighbor search in large metric spaces. The VLDB Journal, 574–584 (1995)
4. Ciaccia, P., Patella, M., Zezula, P.: M-tree: An efficient access method for similarity search in metric spaces. In: Jarke, M., Carey, M.J., Dittrich, K.R., Lochovsky, F.H., Loucopoulos, P., Jeusfeld, M.A. (eds.) VLDB 1997, Proceedings of 23rd International Conference on Very Large Data Bases, Athens, Greece, August 25-29, pp. 426–435. Morgan Kaufmann, San Francisco (1997)
5. Gillet, V.J., Willett, P., Bradshaw, J.: Similarity searching using reduced graphs. Journal of Chemical Information and Computer Sciences 43(2), 338–345 (2003)
6. Huafeng, X., Agrafiotis, D.K.: Nearest neighbor search in general metric spaces using a tree data structure with a simple heuristic. Journal of Chemical Information and Modeling 43(6), 1933–1941 (2003)
7. Irwin, J.J., Shoichet, B.K.: ZINC: A free database of commercially available compounds for virtual screening. Journal of Chemical Information and Modeling 45(1), 177–182 (2005)
8. Kristensen, T.G., Nielsen, J., Pedersen, C.N.S.: A tree-based method for the rapid screening of chemical fingerprints. Algorithms for Molecular Biology 5(1), 9 (2010)
9. Kristensen, T.G.: Transforming Tanimoto queries on real valued vectors to range queries in Euclidian space. Journal of Mathematical Chemistry (March 2010)
10. Leach, A.R., Gillet, V.J.: An Introduction to Chemoinformatics, rev. ed edn. Kluwer Academic Publishers, Dordrecht (2007)
11. Lipkus, A.H.: A proof of the triangle inequality for the Tanimoto distance. Journal of Mathematical Chemistry 26(1-3), 263–265 (1999)
12. Molegro: Molegro Virtual Docker User Manual version 3.0.0 (2008)
13. Späth, H.: Cluster Analysis Algorithms for Data Reduction and Classification of Objects. Ellis Horwood (1980)
14. Steinbeck, C., Han, Y., Kuhn, S., Horlacher, O., Luttmann, E., Willighagen, E.: The Chemistry Development Kit (CDK): an open-source Java library for chemo- and bioinformatics. Journal of Chemical Information and Computer Sciences 43(2), 493–500 (2003)
15. Swamidass, S.J., Baldi, P.: Bounds and algorithms for fast exact searches of chemical fingerprints in linear and sublinear time. Journal of Chemical Information and Modeling 47(2), 302–317 (2007)
16. Weber, R., Schek, H.J., Blott, S.: A quantitative analysis and performance study for similarity-search methods in high-dimensional spaces. In: VLDB 1998: Proceedings of the 24rd International Conference on Very Large Data Bases, pp. 194–205. Morgan Kaufmann Publishers Inc., San Francisco (1998)

17. Willett, P.: Similarity-based approaches to virtual screening. Biochemical Society Transactions 31(Pt 3), 603–606 (2003)
18. Willett, P., Barnard, J.M., Downs, G.M.: Chemical similarity searching. Journal of Chemical Information and Computer Sciences 38(6), 983–996 (1998)
19. Yianilos, P.N.: Data structures and algorithms for nearest neighbor search in general metric spaces. In: Proceedings of the Fourth ACM-SIAM Symposium on Discrete Algorithms (1993)

Sparsification of RNA Structure Prediction Including Pseudoknots

Mathias Möhl[1,⋆], Raheleh Salari[2,⋆], Sebastian Will[1,3,⋆], Rolf Backofen[1,⋆⋆], and S. Cenk Sahinalp[2,⋆⋆]

[1] Bioinformatics, Institute of Computer Science, Albert-Ludwigs-Universität, Freiburg, Germany
[2] Lab for Computational Biology, School of Computing Science, Simon Fraser University, Burnaby, BC, Canada
[3] Computation and Biology Lab, CSAIL, MIT, Cambridge MA, USA

Abstract. Although many RNA molecules contain pseudoknots, computational prediction of pseudoknotted RNA structure is still in its infancy due to high running time and space consumption implied by the dynamic programming formulations of the problem. In this paper, we introduce sparsification to significantly speedup the dynamic programming approaches for pseudoknotted RNA structure prediction, which also lower the space requirements. Although sparsification has been applied to a number of RNA-related structure prediction problems in the past few years, we provide the first application of sparsification to pseudoknotted RNA structure prediction specifically and to handling gapped fragments more generally - which has a much more complex recursive structure than other problems to which sparsification has been applied. We show that sparsification, when applied to the fastest, as well as the most general pseudoknotted structure prediction methods available, - respectively the Reeder-Giegerich algorithm and the Rivas-Eddy algorithm - reduces the number of "candidate" substructures to be considered significantly. In fact, experimental results on the sparsified Reeder-Giegerich algorithm suggest a linear speedup over the unsparsified implementation.

1 Introduction

Recently discovered catalytic and regulatory RNAs [1,2], exhibit their functionality due to specific secondary and tertiary structures [3,4]. The vast majority of computational analysis of non-coding RNAs have been restricted to nested secondary structures, neglecting pseudoknots - which are "among the most prevalent RNA structures" [5]. For example, Xayaphoummine et al. [6] estimated that up to 30% of the base pairs in G+C-rich sequences form pseudoknots.

However the general problem of pseudoknotted RNA structure prediction is NP-hard. As a result, a number of approaches have been introduced for handling restricted classes of pseudoknots [7,8,9,10,11,12,13]. Condon *et al.* [14] give an

⋆ Joint first authors.

⋆⋆ To whom correspondence should be addressed.

V. Moulton and M. Singh (Eds.): WABI 2010, LNBI 6293, pp. 40–51, 2010.

overview of their structure classes and the algorithm-specific restrictions and Möhl *et al.* [15] develop a general framework showing that all these algorithms follow a general scheme, which they use for efficient alignment of pseudoknotted RNA.

The most general algorithm (with respect to the pseudoknot classes handled) among the above by Rivas and Eddy (R&E) has a running time of $O(n^6)$ time and space consumption of $O(n^4)$. It is therefore too expensive to directly apply this algorithm for large scale data analysis. Unfortunately, even the most efficient algorithm by Reeder and Giegerich (R&G) still has a high running time of $O(n^4)$, although it strongly restricts the class of predictable pseudoknots.

In this paper we introduce the technique of sparsification to the problem of pseudoknotted RNA structure prediction. Sparsification improves the expected running time and space usage of a dynamic programming based structure prediction algorithm without introducing additional restrictions on the structure class handled or compromising the optimality of solutions. Sparsification has been recently applied to improve time and space complexity of various existing RNA-related structure prediction algorithms. In particular, it turned out to be successful for RNA folding for pseudoknot-free structures [16,17], simultaneous alignment and folding [18] as well as RNA RNA interaction prediction [19].

Contributions. We study sparsification of pseudoknotted RNA structure prediction. Algorithms developed for this problem differ from the previously sparsified algorithms by their use of gapped fragments and their more complex recursion structure. Our main contribution in this paper is the solution to the algorithmic challenges due to this increased complexity. Among all DP based pseudoknot prediction algorithms, we focus on the fastest algorithm (R&G) and the most general one (R&E) and develop sparse variants of these dynamic programming algorithms. Due to sparsification, the resulting algorithms need to consider only a limited number of candidates substructures compared to the original algorithms. As a result, we analyze the theoretical worst case complexities in terms of the number of candidate substructures. We also present experimental results, comparing our implementations of the original and sparsified *R&G* algorithm. These results suggest a significant (roughly a linear factor) reduction in the number of candidates over the original algorithm.

2 Sparsification of the Reeder and Giegerich algorithm

The R&G algorithm [13] predicts the minimum free energy structure allowing canonical pseudoknots for a sequence S of length n. It extends the Zuker algorithm by adding one more matrix K (for knot), where $K(i,j)$ denotes the energy for the best *canonical* pseudoknot that starts at position i and ends at position j.[1] Canonical pseudoknots are defined as follows. Each pair of base pairs $p_1 = (i,i')$ and $p_2 = (j',j)$ with $i < j' < i' < j$ induces one canonical pseudoknot that consists of two crossing stems $\{(i,i'),(i+1,i'-1),\ldots,(i+d_{i,i'}-1,i'-d_{i,i'}+1)\}$

[1] The original presentation of the algorithm in terms of the ADP framework does not explicitly consider a matrix K but only a motif *knot*.

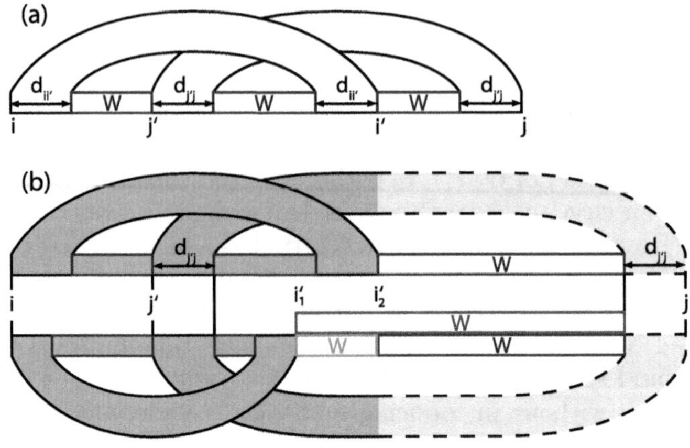

Fig. 1. Recursion for canonical pseudoknots (a) and their sparsification (b)

and $\{(j', j), (j'+1, j-1), \ldots, (j'+d_{j',j}-1, j-d_{j',j}+1)\}$ where the stacking length of the two stems, $d_{i,i'}$ and $d_{j',j}$, respectively, is chosen as large as possible such that still all base pairs are valid Watson-Crick base pairs.

To allow for sparsification, we restrict the scoring scheme slightly such that the energy of a canonical pseudoknot only depends on the left ends of its base pairs[2] and hence can be described as PK-Energy $(i, d_{i,i'}, j', d_{j',j})$. Then,

$$K(i,j) = \min_{i',j'} \text{score}\,(i, j', i', j) \tag{1}$$

with score $(i, j', i', j) =$

$$\begin{pmatrix} \text{PK-Energy}\,(i, d_{i,i'}, j', d_{j',j}) + \\ W(i+d_{i,i'}, j'-1) + W(j'+d_{j',j}, i'-d_{i,i'}) + W(i'+1, j-d_{j',j}) \end{pmatrix}. \tag{2}$$

As shown in Fig. 1(a), for each canonical pseudoknot starting at i and ending at j the recursion decomposes into the pseudoknot itself and the three fragments in-between its two crossing stems. Such pseudoknots add one case in the computation of a matrix entry $W(i,j)$, which, as in the Zuker algorithm, contains the optimal energy of a substructure starting at position i and ending at position j. Due to the restriction to canonical pseudoknots, the recursion of R&G minimizes only over all possible instances of i' and j', because the maximal stacking lengths $d_{i,i'}$ and $d_{j',j}$ are uniquely determined once i' and j' are fixed. Furthermore, Reeder and Giegerich note that the maximal stacking length $d_{x,y}$ can be precomputed for all x, y in $O(n^3)$ time and stored in an $O(n^2)$ table.

In order to sparsify the algorithm, we develop an appropriate notion of a *candidate* such that it is not necessary to minimize over all possible i' and j' but only over the candidates.

[2] The restricted scoring scheme does not distinguish between G-C and G-U base pairs in pseudoknot-stems, since their left ends are identical.

Definition 1 (R&G candidate). *Let $i < j' < i'_1 < i'_2$ and $d_{j',j} \leq i'_1 - j'$. Then i'_1 dominates i'_2 with respect to $(i, j', d_{j',j})$ iff*

$$\text{score}_{i'_2}(i, j', i'_2) \geq \text{score}_{i'_2}(i, j', i'_1), \ \textit{where}$$

$$\begin{aligned}\text{score}_{i_c}(i, j', i') := &\ \text{PK-Energy}\,(i, d_{i,i'}, j', d_{j',j}) \\ &+ W(i + d_{i,i'}, j' - 1) + W(j' + d_{j',j}, i' - d_{i,i'}) + W(i' + 1, i'_c).\end{aligned}$$

We say that i'_2 is a candidate with respect to $(i, j', d_{j',j})$ if there does not exist any i'_1 that dominates it.

The notion of a candidate is visualized in Fig. 1(b). There, i'_1 dominates i'_2 if the score for the gray area at the top (including the dashed part whose exact position is not determined) is not better than the score for the corresponding gray area at the bottom plus the green part. Note that these scores (and hence the candidate i') depend only on i, j', and $d_{j',j}$ and are independent of $d_{i,i'}$ and j. The following lemma shows that the notion of a candidate given in Def. 1 is suitable for sparsification, i.e. some i' needs to be considered in the recursion (for all j) only if it is a candidate, because otherwise it is dominated by a candidate that yields a better score.

Lemma 1 (R&G sparsification). *Let i'_2 be dominated by i'_1 with respect to some $(i, j', d_{j',j})$. Then for all j it holds* $\text{score}\,(i, j', i'_1, j) \leq \text{score}\,(i, j', i'_2, j)$.

Proof. We start with the inequality of Def. 1 and add $W(i'_2+1, j-d_{j',j})$ on both sides. Then the claim follows immediately from $W(i'_1 + 1, j - d_{j',j}) \leq W(i'_1 + 1, i'_2) + W(i'_2+1, j-d_{j',j})$. In Fig. 1(b) this corresponds to the fact that the score for the red box is at least as good as the score from the green and the blue box together. This triangle inequality holds by the correctness of the (unsparsified) algorithm: For all $x < y < z$ we have $W(x, y) + W(y + 1, z) \leq W(x, z)$ since the concatenation of the best structures for the ranges (x, y) and (y, z) always forms a valid structure for the range (x, z) with score $W(x, y) + W(y + 1, z)$ which is hence never better than the optimal score $W(x, z)$ for that range. □

The sparsified algorithm maintains lists L_i of candidates for each pair $(j', d_{j',j})$ since only the lists for one i need to be maintained in memory at the same time. Whenever in the computation of some score(i, j', i', j) the i' is considered the first time for this i and j', it is checked whether it is a candidate and if so, it is added to the respective list. For all other instances of j, i' is then considered only if it is contained in the list. The sparsified algorithm is given by the following pseudo-code ($n := |S|$).

for $i := n$ to 1 **do**
    for all $d_{j',j}, j' \leq n$ **do** $L_i(j', d_{j',j})$:=empty list;
    for $j := i + 3$ to n **do**
        $K(i, j) := \infty$
        for $j' := i + 1$ to $j - 2$ **do**
            // check new elements for candidacy

7: **for** $i_c := \max\{j' + d_{j',j}, \text{checked}_{i,j',d_{j',j}} + 1\}$ to $j - d_{j'j}$ **do**
8: **if** $\text{score}_{i_c}(i, j', i_c) < \text{score}_{i_c}(i, j', i')$ for all $i' \in L_i(j', d_{j',j})$ **then**
9: add i_c to $L_i(j', d_{j',j})$
10: **end if**
11: **end for**
12: $\text{checked}_{i,j',d_{j',j}} := \max(\text{checked}_{i,j',d_{j',j}}, j - d_{j',j})$
13: // *iterate over all candidates*
14: $K_{i,j',j} := \infty$
15: **for all** $i' \in L_i(j', d_{j',j})$ **do**
16: $K_{i,j',j} := \min\{K_{i,j',j}, \text{score}(i, j', i', j)\}$
17: **end for**
18: $K(i,j) := \min\{K(i,j), K_{i,j',j}\}$
19: **end for**
20: compute matrix entries $V(i,j)$ and $W(i,j)$ as in Wexler *et al.*
21: $W(i,j) := \min(W(i,j), K(i,j))$
22: **end for**
23: **end for**

The candidate lists are initialized in line 2. In lines 7 to 11 all new values i_c that have not been considered so far, are tested for candidacy. Here, $\text{checked}_{i,j',d_{j',j}}$ denotes the largest i' that has been checked for candidacy in list $L_i(j', d_{j',j})$.

Lines 14 to 17 compute scores $\text{score}(i, j', i', j)$ for all candidates i'. In line 20, we compute $W(i,j)$ and $V(i,j)$ as in the sparsified pseudoknot-free structure prediction approach due to Wexler *et al.* [16]. The computation of matrices K and W is interleaved such that all entries $K(i,j)$ and $W(i,j)$ are computed before all entries $K(i',j')$ and $W(i',j')$ for $i \leq i' \leq j' \leq j$ and $i \neq i'$ or $j \neq j'$.

Complexity Analysis. Whereas the original algorithm requires $O(n^4)$ time (for $n = |S|$), the sparsified variant requires $O(n^3 L)$ time where L is the total size for all candidate lists of some i i.e. $L := \max_i \sum_{j',d_{j',j}} |L_i(j', d_{j'.j})|$. Obviously, $L \leq n$. In order to maintain the asymptotic space complexity $O(n^2)$ of the original algorithm, we do not maintain all lists $L_i(j', d_{j',j})$ in memory but only the lists with $d_{j',j} \leq k$ where $k > 0$ is a small constant. Please note that to keep presentation simple, we didn't make this explicit in the pseudo-code. Since the maximal stacking length is usually small, there are only very few instances of j with $d_{j',j} > k$ such that for those few j it is cheap to consider all i' as candidates. Hence, we store $O(kn) = O(n)$ candidate lists each requiring at most $O(n)$ space.

3 Sparsification of the Rivas and Eddy Algorithm

The class of structures predicted by the R&E algorithm [8], here called class of R&E structures, is the most general RNA secondary structure prediction algorithm described in the literature [14]. To keep presentation simple we explain the sparsification strategy for a base-pair maximization algorithm that handles the R&E structure class. Finally, we motivate that sparsification can be transferred to the R&E energy minimization algorithm.

First, we give recursions of base pair maximization for R&E structures. Note that the recursions are intentionally very close to the recursions of the R&E energy minimization algorithm. After initialization for $i \geq j$ and $k \geq l$

$$W(i,j) = \begin{cases} 0 & \text{if } i = j \text{ or } i = j+1 \\ -\infty & \text{if } i > j+1 \end{cases} \quad \text{and} \quad \begin{array}{l} W(i,j;k,l) = -\infty \text{ if } j < i \text{ or } l < k \\ W(i,i;k,k) = \text{bp}(i,k) \end{array}$$

where $\text{bp}(i,j) = \begin{cases} 1 & \text{if } S_i, S_k \text{ complementary} \\ -\infty & \text{otherwise,} \end{cases}$ is the *base pair contribution*, the recursions (R&E recursions) are given for $1 \leq i < j < k < l \leq |S|$ as

$$W(i,j) = \max \begin{cases} W(i,j-1) & \texttt{(12')} \\ \text{bp}(i,j) + W(i+1,j-1) & \texttt{(1'21')} \\ \max_{j'} W(i,j'-1) + W(j',j) & \texttt{(12)} \\ \max_{j',k',l'} W(i,j'-1;k'+1,l'-1) + W(j',k';l',j) & \texttt{(1212)} \end{cases}$$

$$W(i,j;k,l) = \max \begin{cases} W(i+1,j;k,l) & \texttt{(1'2G2)} \\ W(i,j-1;k,l) & \texttt{(12'G1)} \\ W(i,j;k+1,l) & \texttt{(1G2'1)} \\ W(i,j;k,l-1) & \texttt{(1G12')} \\ \max_{j'} W(i,j') + W(j'+1,j;k,l) & \texttt{(12G2)} \\ \max_{j'} W(i,j'-1,j;k,l) + W(j',j) & \texttt{(12G1)} \\ \max_{l'} W(i,j;l'+1,l) + W(k,l') & \texttt{(1G21)} \\ \max_{l'} W(i,j;k,l'-1) + W(l',l) & \texttt{(1G12)} \\ \max_{j',k'} W(i,j'-1;k'+1,l) + W(j',j;k,k') & \texttt{(12G21)} \\ \max_{j',k'} W(i,j'-1;k,k'-1) + W(j',j;k',l) & \texttt{(12G12)} \\ \max_{k',l'} W(i,j;k'+1,l'-1) + W(k,k';l',l) & \texttt{(1G212)} \\ \max_{i',j'} W(i,i'-1;j'+1,j) + W(i',j';k,l) & \texttt{(121G2)}. \end{cases}$$

It is easy to check that $W(1,|S|)$ is the maximal number of base pairs in a *R&E* structure of S, because the recursions perform the same decompositions as the original R&E recursions. Note that $W(i,j;k,l)$ is the maximal number of base pairs in structures with at least one base pair that spans the gap. We label each recursion case in a way that illustrates the type of the decomposition of this case. The idea of these labels is taken from Möhl *et al.* [15], where we developed a type system for decompositions, which there are called splits. For this reason, we call these labels split types, however, we won't need any details of the typing system. The decomposition by R&E is illustrated in Figure 2.

A *fragment* is defined as a set of positions of the fixed sequence S. The fragments corresponding to matrix entries in the *R&E* recursion can be described conveniently by their boundaries. We distinguish *ungapped fragments* $F = \{i,\ldots,j\}$, written (i,j), and *1-gap fragments* $F' = \{i,\ldots,j\} \cup \{k,\ldots,l\}$, written $(i,j;k,l)$ where i, j, k, l, are called *boundaries* of respective F or F'. A *split* of a fragment F is a tuple (F_1, F_2) such that $F = F_1 \cup F_2$ and $F_1 \cap F_2 = \emptyset$.

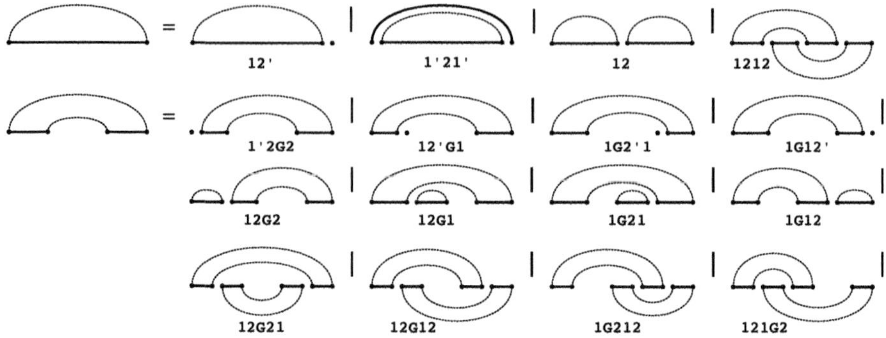

Fig. 2. Decomposition for R&E base pair maximization annotated with labels, i.e. split types, of the corresponding recursion cases

For our sparsification approach, we will show that in each recursion case, certain optimally decomposable fragments do not have to be considered for computing an optimal solution, because each decomposition using these fragments can be replaced by a decomposition using a smaller fragment. We define optimal decomposability with respect to the split type of a R&E recursion case.

Definition 2 (Optimally decomposable). *A fragment F is* optimally decomposable by a split of type T (T-OD) *iff there is a split (F_1, F_2) that occurs in recursion case T and $W(F_1) + W(F_2) \geq W(F)$.*

A fragment F is optimally decomposable w.r.t a set of split types $\mathcal{T}$ ($\mathcal{T}$-OD) *iff F is T-OD for some $T \in \mathcal{T}$.*

Here, we emphasize that testing T-OD for a fragment F is simple in a run of the DP algorithm. After evaluating the case T in the computation of $W(F)$, one compares the maximum of the case to $W(F)$. For example, a fragment $(i, j; k, l)$ is 12G21-OD iff $W(i, j; k, l) = \max_{j',k'} W(i, j' - 1; k' + 1, l) + W(j', j; k, k')$.

In the following we show that for the maximization in a recursion case T, we do not need to consider T'-OD fragments as second fragment of the split, where T' is from a T-specific set of split types. As an example consider the recursion case 12G21, which splits fragments $(i, j; k, l)$ into $F_1 = (i, j' - 1; k' + 1, l)$ and $F_2 = (j', j; k, k')$. Assume that F_2 is 12G21-OD. Then we can show that every evaluation of $W(F)$ where $W(F) = W(F_1) + W(F_2)$ can be replaced by another at least equally good evaluation that splits F into F_1' and $F_2' \subset F_2$, where F_2' is the second fragment in the 12G21-split of F_2. However, note that the argument is split type specific and cannot be applied e.g. when F_2 is 12G12-OD.

For sparsifying R&E, we define the following sets of split types.

$$\mathcal{T}^{\text{RE}}_{\texttt{12}} = \{\texttt{12}\} \qquad \mathcal{T}^{\text{RE}}_{\texttt{1212}} = \{\texttt{12G2}, \texttt{12G1}, \texttt{1G21}\}$$

$$\mathcal{T}^{\text{RE}}_{\texttt{12G1}} = \mathcal{T}^{\text{RE}}_{\texttt{1G12}} = \mathcal{T}^{\text{RE}}_{\texttt{1G21}} = \{\texttt{12}\} \qquad \mathcal{T}^{\text{RE}}_{\texttt{12G2}} = \{\texttt{12G2}\}$$

$$\mathcal{T}^{\text{RE}}_{\texttt{12G21}} = \{\texttt{12G2}, \texttt{1G12}, \texttt{12G21}\} \qquad \mathcal{T}^{\text{RE}}_{\texttt{12G12}} = \{\texttt{12G2}, \texttt{1G21}, \texttt{12G12}\}$$

$$\mathcal{T}^{\text{RE}}_{\texttt{1G212}} = \{\texttt{12G1}, \texttt{1G21}, \texttt{12G21}\} \qquad \mathcal{T}^{\text{RE}}_{\texttt{121G2}} = \{\texttt{12G2}, \texttt{12G1}, \texttt{121G2}\}$$

These sets are defined such that in a recursion case T, whenever the second fragment of a split (F_1, F_2) of F can be optimally decomposed by a split of a type in $\mathcal{T}_T^{\text{RE}}$, a different split (F_1', F_2') of type T can be applied to F, where $F_2' \subset F_2$. As we show later, this split will be just as good as (F_1, F_2) for computing $W(F)$.

Then, one systematically obtains sparsified recursion equations $W'(i,j)$ and $W'(i,j;k,l)$ from the equations for $W(i,j)$ and $W(i,j;k,l)$ by replacing symbol W by W' and modifying them in the following way. For each case T in the recursion of $W(i,j)$ and $W(i,j;k,l)$ that maximizes over $W(F_1) + W(F_2)$ for respective splits of the fragment $F = (i,j)$ or $F = (i,j;k,l)$, maximize only over fragments F_2 that are not $\mathcal{T}_T^{\text{RE}}$-OD. In an algorithm that evaluates the sparsified recursion, such non-$\mathcal{T}_T^{\text{RE}}$-OD fragments correspond to entries of candidate lists. For example, case 12G21 of W is modified in the equation for $W'(i,j;k,l)$ to

$$\max_{j',k',\ (j',j;k,k') \text{ not } \mathcal{T}_{\texttt{12G21}}^{\text{RE}}\text{-OD}} W'(i,j'-1;k'+1,l) + W'(j',j;k,k') \qquad (\texttt{12G21} \textbf{ of W'}).$$

Theorem 1. *Let W be the matrix of the R&E recursion and W' its sparsified variant, then $W(1,|S|) = W'(1,|S|)$.*

Proof. We show for all $1 \leq i,j,k,l \leq |S|$, $W(i,j) = W'(i,j)$ and $W(i,j;k,l) = W'(i,j;k,l)$. First note that it holds that $W(i,j) \geq W'(i,j)$ and $W(i,j;k,l) \geq W'(i,j;k,l)$. The claim is shown by induction on the fragment size and a case distinction over recursion cases. For the case of split type 12, we show that

$$\max_{j'} W(i,j'-1) + W(j',j) = \max_{j',\ (j',j) \text{ not } \mathcal{T}_{12}^{\text{RE}}\text{-OD}} W'(i,j'-1) + W'(j',j).$$

Let (j',j) be 12-OD for some $j' : i \leq j' \leq j$. By IH, it suffices to find a (smaller) fragment (j'',j), where $j'' > j$ and $W(i,j''-1) + W(j'',j) \geq W(i,j'-1) + W(j',j)$. Either (j',j) is not 12-OD or there is a j'', such that $W(j',j) = W(j',j''-1) + W(j'',j)$ and thus $W(i,j''-1) + W(j'',j) \geq W(i,j'-1) + W(j',j)$ because

$$\begin{aligned} W(i,j''-1) + W(j'',j) &\geq_{\Delta\text{-ineq}} W(i,j'-1) + W(j',j''-1) + W(j'',j) \\ &=_{\text{12-OD}} W(i,j'-1) + W(j',j). \end{aligned}$$

The triangle inequality (Δ-ineq) is an immediate consequence of the correctness of the recursion for W. Thus, for the decompositions of all recursion cases there holds such a corresponding inequation. Analogous arguments can be given for all other modified recursion cases. Exemplarily, we elaborate the argument for the complex case 12G21. Let $F_1 = (i,j'-1;k'+1,l)$ and $F_2 = (j',j;k,k')$, such that (F_1, F_2) is a split of type 12G21 of $(j,j;k,k)$. We need to show for all $\mathcal{T}_{\texttt{12G21}}^{\text{RE}}$-OD fragments F_2 there are non-empty ungapped or 1-gap fragments F_1' and F_2', where $F_1' \cup F_2' = F_2$, $F_1' \cap F_2' = \emptyset$, and $W(F_1 \cup F_1') + W(F_2') \geq W(F_1) + W(F_2)$ and the split $(F_1 \cup F_1', F_2')$ occurs in a recursion case of R&E. Again, either F_2 is not $\mathcal{T}_{\texttt{12G21}}^{\text{RE}}$-OD or one of the following cases applies. Case 1 (12G2): for some j'', $W(j',j;k,k') = W(j',j''-1) + W(j'',j;k,k')$. Then, the claim holds for $F_1' = (j',j''-1)$ and $F_2' = (j'',j;k,k')$ by triangle inequality and split $(F_1 \cup F_1', F_2')$ occurs in recursion case 12G21. Case 2 (2G21): for some k'', $W(j',j;k,k') =$

$W(j', j; k, k'') + W(k'' + 1, k')$. The claim holds for $F_2' = (j', j; k, k'')$. Case 3 (12G21): for some j'', k'', $W(j', j; k, k') = W(j', j''-1; k''+1, k') + W(j'', j; k, k'')$. Again, this satisfies the claim by triangle inequality. □

Algorithm. The recursion equation W' tailors a sparsified dynamic programming algorithm for the evaluation of $W'(1, |S|)$ with very limited overhead. We maintain separate candidate lists for each sparsified recursion case. As already mentioned, the T-OD properties of each fragment F can be easily checked after evaluation of each case of $W(F)$. A fragment is added to a candidate list for recursion case T iff it is not T_T^{RE}-OD. The maximizations are restricted to run only over the candidates in the respective candidate list. Their intended use dictates the exact nature of such candidate lists. For a case T, which splits a fragments T into T_1 and T_2, there are candidate lists for all boundaries of a fragment T_2 that are not adjacent to boundaries of T_1 due to split type T. The list entries are tuples of the adjacent boundaries and the fragment score for T_2. In order to profit from a reduced number of candidates in space, we maintain two three-dimensional slices of the matrix for $W(i, j; k, l)$, storing entries only for the current i and $i+1$. Scores $W(i, j; k, l)$ for larger i are stored for candidates only.

R&E Free Energy Minimization. Sparsification is analogously applied to the energy minimizing R&E algorithm. This algorithm distinguishes several additional matrices that contain minimal energies for fragments (i, j) or $(i, j; k, l)$ under the condition that respectively the base pair (i, j) or base pairs (i, l) and (j, k) or one of them exist. Almost all decompositions in the recursion for these matrices are of discussed split types and are sparsified analogously. The only notable exception is due to internal loops. Internal loops require minimizing over all possible positions of the inner loop base pair, where commonly the loop size is restricted by a constant K such that minimizing takes constant time. However, handling inner loops requires access to entries of non-candidate fragments $(i', j'; k', l')$ for $i \leq i' \leq i+K+2$. This is handled by maintaining matrix slices for i to $i+K+2$ in $O(n^3)$ space, which preserves total space complexity.

Complexity Analysis. The described algorithm profits from sparsification in time and space. Compared to $O(n^6)$ time and $O(n^4)$ space of the unsparsified algorithm (for $n = |S|$), we obtain complexities in the number of candidates. Let Z_T denote the maximal length of a candidate lists for case T and Z denote the total number of entries in all lists. Then, the time complexity is $O(n^2(Z_{12} + Z_{1212}) + n^4(Z_{12G2} + Z_{12G1} + Z_{1G21} + Z_{1G12} + Z_{12G21} + Z_{12G12} + Z_{1G212} + Z_{121G2}))$ and space complexity is $O(n^3 + Z)$. In the worst case, Z_{12}, Z_{12G2}, Z_{12G1}, Z_{1G21} and Z_{1G12} are $O(n)$, Z_{12G21}, Z_{12G12}, Z_{1G212}, Z_{121G2} are $O(n^2)$, and Z_{1212} is $O(n^3)$; finally Z is $O(n^4)$ in the worst case.

4 Experimental Results

In order to evaluate the effect of sparsification on pseudoknotted RNA secondary structure prediction, we implemented original and sparsified variants of the Reeder and Giegerich (R&G) algorithm.

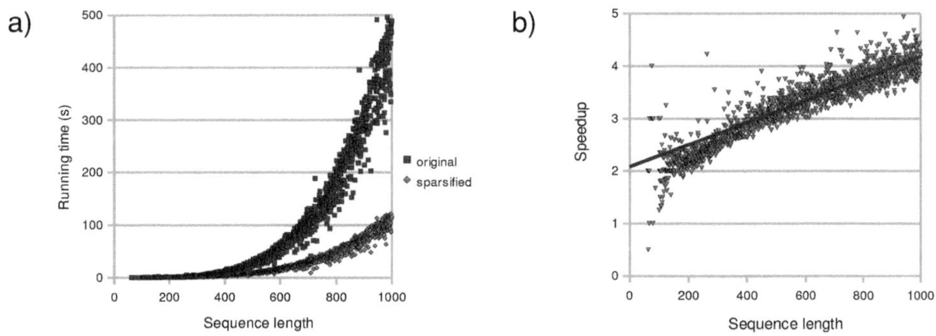

Fig. 3. Running times of the original and sparsified variants of the R&G algorithm

Data Set. We obtained all RNA sequences from PseudoBase[20], which are known to have some pseudoknots in their secondary structures. This set contains 294 sequences that their length is distributed between 76nt and 93399nt. We randomly divided all long sequences into subsequences shorter than 1000nt. Therefore the data set that we used in our experiments contains 1563 sequences with length between 76nt and 1000nt.

Performance. We applied both variants of the R&G algorithm to our data set. Fig. 3 shows the running time of the algorithms on a server with Intel Core Duo CPU at 2.53GHz and 4GB RAM. The results in Fig. 3 show that sparsification significantly improves the running time of the R&G algorithm. As the RNA sequences get longer, the relative performance of the sparsified algorithm (with respect to the non-sparsified ones) improves. Fig. 3.(b) shows the speedup of the sparsified algorithm, which fits well to a linear regression ($R^2 = 0.84$).

Number of candidates. For a better understanding of the effect of sparsification on the R&G algorithm, we measured the number of (i', j') pairs which are checked in each fragment $[i, j]$ in both original and sparsified variants of the algorithm. Note that the number of (i', j') pairs is in order of $O((j - i)^2)$ in the worst case. Fig. 4 shows the average number of (i', j') pairs on fragments of equal length which are checked by the two variants of the algorithm. As expected, this amount is significantly smaller for the sparsified algorithm compared to the original one. Moreover, we observe that as the fragments get longer, the difference between the average number of (i', j') pairs in the sparsified and the original algorithm increases. We define the work load per each fragment $[i, j]$ as the number of candidate (i', j') pairs. Figure 4(b), shows a significant reduction of the work load in the sparsified algorithms. As it can be seen for subsequences of length 1000nt, the work load by the sparsified algorithm is reduced by a factor of about 10 compared to the original algorithm. Note that the work load reduction at fragment length 1000nt does not yield the same speedup for sequences of length 1000nt (here this speedup is about 3.5, confer Fig.3(b)), because for a sequence of length n, all fragments of smaller length are processed by the algorithm.

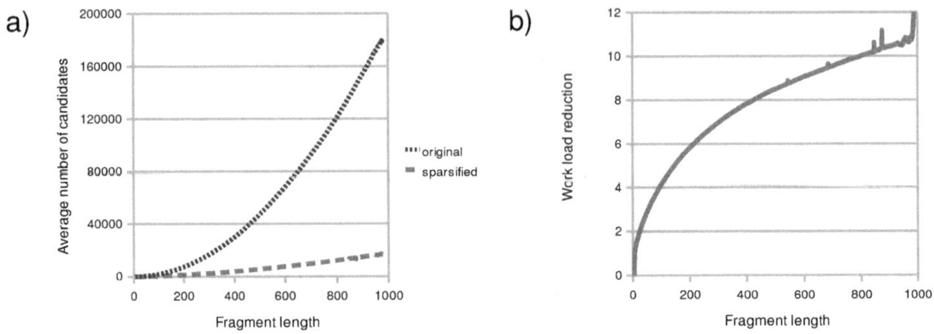

Fig. 4. Average number of (i', j') candidates in the original and sparsified variants of the R&G algorithm

5 Conclusion

The presented work gives two examples for sparsification in the context of gap fragments and a complex recursion structure. Since we successfully sparsified the fastest and the most complex pseudoknot structure prediction algorithm for RNA, it is likely that all other DP-based pseudoknot-algorithm can be sparsified. Thus, the paper motivates further generalization of sparsification for systematic application to complex DP-algorithms as RNA structure prediction algorithms. Even more, by providing detailed examples the paper directly prepares such generalization. Our results from an implementation of the sparsified Reeder and Giegerich algorithm show a significant, presumably even linear, expected work load reduction due to sparsification.

Acknowledgments. This work is partially supported by DFG grants WI 3628/1-1 and BA 2168/3-1 R. Salari was supported by SFU-CTEF funded Bioinformatics for Combating Infectious Diseases Project co-lead by Sahinalp. S.C. Sahinalp was supported by MITACS, NSERC, the CRC program and the Michael Smith Foundation for Health Research.

References

1. Sharp, P.A.: The centrality of RNA. Cell 136(4), 577–580 (2009)
2. Amaral, P.P., Dinger, M.E., Mercer, T.R., Mattick, J.S.: The eukaryotic genome as an RNA machine. Science 319(5871), 1787–1789 (2008)
3. Washietl, S., Pedersen, J.S., Korbel, J.O., Stocsits, C., Gruber, A.R., Hackermuller, J., Hertel, J., Lindemeyer, M., Reiche, K., Tanzer, A., Ucla, C., Wyss, C., Antonarakis, S.E., Denoeud, F., Lagarde, J., Drenkow, J., Kapranov, P., Gingeras, T.R., Guigo, R., Snyder, M., Gerstein, M.B., Reymond, A., Hofacker, I.L., Stadler, P.F.: Structured RNAs in the ENCODE selected regions of the human genome. Genome Res. 17(6), 852–864 (2007)
4. Mattick, J.S., Makunin, I.V.: Non-coding RNA. Hum. Mol. Genet. 15 Spec No. 1, R17–R29 (2006)

5. Staple, D.W., Butcher, S.E.: Pseudoknots: RNA structures with diverse functions. PLoS Biol. 3(6), e213 (2005)
6. Xayaphoummine, A., Bucher, T., Thalmann, F., Isambert, H.: Prediction and statistics of pseudoknots in RNA structures using exactly clustered stochastic simulations. Proc. Natl. Acad. Sci. USA 100(26), 15310–15315 (2003)
7. Lyngso, R.B., Pedersen, C.N.S.: Pseudoknots in RNA secondary structures. In: Proc. of the Fourth Annual International Conferences on Computational Molecular Biology (RECOMB 2000). ACM Press, New York (2000) (BRICS Report Series RS-00-1)
8. Rivas, E., Eddy, S.R.: A dynamic programming algorithm for RNA structure prediction including pseudoknots. Journal of Molecular Biology 285(5), 2053–2068 (1999)
9. Uemura, Y., Hasegawa, A., Kobayashi, S., Yokomori, T.: Tree adjoining grammars for RNA structure prediction. Theoretical Computer Science 210, 277–303 (1999) (Paper as Print Copy)
10. Akutsu, T.: Dynamic programming algorithms for RNA secondary structure prediction with pseudoknots. Discrete Applied Mathematics 104, 45–62 (2000)
11. Deogun, J.S., Donis, R., Komina, O., Ma, F.: RNA secondary structure prediction with simple pseudoknots. In: APBC 2004: Proceedings of the second conference on Asia-Pacific bioinformatics, pp. 239–246. Australian Computer Society, Inc., Darlinghurst (2004)
12. Dirks, R.M., Pierce, N.A.: A partition function algorithm for nucleic acid secondary structure including pseudoknots. J. Comput. Chem. 24(13), 1664–1677 (2003)
13. Reeder, J., Giegerich, R.: Design, implementation and evaluation of a practical pseudoknot folding algorithm based on thermodynamics. BMC Bioinformatics 5, 104 (2004)
14. Condon, A., Davy, B., Rastegari, B., Zhao, S., Tarrant, F.: Classifying RNA pseudoknotted structures. Theoretical Computer Science 320(1), 35–50 (2004)
15. Möhl, M., Will, S., Backofen, R.: Lifting prediction to alignment of RNA pseudoknots. Journal of Computational Biology (2010) (accepted)
16. Wexler, Y., Zilberstein, C.B.Z., Ziv-Ukelson, M.: A study of accessible motifs and rna folding complexity. In: Apostolico, A., Guerra, C., Istrail, S., Pevzner, P.A., Waterman, M.S. (eds.) RECOMB 2006. LNCS (LNBI), vol. 3909, pp. 473–487. Springer, Heidelberg (2006)
17. Backofen, R., Tsur, D., Zakov, S., Ziv-Ukelson, M.: Sparse RNA folding: Time and space efficient algorithms. In: Kucherov, G., Ukkonen, E. (eds.) CPM 2009. LNCS, vol. 5577, pp. 249–262. Springer, Heidelberg (2009)
18. Ziv-Ukelson, M., Gat-Viks, I., Wexler, Y., Shamir, R.: A faster algorithm for RNA co-folding. In: Crandall, K.A., Lagergren, J. (eds.) WABI 2008. LNCS (LNBI), vol. 5251, pp. 174–185. Springer, Heidelberg (2008)
19. Salari, R., Möhl, M., Will, S., Sahinalp, S.C., Backofen, R.: Time and space efficient RNA-RNA interaction prediction via sparse folding. In: Berger, B. (ed.) RECOMB 2010. LNCS, vol. 6044, pp. 473–490. Springer, Heidelberg (2010)
20. van Batenburg, F.H., Gultyaev, A.P., Pleij, C.W., Ng, J., Oliehoek, J.: Pseudobase: a database with RNA pseudoknots. Nucleic Acids Research 28(1), 201–204 (2000)

Prediction of RNA Secondary Structure Including Kissing Hairpin Motifs

Corinna Theis, Stefan Janssen, and Robert Giegerich

Faculty of Technology, Bielefeld University
33501 Bielefeld, Germany
robert@techfak.uni-bielefeld.de

Abstract. We present three heuristic strategies for folding RNA sequences into secondary structures including kissing hairpin motifs. The new idea is to construct a kissing hairpin motif from an overlay of two simple canonical pseudoknots. The difficulty is that the overlay does not satisfy Bellman's Principle of Optimality, and the kissing hairpin cannot simply be built from *optimal* pseudoknots. Our strategies have time/space complexities of $O(n^4)/O(n^2)$, $O(n^4)/O(n^3)$, and $O(n^5)/O(n^2)$. All strategies have been implemented in the program *pKiss* and were evaluated against known structures. Surprisingly, our simplest strategy performs best. As it has the same complexity as the previous algorithm for simple pseudoknots, the overlay idea opens a way to construct a variety of practically useful algorithms for pseudoknots of higher topological complexity within $O(n^4)$ time and $O(n^2)$ space.

1 Introduction

1.1 Biological Relevance of Pseudoknots in RNA Structure

RNA is a chain molecule, the activated form of genetic information in all living organisms. Folding back onto itself, RNA forms secondary structure via base pairing of complementary nucleotides. Stacks of base pairs form helices, akin to the Watson-Crick helix of DNA, but with base pairs A-U, G-C, G-U, and occasionally some non-standard pairs. Ultimately, a tertiary (spatial) structure forms which is essential for biological function. *Pseudoknots* are structural motifs also defined via base pairing patterns, but, as they form late in the folding process, are generally considered as elements of tertiary structure.

Kissing hairpins are a common RNA folding motif belonging to the class of pseudoknots. The unpaired bases of a secondary structure build crossing base pairs by loop-loop interactions (the "kiss") and form a stable tertiary structure motif. Although these motifs have been known for over fifteen years, our understanding of kissing hairpins is still small. Especially viral genomes have been investigated for kissing hairpins, but also bacterial and eukaryotic ones. Researchers showed that kissing hairpins have important duties in a wide variety of RNA mediated processes. For example, they contribute extensively in stabilizing the structure and also play a role in viral plasmid DNA replication [5]

V. Moulton and M. Singh (Eds.): WABI 2010, LNBI 6293, pp. 52–64, 2010.

or RNA synthesis [19]. Li et al. investigated in 2006 the mechanical unfolding of a minimal kissing complex [15]. They discovered that the loop-loop interaction is exceptionally stable.

1.2 RNA Folding of Nested Structures

In RNA structure prediction, there is a dichotomy between prediction of *nested* and *pseudo-knotted* structures. The former is essentially a solved problem, whereas the latter is an active area of research. A structure holds a pseudoknot, if residues $i-j$ and $k-l$ form base pairs such that $i < k < j < l$. This situation is also called a *crossing* interaction. Without any crossing interaction, a structure is *nested*.

Nested structures can be naturally represented as trees, and they lend themselves to structure prediction in $O(n^3)$ time and $O(n^2)$ space. Early algorithms used a simple optimization criterion such as base pair maximization, while today's algorithms of practical relevance [27,14,17] use free energy minimization under an experimentally established thermodynamic model [18]. An improvement to $O(n^3/\log n)$ time for folding of nested structures has recently been contributed by Frid et al. [9], but this approach is not easily adapted to the established energy model. Recent progress in the field of nested structure prediction has been made mostly in the area of a more comprehensive analysis of the folding space [26,4], comparative prediction from multiple sequences [8], or trading the thermodynamic model for machine learning techniques [2].

1.3 Folding Pseudoknots

Structures with pseudoknots are much more difficult to predict. Even under energy models much simpler than what we use in practice, prediction of the optimal pseudo-knotted structure has been shown to be NP-hard [16,1]. This has generated considerable interest in algorithms that solve the problem in polynomial time for restricted topologies of pseudoknots – see the review by Condon and Jabbari [7]. In an investigation of pseudoknot topologies [23], Rødland argues that the full topological complexity of pseudoknots is probably not needed in practical applications. For reasons of space, in the sequel we focus on those approaches which have resulted in realistic programs.

Pseudoknot folding using the established energy model was pioneered by Rivas and Eddy [22]. They presented an $O(n^6)$ time, $O(n^4)$ space algorithm for a fairly general class of pseudoknots. The high effort allows to fold only rather short sequences, and hence, the generality of the algorithm cannot really be exploited. A pragmatic approach was chosen by Reeder and Giegerich with the program *pknotsRG* [20]. They restricted the analysis to the class of *canonical simple recursive* pseudoknots, achieving $O(n^4)$ time, $O(n^2)$ space, and leading to a program widely used [1] today. The program *HotKnots* [21] uses a heuristics to assemble pseudoknots from low-energy helices.

[1] Counting over 200 downloads and over 4,000 submissions per year according to `http://bibiserv.techfak.uni-bielefeld.de/statistics/`

Quite recently, a new algorithm has been published in [6], but at the point of this writing, an implementation was not yet available. Our new approach presented here is an extension of the ideas used with *pknotsRG*, which we will review in necessary detail in Section 2.1.

1.4 Typology of Structures

Notation. Dynamic programming over sequences leads to a decomposition of the given sequence into subwords, typically in all possible ways. Let $S = {}_0s_1 \ldots s_n$ be a sequence over the RNA alphabet $\{A, C, G, U\}$. The use of a fictitious 0-position at the start of S allows us to describe subwords by their bounding positions. For example, subword $(0, n)$ is S and subword $(2, 4)$ is ${}_2s_3s_4$. A subword (i, j) has length $j - i$ and splits seamlessly into subwords (i, k) and (k, j) for $i \leq k \leq j$. This convention avoids a lot of fiddling with ± 1.

We write $s = xyz$ to indicate that s is split into subwords x, y, z. The notation $s = {}_ix_ky_lz_j$ indicates, more concretely, that s is itself a subword of the overall input sequence S with boundaries i and j, and k, l denote the subword boundaries between x, y, z. If all boundaries are independent, a Dynamic Programming algorithm investigating all possible decompositions of this type has at least $O(n^4)$ steps, iterating over all $0 \leq i \leq k \leq l \leq j \leq n$.

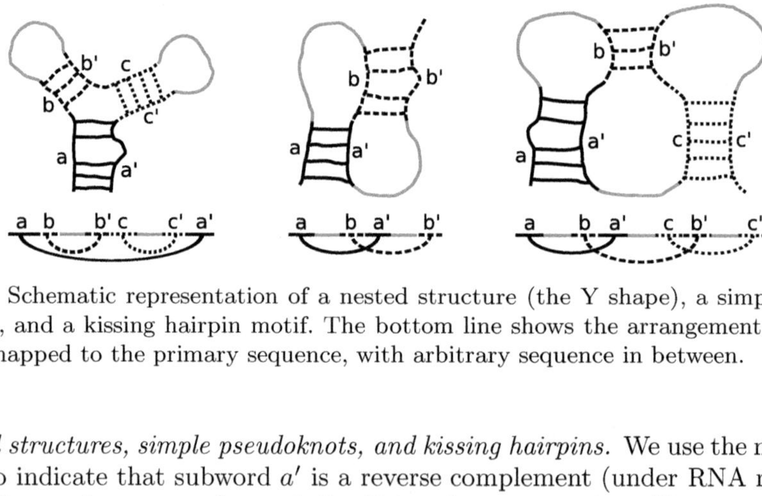

Fig. 1. Schematic representation of a nested structure (the Y shape), a simple pseudoknot, and a kissing hairpin motif. The bottom line shows the arrangement of helix parts mapped to the primary sequence, with arbitrary sequence in between.

Nested structures, simple pseudoknots, and kissing hairpins. We use the notation axa' to indicate that subword a' is a reverse complement (under RNA rules) of a, and hence the two can form a helix. Using these conventions, Figure 1 sketches three types of RNA structures, together with their associated sequence decomposition. The first is a nested structure, the so-called Y-shape, the second a simple pseudoknot (sometimes called H-type), and the third is a kissing hairpin structure, which is our specific concern here. We shall reserve the word "pseudoknot" for simple pseudoknots here, to distinguish them from kissing hairpins. When we allude to pesudoknots with a more complex topology than these two classes, we shall explicitly say so.

To evaluate the folding energy of a kissing hairpin motif on subword s, we need to split $s = aubva'wcxb'yc'$. The subwords named u, v, w, x, y can attain arbitrary (sub)structures, so kissing hairpins (as well as pseudoknots) may be embedded within each other.

2 Three Strategies for Kissing Hairpin Prediction

2.1 The Combined Power of Canonization Rules and Non-ambiguous Dynamic Programming

Canonization. The algorithm of *pknotsRG* reduces computational complexity by imposing three canonization rules on the pseudoknots it considers:

Rule 1: In a helix $s = aua'$, a and a' are perfect helices.

Rule 2: In a helix $s = aua'$, a and a' extend towards each other maximally according to the rules of base pairing, except the following case:

Rule 3: With crossing helices as in $aubva'wb'$, Rule 2 might imply a negative length of v. We set $v = \varepsilon$ and both helices meet at an arbitrary position.

Note that these rules are imposed on pseudoknots only, the search space of nested structures remains untouched. The beneficial effect of these rules is that maximal helices of form ${}_iaza'_j$ can be precomputed, and a canonical split into a pseudoknot of form $s = aubva'wb'$ is uniquely characterized by four moving boundaries only, more precisely as $s = {}_iau_kbva'_lwb'_j$. This is the key to achieve $O(n^4)$ time, $O(n^2)$ space efficiency. For details, we refer to [20]. There, it is shown that while an optimal, pseudoknotted structure P may not satisfy the canonicity constraints, there is a near-optimal pseudoknot P_{can} which does. However, minimum free energy folding might deliver an unknotted structure U with free energy such that $E(P) \leq E(U) \leq E(P_{can})$. U will be returned without a hint to P_{can}, and hence to the potential existence of P. At this point, computing with canonical pseudoknots seems but another heuristic approach.

Semantic non-ambiguity. A Dynamic Programming Algorithm is called *semantically ambiguous* [10,11], if it examines an object of interest in its search space more than once. This typically leads to exponential explosion of redundant solution candidates. For finding a single, optimal solution in a Dynamic Programming Algorithm, such redundancy does not matter, but it renders the algorithm useless for producing near-optimals. The *pknotsRG* program is implemented in a non-ambiguous way.

Combining canonicity with a non-ambiguous algorithm allows the program to return suboptimals. In particular, we can ask the best canonical pseudoknot from the near-optimal search space, even when the minimum free energy structure comes out unknotted. The best canonical pseudoknot P_{can} may be checked for potential extension to a non-canonical structure P of even lower energy. In this sense, the heuristic constraint of canonization appears tolerable. Our algorithms presented here adhere to the same idea. All considered structures are canonical, and there will be only one situation where a structure is considered twice.

2.2 Decomposition Alternatives of the Kissing Hairpin Motif

An elementary decomposition of a kissing hairpin leads to three helices ($a-a', b-b', c-c'$) with intervening sequences u, v, w, x, y, folded in arbitrary ways, with the overall arrangement $aubva'wcxb'yc'$. See Figure 2 for an illustration. Such a decomposition, in full generality, leads to 12 moving boundaries, and makes us resort to canonization. Rule 2 of our canonization constraints eliminates six moving boundaries – the inner endpoints of three helices, which are now fixed by the helix maximality rule. The remaining boundaries are the outer endpoints of the three helices. Iterating over these six boundaries would lead to an $O(n^6)$ time, $O(n^2)$ space strategy. Our goal is to do better than this.

Our key idea is the view of the kissing hairpin motif as an overlay of two simple pseudoknots (Figure 2). Given that we already know how to compute optimal simple pseudoknots for the overlapping subwords $aubva'zb'$ and $btcxb'yc'$, can we find their optimal overlay such that $z = wcx$ and $t = va'w$, thus defining the overall optimal decomposition into $aubva'wcxb'yc'$? Can we find its optimal energy as the sum from its two constituents?

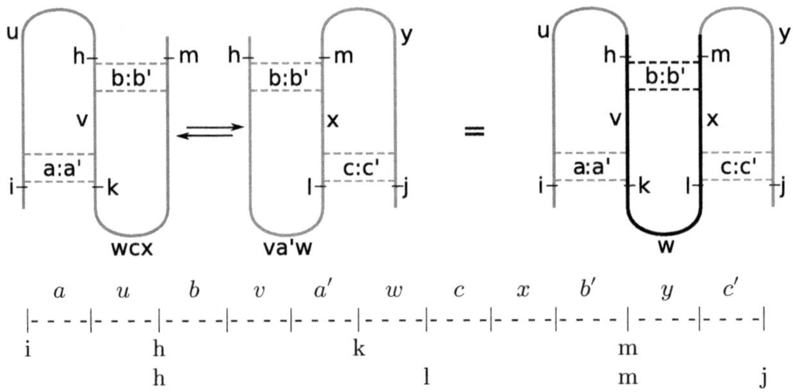

Fig. 2. The composition of two pseudoknots leading to a kissing hairpin motif with the overlay of parts of the sequence and the moving boundaries i, h, k, l, m, and j on top. The linear form of the sequence below shows 12 moving boundaries (vertical lines). With the canonization rules, only six boundaries (labeled lines) remain.

Simple as it seems, there is a problem. First, if $w = \varepsilon$, the optimal choice of a' (with respect to a and b') may conflict with the optimal choice of c (with respect to b and c'). Moreover, in the overlay, the energy contribution of the middle helix ($b - b'$) and the structure for v, w, and x embedded within both pseudoknots are accounted for twice, and must be subtracted from the energy sum of both parts. This violates the monotonicity requirement for dynamic programming known as Bellman's Principle: for the overlay, the energy function is non-monotonic, and as a consequence, an optimal kissing hairpin motif may arise as an overlay of sub-optimal pseudoknots.

We will present three, increasingly complex strategies A, B, and C, such that their search spaces are properly included in the form $Searchspace_A \subseteq Searchspace_B \subseteq Searchspace_C \subset Searchspace_{KH}$. This relation will allow us to evaluate whether the expense for a more general strategy pays off in practice, but we will not be able to relate our results to an evaluation of the complete search space $Searchspace_{KH}$ of all (non-canonical) structures.

2.3 Strategy A – An $O(n^4)$ Time, $O(n^2)$ Space Algorithm

Strategy A makes the optimistic assumption that at least one of the pseudoknots is the optimal structure for its underlying subword. This fixed, we choose the rest of the motif in the best possible way.

(1) For all subwords p, find the optimal pseudoknot such that $p = aubva'zb'$. Store results in a table of size $O(n^2)$.

(2) For all subwords s, split in all ways $s = pt$ and look up the optimal decomposition $p = aubva'zb'$.

(3) For all s of Step 2, use $s = auq$ and find the pseudoknot decomposition such that $q = brcxb'yc'$ and $r = va'w$, to complete the kissing hairpin decomposition $s = aubva'wcxb'yc'$. This pseudoknot must be chosen such that c lies strictly to the right of a', hence this is not, in general, the optimal pseudoknot over its underlying subword q. Record the decomposition of lowest free energy.

(4 - 6) Apply symmetric steps starting from an optimal choice for the right pseudoknot in the overlay.

(7) Choose lower energy value from (3) and (6); store it in a table of size $O(n^2)$.

The symmetry of (1-3) and (4-6) leads to the only case of ambiguity in our approach: If the two locally optimal pseudoknots make a perfect overlay as a kissing hairpin, this (optimal) structure will be found twice.

Efficiency: (1) takes $O(n^4)$ steps as with *pknotsRG*. (2) takes $O(n^3)$ steps, as the decomposition of p is already computed. (3) takes also $O(n^4)$, because it inherits $O(n^3)$ from Step 2 for all splits of s, which determine au and hence, the split auq. (Only) one extra factor of n arises from the split rc, which in turn determines the inner endpoints of helix $(c - c')$ due to the maximality rule, and hence implies the split yc'. (4-6) take $O(n^4)$ steps for symmetry reasons. (7) takes $O(n^2)$ steps. Postponing implementation details, we see that this yields an algorithm with $O(n^4)$ time, $O(n^2)$ space requirements.

Note that Strategy A does some redundant work – the right pseudoknot determined in Step 3 has already been considered as a (generally sub-optimal) pseudoknot in Step 1.

2.4 Strategy B – An $O(n^4)$ Time, $O(n^3)$ Space Algorithm

Strategy B avoids the redundant work of Strategy A, and also enlarges the search space. We spend extra space in Step 1 to store results about sub-optimal pseudoknots.

(1) For $p = aubva'zb'$, and for each choice of b therein, we record the optimal choice of a'. Conversely, for each choice of a', we store the optimal choice of b. This requires two tables of size $O(n^3)$.
(2) For the kissing hairpin motif, we first choose a, b, b', and c', which costs $O(n^4)$, and use the stored information to optimally determine the other bounds for a' and c by lookup with $O(1)$.
(3) Unfortunately, the stored information may suggest that with an optimal choice, a' and c would overlap (and w have negative length).We correct this by a heuristic decision – selecting an a' further to the left and a c further to the right. This decision will also be based on precomputed information in order to retain a runtime of $O(n^4)$.
(4) We minimize over all cases considered.

The overall efficiency is $O(n^4)$ time and $O(n^3)$ space. Note that the search space here is more general than with strategy A, as neither pseudoknot needs to be optimal with respect to its underlying subword. This generalization lies with Step 1. In Strategy A, only the optimal choice of b within p is considered for overlay, while here, all possible choices of b are tried.

2.5 Strategy C – An $O(n^5)$ Time, $O(n^2)$ Space Algorithm

Strategy C avoids the extra storage required by Strategy B. The necessary information is re-computed on demand, after choosing a, b, b' and c'. This increases runtime, but also allows us to avoid the heuristic decision when a' and c would overlap. For each choice of a', we compute the best choice of c strictly to its right. This threatens to raise time complexity to $O(n^6)$, but with a clever arrangement of computations and an extra table of size $O(n)$, we can keep it at $O(n^5)$.

The optimal choice of l with respect to (h, j) as a pseudoknot is a heuristics with respect to (i, j) as a kissing hairpin (see Figure 3). It assumes that $va'w$ can fold optimally. For the kiss, however, v and w can only fold individually, as they are separated by a', which is the partner of a. Thus, l need not be optimal for (i, j) as a kissing hairpin.

3 Algorithms

3.1 Algorithmic Subtleties

Annotated energies. When computing minimum free energies from pseudoknots, we will need to also record the internal boundaries of the given subword which achieved optimal energy. These will be data of the form (E, h, k). When we minimize over these tuples, we do this with a lexicographic ordering. This is consistent with mimimizing over energies alone. When two structures have the same energy, then the choice is arbitrary and remains unspecified.

Exact subword boundaries in the input decomposition. Substructures have certain minimal sizes. For example, we forbid lonely pairs, i.e. helices of length 1. Therefore, in ${}_ia_kza'_j$, we do not iterate k over $i \leq k \leq j$, but only over $i+2 \leq k \leq j-2$.

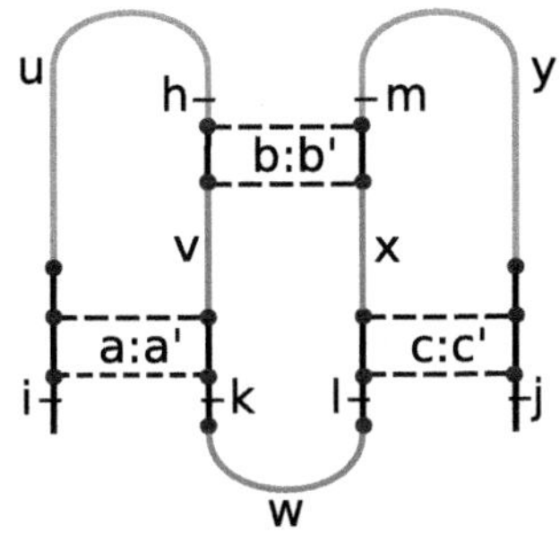

Fig. 3. The graphic shows the mandatory bases (black dots) of a kissing hairpin and the indices i, h, k, l, m, and j determining the start and end points of the helices (black tics). Gray regions u, v, w, x, and y can fold in an arbitrary way.

This does not affect the asymptotics, but saves substantial time in practice. The minimal subword sizes used are two base pairs for each helix, loop u and y have one unpaired base. Loop w has two single bases ($k+2 \leq l$). The size of loop v and x is ≤ 0, because we want to keep the possibility of coaxially stacking of the helices. With that, we get a minimal sequence length of 16 bases to form a kissing hairpin (see Figure 3).

To be concrete in the following recurrences, we use the precise boundaries consistent with our implementation. But for understanding the essentials of the algorithms, the reader may choose to ignore them.

3.2 Pseudoknot-Recurrence of *pknotsRG* – csrPK

Due to the canonization of *pknotsRG*, the calculation of a canonical simple recursive pseudoknot (csrPK) for a given subword needs two boundaries in addition to (i, j): h, the start position of the $b - b'$ helix, and k, the end position of the $a - a'$ helix. The recurrence of a csrPK for a subword (i, j) is:

$$\mathrm{csrPK}\,(i,j) = \min_{\substack{i+3 \,\leq\, h \,\leq\, j-8 \\ h+4 \,\leq\, k \,\leq\, j-4}} E_{\mathrm{csrPK}}\left({}_{i}au_{h}bva'_{k}rb'_{j}\right)$$

The energy function E_{csrPK} makes use of a precomputed table to determine the inner endpoints of the helices in a unique, maximal and non-overlapping fashion. With these boundaries fixed, the energy value is the sum of stabilizing energies of both helices + energy contributions of the arbitrary folded regions u, v and w + contributions from bases which dangle onto the helices from inside the csrPK + penalties for explicitly unpaired bases in front of u and b'. For later use, we adapt E_{csrPK} to additionaly store h and k, which can be retrieved by the functions $\mathrm{boundary}_{\mathrm{left}}$ and $\mathrm{boundary}_{\mathrm{right}}$.

3.3 Recurrences of Strategy A – csrKH$_A$

For Strategy A we make two strong assumptions. (1) Helices $a - a'$ and $b - b'$ of an optimal csrPK, starting at i and ending at m, can be adopted for the overall csrKH and thus determine the boundaries h and k. We can look up these values via the table csrPK. (2) The remaining boundary l, the starting point for the $c - c'$ helix, can be determined by using the energy of a second csrPK as an objective function. This second csrPK must start at h, end at j and have its end position of the left helix $b : b'$ at m, thus overlaying a part of the first csrPK:

$$\begin{aligned}
\text{left}(i,j) &= \min_{i+13 \leq m \leq j-3} E_{\text{csrKH}}\left({}_i a u_h b v a'_k w_l c x b'_m y c'_j\right), \quad \text{where} \\
h &= \text{boundary}_{\text{left}}\left(\text{csrPK}(i,m)\right), \\
k &= \text{boundary}_{\text{right}}\left(\text{csrPK}(i,m)\right), \\
l &= \text{boundary}_{\text{left}}\left(\min_{k+2 \leq d \leq m-4} E_{\text{csrPK}}\left({}_h b v a' w_d c x b'_m y c'_j\right)\right)
\end{aligned}$$

A csrKH may alternatively arise from the opposite direction, i.e. an optimal csrPK on its right half overlaying a suboptimal csrPK at its left:

$$\begin{aligned}
\text{right}(i,j) &= \min_{i+3 \leq h \leq j-13} E_{\text{csrKH}}\left({}_i a u_h b v a'_k w_l c x b'_m y c'_j\right), \quad \text{where} \\
l &= \text{boundary}_{\text{left}}\left(\text{csrPK}(h,j)\right), \\
m &= \text{boundary}_{\text{right}}\left(\text{csrPK}(h,j)\right), \\
k &= \text{boundary}_{\text{right}}\left(\min_{h+4 \leq d \leq l-2} E_{\text{csrPK}}\left({}_i a u_h b v a'_d w c x b'_m\right)\right)
\end{aligned}$$

The optimal csrKH with Strategy A is:

$$\text{csrKH}_A(i,j) = \min\left(\text{left}(i,j), \text{right}(i,j)\right)$$

3.4 Recurrences of Strategy B – csrKH$_B$

Since Strategy B has to store the optimal choice of a' for every given b for csrPKs on the left side and the optimal b for every given a' for csrPKs on the right side of the csrKH, we have to replace the function csrPK with lpk and rpk. A csrPK for a subword (i,j) can now be determined by minimizing over $lpk(i,h,j)$ and $rpk(i,k,j)$:

$$\begin{aligned}
lpk(i,h,j) &= \min_{h+4 \leq k \leq j-4} E_{\text{csrPK}}\left({}_i a u_h b v a'_k r b'_j\right) \\
rpk(i,k,j) &= \min_{i+3 \leq h \leq k-4} E_{\text{csrPK}}\left({}_i a u_h b v a'_k r b'_j\right)
\end{aligned}$$

An overlay of csrPKs from lpk and rkp might overlap in region w of the csrKH, when building it. We can overcome this obstacle in a heuristic way by introducing an artifical border ξ:

$$\begin{aligned}
lpk_{\text{heuristic}}(i,h,j) &= \min_{h+4 \leq k \leq \xi} E_{\text{csrPK}}\left({}_i a u_h b v a'_k r b'_j\right) \\
rpk_{\text{heuristic}}(i,k,j) &= \min_{\xi \leq h \leq k-4} E_{\text{csrPK}}\left({}_i a u_h b v a'_k r b'_j\right)
\end{aligned}$$

Thus we can construct a csrKH with Strategy B by first iterating over the outer endpoints of helix $b - b'$, namely m and h. Second, we choose the energetically optimal combination of k and l by overlaying all csrPKs from $lpk\,(i,h,m)$ and $rpk\,(h,m,j)$, as well as their heuristic counterparts $lpk_{\text{heuristic}}\,(i,h,m)$ and $rpk_{\text{heuristic}}\,(h,m,j)$ to guarantee at least one feasible overlay:

$$\text{csrKH}_B(i,j) = \min_{\substack{i+13 \,\le\, m \,\le\, j-3 \\ i+3 \,\le\, h \,\le\, m-10}} E_{\text{csrKH}}\left({}_iau_hbva'_kw_lcxb'_myc'_j\right), \quad \text{where}$$

$$k \in \text{boundary}_{\text{right}}\left\{lpk\,(i,h,m)\,,\ lpk_{\text{heuristic}}\,(i,h,m)\right\}$$
$$l \in \text{boundary}_{\text{left}}\left\{rpk\,(h,m,j)\,,\ rpk_{\text{heuristic}}\,(h,m,j)\right\}$$

3.5 Recurrences of Strategy C – csrKH$_C$

We start with Strategy C identical to Strategy B, by iterating over m and h. But instead of retrieving k and l from precomputed csrPK tables, we now also iterate k to determine a' and look up the optimal choice for l depending on k in a one dimensional table rpk:

$$\text{csrKH}_C(i,j) = \min_{\substack{i+13 \,\le\, m \,\le\, j-3 \\ i+3 \,\le\, h \,\le\, m-10 \\ h+4 \,\le\, k \,\le\, m-6 \\ l \,=\, \text{boundary}_{\text{left}}(rpk(k))}} E_{\text{csrKH}}\left({}_iau_hbva'_kw_lcxb'_myc'_j\right)$$

When iterating over k, we go from right to left. Thus we have a growing subword (k,m). While shifting k one position to the left, the function $rpk(k)$ also determines the optimal csrPK that begins at h, ends at j, has its b' at m and its c somewhere in the subword (k,m). Since we temporarily store the results for $rpk(k)$, it can be calculated in $O(1)$ time. We just compare the existing result for the one letter shorter subword $rpk(k+1)$ with one new csrPK, whose boundaries are at $h, k+2, m, j$:

$$rpk(k) = \min\left(E_{\text{csrPK}}\left({}_hbva'w_{k+2}cxb'_myc'_j\right), rpk(k+1)\right)$$

3.6 Implementation via Algebraic Dynamic Programming

Alike *pknotsRG*, *pKiss* is implemented with the algebraic dynamic programming technique [12]. This makes it easy to add and combine different types of analysis. Currently, we compute optimal and suboptimal structures. We plan to add shape abstraction and computation of best knotted and un-knotted folding.

4 Evaluation of Strategies A, B, and C

A piece of anecdotal evidence. The RNA polymerase gene (gene 1) of the human coronavirus 229E is a good example for the usefulness of improved secondary structure prediction tools. Analyzing the genome of the human coronavirus, Herold and Siddell [13] guessed, that a "slippery site" together with an H-type pseudoknot acts as a frameshift inducing structure. Extensive mutational analyses showed that a kissing hairpin is required for high frequency frameshifts. Their work implied computer-assisted modeling, but prior prediction tools could not detect kissing hairpin motifs. *pKiss* finds the proper kissing hairpin.

Available test data. Verified structures holding pseudoknots and kissing hairpins are rare. We collected a dataset of 61 pseudoknotted structures include 6 kissing hairpins, one "double" pseudoknot with topology $a\_b\_c\_d\_c'\_a'\_d'\_b'$ and 5 simple pseudoknots with nested sub-structures (see Appendix). The sequence length varies from 28 to 115 nt. The sequence types consist of viral ribosomal frame shifting or readthrough, mRNA, tmRNA, viral 3' UTR, ribozymes, signal recognition particle RNA [25], sequences with high affinity to HIV-1-RT [24] and viral RNA. These well-studied structures are subsequently called the true structures.

Comparison of the Strategies A, B, and C. On 57 out of 61 sequences, Strategies A, B, and C agree. B finds a structure of lower energy than A in two cases, and C in the same two cases and two further ones. This is consistent with the hierarchy of search space inclusion, but the small disagreement is surprising.

Positive and negative test cases. For a true positive prediction, we require the structure with the right topology in the right sequence position, but allow for a few missing base pairs (the price of canonization) or extra base pairs when they are consistent with the true structure. All 6 true kissing hairpins are precisely predicted by each strategy. Overall, 46 structures (75.4%) are correctly predicted while 15 sequences (24.6%) deviate from the true structure. These negative cases contain the complex pseudoknot which is beyond the class of kissing hairpins, but the helices actually predicted are correct. In seven cases, a kissing hairpin is predicted instead of a simple pseudoknot. One cannot exclude that this kissing hairpin is actually correct, but has not been detected before due to the lack of appropriate tools.

Further evaluations. Comparing *pKiss* to the program by Rivas and Eddy brought little insight, as the program solves a more general problem and, as expected from their asymptotics, is much slower and greedy for space. Comparing *pKiss* to the most recent version of *HotKnots* [3] on our data set, we find the following: *HotKnots* currently provides four different parameter sets. Choosing the best prediction from those four in each case, it agrees with Strategy A in 3 out of our 6 positive test cases. On the larger data set of simple pseudoknots, there is more agreement between the methods. Execution time for a single parameter choice is generally lower than for *pKiss* by a factor of 3 – 6. We have also evaluated *pKiss* on random data and tested the robustness of predictions under varied energy parameters for kissing hairpin initiation. All evaluation data, as well as the first author's M.Sc. thesis, can be obtained from our website at `http://bibiserv.techfak.uni-bielefeld.de/pkiss/`.

5 Conclusion

Should the observations from our evaluation on sparse data generalize, interesting algorithmic perspectives open up. Strategy A evaluates a more complex motif than simple pseudoknots – without increasing asymptotic complexity. Unexpectedly, Strategy A performs best among A, B, and C – it is faster, agrees on the true positives, and has fewer false negatives. Closer inspection showed that it is always the left pseudoknot of the overlay which was chosen optimally. One may speculate that this is because the strategy is consistent with the hierarchic folding path during transcription. Boldly dropping the symmetric computation starting from the right pseudoknot reduces work in the innermost loop and may provide a speed-up factor close to 2.

The more exciting perspective is the extension of the overlay idea to more complex structures. A motif of four hairpins with two kissing interactions, for example, can be overlaid as $a\_b\_a'\_c\_b'\_c'$ and $b\_c\_b'\_d\_c'\_d'$. Using ideas of Strategy A, this can, again, be achieved in $O(n^4)$ time and $O(n^2)$ space! Additionally, alternative decompostions, say $a\_b\_a'\_c\_b'\_c'$ with $c\_d\_c'\_d'$ (a kissing hairpin overlaid with a simple pseudoknot) may be investigated, without raising the asymptotics. Furthermore, two such double kissing structures can form an overlay, and so on. It appears that one can construct a variety of practically useful, albeit increasingly heuristic, programs for pseudoknotted motifs of increasingly complex topologies within $O(n^4)$ time and $O(n^2)$ space.

Acknowledgement. RG thanks A. Condon and H. Jabbari for discussion of the *pKiss* ideas in their early state.

References

1. Akutsu, T.: Dynamic programming algorithms for RNA secondary structure prediction with pseudoknots. Discrete Appl. Math. 104(1-3), 45–62 (2000)
2. Andronescu, M.S., Condon, A.E., Hoos, H.H., Mathews, D.H., Murphy, K.P.: Efficient parameter estimation for RNA secondary structure prediction. Bioinformatics 23, 19–28 (2007)
3. Andronescu, M.S., Pop, C., Condon, A.E.: Improved free energy parameters for RNA pseudoknotted secondary structure prediction. RNA 16(1), 26–42 (2010)
4. Chan, C.Y., Lawrence, C.E., Ding, Y.: Structure clustering features on the Sfold Web server. Bioinformatics 21(20), 3926–3928 (2005)
5. Chang, K.Y., Tinoco, I.: Characterization of a kissing hairpin complex derived from the human immunodeficiency virus genome. Proc. Natl. Acad. Sci. USA 91(18), 8705–8709 (1994)
6. Chen, H.L., Condon, A.E., Jabbari, H.: An $O(n^5)$ Algorithm for MFE Prediction of Kissing Hairpins and 4-Chains in Nucleic Acids. J. Comput. Biol. 16(6), 803–815 (2009)
7. Condon, A.E., Jabbari, H.: Computational prediction of nucleic acid secondary structure: Methods, applications, and challenges. Theoretical Computer Science 410(4-5), 294–301 (2009)
8. Deblasio, D., Bruand, J., Zhang, S.: PMFastR: A New Approach to Multiple RNA Structure Alignment. In: Salzberg, S.L., Warnow, T. (eds.) WABI 2009. LNCS, vol. 5724, pp. 49–61. Springer, Heidelberg (2009)

9. Frid, Y., Gusfield, D.: A simple, practical and complete $O(n^3/\log n)$-time Algorithm for RNA folding using the Four-Russians Speedup. Algorithms Mol. Biol. 5(1), 13 (2010)
10. Giegerich, R.: Explaining and Controlling Ambiguity in Dynamic Programming. In: Giancarlo, R., Sankoff, D. (eds.) CPM 2000. LNCS, vol. 1848, pp. 46–59. Springer, Heidelberg (2000)
11. Giegerich, R., Hoener, C., Siederdissen, z.: Semantics and Ambiguity of Stochastic RNA Family Models. IEEE/ACM Transactions on Computational Biology and Bioinformatics 99(PrePrints) (2010)
12. Giegerich, R., Meyer, C., Steffen, P.: A discipline of dynamic programming over sequence data. Science of Computer Programming 51(3), 215–263 (2004)
13. Herold, J., Siddell, S.G.: An 'elaborated' pseudoknot is required for high frequency frameshifting during translation of HCV 229E polymerase mRNA. Nucl. Acids Res. 21(25), 5838–5842 (1993)
14. Hofacker, I.L., Fontana, W., Stadler, P.F., Bonhoeffer, S.L., Tacker, M., Schuster, P.: Fast Folding and Comparison of RNA Secondary Structures. Monatsh. Chem. 125, 167–188 (1994)
15. Li, P.T.X., Bustamante, C., Tinoco, I.: Unusual mechanical stability of a minimal RNA kissing complex. Proc. Natl. Acad. Sci. USA 103(43), 15847–15852 (2006)
16. Lyngsø, R.B., Pedersen, C.N.S.: RNA Pseudoknot Prediction in Energy-Based Models. J. Comput. Biol. 7(3-4), 409–427 (2000)
17. Mathews, D.H., Disney, M.D., Childs, J.L., Schroeder, S.J., Zuker, M., Turner, D.H.: Incorporating chemical modification constraints into a dynamic programming algorithm for prediction of RNA secondary structure. Proc. Natl. Acad. Sci. USA 101(19), 7287–7292 (2004)
18. Mathews, D.H., Turner, D.H.: Prediction of RNA secondary structure by free energy minimization. Curr. Opin. Struct. Biol. 16(3), 270–278 (2006)
19. Melchers, W.J.G., Hoenderop, J.G.J., Slot, H.J.B., Pleij, C.W.A., Pilipenko, E.V., Agol, V.I., Galama, J.M.D.: Kissing of the two predominant hairpin loops in the coxsackie B virus 3' untranslated region is the essential structural feature of the origin of replication required for negative-strand RNA synthesis. J. Virol. 71(1), 686–696 (1997)
20. Reeder, J., Giegerich, R.: Design, implementation and evaluation of a practical pseudoknot folding algorithm based on thermodynamics. BMC Bioinformatics 5(1), 104 (2004)
21. Ren, J., Rastegari, B., Condon, A.E., Hoos, H.H.: HotKnots: Heuristic prediction of RNA secondary structures including pseudoknots. RNA 11(10), 1494–1504 (2005)
22. Rivas, E., Eddy, S.R.: A dynamic programming algorithm for RNA structure prediction including pseudoknots. J. Mol. Biol. 285(5), 2053–2068 (1999)
23. Rødland, E.A.: Pseudoknots in RNA Secondary Structures: Representation, Enumeration, and Prevalence. J. Comput. Biol. 13(6), 1197–1213 (2006)
24. Tuerk, C., MacDougal, S., Gold, L.: RNA pseudoknots that inhibit HIV type 1 reverse transcriptase. Proc. Natl. Acad. Sci. USA 89(15), 6988–6992 (1992)
25. van Batenburg, F.H.D., Gultyaev, A.P., Pleij, C.W.A.: PseudoBase: structural information on RNA pseudoknots. Nucl. Acids Res. 29(1), 194–195 (2001)
26. Wuchty, S., Fontana, W., Hofacker, I.L., Schuster, P.: Complete suboptimal folding of RNA and the stability of secondary structures. Biopolymers 49(2), 145–165 (1999)
27. Zuker, M., Stiegler, P.: Optimal computer folding of large RNA sequences using thermodynamics and auxiliary information. Nucl. Acids Res. 9(1), 133–148 (1981)

Reducing the Worst Case Running Times of a Family of RNA and CFG Problems, Using Valiant's Approach

Shay Zakov, Dekel Tsur, and Michal Ziv-Ukelson

Department of Computer Science, Ben-Gurion University of the Negev, Israel
{zakovs,dekelts,michaluz}@cs.bgu.ac.il

Abstract. We study Valiant's classical algorithm for Context Free Grammar recognition in sub-cubic time, and extract features that are common to problems on which Valiant's approach can be applied. Based on this, we describe several problem templates, and formulate generic algorithms that use Valiant's technique and can be applied to all problems which abide by these templates. These algorithms obtain new worst case running time bounds for a large family of important problems within the world of RNA Secondary Structures and Context Free Grammars.

1 Introduction

Computational prediction of RNA structures serves as the basis of many approaches related to RNA functional analysis [1]. Most computational tools for RNA structural prediction focus on RNA *secondary structures* — a reduced structural representation of RNA molecules which describes a set of paired nucleotides, through hydrogen bonds, in an RNA sequence. Over the last decades, several variants of RNA secondary structure prediction problems were defined, to which polynomial algorithms have been designed [2,3,4,5,6,7,8,9]. The computational feasibility of these variants (as opposed to three-dimensional structure prediction), combined with the fact that secondary structures still reveal important information about the functional behavior of RNA molecules, account for the high popularity of state-of-the-art tools for RNA secondary structure prediction.

Sakakibara et al. [10] noticed that the basic variants of *RNA Secondary Structure Prediction* problems [2,3] are in fact special cases of the *Weighted Context Free Grammar (WCFG) Parsing* problem [11]. This approach was then followed by Dowell and Eddy [12], Do et al. [13], and others, who studied different aspects of the relationship between these two domains. The WCFG Parsing problem is a generalization of the simpler *Context Free Grammar (CFG) Parsing* problem, where both problems can be solved by the Cocke-Kasami-Younger (CKY) algorithm [14,15,16], whose running time is cubic in the number of words in the input sentence (or the number of nucleotides in the input RNA sequence).

The CFG literature describes two improvements which allow to obtain a sub-cubic time for the CKY algorithm. The first among these improvements was a

V. Moulton and M. Singh (Eds.): WABI 2010, LNBI 6293, pp. 65–77, 2010.

Table 1. Time complexities of several VMT problems

	Problem	Standard DP running time	VMT algorithm running time
Results previously published	CFG Recognition / Parsing	$\Theta(n^3)$ [15, 16, 14]	$\Theta(BS(n))$ [17]
	WCFG Parsing	$\Theta(n^3)$ [11]	$\Theta(MP(n))$ [19, 20]
	RNA Base-Pairing Maximization	$\Theta(n^3)$ [2]	$\Theta(MP(n))$ [19]
	RNA Energy Minimization	$\Theta(n^3)$ [3]	$\Theta(MP(n))$ [19]
In this paper	WCFG Inside-Outside	$\Theta(n^3)$ [33]	$\Theta(BS(n))$
	RNA Partition Function	$\Theta(n^3)$ [5]	$\Theta(BS(n))$
	RNA Simultaneous Alignment and Folding	$\Theta(n^{3m})$ [9]	$\Theta(MP(n^m))$
	RNA-RNA Interaction	$\Theta(n^6)$ [4]	$\Theta(MP(n^2))$
	RNA-RNA Interaction Partition Function	$\Theta(n^6)$ [7]	$\Theta(BS(n^2))$
	RNA Sequence to Structured-Sequence Alignment	$\Theta(n^4)$ [8]	$\Theta(nMP(n))$

technique suggested by Valiant [17], who showed that the CFG parsing problem can be solved in a running time which matches the running time of a *Boolean Matrix Multiplication* of two $n \times n$ matrices. The currently fastest algorithm for this variant of matrix multiplication runs in $O(n^{2.376})$ time [18]. In [19], Akutsu argued that the algorithm of Valiant can be modified to deal also with WCFG Parsing, and consequentially with RNA Folding (this extension is described in more details in [20]). The running time of the adapted algorithm is different from that of Valiant's algorithm, and matches the running time of a *Max-Plus Matrix Multiplication* of two $n \times n$ matrices. The currently fastest algorithm for this variant of matrix multiplication runs in $O(\frac{n^3 \log^3 \log n}{\log^2 n})$ time [21]. The second improvement was introduced by Graham et al. [22], who applied the *Four Russians* technique [23] to the CFG parsing problem, and obtained an $O\left(\frac{n^3}{\log n}\right)$ running time algorithm. To the best of our knowledge, no extension of this approach to the WCFG Parsing problem has been described. Recently, Frid and Gusfield [24] showed how to apply the *Four Russians* technique to the RNA Energy Minimization problem (under the assumption of a discrete scoring scheme), obtaining the same running time of $O\left(\frac{n^3}{\log n}\right)$. Several other techniques have been previously developed to accelerate the practical running times of different variants of CFG and RNA related algorithms. Nevertheless, these techniques either retain the same worst case running times of the standard algorithms [22,25,26,27,28,29,30], or apply heuristics which compromise the optimality of the obtained solutions [31,32].

In this extended abstract, we present three template formulations, entitled *Vector Multiplication Templates (VMTs)*, which abstract the essential properties that characterize problems for which a Valiant-like algorithmic approach can be applied. Then, we exemplify how Valiant's approach can be simplified and applied in order to describe generic algorithms for all problems sustaining these templates. Table 1 lists some examples of VMT problems. The table compares between the running times of standard dynamic programming (DP) algorithms, and the VMT algorithms presented here. In the single string problems, n denotes the length of the input string. In the double-string problems [4,7,8], we assume that both input strings are of length n. For the *RNA Simultaneous Alignment and Folding* problem, m denotes the number of input strings and n is the length of each string. $BS(n)$ denotes the time complexity of a Scalar or a Boolean matrix multiplication of two $n \times n$ matrices, for which the current best result is $O(n^{2.376})$, due to [18]. $MP(n)$ denotes the time complexity of a Min-Plus or a Max-Plus matrix multiplication of two $n \times n$ matrices, for which the current best result is $O(\frac{n^3 \log^3 \log n}{\log^2 n})$, due to [21]. For most of the problems, the algorithms presented here obtain lower running time bounds than the best previous algorithms.

2 Preliminaries

2.1 Interval and Matrix Notations

For two integers a, b, denote by $[a, b]$ the interval which contains all integers q such that $a \leq q \leq b$. For two intervals $I = [i_1, i_2]$ and $J = [j_1, j_2]$, define the following intervals: $[I, J] = \{q : i_1 \leq q \leq j_2\}$, $(I, J) = \{q : i_2 < q < j_1\}$, $[I, J) = \{q : i_1 \leq q < j_1\}$, and $(I, J] = \{q : i_2 < q \leq j_2\}$. When an integer r replaces one of the intervals I or J in the notation above, we regard it as if it was the interval $[r, r]$. For example, $[0, I) = \{q : 0 \leq q < i_1\}$, and $(i, j) = \{q : i < q < j\}$. For two intervals $I = [i_1, i_2]$ and $J = [j_1, j_2]$ such that $j_1 = i_2 + 1$, define IJ to be the *concatenation* of I and J, i.e. the interval $[i_1, j_2]$.

Let X be an $n_1 \times n_2$ matrix, with rows indexed with $0, 1, \ldots, n_1 - 1$ and columns indexed with $0, 1, \ldots, n_2 - 1$. Denote by $X_{i,j}$ the element in the ith row and jth column of X. For two intervals $I \subseteq [0, n_1)$ and $J \subseteq [0, n_2)$, let $X_{I,J}$ denote the sub-matrix of X obtained by projecting it onto the subset of rows I and subset of columns J. Denote by $X_{i,J}$ the sub-matrix $X_{[i,i],J}$, and by $X_{I,j}$ the sub-matrix $X_{I,[j,j]}$.

Let $\mathcal{D}$ be a domain of elements, and $\otimes$ and $\oplus$ two binary operations on $\mathcal{D}$. We assume that (1) $\oplus$ is associative (i.e. for three elements a, b, c in the domain, $(a \oplus b) \oplus c = a \oplus (b \oplus c)$), and (2) there exists a *zero* element ϕ in $\mathcal{D}$, such that for every element $a \in \mathcal{D}$, $a \oplus \phi = \phi \oplus a = a$ and $a \otimes \phi = \phi \otimes a = \phi$.

Let X and Y be a pair of matrices of sizes $n_1 \times n_2$ and $n_2 \times n_3$, respectively, which elements are taken from $\mathcal{D}$. Define the result of the *matrix multiplication* $X \otimes Y$ to be the matrix Z of size $n_1 \times n_3$, where each entry $Z_{i,j}$ is given by

$$Z_{i,j} = \oplus_{q \in [0,n_2)} (X_{i,q} \otimes Y_{q,j}) = (X_{i,0} \otimes Y_{0,j}) \oplus (X_{i,1} \otimes Y_{1,j}) \oplus \ldots \oplus (X_{i,n_2-1} \otimes Y_{n_2-1,j}).$$

In the special case where $n_2 = 0$, define the result of the multiplication Z to be an $n_1 \times n_3$ matrix in which all elements are ϕ. In the special case where $n_1 = n_3 = 1$, the matrix multiplication $X \otimes Y$ is also called a *vector multiplication* (where the resulting matrix Z contains a single element).

Let X and Y be two matrices. When X and Y are of the same size, define the result of the *matrix addition* $X \oplus Y$ to be the matrix Z of the same size as X and Y, where $Z_{i,j} = X_{i,j} \oplus Y_{i,j}$. When X and Y have the same number of columns, denote by $\begin{bmatrix} X \\ Y \end{bmatrix}$ the matrix obtained by concatenating Y below X. When X and Y have the same number of rows, denote by $[XY]$ the matrix obtained by concatenating Y to the right of X. The following properties are well known, and can be easily deduced from the definition of matrix multiplication and the associativity of the $\oplus$ operation. In each property we assume that the participating matrices are of the appropriate sizes.

$$X \otimes \left[Y^1 Y^2\right] = \left[(X \otimes Y^1)(X \otimes Y^2)\right] \tag{2.1}$$

$$(X^1 \otimes Y^1) \oplus (X^2 \otimes Y^2) = \left[X^1 X^2\right] \otimes \begin{bmatrix} Y^1 \\ Y^2 \end{bmatrix} \tag{2.2}$$

Under the assumption that the operations $\otimes$ and $\oplus$ between two domain elements consume $\Theta(1)$ computation time, a straightforward implementation of a matrix multiplication between two $n \times n$ matrices can be computed in $\Theta(n^3)$ time. Nevertheless, for some variants of multiplications, sub-cubic algorithms for square matrix multiplications are known. Here, we consider three such variants, which will be referred to as *standard multiplications* in the rest of this paper:

- *Scalar* multiplication: The matrices hold numerical elements, $\otimes$ stands for number multiplication $(\cdot)$ and $\oplus$ stands for number addition $(+)$. The *zero* element is 0. The running time of the currently fastest algorithm for this variant is $O(n^{2.376})$ [18].
- *Min-plus/Max-plus* multiplication: The matrices hold numerical elements, $\otimes$ stands for number addition and $\oplus$ stands for min or max (where $a \min b$ is the minimum between a and b, and similarly for max). The *zero* element is ∞ for the min-plus variant and $-\infty$ for the max-plus variant. The running time of the currently fastest algorithm for these variants is $O(\frac{n^3 \log^3 \log n}{\log^2 n})$ [21].
- *Boolean* multiplication: The matrices hold boolean elements, $\otimes$ stands for *boolean AND* $(\wedge)$ and $\oplus$ stands for *boolean OR* $(\vee)$. The *zero* element is the *false* value. Boolean matrix multiplication is computable with the same complexity as the scalar multiplication, having the running time of $O(n^{2.376})$ [18].

2.2 String Notations

Let $s = s_0 s_1 \ldots s_{n-1}$ be a string of length n over some alphabet. Let $s_{i,j}$ denote the substring of s between indices i (inclusive) and j (exclusive). In a case where $i = j$, $s_{i,j}$ corresponds to an empty string, and for $i > j$, $s_{i,j}$ does not correspond to a valid string. An *inside value* $\beta_{i,j}$ is a value which reflects some property dependent only on the substring $s_{i,j}$. In the context of RNA, an input string usually represents a sequence of nucleotides, where in the context of CFGs, it usually represents a sequence of words. Examples of inside values in the world of RNA problems are the maximum number of base-pairs in a secondary structure of

$s_{i,j}$ [2], the minimum free energy of a secondary structure of $s_{i,j}$ [3], the sum of weights of all secondary structures of $s_{i,j}$ [5], etc. In CFGs, inside values can be boolean values which state whether the sub-sentence can be derived from some non-terminal symbol of the grammar, or numeric values corresponding to the probability of (all or best) such derivations [14,15,16,11].

In the rest of this paper, the notation β will be used to denote the set of all values of the form $\beta_{i,j}$ with respect to substrings $s_{i,j}$ of some given string s. It is convenient to visualize β as an upper triangle of an $(n+1) \times (n+1)$ matrix, where n denotes the length of s and the (i,j)-th entry in the matrix contains the value $\beta_{i,j}$. We will also use notations such as $\beta_{I,J}$, $\beta_{i,J}$, and $\beta_{I,j}$ to denote the corresponding sub-matrices of β, as defined in Section 2.1.

3 The Inside Vector Multiplication Template

In this section we describe a class of problems which are called *Inside VMT* problems. We start by giving a simple motivating example in Section 3.1. Then, we formally define the class of Inside VMT problems in Section 3.2, and in Section 3.3 we formulate an efficient generic algorithm for all Inside VMT problems.

3.1 Example: RNA Base-Pairing Maximization

The *RNA Base-Pairing Maximization* problem [2] is a simple problem which exhibits the main characteristics of Inside VMT problems. In this problem, an input string $s = s_0 s_1 \cdots s_{n-1}$ represents a string of *bases* over the alphabet A, C, G, U. The goal is to compute the maximum number of nested complementary base-pairs in a secondary structure of s (we refer the reader to [2] for the formal problem definition). We call such a number the *solution* for the instance s, where $\beta_{i,j}$ denotes the solution for the substring $s_{i,j}$. The value $\beta_{i,j}$ is 0 for $j - i \leq 1$, and otherwise can be computed according to the following recurrence:

$$\beta_{i,j} = \max \left\{ \begin{array}{l} (I)\ \beta_{i+1,j-1} + \delta_{i,j-1}, \\ (II)\ \max\limits_{q \in (i,j)} \{\beta_{i,q} + \beta_{q,j}\} \end{array} \right\},$$

where $\delta_{i,j-1} = 1$ if s_i and s_{j-1} are complementary bases, and otherwise $\delta_{i,j-1} = -\infty$. This recursive computation can be efficiently implemented using *dynamic programming* (DP). For an input string s of length n, the DP algorithm maintains the upper triangle of an $(n+1) \times (n+1)$ matrix B, where each entry $B_{i,j}$ in B corresponds to a solution $\beta_{i,j}$. The entries in B are filled, starting from short base-case entries of the form $B_{i,i}$ and $B_{i,i+1}$, and continuing by computing entries corresponding to substrings with increasing lengths. In order to compute a value $\beta_{i,j}$, the algorithm needs to examine only values of the form $\beta_{i',j'}$ such that $s_{i',j'}$ is a strict substring of $s_{i,j}$. Due to the bottom-up computation, these values are already computed and stored in B, and can be obtained in $\Theta(1)$ time.

Upon computing some value $\beta_{i,j}$, the algorithm needs to compute term (II) of the recurrence. This computation is of the form of the *vector multiplication operation* $\oplus_{q\in(i,j)}\left(\beta_{i,q}\otimes\beta_{j,q}\right)$, where the multiplication variant is the *Max Plus* multiplication. Since all relevant values in B are computed, this computation can be implemented by computing $B_{i,(i,j)}\otimes B_{(i,j),j}$ (see Fig. 1a), which takes $\Theta(j-i)$ running time. After computing term (II), the algorithm needs to perform additional operations for computing $\beta_{i,j}$ which take $\Theta(1)$ running time (computing term (I), and taking the maximum between the results of the two terms). Thus, on average, the running time for computing each value $\beta_{i,j}$ is $\Theta(n)$, and the overall running time for computing all $\Theta(n^2)$ values $\beta_{i,j}$ is $\Theta(n^3)$. Upon termination, the computed matrix B equals to the matrix β, and the required result $\beta_{0,n}$ is found in the entry $B_{0,n}$.

Note that reducing vector multiplication computation time would reduce the overall running time of the described algorithm. This observation holds in general for all VMT problems, as we show below. The algorithms we develop follow the approach of Valiant, which organizes the required vector multiplication operations in a manner that allows to compute them via the usage of fast matrix multiplication algorithms, and hence reduces the amortized running time of each vector multiplication to sub-linear time.

3.2 Inside VMT Definition

In this section we characterize the class of *Inside VMT* problems. The *RNA Base-Paring Maximization* problem, which was presented in the previous section, exhibits a simple special case of an Inside VMT problem, in which the goal is to compute a single inside value for a given input string. Note that this requires the computation of such inside values for all substrings of the input. In other Inside VMT problems the case is similar, hence we will assume that the goal of Inside VMT problems is to compute inside values for *all* substrings of an input string. In the more general case, an Inside VMT problem defines several inside values for each substring, where the computation of these values is formulated in a mutually recursive manner. Examples of such a problems are the *RNA Energy Minimization* problem [3] and the *CFG Parsing* problem [14,15,16]. A common property of all Inside VMT problems is that the computation of at least one type of the inside values requires the result of a vector multiplication operation, which is of similar structure to the multiplication presented for the RNA Base-Paring Maximization problem.

Definition 1. *A problem is considered an* Inside VMT *problem if it fulfills the following requirements:*

1. *The instances of the problem are strings, and the goal of the problem is to compute for every substring $s_{i,j}$ of an input string s,* a series *of inside values $\beta^1_{i,j},\beta^2_{i,j},\ldots,\beta^K_{i,j}$.*
2. *Let n denote the length of s, and let $0\leq i\leq j\leq n$ and $1\leq k\leq K$. Let $\mu^k_{i,j}$ be a result of a standard vector multiplication of the form $\mu^k_{i,j}=\oplus_{q\in(i,j)}\left(\beta^{k'}_{i,q}\otimes\beta^{k''}_{q,j}\right)$, where $1\leq k',k''\leq K$. Assume that the following values can be obtained in $\Theta(1)$ running time: $\mu^k_{i,j}$, all values $\beta^{k'}_{i',j'}$ for $1\leq k'\leq$*

K and $s_{i',j'}$ a strict substring of $s_{i,j}$, and all values $\beta^{k'}_{i,j}$ for $1 \leq k' < k$. Then, the computation of $\beta^k_{i,j}$ can be performed in $o\left(\frac{M(n)}{n^2}\right)$ running time, where $M(n)$ is the running time of the matrix multiplication algorithm which corresponds to the multiplication variant for computing $\mu^k_{i,j}$.

3.3 Inside VMT Algorithm

We next describe a generic algorithm, based on Valiant's algorithm [17], for solving problems sustaining the Inside VMT requirements. For simplicity, we assume that a single value $\beta_{i,j}$ needs to be computed for each substring $s_{i,j}$ of the input string s. The new algorithm also maintains the matrix B as defined in Section 3.1. At each stage of the run, each entry $B_{i,j}$ either contains the value $\beta_{i,j}$, or some intermediate result in the computation of $\mu_{i,j}$. Note that only the upper triangle of B corresponds to valid substrings of the input. Nevertheless, our formulation handles all entries uniformly, implicitly ignoring values in entries $B_{i,j}$ when $j < i$. At each stage, the algorithm computes the values in a sub-matrix $B_{I,J}$ for some pair of intervals $I, J \subseteq [0, n]$. The following pre-condition is maintained at the beginning of the stage (Fig. 1b):

1. Each entry $B_{i,j} \in B_{[I,n],[0,J]}$, such that $B_{i,j} \notin B_{I,J}$, contains the value $\beta_{i,j}$.
2. Each entry $B_{i,j} \in B_{I,J}$ contains the value $\oplus_{q\in(I,J)} \left(\beta_{i,q} \otimes \beta_{j,q}\right)$. In other words, $B_{I,J} = \beta_{I,(I,J)} \otimes \beta_{(I,J),J}$.

Upon initialization, $I = J = [0, n]$, and all values in B are set to ϕ. Observe that at this stage the pre-condition is met. Now, consider a call to the algorithm with some pair of intervals I, J. If $I = [i, i]$ and $J = [j, j]$, then from the pre-condition we have that all values $\beta_{i',j'}$ which are required for the computation of $\beta_{i,j}$ are computed and stored in B, and $B_{i,j} = \mu_{i,j}$ (Fig. 1a). Thus, $\beta_{i,j}$ can be evaluated in $o\left(\frac{M(n)}{n^2}\right)$ running time, and be stored in $B_{i,j}$.

Else, either $|I| > 1$ or $|J| > 1$ (or both), and the algorithm partitions $B_{I,J}$ into two sub-matrices of approximately equal sizes, and computes each part recursively. This partition is described next.

In the case where $|I| \leq |J|$, $B_{I,J}$ is partitioned "vertically" (Fig. 1,(b),(c),(d)): Let J_1 and J_2 be two column intervals such that $J = J_1J_2$ and $|J_1| = \lfloor|J|/2\rfloor$. Since J and J_1 start at the same position, $(I, J) = (I, J_1)$. Thus, from the pre-condition and Equation 2.1, $B_{I,J_1} = \beta_{I,(I,J_1)} \otimes \beta_{(I,J_1),J_1}$. Therefore, the pre-condition with respect to the sub-matrix B_{I,J_1} is met, and the algorithm computes this sub-matrix recursively. After B_{I,J_1} is computed, the first part of the pre-condition with respect to B_{I,J_2} is met, i.e. all necessary solutions, except for those in B_{I,J_2} are computed and stored in B. In addition, at this stage $B_{I,J_2} = \beta_{I,(I,J)} \otimes \beta_{(I,J),J_2}$. Let L be the interval such that $(I, J_2) = (I, J)L$. Observe that $L = J_1$ if the last index in I is smaller than the first index in J, or otherwise L is some (possibly empty) suffix of J_1. To meet the full pre-condition requirements with respect to I and J_2, B_{I,J_2} is updated using Equation 2.2

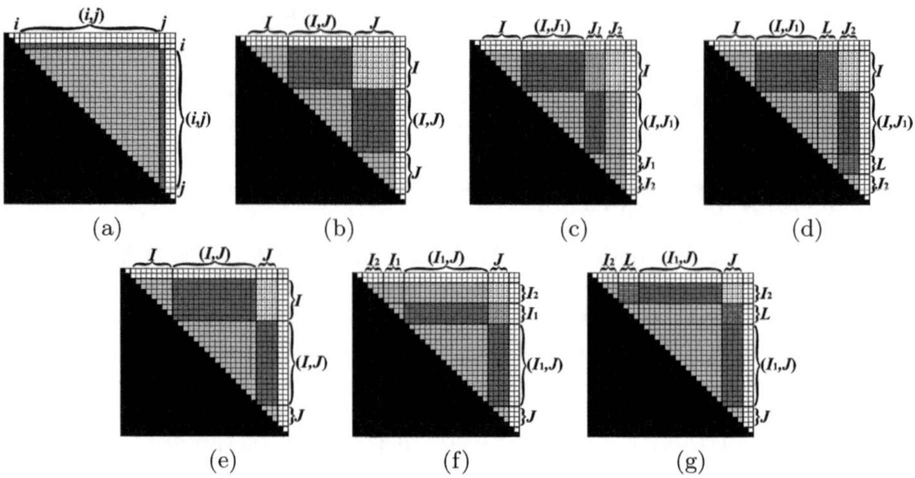

Fig. 1. An exemplification of the matrix B maintained by the Inside VMT algorithm

to be $B_{I,J_2} \oplus (B_{I,L} \otimes B_{L,J_2}) = \left(\beta_{I,(I,J)} \otimes \beta_{(I,J),J_2}\right) \oplus (\beta_{I,L} \otimes \beta_{L,J_2}) = \beta_{I,(I,J_2)} \otimes \beta_{(I,J_2),J_2}$. Now, the pre-condition with respect to B_{I,J_2} is established, and the algorithm computes B_{I,J_2} recursively.

In the case where $|I| > |J|$, $B_{I,J}$ is partitioned "horizontally", in a symmetric manner to the vertical partition. Fig. 1,(e),(f),(g) exemplifies this partition, where further technical details are excluded.

The extension of the algorithm to the general case, where the goal is to compute a series of inside value-sets $\beta^1, \beta^2, \ldots, \beta^K$, is implemented by maintaining K matrices instead of a single matrix, and applying the recurrence over all K matrices simultaneously. Time complexity analysis of this algorithm is excluded from this abstract. It is similar to that of Valiant's algorithm (presented in [17]), showing that the time complexity of the Inside VMT algorithm over strings of length n matches the time complexity of multiplying two $n \times n$ matrices, with the corresponding variant of matrix multiplication.

Theorem 1. *For every Inside VMT problem there is an algorithm whose running time is $\Theta(M(n))$, where n is the length of the input string and $M(n)$ is the running time of the corresponding matrix multiplication algorithm over two $n \times n$ matrices.*

4 Additional Vector Multiplication Templates

This section presents two additional vector multiplication templates: *Outside VMT* and *Multiple String VMT*. These templates are sustained by several important problems which do not follow the Inside VMT requirements, yet share similar properties. Algorithms for these VMT problems may also be accelerated by incorporating fast matrix multiplication sub-routines.

4.1 Outside VMT

Let s be a string over some alphabet. An *outside value* $\alpha_{i,j}$ is a value that reflects some property of the string obtained by removing the substring $s_{i,j}$ from s. Similarly to Inside VMT problems, the goal of *Outside VMT problems* is to compute a set of outside values with respect to a given input string. Examples of problems which require outside value computations and adhere to the VMT requirements are the *RNA Partition Function* problem [5] and the *WCFG Inside-Outside* problem [33]. In both problems, the computation of outside values requires a set of pre-computed inside values, where these inside values can be computed with the Inside VMT algorithm. In such cases, we call the problems *Inside-Outside VMT* problems.

Definition 2. *A problem is considered an* Outside VMT *problem if it fulfills the following requirements:*

1. *The instances of the problem are strings, and the goal of the problem is to compute for every substring $s_{i,j}$ of an input string s, a series of outside values $\alpha^1_{i,j}, \alpha^2_{i,j}, \ldots, \alpha^K_{i,j}$.*
2. *Let n denote the length of s, and let $0 \leq i \leq j \leq n$ and $1 \leq k \leq K$. Let $\beta^1, \beta^2, \ldots, \beta^K$ be a set of pre-computed inside value matrices for s, and let $\mu^k_{i,j}$ be an expression of the form $\mu^k_{i,j} = \oplus_{q\in[0,i)} \left(\beta^k_{q,i} \otimes \alpha^{k'}_{q,j}\right)$ or of the form $\mu^k_{i,j} = \oplus_{q\in(j,n]} \left(\alpha^{k'}_{i,q} \otimes \beta^k_{j,q}\right)$, where $1 \leq k' \leq K$. Assume that the following values can be obtained in $\Theta(1)$ running time: $\mu^k_{i,j}$, all values $\alpha^{k'}_{i',j'}$ for $1 \leq k' \leq K$ and $s_{i,j}$ a strict substring of $s_{i',j'}$, and all values $\alpha^{k'}_{i,j}$ for $1 \leq k' < k$. Then, the computation of $\alpha^k_{i,j}$ can be performed in $o\left(\frac{M(n)}{n^2}\right)$ running time, where $M(n)$ is the running time of the matrix multiplication algorithm which corresponds to the multiplication variant for computing $\mu^k_{i,j}$.*

A generic algorithm for Outside VMT problems, which is similar to the Inside VMT algorithm presented in the previous section, can be described. This algorithm obtains the same running time as that of the Inside VMT algorithm. Due to its technicality and the space requirements, the description of the Outside VMT algorithm is excluded from this abstract, and will be presented in an extended version of this paper.

Theorem 2. *For every Outside VMT problem there is an algorithm whose running time is $\Theta(M(n))$, where n is the length of the input string and $M(n)$ is the running time of the corresponding matrix multiplication algorithm over two $n \times n$ matrices.*

4.2 Multiple String VMT

In this section we describe another extension to the VMT framework, intended for problems whose instances are sets of strings, rather than a single string.

Examples of such problems are the *RNA Simultaneous Alignment and Folding* problem [9,31], and the *RNA-RNA Interaction* problem [4]. Additional problems which exhibit slight divergences of the presented template, such as the *RNA-RNA Interaction Partition Function* problem [7] and the *RNA Sequence to Structures-Sequence Alignment* problem [8], can be solved in similar manners.

To define the Multiple String VMT variant in a general manner, we first give some related notation. An *instance* of a Multiple String VMT problem is a set of strings $S = \left(s^0, s^1, \ldots, s^{m-1}\right)$. For simplicity, we will assume that all strings $s^p \in S$ are of the same length n, though this assumption can be easily relaxed. A *position* in S is a set of indices $X = (i_0, i_1, \ldots, i_{m-1})$, where each index $i_p \in X$ is in the range $0 \leq i_p \leq n$. Let $X = (i_0, i_1, \ldots, i_{m-1})$ and $Y = (j_0, j_1, \ldots, j_{m-1})$ be two positions in S. Say that $X \leq Y$ if $i_p \leq j_p$ for every $0 \leq p < m$, and say that $X < Y$ if $X \leq Y$ and $X \neq Y$. When $X \leq Y$, denote by $S_{X,Y}$ the *sub-instance* $S_{X,Y} = \left(s^0_{i_0,j_0}, s^1_{i_1,j_1}, \ldots, s^{m-1}_{i_{m-1},j_{m-1}}\right)$ of S. Denote by (X, Y) the *set* of all positions Q such that $X < Q < Y$.

An inside value $\beta_{X,Y}$ is a value which reflects some property that depends only on the sub-instance $S_{X,Y}$, where an outside value $\alpha_{X,Y}$ reflects some property that depends on the instance S, after excluding from each string in S the corresponding substring in $S_{X,Y}$. We next define *Multiple String Inside VMT* problems. The "outside" variant can be formulated in a similar manner.

Definition 3. *A problem is considered a* Multiple String Inside VMT *problem if it fulfills the following requirements:*

1. *The instances of the problem are sets of strings, and the goal of the problem is to compute for every sub-instance $S_{X,Y}$ of an input instance S, a series of inside values $\beta^1_{X,Y}, \beta^2_{X,Y}, \ldots, \beta^K_{X,Y}$.*
2. *Let X and Y be two positions in S such that $X \leq Y$, and let $1 \leq k \leq K$. Let $\mu^k_{X,Y}$ be a value of the form $\mu^k_{X,Y} = \oplus_{Q \in (X,Y)} \left(\beta^{k'}_{X,Q} \otimes \beta^{k''}_{Q,Y}\right)$, where $1 \leq k', k'' \leq K$. Assume that the following values can be obtained in $\Theta(1)$ running time: $\mu^k_{X,Y}$, all values $\beta^{k'}_{X',Y'}$ for $1 \leq k' \leq K$ and $S_{X',Y'}$ a strict sub-instance of $S_{X,Y}$, and all values $\beta^{k'}_{X,Y}$ for $1 \leq k' < k$. Then, the computation of $\beta^k_{X,Y}$ can be performed in $o\left(\frac{M((n+1)^m)}{(n+1)^{2m}}\right)$ running time, where $M(n)$ is the running time of the matrix multiplication algorithm which corresponds to the multiplication variant for computing $\mu^k_{i,j}$.*

In order to solve Multiple String Inside or Outside VMT problems, we reduce them to standard single string Inside or Outside VMT problems. The reduction maps a Multiple String VMT instance S composed of m strings of length n each, to a single string s of length $(n+1)^m$. Each sub-instance $S_{X,Y}$ of S corresponds to a substring $s_{i,j}$ of s, and expressions of the form $\oplus_{Q \in (X,Y)} \left(\beta^{k'}_{X,Q} \otimes \beta^{k''}_{Q,Y}\right)$ can be computed by computing expressions of the form $\oplus_{q \in (i,j)} \left(\beta^{k'}_{i,q} \otimes \beta^{k''}_{q,j}\right)$ (and similarly for expressions required for the computation of outside values). Due to the space requirements, the description of this reduction is excluded from this abstract, and will be presented in an extended version of this paper.

Theorem 3. *Let P be a Multiple String (Inside or Outside) VMT problem, and let S be an input instance of P composed of m strings of length n each. Then, a solution for S with respect to P can be computed in the same running time as that of computing a solution for a string of length $(n+1)^m$ with respect to a regular (single string) VMT problem and the same vector multiplication variant.*

5 Concluding Discussion

This paper presents a simplification and a generalization of Valiant's technique, which speeds up a family of algorithms by incorporating fast matrix multiplication procedures. We suggest generic templates that identify problems for which this approach is applicable, where these templates are based on general recursive properties of the problems (rather than on their specific algorithms). An explicit algorithm is described for the Inside Vector Multiplication Template, where algorithms for other variants of problems can be similarly designed.

The presented framework yields new worst case running time bounds for a family of important problems. The examples given here come from the fields of RNA secondary structure prediction and CFG parsing, yet it is possible that problems from other domains can be similarly accelerated. While previous works describe other practical acceleration techniques for some of these problems, Valiant's technique, along with the Four Russians technique [24], are the only two techniques which currently allow to reduce the worst case running times of the standard algorithms, without compromising the correctness of the computations.

Valiant's technique has several advantages over the Four Russians technique. First, the running times obtained by applying it are faster than those obtained by applying the Four Russians technique (for example, the running times of algorithms for the RNA Energy Minimization problems which apply these techniques are $O(\frac{n^3 \log^3 \log(n)}{\log^2(n)})$ and $O(\frac{n^3}{\log(n)})$, respectively). Second, it does not require the enumeration of sub-matrices of the dynamic programming matrix, which may be done only in the case of discrete scoring schemes. In addition, Valiant's technique extends in a natural manner to a whole set of problems, without the need of taking into consideration many problem-specific aspects.

Many of the previous acceleration techniques for RNA and CFG related algorithms are based on sparsification, and are applicable only to optimization problems. Another important advantage of the technique presented here over previous ones is that it is the first technique which reduces the running times of algorithms for the non-optimization problem variants, such as RNA Partition Function related problems [5,7] and the PCFG Inside-Outside algorithm [33] (in which the goals are to sum the scores of all solutions of the input, instead of computing the score of an optimal solution).

The time complexities of the VMT algorithms we describe here are dictated by the time complexities of matrix multiplication algorithms. As matrix multiplication variants are essential operations in many computational problems, much work has been done to improve both the theoretical and the practical running times of these operations, including many recent achievements (see e.g. [21,34,35]). It is expected that even further improvements in this domain will be developed in the future, due to its importance.

Acknowledgments. This work was partially supported by the Frankel Center for Computer Science at Ben Gurion University of the Negev.

References

1. Consortium, A.F.B., Backofen, R., Bernhart, S.H., Flamm, C., Fried, C., Fritzsch, G., Hackermuller, J., Hertel, J., Hofacker, I.L., Missal, K., Mosig, A., Prohaska, S.J., Rose, D., Stadler, P.F., Tanzer, A., Washietl, S., Will, S.: RNAs everywhere: genome-wide annotation of structured RNAs. J. Exp. Zoolog. B. Mol. Dev. Evol. 308, 1–25 (2007)
2. Nussinov, R., Jacobson, A.B.: Fast algorithm for predicting the secondary structure of single-stranded RNA. PNAS 77, 6309–6313 (1980)
3. Zuker, M., Stiegler, P.: Optimal computer folding of large RNA sequences using thermodynamics and auxiliary information. Nucleic Acids Research 9, 133–148 (1981)
4. Alkan, C., Karakoç, E., Nadeau, J.H., Sahinalp, S.C., Zhang, K.: RNA-RNA interaction prediction and antisense RNA target search. Journal of Computational Biology 13, 267–282 (2006)
5. McCaskill, J.S.: The equilibrium partition function and base pair binding probabilities for RNA secondary structure. Biopolymers 29, 1105–1119 (1990)
6. Bernhart, S., Tafer, H., Mückstein, U., Flamm, C., Stadler, P., Hofacker, I.: Partition function and base pairing probabilities of RNA heterodimers. Algorithms for Molecular Biology 1, 3 (2006)
7. Chitsaz, H., Salari, R., Sahinalp, S.C., Backofen, R.: A partition function algorithm for interacting nucleic acid strands. Bioinformatics 25, i365–i373 (2009)
8. Zhang, K.: Computing similarity between RNA secondary structures. In: INTSYS 1998: Proceedings of the IEEE International Joint Symposia on Intelligence and Systems, p. 126. IEEE Computer Society, Washington (1998)
9. Sankoff, D.: Simultaneous solution of the RNA folding, alignment and protosequence problems. SIAM Journal on Applied Mathematics 45, 810–825 (1985)
10. Sakakibara, Y., Brown, M., Hughey, R., Mian, I., Sjolander, K., Underwood, R., Haussler, D.: Stochastic context-free grammers for tRNA modeling. Nucleic Acids Research 22, 5112 (1994)
11. Teitelbaum, R.: Context-free error analysis by evaluation of algebraic power series. In: STOC, pp. 196–199. ACM, New York (1973)
12. Dowell, R., Eddy, S.: Evaluation of several lightweight stochastic context-free grammars for RNA secondary structure prediction. BMC bioinformatics 5, 71 (2004)
13. Do, C.B., Woods, D.A., Batzoglou, S.: CONTRAfold: RNA secondary structure prediction without physics-based models. Bioinformatics 22, e90–e98 (2006)
14. Cocke, J., Schwartz, J.T.: Programming Languages and Their Compilers. Courant Institute of Mathematical Sciences, New York (1970)
15. Kasami, T.: An efficient recognition and syntax analysis algorithm for context-free languages. Technical Report AFCRL-65-758, Air Force Cambridge Res. Lab., Bedford Mass. (1965)
16. Younger, D.H.: Recognition and parsing of context-free languages in time n^3. Information and Control 10, 189–208 (1967)

17. Valiant, L.: General context-free recognition in less than cubic time. Journal of Computer and System Sciences 10, 308–315 (1975)
18. Coppersmith, D., Winograd, S.: Matrix multiplication via arithmetic progressions. J. Symb. Comput. 9, 251–280 (1990)
19. Akutsu, T.: Approximation and exact algorithms for RNA secondary structure prediction and recognition of stochastic context-free languages. Journal of Combinatorial Optimization 3, 321–336 (1999)
20. Benedí, J., Sánchez, J.: Fast Stochastic Context-Free Parsing: A Stochastic Version of the Valiant Algorithm. In: Martí, J., Benedí, J.M., Mendonça, A.M., Serrat, J. (eds.) IbPRIA 2007. LNCS, vol. 4477, pp. 80–88. Springer, Heidelberg (2007)
21. Chan, T.M.: More algorithms for all-pairs shortest paths in weighted graphs. In: STOC 2007: Proceedings of the thirty-ninth annual ACM symposium on Theory of computing, pp. 590–598. ACM, New York (2007)
22. Graham, S.L., Harrison, M.A., Ruzzo, W.L.: An improved context-free recognizer. ACM Transactions on Programming Languages and Systems 2, 415–462 (1980)
23. Arlazarov, V.L., Dinic, E.A., Kronod, M.A., Faradzev, I.A.: On economical construction of the transitive closure of an oriented graph. Soviet. Math. Dokl. 11, 1209–1210 (1970)
24. Frid, Y., Gusfield, D.: A simple, practical and complete $O(\frac{n^3}{\log n})$-time algorithm for RNA folding using the four-russians speedup. In: Salzberg, S.L., Warnow, T. (eds.) WABI 2009. LNCS, vol. 5724, pp. 97–107. Springer, Heidelberg (2009)
25. Klein, D., Manning, C.D.: A* parsing: Fast exact viterbi parse selection. In: HLT-NAACL, pp. 119–126 (2003)
26. Jansson, J., Ng, S., Sung, W., Willy, H.: A faster and more space-efficient algorithm for inferring arc-annotations of RNA sequences through alignment. Algorithmica 46, 223–245 (2006)
27. Wexler, Y., Zilberstein, C.B.Z., Ziv-Ukelson, M.: A study of accessible motifs and RNA folding complexity. Journal of Computational Biology 14, 856–872 (2007)
28. Ziv-Ukelson, M., Gat-Viks, I., Wexler, Y., Shamir, R.: A faster algorithm for RNA co-folding. In: Crandall, K.A., Lagergren, J. (eds.) WABI 2008. LNCS (LNBI), vol. 5251, pp. 174–185. Springer, Heidelberg (2008)
29. Backofen, R., Tsur, D., Zakov, S., Ziv-Ukelson, M.: Sparse RNA folding: Time and space efficient algorithms. In: Kucherov, G., Ukkonen, E. (eds.) CPM 2009. LNCS, vol. 5577, pp. 249–262. Springer, Heidelberg (2009)
30. Salari, R., Mohl, M., Will, S., Sahinalp, S., Backofen, R.: Time and Space Efficient RNA-RNA Interaction Prediction via Sparse Folding. In: Berger, B. (ed.) RECOMB 2010. LNCS, vol. 6044, pp. 473–490. Springer, Heidelberg (2010)
31. Havgaard, J., Lyngso, R., Stormo, G., Gorodkin, J.: Pairwise local structural alignment of RNA sequences with sequence similarity less than 40%. Bioinformatics 21, 1815–1824 (2005)
32. Will, S., Reiche, K., Hofacker, I.L., Stadler, P.F., Backofen, R.: Inferring non-coding RNA families and classes by means of genome-scale structure-based clustering. PLOS Computational Biology 3, 65 (2007)
33. Baker, J.K.: Trainable grammars for speech recognition. The Journal of the Acoustical Society of America 65, S132 (1979)
34. Goto, K., Geijn, R.: Anatomy of high-performance matrix multiplication. ACM Transactions on Mathematical Software (TOMS) 34, 1–25 (2008)
35. Robinson, S.: Toward an optimal algorithm for matrix multiplication. News Journal of the Society for Industrial and Applied Mathematics 38 (2005)

Reconstruction of Ancestral Genome Subject to Whole Genome Duplication, Speciation, Rearrangement and Loss

Denis Bertrand[1], Yves Gagnon[1], Mathieu Blanchette[2], and Nadia El-Mabrouk[1]

[1] DIRO, Université de Montréal, H3C 3J7, Canada
bertrden@iro.umontreal.ca, y.gagnon@umontreal.ca,
mabrouk@iro.umontreal.ca
[2] McGill Centre for Bioinformatics, McGill University, H3A 2B4, Canada
blanchem@mcb.mcgill.ca

Abstract. Whole genome duplication (WGD) is a rare evolutionary event that has played a dramatic role in the diversification of most eukaryotic lineages. Given a set of species known to have evolved from a common ancestor through one or many rounds of WGD together with a set of genome rearrangements, and a phylogenetic tree for these species, the goal is to infer the pre-duplication ancestral genomes. We use a two step approach: (1) Compute a score for each possible ancestral adjacency at each internal node of the phylogeny; (2) Combine adjacencies to form ancestral chromosomes. We first apply our method on simulated datasets and show a high accuracy for adjacency prediction. We then infer the pre-duplicated ancestor of a set of 11 yeast species and compare it to a manually assembled ancestral genome obtained by Gordon *et al.* (2009).

1 Introduction

Whole genome duplication (WGD) is a spectacular evolutionary event that has the effect of simultaneously doubling all the chromosomes of a genome. Evidence for WGD events has shown up across the whole eukaryote spectrum, from the protist *Giardia* to the yeast species, including most plant lineages, several insect, fish, amphibians, and even to mammalian species. For some genomes, recent duplication is easily detected by the presence of a nearly complete set of duplicated chromosomes. However, in most cases, due to a series of rearrangements disrupting the initial perfectly doubled structure of the genome, all that we can observe is a set of duplicated blocks (chromosomal segments or genes) representing a high proportion of the genome, scattered throughout the genome.

Studying the evolution of a lineage that have been subject to one or many WGD is challenging due to the high rates of paralogy in their genomes. Inferring the content and chromosome organization of ancestral genomes preceding the WGD is a major step towards solving this difficulty, and also answering biological questions such as the mechanisms of polyploid formation and the consequence of such variations on the genetic and physiological specificities of species.

In 2003, we have presented the first formal result related to genome duplication, which is an exact linear-time algorithm for solving the *genome halving*

V. Moulton and M. Singh (Eds.): WABI 2010, LNBI 6293, pp. 78–89, 2010.

problem [5]: Given a present-day genome G represented as a set of strings (chromosomes) with each block present exactly twice, infer a perfectly duplicated genome H (a genome with exactly two copies of each chromosome) minimizing the rearrangement distance to G (inversions, reciprocal translocations or both). Our results have been reformulated recently by Alekseyev and Pevzner [1] using an alternative representation of the breakpoint graph. Subsequently, Sankoff and colleagues [13,14], and more recently Gavranović and Tannier [6], used variations of the genome halving strategy (Guided Genome Halving or GGH) to find the preduplicated ancestor of a doubled genome in the presence of a non-duplicated outgroup [13,14]. As noticed in [7], the GGH algorithms can hardly be generalized to a complete phylogenetic tree, with more than one WGD event on a path from an extant species to the root of the tree, and an arbitrary number of post-WGD genomes and non-WGD outgroups. Moreover, as for genome halving, GGH algorithms can only consider genes that have retained two copies after the WGD. In the case of reconstructing the ancestor of *Saccharomyces cerevisiae*, Gordon *et al.* [7] have noticed that less than 20% of all genes can be taken into account by the GGH strategy. Subsequent work shows that this limitation can be circumvented by grouping genes into double conserved syntenies [3,6].

In this paper, we consider the general problem of inferring the pre-duplicated genome preceding the first duplication event in a multi-species evolutionary history involving WGDs, rearrangement events, and block losses. The input of our problem is a set of extant genomes, each represented as a set of strings on an alphabet of blocks (each block potentially present more than once in each genome), and a phylogenetic tree representing the evolution of the species, with specific branches marked with WGD events. Such data and phylogenetic information is available for a variety of eukaryotic lineages, such as the yeast species [7], grass genomes [12], and many other lineages. Our approach for ancestral genome prediction is to maximize the conservation of block adjacencies in the phylogeny. We use a two-step methodology: (1) at each node of the phylogeny, compute the adjacency score of each pair of blocks; (2) infer a pre-duplicated ancestral genome by an optimal chaining of adjacencies. The main contribution of our method is the computation of a rigorous score for each potential ancestral adjacency (a, b), reflecting the maximum number of times a and b can be adjacent in the whole phylogeny, for any setting of ancestral genomes. As it is the case for the other local approaches [4], in the absence of a complete set of reliable syntenies, the output of our algorithm is a set of *Contiguous Ancestral Regions fragments* (CAR) [4,10], rather than a completely assembled ancestral genome.

The approaches most comparable to ours are those developed by Ma *et al.* (see the method in [10] for single gene copies, and its generalization to genomes with duplications in [11]). We show that our approach outperforms the former on simulated data. The latter can only be used if accurate gene trees, with branch lengths, are available, which is often limiting. In contrast, our approach works under stronger assumptions but requires only a species tree and extant genomes as input. Our paper is structured as follows: after introducing basic notations, we introduce the notion of adjacency scores, show how to compute it, and how to use it to assemble putative ancestral genomes by solving an instance of the traveling salesperson problem. We then show, using simulated data, that the predicted pre-WGD genomes are highly accurate, even in the presence of a large number of rearrangements. Finally, we apply our approach to the prediction of the ancestral pre-WGD yeast genome and obtain results very similar to the hand-curated ancestral genome produced by Gordon *et al.* [7].

2 Preliminaries

Notation: Let $\mathcal{B}$ be a set of unsigned blocks (e.g. genes, or any other type of genomic markers). A *string* is a sequence of blocks from $\mathcal{B}$, where each block is signed (+ or −) to mark its orientation. A *genome* G is a collection of strings $C_1, C_2, \ldots, C_N$ called its *chromosomes*, where each element of $\mathcal{B}$ may be present more than once. To represent chromosomal ends, we add an artificial block O, which is also added to our alphabet $\mathcal{B}$, at an extremity of each chromosome, and consider each chromosome as circular. We denote by $\Sigma_G \subseteq \mathcal{B}$ the set of blocks present in G (including O), by $mult(a, G)$ the multiplicity of block a in G , and by $\pm\Sigma_G$ the set obtained from Σ_G by considering each block in its positive and negative directions. The artificial block O is always considered positively signed. A *multiset* of $\pm\Sigma_G$ is a subset of $\pm\Sigma_G$ with possibly repeated blocks. Let $a \in \Sigma_G$ and $b \in \pm\Sigma_G$. We say that b is a left-adjacency of a in G iff "$b\ +a$" or "$-a\ b$" is a substring of G. Symmetrically, b is a right-adjacency of a in G iff "$+a\ b$" or "$b\ -a$" is a substring of G. We denote by $LA(G, a)$ and $RA(G, a)$ the *multisets* of left- and right-adjacencies of the one or more copies of a in G.

Evolutionary model: A *Whole Genome Duplication* (or WGD for short) is an event transforming a genome $G = \{C_1, C_2, \ldots, C_N\}$ into a genome G^D containing $2N$ chromosomes, i.e. $G^D = \{C_1, C_1', C_2, C_2', \ldots, C_N, C_N'\}$, where, for each $1 \leq i \leq N$, $C_i = C_i'$. Let $G_1, G_2, \ldots, G_n$ be a set of n related species at the leaves of a species tree T, assumed to have evolved from a common ancestor through WGD events, intra-chromosomal (inversions or transpositions) and inter-chromosomal (reciprocal translocations, fusions, or fissions) rearrangements, and block losses. A phylogeny for $(G_i)_{i=1}^n$ is a tree T with n leaves, where G_i, for $1 \leq i \leq n$, is the label of leaf i, and each internal node (also called *speciation node*) has exactly two children and represents a speciation event.

In our model, WGDs are the only duplication events responsible for block multiplicity (in particular, single-block duplications are not considered). In order to account for those duplication events, we create new internal nodes in T, called *WGD nodes*, and position them appropriately on the edges of T. Contrary to speciation nodes, each WGD node has only a single child. Moreover, if all leaf genomes have multiplicity greater than 1, then we add one or more WGD nodes above the root r of T and we create a new root D, that we call *the duplication root of* T.

Assuming a model with no convergent evolution and minimum losses, the multiset of blocks Σ_u present at node u can be obtained as follows (see Figure 1(a)). Let $A(a)$ be the node of T representing the least common ancestor of the leaves that contain a given block a. Then, we assign a to each node belonging to a path from $A(a)$ to any leaf containing a. In order to define the multiset Σ_u, we also need to know the multiplicity of each block at u. The multiplicity of a in Σ_u is recursively obtained as the maximum of its multiplicities in u's two children.

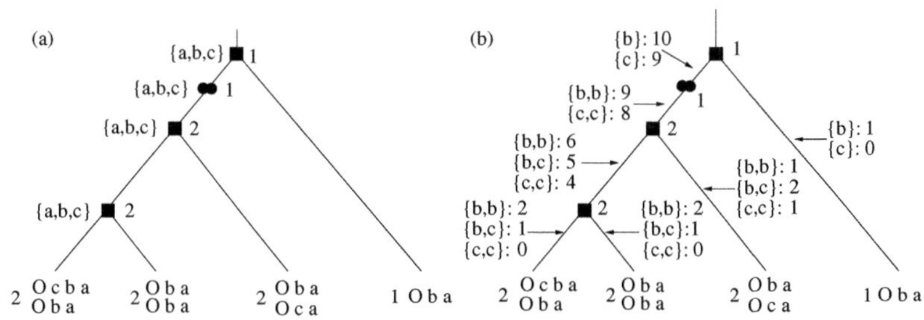

Fig. 1. (a) A species tree with each leaf labeled with its corresponding genome and multiplicity number, and each internal node labeled with the multiplicity and block set of the ancestral genome just preceding the speciation or WGD event. Squares indicate speciation nodes, and the double circle indicates a WGD node. (b) An illustration of the algorithm computing L^{below} for the adjacencies of gene a.

3 Problem Definition

Given a species tree T for the genomes $(G_i)_{i=1}^n$, augmented with WGD nodes as described in the previous section, we want to infer the pre-duplicated ancestral genomes, i.e. the ancestral genomes just preceding the first WGD nodes on the paths from the root D of T to a leaf. We will use a parsimony criteria seeking a solution with a maximum number of adjacency conservations along the branches of T, or, equivalently, a maximum of adjacency conservation.

Ancestral genome assignment. A *genome assignment* $G(u)$ at u is a genome on $\mathcal{B}$ respecting the content and multiplicity constraints given by Σ_u. If u is a WGD node, $G(u)$ must be a duplicated genome. Let u and v be two nodes of T with u being the parent of v. In the case of genomes with single gene copies, it is easy to define the number of adjacencies preserved along branch (u, v) as the number of common substrings of size 2 between $G(u)$ and $G(v)$. This definition is not directly transposable to the case of genomes with multiple gene copies, as the one to one orthology between genes is not set. Instead, for each block a, we compare its left and right-adjacency multisets in $G(u)$ and $G(v)$. More precisely, we define $adjCons(a, G(u), G(v)) = |LA(G(u), a) \cap LA(G(v), a)| + |RA(G(u), a) \cap RA(G(v), a)|$, as the number of left and right conserved adjacencies of a on the branch (u, v), and $adjCons(G(u), G(v)) = \sum_{a \in \Sigma_u \cap \Sigma_v} adjCons(a, G(u), G(v))$ as the total number of left and right conserved adjacencies on the branch (u, v). In both formulas, intersections and cardinalities are taken over multisets. Notice that $adjCons(G(u), G(v))$ accounts for each adjacency conservation twice.

We then define *adjCons*(T) as the maximum number of conserved adjacencies in T, over all possible ancestral genome assignments $G(u_1), \ldots G(u_k)$ at all internal (speciation and WGD) nodes $n_1, \ldots n_k$ of T:

$$adjCons(T) = \max_{G(u_1),\ldots,G(u_k)} \sum_{(u,v)\in E(T)} adjCons(G(u), G(v))$$

Finally, for a given ancestral node u_i with genome assignment $H(u_i)$, define

$$adjCons(T|_{G(u_i)=H(u_i)}) = \max_{G(u_1),\ldots,G(u_k)|G(u_i)=H(u_i)} \sum_{(u,v)\in E(T)} adjCons(G(u), G(v)),$$

which is the maximum number of adjacencies that can be preserved along the branches of T, if the genome at node u_i is set to $H(u_i)$. We can now state our optimization problem precisely.

Ancestral Genome Assignment Problem:
INPUT: A species tree T for the genomes $(G_i)_{i=1}^{n}$ augmented with one or more WGD nodes as described in the previous section; The multiset of blocks at each internal node; A particular WGD node u of interest.
OUTPUT: An ancestral genome assignment $H(u)$ to u such that *adjCons*$(T|_{G(u)=H(u)})$ is maximized.

In this paper, we focus on inferring the pre-duplicated genomes preceding a first WGD event on a branch from the root of T to a leaf. In other words, u is the first WGD node on a branch from the root of T to a leaf.

4 Method

We start by defining an upper bound on our objective function, *adjCons*$(T|_{G(u)=H(u)})$. We define *adjCons*$(a, T|_{\mathcal{C}})$ as the maximum number of left and right adjacencies of a that can be preserved over the whole tree, for any assignment $G(u_1), \ldots, G(u_k)$ of ancestral genomes subject to a set of constraints $\mathcal{C}$. Then, it is straightforward to show that:

$$\begin{aligned} adjCons(T|_{G(u)=H(u)}) &\leq \sum_a adjCons(a, T|_{LA(a,G(u))=LA(a,H(u)),RA(a,G(u))=RA(a,H(u))}) \\ &\leq 1/2 \cdot \sum_a adjCons(a, T|_{LA(a,G(u))=LA(a,H(u))}) + \\ &\qquad adjCons(a, T|_{RA(a,G(u))=RA(a,H(u))}) \end{aligned}$$

Our ancestral reconstruction algorithm thus seeks a genome H such that the above term is maximized. It proceeds in two steps: (1) For each internal node u of the tree (speciation or WGD node), each block $a \in \Sigma_u$, and each multisets X of possible left adjacencies of a at node u, we compute *adjCons*$(a, T|_{LA(a,G(u))=X})$, using a dynamic programming algorithm. We then proceed similarly for right adjacencies. (2) We obtain the desired pre-duplicated genome at WGD node u by chaining the adjacencies at node u in a optimal way.

4.1 Computing Adjacency Scores

We first describe how to compute $adjCons(a, T|_{LA(a,G(u_i))=X})$, for any node u_i, block $a \in \Sigma_{u_i}$, and candidate left-adjacencies X. The algorithm to compute right-adjacencies is very similar. Consider an edge (u, v) in T, where u is the parent of v. Let X be a multisubset of $\pm\Sigma_u$ and $G(u)$ be a genome assignment at node u such that $LA(a, G(u)) = X$. We define $L^{below}_{(u,v)}(a, X)$ as the maximum number (over all possible genome assignments of T's internal nodes) of left-adjacencies involving the copies of a that can be preserved along the branch (u, v) and all the branches of the subtree rooted at node u. Similarly, we define $L^{above}_{(u,v)}(a, X)$ as the maximum number of left-adjacencies involving a that can be preserved, along branch (u, v) and all the branches outside the subtree rooted at node u. Then, for an internal node u with children v and w and parent p, we obtain $adjCons(a, T|_{LA(a,G(u))=X}) = L^{below}_{(u,v)}(a, X) + L^{below}_{(u,w)}(a, X) + L^{above}_{(p,u)}(a, X)$. Notice that, if u is a WGD node, then u has a single child v, and thus the term $L^{below}_{(u,w)}(a, X)$ should be removed from the above formula. Similarly, if u is the root of the tree, then the term $L^{above}_{(p,u)}(a, X)$ should be removed.

Algorithm 1: $L^{below}_{(u,v)}(a, X)$

---

if *v is a leaf* **then**
- **if** *u is a speciation node* **then**
  - $L^{below}_{(u,v)}(a, X) = |X \cap LA(G(v), a)|$;
- **if** *u is duplication node* **then**
  - $L^{below}_{(u,v)}(a, X) = |(X \cup X) \cap LA(G(v), a)|$;

else v is an internal node
- **if** *v is a speciation node with children x and y* **then**
  - **if** *u is a speciation node* **then**
    - $L^{below}_{(u,v)}(a, X) = max_{X'}\{L^{below}_{(v,x)}(a, X') + L^{below}_{(v,y)}(a, X') + |X \cap X'|\}$;
  - **if** *u is a duplication node* **then**
    - $L^{below}_{(u,v)}(a, X) = max_{X'}\{L^{below}_{(v,x)}(a, X') + L^{below}_{(v,y)}(a, X') + |(X \cup X) \cap X'|\}$;
- **else** v is a duplication node with one child w
  - $L^{below}_{(u,v)}(a, X) = max_{X'}\{L^{below}_{(v,x)}(a, X') + |X \cap X'|\}$;

Algorithm 2: $L^{above}_{(p,u)}(a, X)$

---

if *u is the root r of T* **then**
- $p = D$ and $L^{above}_{(p,u)}(z, X) = 0$;

else let p' be the parent of p
- **if** *p is a speciation node and s is the sibling of u* **then**
  - $L^{above}_{(p,u)}(a, X) = max_{X'}\{L^{above}_{(p',p)}(a, X') + L^{below}_{(p,s)}(a, X') + |X \cap X'|\}$;
- **if** *p is a duplication node (its only child is u)* **then**
  - $L^{above}_{(p,u)}(a, X) = max_{X'}\{L^{above}_{(p',p)}(a, X') + |X \cap (X' \cup X')|\}$;

We are thus interested in calculating the tables $L^{below}_{(u,v)}$ and $L^{above}_{(u,v)}$ for each edge (u, v) of T. Those are obtained by the dynamic programming algorithms shown in Algorithms 1 and 2. An illustration of this algorithm is given in Figure 1(b). Although expressed in a recursive manner for simplicity, both algorithms can be re-written using a dynamic programming approach that proceeds in a bottom-up fashion to obtain L^{above} and in a top-down fashion to obtain L^{below}. The running time to compute $adjCons(a, T|_{LA(a,G(u))=X})$ is thus $O(\Sigma_{(u,v)\in T}(| \pm \Sigma_u|^{mult(a,G(u))} \times | \pm \Sigma_v|^{mult(a,G(v))}))$.

4.2 Assembling an Pre-duplication Ancestral Genome

We now seek to build a solution to the Ancestral Genome Assignment Problem, i.e. to chain blocks at u so as to maximize the objective function. We achieve this by solving a Traveling Salesperson Problem (TSP) on a complete undirected graph where vertices correspond to blocks. We initially weighted edges according to our upper bound L_u^{all}. However, this weighting gives too much importance to adjacencies involving blocks with high multiplicity. Thus, we decided to weight the edges according to the ratio $rL_u^{all}(a, X) = L_u^{all}(a, X)/adjConsMax(a, T)$, where $adjConsMax(a, T) = \Sigma_{(u,v)\in E(T)} \min(mult(a, G_u), mult(a, G_v))$. Notice that $adjConsMax(a, T)$ represents the number of conserved adjacencies for the block a in T if a is always adjacent to the same gene in all leaves of T. This ratio allows us to evaluate the confidence of an inferred adjacency (see Figures 3 and 4 top right).

More precisely, we build an undirected graph Q that contains a pair of vertices a^t, a^h for each block $a \in \Sigma_u$, as well as a set of vertices O_1, O_2, O_k marking chromosome ends, where k is chosen to be at least as large as (but possibly larger than) the maximum number of chromosomes in the ancestral genome we seek to infer. Edge weights are chosen as follows, for $a \neq b \in \Sigma_u$, $i \neq j \in \{1, ..., k\}$, and M some large number:

$$w(a^h, b^t) = rR_u^{all}(a, \{+b\}) + rL_u^{all}(b, \{+a\}) \qquad w(a^t, a^h) = M$$
$$w(a^h, b^h) = rR_u^{all}(a, \{-b\}) + rR_u^{all}(b, \{-a\}) \qquad w(O_i, a^t) = 2 \times rL_u^{all}(a, \{O\})$$
$$w(a^t, b^t) = rL_u^{all}(b, \{-a\}) + rL_u^{all}(a, \{-b\}) \qquad w(O_i, a^h) = 2 \times rR_u^{all}(a, \{O\})$$
$$w(a^t, b^h) = rL_u^{all}(a, \{+b\}) + rR_u^{all}(b, \{+a\}) \qquad w(O_i, O_j) = 0$$

Because a^t and a^h are connected by heavy edges, any maximum weight hamiltonian cycle must include all of them. A hamiltonian cycle through Q thus defines a set of strings (chromosomes; delimited by O vertices), with some possibly empty (two consecutive O vertices). Starting from O_1, the cycle visits pairs (a^t, a^h) (corresponding to $+a$) or (a^h, a^t) (corresponding to $-a$). The heaviest hamiltonian cycle through Q thus corresponds to an hypothetical ancestral genome H at u that preserves a large number of adjacencies.

The instance of the TSP we need to solve is a symmetrical weighted graph with $2 \cdot |\Sigma_u| + k$ vertices. In the case of the reconstruction of the ancestral pre-duplication yeast genome, $|\Sigma_u| = 4705$, so the graph is quite large. Although an NP-Complete problem, TSP is one of the best studied algorithmic problems and excellent heuristics exist. We considered two of them. The first is a simple greedy approach that repeatedly selects the heaviest edge remaining unless this results in the premature closing of a cycle. The second is the Chained Lin-Kernighan heuristic [9] implemented in Concorde [2], referred as TSP from now on.

5 Results

A strong prerequisite for reconstructing accurate ancestral genomes is to have enough data on extant species, and sufficient colinearity of gene order among a reasonably large number of related species. Yeast genomes are a perfect example of a data set satisfying all these conditions. Following the extensive work of Wolfe and colleagues during the last decade, it is now almost universally accepted that *Saccharomyces cerevisiae* is the descendant of an ancient whole-genome duplication event. Moreover, the availability of a large number of completely sequenced yeast genomes as well as the Yeast Gene Order Browser [7], provides the material for an accurate ancestral genome reconstruction. We therefore focus, in this paper, on the study of the yeast species data sets.

5.1 Simulated Data Sets

We first test our method on a data set obtained through a simulated evolution that is as close as possible to the one observed for yeast species. The phylogenetic tree given in Figure 2(a) reflects the evolution of the 11 yeast species recorded in the Yeast Gene Order Browser, as given by [8]. Gene sets at leaves are those provided in [7], and gene sets at internal nodes, as well as the number of gene losses on each branch, are directly inferred from those at the leaves.

Based on this tree, we simulate the evolution of 11 genomes, starting from an ancestor with 4705 genes distributed among 8 chromosomes (the number of genes and chromosomes of the pre-duplicated ancestor as predicted by Gordon *et al.*), and performing a certain number of rearrangements and gene losses on each branch of the tree. The number of gene losses is simply the one observed in the phylogenetic tree of Figure 2 . The number of rearrangements is selected randomly from an interval $[\mu/2, \mu]$, where μ is a parameter chosen prior to the generation, and the size of each rearrangement is random. As for the rate of rearrangement operations it is chosen to be similar to that reported for *S. cerevisiae* in [7]. More precisely we choose the rates (Inv : Trans : Fus+Fiss) = (5 : 4 : 1).

Notice that four of the species represented in the phylogenetic tree of Figure 2 (those indicated by $*$), are partially sequenced species for which only scaffolds are available. To account for this specificity of the data, we perform random fissions on four of our simulated genomes. Moreover, as scaffolds just represent parts of chromosomes, adjacencies at the extremities are not relevant to our study and are not taken into account.

Simulations without WGD. In order to measure the efficiency of our approach in absence of WGD events, we compare our results with those obtained by the algorithm of Ma *et al.* [10]. We refer to this software as the Ma method.

We simulate data sets based on the subtree of the yeast phylogeny containing only the six non-duplicated yeast species. Moreover, as the Ma *et al.* algorithm does not support losses, we only consider the set of genes present in all six species. We performed our simulations with 10 different values of μ, varying from 100 to 1000. For each of those 10 μ values, 50 different data sets are obtained, an ancestor is inferred for each dataset and compared to the "true" known ancestor.

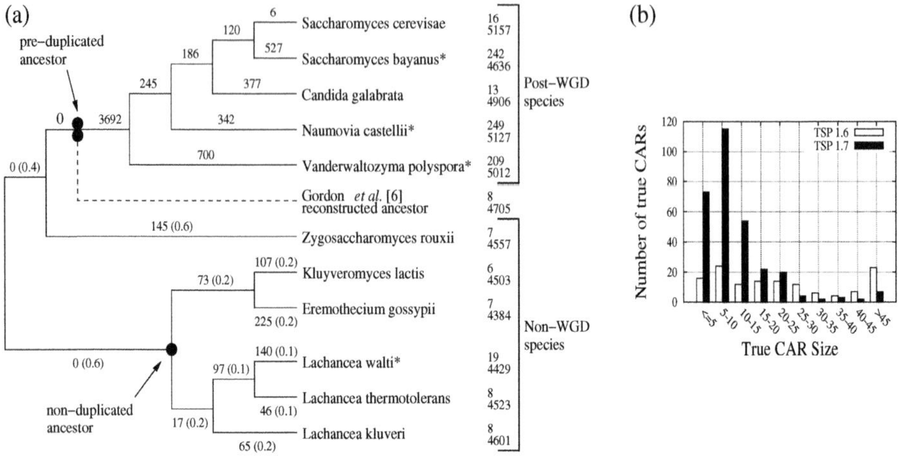

Fig. 2. (a) Yeast phylogenetic tree used for the simulations. The ancestors are represented by black dots. The branch lengths represents the number of gene losses. The ones in parentheses are the distances used for the Ma *et al.* method. * indicates partially sequenced organisms. On leaves, the top number is the number of chromosomes, contigs or scaffolds. The bottom number is the number of genes. (b) Size distribution of true CARs in our inferred pre-duplicated yeast ancestor, considering the Gordon *et al.* [7] ancestor as the "true" ancestor.

Results (error rates and number of preserved CARs) are averaged over all data sets showing a comparable ancestral genome divergence, where the genome divergence of a data set is the fraction of adjacencies in the ancestor that are preserved in at least one leaf of the tree. In Figure 3, the *error rate* is the rate of inferred adjacencies that are not present in the true ancestor, and the proportion of true CARs is the proportion of inferred CARs that are present in the true ancestor. Recall that a CAR is a chromosomal segment inferred by the algorithm, and it is "true" if it is a subsequence of the ancestral genome. We arbitrarily imposed a minimum size of 4 adjacencies to consider a CAR as a true CAR.

Comparing the error rates of the Ma method, and our methods using the greedy or TSP approach (Figure 3 top left), we first notice that the three methods have a good performance (less than 10% errors for genome divergence of 40%), but with the TSP method outperforming the two others. However, a drawback of the TSP approach is the fact that it outputs very few CARs, typically one or two for a genome divergence above 40%. In all cases, our approaches (greedy and TSP) infer fewer CARs than the Ma method (Figure 3 bottom left), and fewer that the actual number of chromosomes of the true ancestor.

In order to improve the proportion of true CARs inferred, we "force" the production of more CARs by defining "TSP τ" which is the TSP method augmented with the procedure of cutting, from the inferred ancestor, all adjacencies with weight less than a certain threshold τ. Figure 3 top right gives the error rate associated to the set of adjacencies of a given weight (rate of such adjacencies in our results that are false predictions). Based on this figure, we choose $\tau = 1.4$. As observed in Figure 3 bottom left, the proportion of true CARs inferred is greatly improved compared to the TSP approach without edge cut, but more interestingly compared to the greedy approach and the Ma method. However whereas the "true adjacencies" of the TSP approach where covering more that 90% of the genome, the number of genes covered by the "true CARs" is less than 40% for a genome divergence of more than 40%. However, this gene coverage remains higher than that of the Ma method (Figure 3 bottom right).

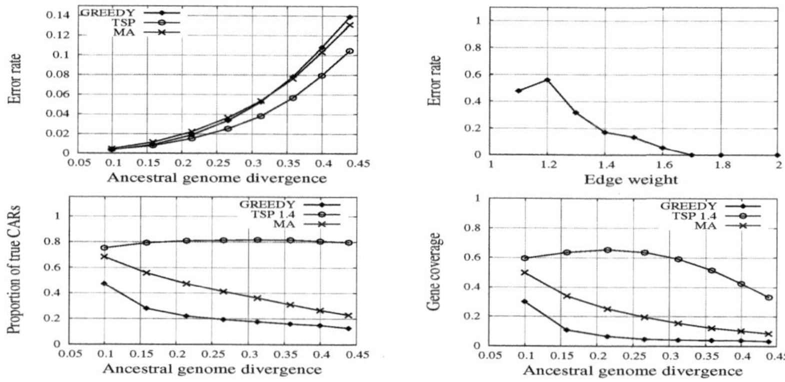

Fig. 3. Simulations for a tree without WGD. Top left: Error rate of the inferred ancestral genomes. Top right: Error rate of adjacencies depending on their weight. Bottom left: Proportion of true CARs inferred. Bottom right: Proportion of genes in the ancestral genome that are covered by the inferred true CARs. See the text for explanation about "Ancestral genome divergence", "Error rate", "TSP", "TSP τ" and "MA".

Simulations with WGD. We now simulate datasets based on the whole yeast tree (Figure 2(a)). Sets of genomes have been generated, with μ varying from 0 to 500, with increments of 50. For each of those 11 μ values, 50 data sets have been generated. Notice first that the addition, in our simulations, of gene loss, increases the ancestral genome divergence. In this case, the TSP approach clearly infers fewer false adjacencies than the greedy approach, regardless of the ancestral genome divergence. Its error rate remains under 10% for ancestors with divergence under 50%. Based on Figure 4 top right, we choose two thresholds for edge-cut $\tau = 1.6$ and $\tau = 1.7$. We observe from Figure 4 (bottom left and bottom right) that the proportion of true CARs inferred by TSP1.7 is over 80% for a gene divergence under 0.3 with a gene coverage over 60%, and over 40% for a genome divergence under 0.5, but with a significantly lower gene coverage (only over 20%).

5.2 Comparison with the Gordon *et al.* Ancestor

We applied our method to the yeast species tree (Figure 2(a)) with the gene datasets of the *Yeast Gene Order Browser* [7] described above, to infer the preduplicated ancestral genome of *Sccharomyces cerevisiae*. Compared with the

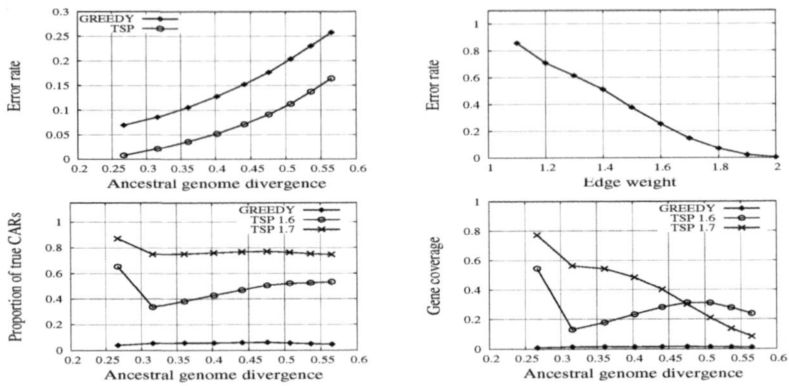

Fig. 4. Simulations for a tree with WGD. See Figure 3.

ancestral genome manually inferred by Gordon *et al.* [7], about 98% of the adjacencies inferred by our method are also present in the Gordon *et al.* ancestor. However, our TSP approach without edge-cut leads to only 4 CARs compared to the 8 likely ancestral chromosomes. We tried two cutoff values for edge weight to decrease the number of incorrect adjacencies. With a cutoff value of 1.6, we obtain smaller CARs (average length 26), 84% of them (covering 79% of the genes) being "true" CARs of the Gordon *et al.* ancestor. With a cutoff value of 1.7, CARs are even smaller (12 in average) with 95% true CARs, covering 75% of the genes. Figure 2(b) illustrates the size distribution of true CARs with the different TSP strategies.

6 Conclusion

We have developed a general method for inferring the ancestral pre-duplicated genomes of a lineage known to have evolved through one or many rounds of whole genome duplication, interspersed with genome rearrangements and gene losses. The input to our method is a phylogenetic tree representing the evolution of the species, with positions of the WGD events, and genomes represented as ordered sets of oriented blocks, each block appearing in one or many copies in each genome. We developed a local approach consisting in inferring ancestral adjacencies and then chaining them in an optimal way. The originality of this method is the computation of a rigorous score for each ancestral set of adjacencies, reflecting the maximum number of conservation of this set of adjacencies among the whole tree. This is done by a rigorous dynamic programming algorithm, which is sufficiently fast to run on large data sets. Chaining adjacencies is then performed using a traveling salesperson strategy on a graph representation of all possible adjacencies.

Applying our method, first on simulated datasets and then on the yeast genomes, reveals a high accuracy for adjacency prediction. However, the number of inferred CARs strongly depends on the cutoff value used to separate good adjacencies from noise. Although the TSP strategy seems appropriate, other chaining strategies may be considered and may improve the quality of our results.

In this paper, we focused on inferring the ancestral genomes preceding a first WGD event on the tree. In other words, the inferred genome has a single copy of each chromosome. This restriction is only required for the chaining part of the method, as the first step that consists in computing the score of each set of adjacencies at each internal node of the tree is general. However, if the ancestor of interest has more than one copy of each gene, it is not clear how to assemble a set of relevant adjacencies to form CARs as the TSP representation breaks down. This is one of the future directions to our project that we aim to pursue.

References

1. Alekseyev, M.A., Pevzner, P.A.: Colored de bruijn graphs and the genome halving problem. IEEE/ACM TCBB 4(1), 98–107 (2007)
2. Applegate, D., Bixby, R., Chvatal, V., Cook, W.: Concorde TSP solver (2006), `http://www.tsp.gatech.edu/concorde/`
3. Chauve, C., Gavranović, H., Ouangraoua, A., Tannier, E.: Yeast ancestral genome reconstructions: the possibilities of computational methods II. Journal of Computational Biology (2010) (in press)
4. Chauve, C., Tannier, E.: A methodological framework for the reconstruction of contiguous regions of ancestral genomes and its application to mammalian genomes. Plos Computational Biology 4(11), e1000234 (2008)
5. El-Mabrouk, N., Sankoff, D.: The reconstruction of doubled genomes. SIAM Journal on Computing 32(1), 754–792 (2003)
6. Gavranović, H., Tannier, E.: Guided genome halving: probably optimal solutions provide good insights into the preduplication ancestral genome of Saccharomyces cerevisiae. In: Pacific Symposium on Biocomputing, vol. 15, pp. 21–30 (2010)
7. Gordon, J.L., Byrne, K.P., Wolfe, K.H.: Additions, losses, and rearrangements on the evolutionary route from a reconstructed ancestor to the modern saccharomyces cerevisiae genome. PloS Genetics 5(5), e1000485 (2009)
8. Hedtke, S.M., Townsend, T.M., Hillis, D.M.: Resolution of phylogenetic conflict in large data sets by increased taxon sampling. Syst. Biol. 55, 522–529 (2006)
9. Lin, S., Kernighan, B.W.: An effective heuristic algorithm for the traveling salesman problem. Operations Research 21, 498–516 (1973)
10. Ma, J., et al.: Reconstructing contiguous regions of an ancestral genome. Genome Research 16, 1557–1565 (2007)
11. Ma, J., et al.: Dupcar: Reconstructing contiguous ancestral regions with duplications. Journal of Computational Biology 15(8), 1–21 (2008)
12. Salse, J., et al.: Reconstruction of monocotelydoneous proto-chromosomes reveals faster evolution in plants than in animals. PNAS 106(35), 14908–14913 (2009)
13. Zheng, C., Zhu, Q., Adam, Z., Sankoff, D.: Guided genome halving: hardness, heuristics and the history of the hemiascomycetes. In: ISMB, pp. 96–104 (2008)
14. Zheng, C., Zhu, Q., Sankoff, D.: Descendants of whole genome duplication within gene order phylogeny. Journal of Computational Biology 15(8), 947–964 (2008)

Genomic Distance with DCJ and Indels

Marília D.V. Braga, Eyla Willing, and Jens Stoye

Technische Fakultät, Universität Bielefeld, Germany
{mbraga,eyla}@cebitec.uni-bielefeld.de, stoye@techfak.uni-bielefeld.de

Abstract. The double cut and join (DCJ) operation, introduced by Yancopoulos, Attie and Friedberg in 2005, allows one to represent most rearrangement events in genomes. However, a DCJ cannot perform an insertion or a deletion and most approaches under this model consider only genomes with the same content and without duplications, including the linear time algorithms to compute the DCJ distance and to find an optimal DCJ sorting sequence. In this work, we compare two genomes with unequal content, but still without duplications, and present a new linear time algorithm to compute the genomic distance, considering DCJ and indel operations. With this method we find preliminary evidence of the occurrence of clusters of deletions in the *Rickettsia* bacterium.

1 Introduction

The double cut and join (DCJ) is an abstract rearrangement operation, introduced by Yancopoulos *et al.* in 2005 [5], that allows to represent most large scale mutation events, such as inversions, translocations, fusions and fissions, that can occur in genomes. No restriction on the genome structure considering linear and circular chromosomes is imposed. An advantage of this general model is that it leads to considerable algorithmic simplifications. However, a DCJ cannot perform an insertion or a deletion and most studies concerning DCJ consider genomes with the same content and without duplications. With these restrictions, linear time algorithms have been proposed to compute the DCJ distance and to find an optimal DCJ sorting sequence [1].

In 2008, Yancopoulos and Friedberg [4] proposed an extension of the DCJ paradigm, to include operations performing insertions and deletions, in order to deal with genomes having unequal content. The authors introduced some concepts, but left open the design of an algorithm to handle this problem, which is the subject of this study. We propose an approach in which the cost of an insertion or deletion is the same as that of a DCJ, where several consecutive markers can be inserted or deleted in a single event. Therefore, we generalize the adjacency graph, introduced by Bergeron *et al.* [1], by incorporating the representation of the markers that occur in only one of the two genomes. We then design a linear time algorithm to compute the distance between two genomes with unequal content, but still without duplications, taking into consideration DCJ operations, insertions and deletions. We used this method to do an interesting analysis of a group of bacterial genomes, as described in the last section.

V. Moulton and M. Singh (Eds.): WABI 2010, LNBI 6293, pp. 90–101, 2010.

2 DCJ, Adjacency Graph and Indels

In this work, duplications are not allowed. Thus, given a genome A over a set of markers $\mathcal{G}_A$, each g in $\mathcal{G}_A$ occurs exactly once in A. Furthermore, each marker g is a DNA fragment and can be either read in direct orientation and represented by the symbol g, or read in reverse orientation and represented by the symbol $\overline{g}$. We have $\overline{\overline{g}} = g$ and, for any set $\mathcal{F}$, we define $\widehat{\mathcal{F}} = \mathcal{F} \cup \overline{\mathcal{F}}$, where $\overline{\mathcal{F}} = \{\overline{f} \mid f \in \mathcal{F}\}$.

Let A be a genome, possibly composed of linear and circular chromosomes. From each chromosome $\mathcal{C}$ of A we can build a string s over $\widehat{\mathcal{G}_A}$, obtained by the concatenation of all symbols in $\mathcal{C}$, read in any of the two directions. Each end of a linear chromosome is called a *telomere*, represented by the symbol $\circ$. Thus, if $\mathcal{C}$ is linear, it is represented by $\circ s \circ$. If $\mathcal{C}$ is circular, it is simply represented by s (we can start to build s in any symbol of $\mathcal{C}$). A genome A with k chromosomes can be represented by a set of k strings and an example is given in Fig. 1.

Fig. 1. In this graphic representation of genome $A = \{\circ aexc\circ, \circ d\overline{y}b\circ, \circ z\overline{w}\circ\}$, composed of three linear chromosomes, each arrow represents a marker and its orientation

In the following we generalize definitions introduced by Bergeron *et al.* [1].

Given a genome A over $\mathcal{G}_A$ and a subset $\mathcal{G} \subseteq \mathcal{G}_A$, for each $g \in \mathcal{G}$ we denote its two extremities by g^t (tail) and g^h (head). A $\mathcal{G}$-*adjacency* is in general a linear string $v = \gamma_1 \ell \gamma_2$, such that γ_1 and γ_2 are telomeres or extremities of markers in $\mathcal{G}$ and ℓ, the substring composed of the markers that are between γ_1 and γ_2 in A, contains no marker that also belongs to $\mathcal{G}$. The substring ℓ is said to be the *label* of v, and the extremities γ_1 and γ_2 are said to be $\mathcal{G}$-*adjacent.* If ℓ is a non-empty string, v is said to be *labeled*, otherwise v is said to be *clean.* Observe that a $\mathcal{G}$-adjacency $\gamma_1 \ell \gamma_2$ can also be represented by $\gamma_2 \overline{\ell} \gamma_1$. Moreover, a labeled $\mathcal{G}$-adjacency $u = \circ \ell \circ$ indicates that A contains a linear chromosome composed only of markers that are not in $\mathcal{G}$, that is, u corresponds to a whole linear chromosome. In the same way, if s is a circular chromosome in A composed only of markers that are not in $\mathcal{G}$, then s is also a $\mathcal{G}$-adjacency. This is the only special case of $\mathcal{G}$-adjacency in which we have a circular instead of a linear string.

A genome A can be then represented by the set $V_{\mathcal{G}}(A)$ containing its $\mathcal{G}$-adjacencies. For example, if $\mathcal{G} = \{a, b, c, d, e\}$, the genome in Fig. 1 has the representation $V_{\mathcal{G}}(A) = \{\circ a^t, a^h e^t, e^h x c^t, c^h \circ, \circ d^t, d^h \overline{y} b^t, b^h \circ, \circ z \overline{w} \circ\}$. However, for $\mathcal{G}' = \mathcal{G}_A = \{a, b, c, d, e, x, y, z, w\}$, we have only clean adjacencies in $V_{\mathcal{G}'}(A) = \{\circ a^t, a^h e^t, e^h x^t, x^h c^t, c^h \circ, \circ d^t, d^h y^h, y^t b^t, b^h \circ, \circ z^t, z^h w^h, w^t \circ\}$.

A *cut* performed on a genome A separates two adjacent markers of A. A cut affects a $\mathcal{G}$-adjacency v of $V_{\mathcal{G}}(A)$ as follows: if v is linear, the cut is done between two symbols of v, creating two open ends in two separate linear strings; if v is circular, the cut creates two open ends in one linear string. A *double-cut and join* or DCJ applied on a genome A is the operation that performs two cuts in $V_{\mathcal{G}}(A)$, creating four open ends, and joins these open ends in a different way.

As an example, considering the genome A from Fig. 1 and $\mathcal{G} = \{a, b, c, d, e\}$, if we apply a DCJ on $a^h e^t$ and $d^h \overline{y} b^t$ of $V_{\mathcal{G}}(A)$ we can create $a^h b^t$ and $d^h \overline{y} e^t$.

Observe that, if the two original $\mathcal{G}$-adjacencies are $\gamma_1 \ell_1 \circ$ and $\gamma_2 \ell_2 \circ$, we can create $\gamma_1 \ell_1 \overline{\ell_2} \gamma_2$ and $\circ\circ$. Conversely, a DCJ operation can be applied to $\circ\circ$ and $\gamma_1 \ell \gamma_2$, creating, for example, $\gamma_1 \ell \circ$ and $\gamma_2 \circ$. These are special cases of the DCJ operation, and the clean $\mathcal{G}$-adjacency $\circ\circ$ is also called *null linear chromosome* [5]. There are also two special cases of a DCJ that change only labels and circular $\mathcal{G}$-adjacencies: when one of the two cuts is on a circular $\mathcal{G}$-adjacency s, the result will be a single $\mathcal{G}$-adjacency v and s will be integrated into the label of v. Conversely, when both cuts of a DCJ are applied to the same $\mathcal{G}$-adjacency, we would have either an inversion in its label or an excision of a circular $\mathcal{G}$-adjacency. With respect to the structure of the involved chromosomes, a DCJ operation can correspond to several events, such as an inversion, a translocation, a fusion, or a fission. In addition a DCJ can also correspond to an excision or an integration of a circular chromosome [1].

Given a genome A over $\mathcal{G}_A$ and a genome B over $\mathcal{G}_B$ with $\mathcal{G} = \mathcal{G}_A \cap \mathcal{G}_B$, the *adjacency graph* $AG(A, B)$ is the graph that has a vertex for each $\mathcal{G}$-adjacency in $V_{\mathcal{G}}(A)$ and a vertex for each $\mathcal{G}$-adjacency in $V_{\mathcal{G}}(B)$. Then, for each $g \in \mathcal{G}$, we have one edge connecting the vertex in $V_{\mathcal{G}}(A)$ and the vertex in $V_{\mathcal{G}}(B)$ that contain g^h and one edge connecting the vertex in $V_{\mathcal{G}}(A)$ and the vertex in $V_{\mathcal{G}}(B)$ that contain g^t. Due to the 1-to-1 correspondence between the vertices of $AG(A, B)$ and the $\mathcal{G}$-adjacencies in $V_{\mathcal{G}}(A)$ and $V_{\mathcal{G}}(B)$, we can identify each adjacency with its corresponding vertex.

We know that $AG(A, B)$ is composed of two types of connected components, cycles and paths, alternating vertices in $V_{\mathcal{G}}(A)$ and $V_{\mathcal{G}}(B)$ [1]. A path that has one endpoint in $V_{\mathcal{G}}(A)$ and the other in $V_{\mathcal{G}}(B)$ is called an *AB-path.* In the same way, both endpoints of an *AA-path* are in $V_{\mathcal{G}}(A)$, as well as both endpoints of a *BB-path* are in $V_{\mathcal{G}}(B)$. Furthermore, the adjacency graph can have two extra types of components: each $\mathcal{G}$-adjacency that corresponds to a linear (respect. circular) chromosome is a *linear* (respect. *circular*) *singleton.* Observe that linear singletons are particular cases of AA-paths and BB-paths. If $\mathcal{G}_A = \mathcal{G}_B = \mathcal{G}$, the adjacency graph is composed only of clean $\mathcal{G}$-adjacencies, has no singletons and is said to be *clean.* An example of an adjacency graph is given in Fig. 2.

Singletons, AB-paths composed of one single edge, and cycles composed of two edges are said to be *DCJ-sorted.* Longer paths and cycles are said to be *DCJ-unsorted.* We call *DCJ-sorting* of A into B the procedure of using DCJ operations to turn $AG(A, B)$ into DCJ-sorted components. The *DCJ distance*

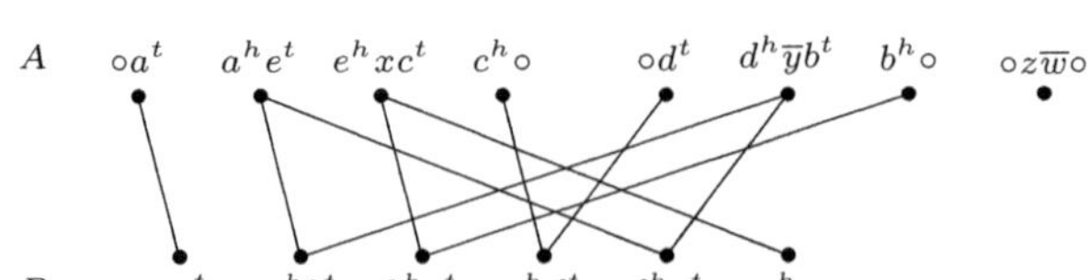

Fig. 2. For genomes $A = \{\circ aexc\circ, \circ d\overline{y}b\circ, \circ z\overline{w}\circ\}$ and $B = \{\circ abcde\circ\}$, the adjacency graph contains one cycle, two AA-paths (one is a linear singleton) and two AB-paths

of A and B, denoted by $d_{DCJ}(A,B)$, corresponds to the minimum number of steps required to do a DCJ-sorting of A into B and can be easily obtained:

Theorem 1 ([1]). *Given a genome A over $\mathcal{G}_A$ and a genome B over $\mathcal{G}_B$, we have $d_{DCJ}(A,B) = n - c - \frac{b}{2}$, where n is the number of markers in $\mathcal{G} = \mathcal{G}_A \cap \mathcal{G}_B$, and c and b are, respectively, the number of cycles and AB-paths in $AG(A,B)$.*

Bergeron *et al.* [1] observed that the number of AB-paths in $AG(A,B)$ is even and that an *optimal* DCJ operation either increases the number of cycles by one, or the number of AB-paths by two (decreasing the DCJ distance by one). In the same way, a *neutral* operation does not affect the number of cycles and AB-paths in the graph, while a *counter-optimal* operation either decreases the number of cycles by one, or the number of AB-paths by two. The problem of finding an optimal sequence of operations that do a DCJ-sorting of A into B can be solved with a simple greedy linear time algorithm [1].

Now let $\mathcal{A}$ be the set of markers that occur only in genome A and let $\mathcal{B}$ be the set of markers that occur only in genome B, that is, $\mathcal{A} = \mathcal{G}_A \setminus \mathcal{G}_B$ and $\mathcal{B} = \mathcal{G}_B \setminus \mathcal{G}_A$. The markers in $\mathcal{A}$ and $\mathcal{B}$ are represented in $AG(A,B)$ as labels and singletons, but they are simply ignored by the approaches to compute the DCJ distance and sorting sequence, mentioned above. However, in order to completely sort A into B, the markers in $\mathcal{A}$ have to be deleted, while the markers in $\mathcal{B}$ have to be inserted. No DCJ operation is actually able to do an insertion or a deletion. Moreover, no operation is able to delete and insert at the same time (such an event would be a replacement, which is not accepted in the model we consider). Thus, for the purpose of this study, an operation is either a DCJ operation, or an *insertion*, or a *deletion*. We will refer to insertions and deletions as *indel* operations. A DCJ and an indel operation have the same cost and we define the *DCJ-indel distance* of A and B, denoted by $d_{DCJ}^{id}(A,B)$, as the minimum number of DCJ and indel operations required to transform A into B.

We can then establish a first simple upper bound for the DCJ-indel distance:

Observation 1. *Given a genome A over $\mathcal{G}_A$ and a genome B over $\mathcal{G}_B$, we have $d_{DCJ}^{id}(A,B) \leq d_{DCJ}(A,B) + |\mathcal{A}| + |\mathcal{B}|$, where $\mathcal{A} = \mathcal{G}_A \setminus \mathcal{G}_B$ and $\mathcal{B} = \mathcal{G}_B \setminus \mathcal{G}_A$.*

3 Accumulating Runs with Optimal DCJ Operations

Observe that a $\mathcal{G}$-adjacency with a non-empty label ℓ can be cut in at least two different positions, either before or after ℓ. Since the position of the cut does not change the effect of the DCJ operation on $d_{DCJ}(A,B)$, we can choose to cut at positions that allow the concatenation of the labels of the original $\mathcal{G}$-adjacencies. As a consequence, a set of labels in $\mathcal{G}$-adjacencies of genome A can be first *accumulated* with DCJ operations and later deleted at once. In the same way, a set of labels in $\mathcal{G}$-adjacencies of genome B can be first inserted at once as a *cluster* and later split with DCJ operations, as we can see in Fig. 3.

Due to the following observation, without loss of generality, we allow operations on both genomes A and B, in order to be able to concatenate labels in

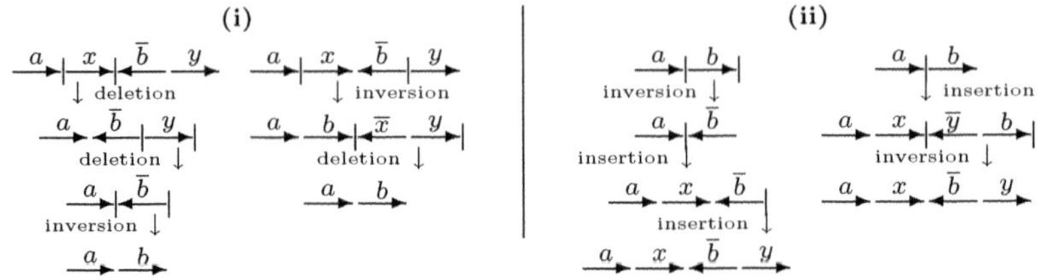

Fig. 3. (i) Two different scenarios sorting $\{\circ a x \overline{b} y \circ\}$ into $\{\circ a b \circ\}$. In the left we have two separate deletions and an optimal DCJ (inversion). In the right, we first perform the inversion, accumulating x and y, so that they can be deleted at once, saving one step. (ii) Conversely, while sorting $\{\circ a b \circ\}$ into $\{\circ a x \overline{b} y \circ\}$, instead of two insertions (left), we can insert a cluster at once and later split it with an inversion (right), saving one step.

$\mathcal{G}$-adjacencies of both genomes. Regarding the operations applied on genome B, this approach can be seen as a backtracing to find the best moment to do a cluster insertion in genome A. An algorithm sorting genome A into B can be derived from this approach.

Observation 2. *Given two genomes A and B, any pair of sequences s_1 and s_2 composed of DCJ and indel operations acting on both genomes A and B, transforming respectively A and B into an intermediate genome I, has a corresponding sequence acting only on genome A, that is, transforming A into B, with length $|s_1| + |s_2|$.*

Given a component C of $AG(A, B)$, we can obtain a string $\ell(C)$ by the concatenation of the labels of the $\mathcal{G}$-adjacencies of C in the order in which they appear. Cycles, AA-paths and BB-paths can be read in any direction, but AB-paths should always be read from A to B. If C is a cycle and has labels in both genomes A and B, we should start to read in a labeled $\mathcal{G}$-adjacency v of genome A, such that the first labeled vertex before v is a $\mathcal{G}$-adjacency in genome B; otherwise C has labels in at most one genome and we can start anywhere. Each maximal substring of $\ell(C)$ in $\widehat{\mathcal{A}}^+$ (respectively in $\widehat{\mathcal{B}}^+$) is called an *$\mathcal{A}$-run* (respectively a *$\mathcal{B}$-run*). Each $\mathcal{A}$-run or $\mathcal{B}$-run can be simply called a *run*. A component composed only of clean $\mathcal{G}$-adjacencies has no run and is said to be *clean*, otherwise the component is *labeled*. We denote by $\Lambda(C)$ the number of runs in a component C. A path can have any number of runs, while a cycle has zero, one, or an even number of runs. Fig. 4 shows a BB-path with 4 runs.

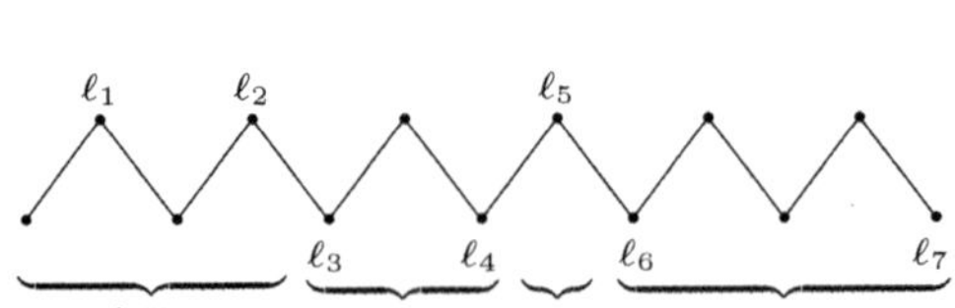

Fig. 4. A BB-path with 4 runs. Only the labels of the $\mathcal{G}$-adjacencies are represented.

Proposition 1. *If $\gamma_1\gamma_2$ is a clean $\mathcal{G}$-adjacency in a DCJ-unsorted component C of $AG(A, B)$, such that neither γ_1 nor γ_2 are telomeres, then it is always possible to extract a clean cycle from C with an optimal DCJ operation.*

Proof. If $\gamma_1\gamma_2$ is in $V_{\mathcal{G}}(B)$, we apply a DCJ on the two vertices $\gamma_1\ell_1\gamma_3$ and $\gamma_2\ell_2\gamma_4$ of $V_{\mathcal{G}}(A)$ that are neighbors of $\gamma_1\gamma_2$, creating the two new vertices $\gamma_3\overline{\ell_1}\ell_2\gamma_4$ and $\gamma_1\gamma_2$. Observe that the vertex $\gamma_1\gamma_2$ in $V_{\mathcal{G}}(B)$ and the new vertex $\gamma_1\gamma_2$ in $V_{\mathcal{G}}(A)$ are extracted into a clean cycle. Analogously, if $\gamma_1\gamma_2$ is in $V_{\mathcal{G}}(A)$, we do the same procedure using the two vertices of $V_{\mathcal{G}}(B)$ that are neighbors of $\gamma_1\gamma_2$. □

Proposition 2. *A run can be entirely accumulated in the label of one single $\mathcal{G}$-adjacency with optimal DCJ operations.*

Proof. A run that is not yet accumulated is distributed over two or more $\mathcal{G}$-adjacencies in one genome. The $\mathcal{G}$-adjacencies in the other genome within the run are clean. We can thus apply optimal DCJs that extract clean cycles (Proposition 1) and accumulate the entire run in the label of one $\mathcal{G}$-adjacency. □

Proposition 2 immediately gives a tighter upper bound for the distance:

Lemma 1. *Given two genomes A and B without duplications, we have*

$$d_{DCJ}^{id}(A, B) \leq d_{DCJ}(A, B) + \sum_{C \in AG(A,B)} \Lambda(C).$$

4 Merging Runs in One Component

For some instances of A and B, the upper bound of Lemma 1 gives the exact number of steps required to sort A into B. However, since two runs can be *merged* together with a DCJ operation, the DCJ-indel distance is often smaller than this upper bound. Given a DCJ operation ρ, let Λ_0 and Λ_1 be, respectively, the number of runs in $AG(A, B)$ before and after ρ. We define $\Delta\Lambda(\rho) = \Lambda_1 - \Lambda_0$.

Proposition 3. *Given any DCJ operation ρ, we have $\Delta\Lambda(\rho) \geq -2$.*

Proof. If ρ cuts between an $\mathcal{A}$-run r_1 and a $\mathcal{B}$-run r_2 and between an $\mathcal{A}$-run r_3 and a $\mathcal{B}$-run r_4, with $r_1 \neq r_3$ and $r_2 \neq r_4$, and joins r_1 with r_3 and r_2 with r_4, then $\Delta\Lambda(\rho) = -2$. As a DCJ has at most two cuts and two joins, it is not possible to do better, that is $\Delta\Lambda(\rho) \geq -2$. □

In order to obtain the exact formula for the DCJ-indel distance, we will first analyze the components of the adjacency graph separately. Given two genomes A and B and a component $C \in AG(A, B)$, we denote by $d_{DCJ}(C)$ the minimum number of DCJ operations required to do a separate DCJ-sorting in C, applying DCJs only on vertices of C (or vertices that result from DCJs applied on vertices that were in C). From [3], we know that it is possible to do a separate DCJ-sorting using only optimal DCJs in any component of $AG(A, B)$, or, in other words, $d_{DCJ}(A, B) = \sum_{C \in AG(A,B)} d_{DCJ}(C)$. Moreover, we denote by $\lambda(C)$ the minimum number of runs that we can obtain doing a separate DCJ-sorting in C with optimal DCJ operations. We then have:

Proposition 4. *Given a component C in $AG(A,B)$, we have $\lambda(C) = \lceil \frac{\Lambda(C)+1}{2} \rceil$, if $\Lambda(C) \geq 1$. Otherwise $\lambda(C) = 0$.*

Proof. The proof is by induction on $i = \Lambda(C)$ and the hypothesis is $T(i) = \lceil \frac{i+1}{2} \rceil$. A labeled DCJ-sorted component can have one or two runs, thus we need two base cases, $T(1) = 1$ and $T(2) = 2$. These cases can be easily verified. More intricate is the inductive step, for $i \geq 3$.

When $i \geq 3$ is odd, we can merge the first and the last runs with an optimal DCJ, obtaining a cycle with $i-1$ runs. This gives $T(i) = T(i-1) = \frac{i-1+2}{2} = \lceil \frac{i+1}{2} \rceil$. If $i \geq 5$, we can also do an optimal DCJ that has $\Delta\Lambda = -2$, extracting a cycle with even $i' \geq 2$ runs and leaving the path with odd $i'' \geq 1$ runs, such that $i = i' + i'' + 2$ and $T(i) = T(i') + T(i'') = \frac{i'+2}{2} + \frac{i''+1}{2} = \frac{i+1}{2} = \lceil \frac{i+1}{2} \rceil$.

When $i \geq 4$ is even, any optimal DCJ merging runs would extract a cycle with even $i' \geq 2$ runs and leave the other component with $i'' \geq 1$ runs. One way is to do a DCJ that has $\Delta\Lambda = -1$, such that $i = i' + i'' + 1$ and $i'' \geq 1$ is odd. This gives $T(i) = \frac{i'+2}{2} + \frac{i''+1}{2} = \frac{i+2}{2} = \lceil \frac{i+1}{2} \rceil$. If $i \geq 6$, it is also possible to do a DCJ that has $\Delta\Lambda = -2$, such that $i = i' + i'' + 2$ and $i'' \geq 2$ is even. We then have $T(i) = \frac{i'+2}{2} + \frac{i''+2}{2} = \frac{i+2}{2} = \lceil \frac{i+1}{2} \rceil$. (All other optimal DCJs applied between runs of components with $\Lambda \geq 3$ would lead to greater values of λ.) □

If λ_0 and λ_1 are, respectively, the sum of the number λ for the components of the adjacency graph before and after ρ, we define $\Delta\lambda(\rho) = \lambda_1 - \lambda_0$. By the definition of λ, any optimal DCJ ρ acting on a single component has $\Delta\lambda(\rho) \geq 0$. However, considering the case in which only one component is affected by ρ, we still need to investigate $\Delta\lambda(\rho)$ when ρ is counter-optimal or neutral.

Proposition 5. *Given a DCJ operation ρ acting on a single component, we have $\Delta\lambda(\rho) \geq 0$, if ρ is counter-optimal, or $\Delta\lambda(\rho) \geq -1$, if ρ is neutral.*

Proof. The linearization of a cycle is the only counter-optimal DCJ that acts on a single component. This can decrease neither Λ, nor λ. Moreover, when $\Lambda \leq 2$, it is not possible to decrease the number λ with any DCJ. When the component has $\Lambda = 3$, the best we can get with a neutral ρ is $\Delta\Lambda(\rho) = -1$. This gives $\lambda_1 = \lceil \frac{(3-1)+1}{2} \rceil = \lceil \frac{3}{2} \rceil = \lceil \frac{3+1}{2} \rceil = \lambda_0$, that is, $\Delta\lambda(\rho) = 0$. And when the component has $\Lambda \geq 4$, we can get $\Delta\Lambda(\rho) = -2$ with a neutral ρ, resulting in $\lambda_1 = \lceil \frac{(\Lambda(C)-2)+1}{2} \rceil = \lceil \frac{\Lambda(C)+1}{2} \rceil - 1 = \lambda_0 - 1$, that is, $\Delta\lambda(\rho) = -1$. □

We denote by $d_{DCJ}^{id}(C)$ the minimum number of DCJ and indel operations required to sort separately a component C of $AG(A,B)$.

Proposition 6. *If C is a component of $AG(A,B)$, then we have $d_{DCJ}^{id}(C) = d_{DCJ}(C) + \lambda(C)$.*

Proof. By the definition of λ, the best we can do with optimal DCJs is $d_{DCJ}(C) + \lambda(C)$. From Proposition 5, we know that $\Delta\lambda(\rho) \geq 0$ if ρ is a counter-optimal DCJ, thus we can only get longer sorting scenarios if we use such operations. We also know that $\Delta\lambda(\rho) \geq -1$ if ρ is neutral, and, since this kind of operation increases the sorting scenario by one with respect to the scenario with only optimal DCJs, this gives at least $d_{DCJ}(C) + \lambda(C)$. □

Proposition 6 gives a new upper bound for the DCJ-indel distance:

Lemma 2. *Given two genomes A and B without duplications, we have*

$$d_{DCJ}^{id}(A,B) \leq d_{DCJ}(A,B) + \sum_{C \in AG(A,B)} \lambda(C).$$

Proof. We can sort the components separately with $\sum_{C \in AG(A,B)} d_{DCJ}^{id}(C)$ steps, which corresponds exactly to $d_{DCJ}(A,B) + \sum_{C \in AG(A,B)} \lambda(C)$. □

Since $\lambda(C) \leq \Lambda(C)$, the upper bound given by Lemma 2 is tighter than the one given by Lemma 1, but can still be improved. Observe that a parsimonious scenario may not simply consist of optimal DCJ operations, insertions and deletions. Sometimes a neutral DCJ can lead to a shorter sequence of operations sorting one genome into another, as we can see in Fig. 5.

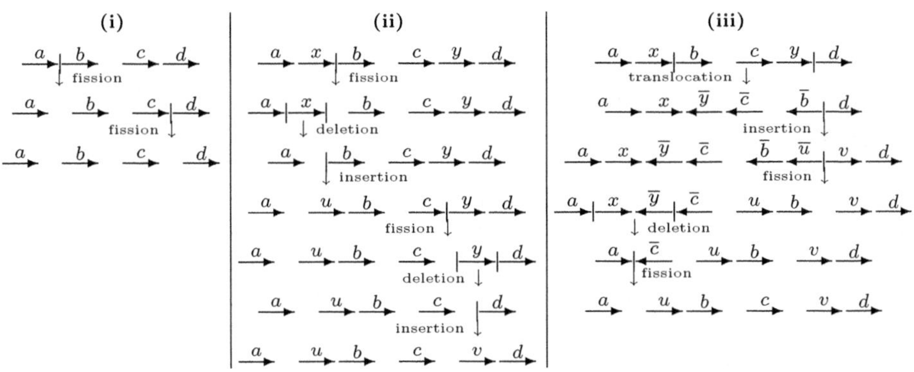

Fig. 5. An optimal scenario sorting $\{\circ ab\circ, \circ cd\circ\}$ into $\{\circ a\circ, \circ b\circ, \circ c\circ, \circ d\circ\}$ (i) and two different scenarios sorting $\{\circ axb\circ, \circ cyd\circ\}$ into $\{\circ a\circ, \circ ub\circ, \circ c\circ, \circ vd\circ\}$. In (ii) in addition to the two optimal DCJ operations (fissions) from (i) we have two insertions and two deletions, using six steps. In (iii) we first use a neutral DCJ operation (translocation) that allows us to do only one deletion and one insertion, achieving a total of five steps.

5 Recombinations and the DCJ-indel Distance

A DCJ operation ρ that acts on two components is called *recombination* and can have $\Delta\lambda(\rho) = -2$. The two components on which the cuts are applied are called *sources* and the components obtained after the joinings are called *resultants* of the recombination.

Proposition 7. *Given any recombination ρ, we have $\Delta\lambda(\rho) \geq -2$.*

Proof. Only the recombinations that decrease or do not change the number of runs ($\Delta\Lambda \leq 0$) have to be analyzed (we can not have $\Delta\lambda \leq -1$ if the number of runs increases). First consider the recombination of two paths with i and j runs, respectively, that result in two new paths with i' and j' runs. Observe that the best we can have is when i and j are even, i' and j' are odd and $\Delta\Lambda = -2$, that gives: $\lambda_1 = \lceil \frac{i'+1}{2} \rceil + \lceil \frac{j'+1}{2} \rceil = \frac{i'+j'+2}{2} = \frac{i+j}{2} = \frac{i}{2} + \frac{j}{2} = \lceil \frac{i+1}{2} \rceil - 1 + \lceil \frac{j+1}{2} \rceil - 1 = \lambda_0 - 2$. The analysis of recombinations involving cycles is analogous. □

Given a recombination ρ, let $\Delta_{dcj}(\rho)$ be respectively 0, +1 and +2 depending whether ρ is optimal, neutral or counter-optimal. Any recombination applied to a vertex of an AA-path and a vertex of a BB-path is optimal [3]. A recombination applied to vertices of two different AB-paths can be either neutral, when the result is also a pair of AB-paths, or counter-optimal, when the result is a pair composed of an AA-path and a BB-path. All other types of path recombinations are neutral. In addition, all recombinations involving at least one cycle are counter-optimal. We define $\Delta d(\rho) = \Delta_{dcj}(\rho) + \Delta\lambda(\rho)$. Any counter-optimal recombination has $\Delta d \geq 0$, thus only path recombinations can have $\Delta d \leq -1$.

Let $\mathcal{A} = \widehat{\mathcal{A}}^+(\widehat{\mathcal{B}}^+\widehat{\mathcal{A}}^+)^*$ (respect. $\mathcal{B} = \widehat{\mathcal{B}}^+(\widehat{\mathcal{A}}^+\widehat{\mathcal{B}}^+)^*$) be a sequence with an odd (≥ 1) number of runs, starting and ending with a run over $\widehat{\mathcal{A}}$ (respect. over $\widehat{\mathcal{B}}$). We can then make any combination of $\mathcal{A}$ and $\mathcal{B}$, such as $\mathcal{AB} = \widehat{\mathcal{A}}^+(\widehat{\mathcal{B}}^+\widehat{\mathcal{A}}^+)^*\widehat{\mathcal{B}}^+$, that is a sequence with an even (≥ 2) number of runs, starting with a run over $\widehat{\mathcal{A}}$ and ending with a run over $\widehat{\mathcal{B}}$. An empty sequence (with no run) is represented by ε. Then each one of the notations AA_ε, $AA_\mathcal{A}$, $AA_\mathcal{B}$, $AA_\mathcal{AB}$, BB_ε, $BB_\mathcal{A}$, $BB_\mathcal{B}$, $BB_\mathcal{AB}$, AB_ε, $AB_\mathcal{A}$, $AB_\mathcal{B}$, $AB_\mathcal{AB}$ and $AB_\mathcal{BA}$ represents a particular type of path (AA, BB or AB) with a particular structure of runs (ε, $\mathcal{A}$, $\mathcal{B}$, $\mathcal{AB}$ or $\mathcal{BA}$). The complete set of path recombinations with $\Delta d \leq -1$ is given in Table 1. In Table 2 we also list recombinations with $\Delta d = 0$ that create at least one source of recombinations of Table 1. We denote by $AB_\bullet$ an AB-path that can not be a source of a recombination in Tables 1 and 2, such as AB_ε, $AB_\mathcal{A}$ and $AB_\mathcal{B}$.

Table 1. Path recombinations that have $\Delta d \leq -1$ and allow the best reuse of the resultants. Optimal recombinations are in the left, neutral recombinations in the right.

sources	resultants	$\Delta\lambda$	Δ_{dcj}	Δd
$AA_\mathcal{AB} + BB_\mathcal{AB}$	$AB_\bullet + AB_\bullet$	−2	0	−2
$AA_\mathcal{A} + BB_\mathcal{AB}$	$AB_\bullet + AB_\mathcal{AB}$	−1	0	−1
$BB_\mathcal{A} + AA_\mathcal{AB}$	$AB_\bullet + AB_\mathcal{BA}$	−1	0	−1
$AA_\mathcal{B} + BB_\mathcal{AB}$	$AB_\bullet + AB_\mathcal{BA}$	−1	0	−1
$BB_\mathcal{B} + AA_\mathcal{AB}$	$AB_\bullet + AB_\mathcal{AB}$	−1	0	−1
$AA_\mathcal{A} + BB_\mathcal{A}$	$AB_\bullet + AB_\bullet$	−1	0	−1
$AA_\mathcal{B} + BB_\mathcal{B}$	$AB_\bullet + AB_\bullet$	−1	0	−1

sources	resultants	$\Delta\lambda$	Δ_{dcj}	Δd
$AA_\mathcal{AB} + AA_\mathcal{AB}$	$AA_\mathcal{A} + AA_\mathcal{B}$	−2	+1	−1
$BB_\mathcal{AB} + BB_\mathcal{AB}$	$BB_\mathcal{A} + BB_\mathcal{B}$	−2	+1	−1
$AA_\mathcal{AB} + AB_\mathcal{AB}$	$AB_\bullet + AA_\mathcal{A}$	−2	+1	−1
$AA_\mathcal{AB} + AB_\mathcal{BA}$	$AB_\bullet + AA_\mathcal{B}$	−2	+1	−1
$BB_\mathcal{AB} + AB_\mathcal{AB}$	$AB_\bullet + BB_\mathcal{B}$	−2	+1	−1
$BB_\mathcal{AB} + AB_\mathcal{BA}$	$AB_\bullet + BB_\mathcal{A}$	−2	+1	−1
$AB_\mathcal{AB} + AB_\mathcal{BA}$	$AB_\bullet + AB_\bullet$	−2	+1	−1

Table 2. Recombinations that have $\Delta d = 0$ and create resultants that can be used in recombinations with $\Delta d \leq -1$

sources	resultants	$\Delta\lambda$	Δ_{dcj}	Δd
$AA_\mathcal{A} + AB_\mathcal{BA}$	$AB_\bullet + AA_\mathcal{AB}$	−1	+1	0
$AA_\mathcal{B} + AB_\mathcal{AB}$	$AB_\bullet + AA_\mathcal{AB}$	−1	+1	0
$BB_\mathcal{A} + AB_\mathcal{AB}$	$AB_\bullet + BB_\mathcal{AB}$	−1	+1	0
$BB_\mathcal{B} + AB_\mathcal{BA}$	$AB_\bullet + BB_\mathcal{AB}$	−1	+1	0

sources	resultants	$\Delta\lambda$	Δ_{dcj}	Δd
$AA_\mathcal{A} + BB_\mathcal{B}$	$AB_\bullet + AB_\mathcal{AB}$	0	0	0
$AA_\mathcal{B} + BB_\mathcal{A}$	$AB_\bullet + AB_\mathcal{BA}$	0	0	0
$AB_\mathcal{AB} + AB_\mathcal{AB}$	$AA_\mathcal{A} + BB_\mathcal{B}$	−2	+2	0
$AB_\mathcal{BA} + AB_\mathcal{BA}$	$AA_\mathcal{B} + BB_\mathcal{A}$	−2	+2	0

Proposition 8. *The recombinations with $\Delta d = 0$ involving cycles or circular singletons cannot create new components that can be used as sources of recombinations listed in Tables 1 and 2.*

Proof. A recombination ρ with $\Delta d = 0$ involving a cycle or a circular singleton C would integrate C to another component C' without changing the type or the structure of runs in C'. Thus, if C' is a source of a recombination in these tables after ρ, C' was already the same type of source before ρ. And if C' was not a source before ρ, C' cannot become a source after ρ. □

With Proposition 8 we already have an exact formula to d_{DCJ}^{id} for a particular set of instances. Given a $\mathcal{G}$-adjacency $\gamma\ell\circ$ of a genome A such that $\gamma \neq \circ$, then γ is said to be a *tail* of a linear chromosome in A. Two genomes are *co-tailed* if their sets of tails are equal (this includes two genomes composed only of circular chromosomes).

Theorem 2. *Given two co-tailed genomes A and B without duplications, we have $d_{DCJ}^{id}(A,B) = d_{DCJ}(A,B) + \sum_{C \in AG(A,B)} \lambda(C)$.*

Proof. The graph $AG(A,B)$ for co-tailed genomes A and B can have only singletons (that could be $AA_{\mathcal{A}}$ and $BB_{\mathcal{B}}$), cycles and AB-paths of one edge. These AB-paths could be $AB_{\mathcal{AB}}$, but never $AB_{\mathcal{BA}}$, thus no recombination listed in Tables 1 and 2 is possible. □

Now we continue the analysis for the general case. The two sources of a recombination can also be called *partners*. Looking at Table 1 we observe that all partners of $AB_{\mathcal{AB}}$ and $AB_{\mathcal{BA}}$ paths are also partners of $AA_{\mathcal{AB}}$ and $BB_{\mathcal{AB}}$ paths, all partners of $AA_{\mathcal{A}}$ and $AA_{\mathcal{B}}$ paths are also partners of $AA_{\mathcal{AB}}$ paths and all partners of $BB_{\mathcal{A}}$ and $BB_{\mathcal{B}}$ paths are also partners of $BB_{\mathcal{AB}}$ paths. Moreover, some resultants of recombinations in Tables 1 and 2 can be used in other recombinations. These observations allow the identification of groups, as listed in Tables 3 and 4.

The deductions shown in Tables 3 and 4 can be computed with an approach that greedily maximizes the number of recombinations in P, Q, T, S, M and N in this order. The P part contains only one operation and is thus very simple. The same happens with Q, since the two groups in this part are exclusive after applying P. The only part that requires more attention is T, in which some combinations of operations can happen at the same time and the order can be relevant. The part S is only the application of all possible remaining operations with $\Delta d = -1$. After S, the two groups in M are exclusive and then the same happens to the six groups in N.

The results presented in this section give rise to the following theorem, that gives the exact formula for the DCJ-indel distance:

Theorem 3. *Given two genomes A and B without duplications, we have*

$$d_{DCJ}^{id}(A,B) = d_{DCJ}(A,B) + \sum_{C \in AG(A,B)} \lambda(C) - 2P - 3Q - 2T - S - 2M - N,$$

where P, Q, T, S, M and N are computed as described above.

Table 3. All recombination groups obtained from Table 1. Observe that the last four groups in T are subsets of groups in Q and the last ten groups in S are subsets of groups in Q and T. The column **scr** indicates the contribution of each path in the distance decrease (the table is sorted in descending order with respect to this column).

	sources	resultants	Δd	scr
P	$AA_{AB} + BB_{AB}$	$2AB_\bullet$	-2	-1
Q	$2AA_{AB} + BB_{A} + BB_{B}$	$4AB_\bullet$	-3	$-3/4$
	$2BB_{AB} + AA_{A} + AA_{B}$	$4AB_\bullet$	-3	$-3/4$
T	$AA_{AB} + BB_{A} + AB_{AB}$	$3AB_\bullet$	-2	$-2/3$
	$AA_{AB} + BB_{B} + AB_{BA}$	$3AB_\bullet$	-2	$-2/3$
	$BB_{AB} + AA_{A} + AB_{BA}$	$3AB_\bullet$	-2	$-2/3$
	$BB_{AB} + AA_{B} + AB_{AB}$	$3AB_\bullet$	-2	$-2/3$
	$2AA_{AB} + BB_{A}$	$2AB_\bullet + AA_{B}$	-2	$-2/3$
	$2AA_{AB} + BB_{B}$	$2AB_\bullet + AA_{A}$	-2	$-2/3$
	$2BB_{AB} + AA_{A}$	$2AB_\bullet + BB_{B}$	-2	$-2/3$
	$2BB_{AB} + AA_{B}$	$2AB_\bullet + BB_{A}$	-2	$-2/3$
S	$AA_{A} + BB_{A}$	$2AB_\bullet$	-1	$-1/2$
	$AA_{B} + BB_{B}$	$2AB_\bullet$	-1	$-1/2$
	$AB_{AB} + AB_{BA}$	$2AB_\bullet$	-1	$-1/2$
	$BB_{AB} + AA_{A}$	$AB_\bullet + AB_{AB}$	-1	$-1/2$
	$AA_{AB} + BB_{A}$	$AB_\bullet + AB_{BA}$	-1	$-1/2$
	$BB_{AB} + AA_{B}$	$AB_\bullet + AB_{BA}$	-1	$-1/2$
	$AA_{AB} + BB_{B}$	$AB_\bullet + AB_{AB}$	-1	$-1/2$
	$AA_{AB} + AB_{AB}$	$AB_\bullet + AA_{A}$	-1	$-1/2$
	$AA_{AB} + AB_{BA}$	$AB_\bullet + AA_{B}$	-1	$-1/2$
	$BB_{AB} + AB_{AB}$	$AB_\bullet + BB_{B}$	-1	$-1/2$
	$BB_{AB} + AB_{BA}$	$AB_\bullet + BB_{A}$	-1	$-1/2$
	$AA_{AB} + AA_{AB}$	$AA_{B} + AA_{A}$	-1	$-1/2$
	$BB_{AB} + BB_{AB}$	$BB_{B} + BB_{A}$	-1	$-1/2$

Table 4. All recombination groups that contain operations from Tables 1 and 2. The groups in N are subsets of the groups in M. The table is sorted in descending order with respect to the contribution of each path in the distance decrease (column **scr**).

	sources	resultants	Δd	scr
M	$2AB_{AB} + AA_{B} + BB_{A}$	$4AB_\bullet$	-2	$-1/2$
	$2AB_{BA} + AA_{A} + BB_{B}$	$4AB_\bullet$	-2	$-1/2$
N	$AB_{AB} + AA_{B} + BB_{A}$	$3AB_\bullet$	-1	$-1/3$
	$AB_{BA} + AA_{A} + BB_{B}$	$3AB_\bullet$	-1	$-1/3$
	$2AB_{AB} + AA_{B}$	$2AB_\bullet + AA_{A}$	-1	$-1/3$
	$2AB_{AB} + BB_{A}$	$2AB_\bullet + BB_{B}$	-1	$-1/3$
	$2AB_{BA} + AA_{A}$	$2AB_\bullet + AA_{B}$	-1	$-1/3$
	$2AB_{BA} + BB_{B}$	$2AB_\bullet + BB_{A}$	-1	$-1/3$

Both $AG(A, B)$ and $d_{DCJ}(A, B)$ can be computed in linear time [1]. The runs can be obtained by a single walk through each component of $AG(A, B)$, which is also linear. The algorithm to compute P, Q, T, S, M and N is a finite sequence of **if** and **else** statements, that depends only on the number of each type of labeled path in $AG(A, B)$, thus the whole procedure takes linear time.

6 Experiments and Discussion

We used our method to analyze the evolution of *Rickettsia*, a group of obligate intracellular parasites that are carried by many vectors (frequently hematophagous arthropods) and occasionally transmitted from the vector to mammalians (including humans), causing several diseases (typhus, spotted fever, etc.) [2]. The genomes of such intracellular parasites are observed to have a reductive evolution, that is, the process by which genomes shrink and undergo extreme levels of gene degradation and loss. There are several completely sequenced *Rickettsia* genomes, and most of them are closely related [2]. The exception is *R. bellii*, which shows a high level of rearrangement with respect to the others. We compared *R. bellii* with six other species of *Rickettsia*, observing in all pairwise analyses a considerable reduction of the indels (see Table 5), when they are grouped into runs (column $\Sigma\Lambda$) and into merged runs (column $\Sigma\lambda$). Although

Table 5. Comparing *R. bellii* (1.52 Mbp) with six other species of *Rickettsia*

species	Mbp	$\lvert\mathcal{A}\rvert + \lvert\mathcal{B}\rvert$	$\Sigma\Lambda$	$\Sigma\lambda$	d_{DCJ}	d_{DCJ}^{id}	species	Mbp	$\lvert\mathcal{A}\rvert + \lvert\mathcal{B}\rvert$	$\Sigma\Lambda$	$\Sigma\lambda$	d_{DCJ}	d_{DCJ}^{id}
R. felis	1.55	333	241	181	312	493	*R. conorii*	1.27	277	192	153	261	414
R. massiliae	1.36	302	218	172	276	448	*R. prowazekii*	1.11	241	130	117	197	314
R. africanae	1.28	290	212	166	260	426	*R. typhi*	1.11	239	126	114	195	309

these are preliminary results, they could suggest that each cluster is composed of genes that have been lost together during the evolution of *Rickettsia*.

Discussion. We propose the first linear time algorithm to compute the distance between two genomes with unequal content, but without duplications, taking into consideration DCJ and indel operations. With this method we analyze a group of bacteria, obtaining interesting results. Due to the lack of available data, we could not yet perform analyses on linear genomes, which would let us test the impact of path recombinations on the distance.

This work opens some perspectives. One is the development of a sorting algorithm that can be derived from the results presented here. Another issue that could be addressed next is the incorporation of replacements in the model, when an insertion and a deletion occur at the same position of the genome.

References

1. Bergeron, A., Mixtacki, J., Stoye, J.: A unifying view of genome rearrangements. In: Bücher, P., Moret, B.M.E. (eds.) WABI 2006. LNCS (LNBI), vol. 4175, pp. 163–173. Springer, Heidelberg (2006)
2. Blanc, G., et al.: Reductive genome evolution from the mother of Rickettsia. PLoS Genetics 3(1), e14 (2007)
3. Braga, M.D.V., Stoye, J.: The solution space of sorting by DCJ. To appear in Journal of Computational Biology (2010)
4. Yancopoulos, S., Friedberg, R.: Sorting Genomes with Insertions, Deletions and Duplications by DCJ. In: Nelson, C.E., Vialette, S. (eds.) RECOMB-CG 2008. LNCS (LNBI), vol. 5267, pp. 170–183. Springer, Heidelberg (2008)
5. Yancopoulos, S., Attie, O., Friedberg, R.: Efficient sorting of genomic permutations by translocation, inversion and block interchange. Bioinformatics 21, 3340–3346 (2005)

Listing All Sorting Reversals in Quadratic Time

Krister M. Swenson[1,2], Ghada Badr[3], and David Sankoff[1]

[1] Department of Mathematics and Statistics, University of Ottawa, Ontario, K1N 6N5, Canada
[2] LaCIM, UQAM, Montréal Québec, H3C 3P8, Canada
[3] SITE, School of Information Technology and Engineering, University of Ottawa, Ontario, K1N 6N5, Canada

Abstract. We describe an average-case $O(n^2)$ algorithm to list all reversals on a signed permutation π that, when applied to π, produce a permutation that is closer to the identity. This algorithm is optimal in the sense that, the time it takes to write the list is $\Omega(n^2)$ in the worst case.

1 Introduction

In 1995 Hannenhalli and Pevzner [9] presented an algorithm to transform one genome into another in a minimum number of biologically plausible moves. They modeled a genome as a signed permutation and the move that they considered was the reversal: the order of a substring of the permutation is reversed, and the sign of each element in the substring is flipped. Since then many refinements and speed improvements have been developed [4,8,10,11,16,18,19].

In 2002 Siepel and Ajana et. al. [1,15] showed how to list every parsimonious scenario of reversals, each scenario being a proposed candidate for the true evolutionary history. Fundamental to their algorithms are $O(n^3)$ techniques for finding all *sorting* reversals; the reversals that at each step produce a permutation that is closer to the target permutation than the last. Ajana et. al. [1] used these results to support the replication-directed reversal hypothesis. Lefebvre et. al. [12] and Sankoff et. al. [14] used similar methodology to gain insight into the distribution of reversal lengths between genomes. Algorithms that attempt to more succinctly represent all shortest-length scenarios [3,6] have also been developed.

In this paper we show how to list all sorting reversals in $O(n^2)$ time on average. This algorithm is optimal in the sense that there are $\Omega(n^2)$ safe cycle-splitting reversals in the worst-case. This affords a significant speedup of the aforementioned methods [1,3,6,12,14,15], since listing all sorting reversals is the kernel of repeated computation in each of them, especially when applied to permutations of sizes 3×10^3 to 3×10^5 (the size of bacterial or mammalian genomes).

2 Background

Take a signed permutation $\pi = \pi_1, \ldots, \pi_n$ on the integers from 1 to n. Define a (signed) *reversal* $\rho(i, j)$ as the signed permutation

$$1, 2, \ldots, (i-1), {}^{-}j, \ldots, {}^{-}i, (j+1), \ldots, n.$$

V. Moulton and M. Singh (Eds.): WABI 2010, LNBI 6293, pp. 102–110, 2010.

That way, applying the reversal $\rho(i,j)$ to permutation π gives

$$\rho(i,j)(\pi) = \pi \circ \rho(i,j) = \pi_1, \ldots, \pi_{i-1}, {}^{-}\pi_j, \ldots, {}^{-}\pi_i, \pi_{j+1}, \ldots, \pi_n.$$

Given signed permutations π_1 and π_2, the *reversal distance* $d(\pi_1, \pi_2)$ is the smallest k such that $\pi_2 = \pi_1 \circ \rho_1 \circ \rho_2 \circ \cdots \circ \rho_k$. Without loss of generality[1] we consider $\pi_2 = I = 1, 2, \ldots, n$ to be the identity permutation. In this paper, we describe our methods using circular permutations (where the leftmost element follows the rightmost element), as any sorting reversal on a circular permutation has its counterpart on a linear version of the permutation.

2.1 All Sorting Reversals

A reversal ρ is a *sorting* reversal on π if $d(\pi \circ \rho) = d(\pi) - 1$. Although the definition is simple, a characterization of all sorting reversals requires effort; to do so we must introduce the *breakpoint graph* [9]. Each element π_i of permutation π has two vertices associated with it denoted by π_i^- and π_i^+ ($\pi^\pm$ can denote either). Embed the graph on a circle as follows: place all $2n$ vertices on the circle so that:

1. π_i^+ and π_i^- are adjacent on the circle,
2. π_i^- is before (in the clockwise direction) π_i^+ if and only if π_i is positive, and
3. a $\pi_i^\pm$ is adjacent to a $\pi_{i+1}^\pm$ if and only if π_i and π_{i+1} are adjacent in π.

For two vertices $v_1 = \pi_i^\pm$ and $v_2 = \pi_j^\mp$ ($i \neq j$) that are adjacent on the circle, add the edge (v_1, v_2)—a *reality* edge (also called a black edge); also add edges (π_i^+, π_{i+1}^-) for all i and (π_n^+, π_1^-)—the *desire* edges (also called gray edges). Figure 1(a) shows the breakpoint graph for $\pi =$ (-1 2 4-5 6 8-7-3). Note that every vertex has indegree 2 and outdegree 2, so the graph has a unique decomposition into cycles of even length (alternating between reality and desire edges).

A reversal $\rho(i,j)$ is said to *act on* the reality edges $(\pi_{i-1}^\pm, \pi_i^\mp)$ and $(\pi_j^\mp, \pi_{j+1}^\pm)$ because these are the only edges in the breakpoint graph of π that are not in the graph of

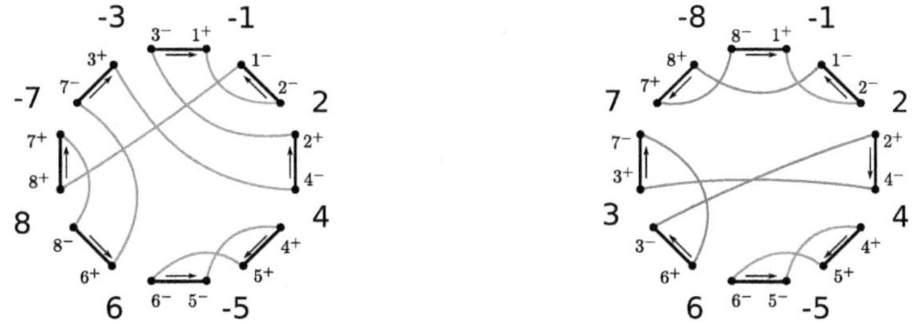

(a) breakpoint graph of $\pi =$ (-1 2 4-5 6 8-7-3) *(b) breakpoint graph of $\pi \circ \rho(6,8)$*

Fig. 1. Two breakpoints graphs. The direction that reality edges are traversed on a tour of the cycles is labeled with arrows. $\rho(6,8)$ is an unsafe reversal on π.

[1] Since $I = \pi_2^{-1} \circ \pi_1 \circ \rho_1 \circ \rho_2 \circ \cdots \circ \rho_k$.

$\pi \circ \rho(i,j)$. In Figure 1, the reversal $\rho(6,8)$ acts on reality edges $(3^-,1^+)$ and $(6^+,8^-)$. Two reality edges on the same cycle are *convergent* if a traversal of their cycle visits each edge in the same direction in the circular embedding; otherwise they are *divergent*. The following definitions classify the action of a reversal on the cycles of the breakpoint graph [9].

Definition 1 (cycle-splitting reversal). *A reversal that acts on a pair of divergent reality edges splits the cycle to which the edges belong, so are called* cycle-splitting *reversals.*

Conversely, no reversal that acts on a pair of convergent reality edges splits their common cycle. A reversal that acts upon a pair of reality edges in two different cycles merges the two cycles. The permutation of Figure 1(a) has 10 cycle-splitting inversions including $\rho(1,2)$, $\rho(4,4)$, and $\rho(6,8)$. Notice that at most one cycle can be created by a reversal, yielding the inequality

$$d(\pi) \geq n - c(\pi), \tag{1}$$

where $c(\pi)$ is the number of cycles in the breakpoint graph. Most cycle-splitting reversals are sorting reversals [17], but not all sorting reversals are cycle-splitting reversals, which indicates a gap between this lower bound and the reversal distance.

We must further explore structure in the permutation that allows us to predict the reversal distance when the lower bound is not realized.

Definition 2 (FCI [5]). *A* framed common interval *(FCI) of a permutation (made circular by considering the first and last elements as being adjacent) is a substring of the permutation, $as_1s_2\ldots s_kb$ or $^-bs_1s_2\ldots s_k{}^-a$ such that*

- *for each i, $1 \leq i \leq k$, $|a| < |s_i| < |b|$, and*
- *for each l, $|a| < l < |b|$, there exists a j with $|s_j| = l$, and*
- *it is not a concatenation of substrings satisfying the previous two properties.*

So the substring $s_1s_2\ldots s_k$ is a (possibly empty) signed permutation of the integers that are greater than a and less than b; a and b are the *frame* elements, while those of $s_1\ldots s_k$ are *trunk* elements if they are not trunk elements of a smaller FCI. The framed interval is said to be common, in that it also exists as an interval $\big(a(a+1)(a+2)\ldots b\big)$ in the identity permutation.

A *component* of a permutation is comprised of the trunk elements of an FCI that are not trunk elements of a shorter FCI, plus the frame elements. The permutation of Figure 1(b) has three components: one framed by elements 2 and 7, another framed by 4 and 6. The third is an interval in the circular sense, framed by elements 7 and 2 with the trunk comprised of elements 8 and 1; in the circular sense we have $7 < 8 < 1 < 2$ here.

Definition 3 (bad component[5]). *A* bad component *of a permutation is a component with at least 4 elements, where the sign of every element is the same.*

In Figure 1(b), the component (2 4 6 3 7) is bad. The existence of one or more bad components in a permutation indicate exactly those situations where the lower bound cannot be met [9]. Siepel's paper [15] describes in detail an $O(n^3)$ algorithm for finding the set of sorting reversals when bad components exist. While further exploration of Siepel's characterization of sorting reversals in the presence of bad components could eventually lead to a worst-case $O(n^2)$ algorithm, we do not address the issue here. Suffice it to say that the average-case complexity is $O(n^2)$ even when the trivial $O(n^3)$ algorithm[2] is used on permutations with bad components; the probability that a permutation chosen uniformly at random has a bad component is $O(n^{-2})$ [7,17] and we can detect the presence of bad components in linear time [2,5].

We focus on the bottleneck of sorting FCIs that do not correspond to bad components: cycle-splitting reversals that create bad components (cycle-splitting reversals that are not sorting reversals).

Definition 4 (bad reversal). *A* bad *reversal is a reversal that creates a bad component.*

Definition 5 (unsafe reversal[9]). *An* unsafe *reversal is a cycle-splitting reversal that is bad.*

In Figure 1(a), the reversal $\rho(6,8)$ is unsafe.

2.2 Outline

Known algorithms that list all sorting reversals check, one by one, if each of the potentially $\Omega(n^2)$ cycle-splitting reversals is unsafe by applying the reversal and then running a linear time check as to whether it produced a bad (unoriented) component [1,15]. Instead of listing all cycle-splitting reversals and then checking them, we do the inverse: we predict which reversals may be *unsafe* (whether cycle-splitting or otherwise) and avoid listing them. We first characterize what we call *ominous* substrings of the permutation, those substrings that could be turned into a bad component with one reversal. Our algorithm searches for ominous substrings by doing the following: for each element of the permutation we posit that it is a smallest element of a potential (after a reversal) bad component and continue by scanning the permutation to detect an ominous substring.

In Section 3 we introduce ominous substrings while Section 4 describes how to detect the set of all ominous substrings efficiently. Section 5 presents the algorithm. Finally Section 6 discusses open problems.

3 Ominous Substrings

Take any unsafe cycle-splitting reversal ρ on permutation π. Since it is unsafe, the permutation $\pi \circ \rho$ has at least one bad component created by ρ. In this section we will show that there exists in π a particular pattern — an *ominous* substring of π — indicating that ρ is unsafe. We first describe ominous substring of permutations with a single component.

---

[2] Each of the $O(n^2)$ reversals can in turn be applied to the permutation and the linear time algorithm for reversal distance [2] can be used to check if it was a sorting reversal.

3.1 Permutation with a Single Component

A substring of a permutation is *ominous* if and only if there exists some elements e and f such that the substring fits one of the following templates (or their reverse):

1. $(eA\underline{X{-f}{-B}})$: where $A, -B$, and X are substrings of the permutation. A has only positive while $-B$ has only negative elements.
2. $(\underline{{-A}{-e}X}Bf)$: where $-A, B$, and X are substrings of the permutation. $-A$ has only negative while B has only positive elements. e is negative.
3. $(eA\underline{-B}Cf)$: where $A, -B$, and C are substrings of the permutation. A and C have only positive while $-B$ has only negative elements.

and A and B (and C if it exists) are comprised of exactly those elements with absolute value i for $e < i < f$.

In template 3, there already exists an FCI with frame elements e and f; the reversal that acts on exactly the elements of B fixes the elements of the interval to have the same sign. In the other two templates, a new interval is created with e and f as the frame elements. For template 1, $\{f\} \cup B \cup X$ are the elements reversed while for template 2, $\{e\} \cup A \cup X$ are the elements reversed. For example, $(-7\ 1\ \underline{-3\ -4\ -5\ -2}\ 6)$ matches template 3 with the unsafe reversal acting upon the elements $\{2,3,4,5\}$; A and C are empty in this case. $(-1\ 2\ 4\ \underline{6\ -5\ -3})$ matches template 1 with the unsafe reversal acting upon the elements $\{3,5,6\}$; $f = -5$, $B = \{-3\}$, and $X = \{6\}$ in this case. $(-2\ \underline{-6\ -8\ -4\ 1}\ 5\ 7\ 9\ -3)$ matches template 2 with the unsafe reversal acting upon the elements $\{1,4,6,8\}$; $A = \{6,8\}$, $B = \{5,7\}$, and $X = \{1\}$ in this case.

We are ready to state the main lemma of this section.

Lemma 1. *There is a one to one correspondence between bad reversals and ominous substrings.*

Proof. By definition, there exists at least one reversal that creates a bad component from an ominous substring. On the other hand, take a permutation $\pi \circ \rho$ that has a bad component — with frame elements e and f — created by the reversal ρ. Say that the elements of the bad component are positive, then e is on the left and f is on the right. If ρ includes both e and f, this implies that the bad component already exists in π, which is a contradiction. Now let us examine the other three possibilities. If ρ does not include e and f, then the ominous substring in π corresponds to template 3. If ρ includes only f, then the ominous substring in π corresponds to template 1. If ρ includes only e, then the ominous substring in π corresponds to the template 2. If the elements of the bad component are negative then the negative analogue holds for each case. Since each ominous substring implies exactly one reversal dictated by the A, B, C, and X, we have the bijection.

3.2 Permutations with Multiple Components

We described ominous substrings on permutations with a single component. Since sorting reversals act only upon adjacencies in a single component [9], we adapt the techniques for single components to the case of multiple components in the following manner.

Consider a component of a permutation with some frame elements of a smaller FCI contained in it. We obtain the *condensed* version of the component by doing the following: for each pair of frame elements a and b (or $\text{-}a$ and $\text{-}b$) of a smaller FCI contained in it, we replace the pair by a (resp. $\text{-}a$) and change the magnitude m of every element $m > b$ in the component to be $m - (b - a)$. The templates can be applied directly to the condensed component. For example, take the component $C = (2\ 4\ 6\ 3\ 7)$ in Figure 1 where the component $(4\ \text{-}5\ 6)$ is contained in it. The condensed version of C is $(2\ 4\ 3\ 5)$. The condensed version of any component can be computed in linear time.

4 Detecting Ominous Substrings

We now turn to the task of detecting an ominous substring associated with a smallest element e. The following methods can be adapted to detect the negative analogue of each template, so we only describe the detection of the templates as they were presented in Section 3.1. The general outline used in each of the following algorithms is the same: we visit the permutation starting with element e, proceeding to element $e+1$, then $e+2$ and so on. At each step we maintain enough information to check whether certain conditions hold that indicate we have found an ominous substring.

Call the set of elements that we visit through the first i steps S_i (those with absolute value in the interval $[e, e+i]$). To check for each template at step i, so that f would be the element $e+i$, we maintain the following values.

- Rightmost positive index visited: $rp = \max(\{|\pi^{-1}(|j|)| \mid j \in S_i, j > 0\})$
- Leftmost positive index visited: $lp = \min(\{|\pi^{-1}(|j|)| \mid j \in S_i, j > 0\})$
- Rightmost negative index visited: $rn = \max(\{|\pi^{-1}(|j|)| \mid j \in S_i, j < 0\})$
- Leftmost negative index visited: $ln = \min(\{|\pi^{-1}(|j|)| \mid j \in S_i, j < 0\})$

Template 1 ($eAX\text{-}f\text{-}B$) exists, with unsafe reversal $\rho(rp+1, rn)$, if and only if the following conditions hold:

1. $lp = \pi^{-1}(|e|)$
   (e is the leftmost element visited)
2. $ln > rp$
   (the negative elements are to the right of the positive)
3. $rn - ln + rp - lp = i - 1$
   (the positive and negative elements are all contiguous)
4. $\pi^{-1}(|e+i|) = ln$
   (the last element visited is the leftmost negative element)
5. $i \geq 3$
   (the FCI has at least 4 elements)

Template 2 ($\text{-}A\text{-}eXBf$) exist, with unsafe reversal $\rho(ln, lp-1)$, if and only if the following conditions hold:

1. $rn = \pi^{-1}(|e|)$
   (e is the rightmost negative element)
2. $lp > rn$
   (the negative elements are to the left of the positive)

3. $rn - ln + rp - lp = i - 1$
   (the positive and negative elements are all contiguous)
4. $\pi^{-1}(|e+i|) = rp$
   (the last element visited is the rightmost element visited)
5. $i \geq 3$
   (the FCI has at least 4 elements)

To check for template 3 we maintain another value $neg = |\{j \mid j \in S_i, j < 0\}|$, the number of negative values visited. We know that we have found template 3 ($eA\text{-}BCf$) with unsafe reversal $\rho(ln, rn)$ if and only if all of the following conditions hold:

1. $lp = \pi^{-1}(|e|)$
   (e is the leftmost element visited)
2. $ln > lp$
   (the negative elements are to the right of some positive)
3. $rp > rn$
   (the negative elements are to the left of some positive)
4. $rp - lp = i$
   (we have visited a contiguous substring)
5. $rn - ln = neg - 1$
   (the negative elements of B are contiguous)
6. $\pi^{-1}(|e+i|) = rp$
   (the last element visited is the rightmost element visited)
7. $i \geq 3$
   (the FCI has at least 4 elements)

Note that if at some iteration i during our scan conditions 1 or 2 for any of the templates are broken, we know that e can no longer match that template.

5 The Algorithm

We begin by proving the following theorem.

Theorem 1. *For a permutation without a bad component, there is an $O(n^2)$ algorithm for listing all sorting reversals.*

Proof. Use the methods of Section 4 to obtain a blacklist of all ominous substrings associated with each possible smallest frame element e. Since the list of all ominous substrings associated with a single smallest frame element is obtained by a linear scan for all possible right endpoints f, the time to build the blacklist is $O(n^2)$. Each element of the list is associated with a bad reversal, the indices of which we mark in an n by n matrix; an entry r at row i and column j indicates that the bad reversal r acts on elements from position i to position j in the permutation. Obtain the list of all cycle-splitting reversals in $O(n^2)$ time using the standard methods [9]. Finally, examine this list one reversal at a time, removing from the list any reversal that has a corresponding entry marked in the matrix.

The methods described so far are applicable to permutations with no bad components. Permutations with bad components can be easily handled by combining our algorithm with that of Siepel [15] in the following way. First make a linear scan of the permutation to detect bad components [2,5]. If there are bad components, use the $O(n^3)$ algorithm of Siepel, otherwise, use our algorithm.

Theorem 2. *Pick a signed permutation uniformly at random, the expected time the above algorithm takes to list all sorting reversals is $O(n^2)$.*

Proof. The probability of seeing a bad component in a permutation taken uniformly at random from the set of all signed permutations is $O(n^{-2})$ [17]. The bound follows since $n^3 \times n^{-2} < n^2$.

6 Conclusions

We presented the first quadratic time algorithm for listing all sorting reversals for a signed permutation. This pattern matching algorithm is simple in that it requires no special data structures. It is optimal in the sense that most permutations can have $\Omega(n^2)$ sorting reversals [20,13]. An improvement on our bound would be an algorithm that runs in $O(n+k)$ time where k is the number of sorting reversals. It is currently unclear how to modify our algorithm to obtain this bound.

References

1. Ajana, Y., Lefebvre, J.-F., Tillier, E.R.M., El-Mabrouk, N.: Exploring the set of all minimal sequences of reversals - an application to test the replication-directed reversal hypothesis. In: Guigó, R., Gusfield, D. (eds.) WABI 2002. LNCS, vol. 2452, pp. 300–315. Springer, Heidelberg (2002)
2. Bader, D.A., Moret, B.M.E., Yan, M.: A linear-time algorithm for computing inversion distance between signed permutations with an experimental study. J. Comput. Biol. 8(5), 483–491 (2001); A preliminary version appeared in WADS 2001, pp. 365–376
3. Baudet, C., Dias, Z.: An Improved Algorithm to Enumerate All Traces that Sort a Signed Permutation by Reversals. In: SIGAPP 2010: Proceedings of the Twenty Fifth Symposium on Applied Computing (2010)
4. Bergeron, A.: A very elementary presentation of the Hannenhalli–Pevzner theory. Discrete Applied Mathematics 146(2), 134–145 (2005)
5. Bergeron, A., Heber, S., Stoye, J.: Common intervals and sorting by reversals: a marriage of necessity. In: Proc. 2nd European Conf. Comput. Biol. ECCB 2002, pp. 54–63 (2002)
6. Braga, M.D.V., Sagot, M., Scornavacca, C., Tannier, E.: The Solution Space of Sorting by Reversals. In: Măndoiu, I.I., Zelikovsky, A. (eds.) ISBRA 2007. LNCS (LNBI), vol. 4463, pp. 293–304. Springer, Heidelberg (2007)
7. Caprara, A.: On the tightness of the alternating-cycle lower bound for sorting by reversals. J. Combin. Optimization 3, 149–182 (1999)
8. Hannenhalli, S., Pevzner, P.A.: Transforming mice into men (polynomial algorithm for genomic distance problems). In: Proc. 36th Ann. IEEE Symp. Foundations of Comput. Sci. (FOCS 1995), pp. 581–592. IEEE Press, Piscataway (1995)
9. Hannenhalli, S., Pevzner, P.A.: Transforming cabbage into turnip: Polynomial algorithm for sorting signed permutations by reversals. J. ACM 46(1), 1–27 (1999)

10. Kaplan, H., Shamir, R., Tarjan, R.E.: Faster and simpler algorithm for sorting signed permutations by reversals. SIAM J. Computing 29(3), 880–892 (1999)
11. Kaplan, H., Verbin, E.: Efficient data structures and a new randomized approach for sorting signed permutations by reversals. In: Baeza-Yates, R., Chávez, E., Crochemore, M. (eds.) CPM 2003. LNCS, vol. 2676, pp. 170–185. Springer, Heidelberg (2003)
12. Lefebvre, J.-F., El-Mabrouk, N., Tillier, E.R.M., Sankoff, D.: Detection and validation of single gene inversions. In: Proc. 11th Int'l. Conf. on Intelligent Systems for Mol. Biol. (ISMB 2003). Bioinformatics, vol. 19, pp. i190–i196. Oxford U. Press (2003)
13. Sankoff, D., Haque, L.: The distribution of genomic distance between random genomes. Journal of Computational Biology 13(5), 1005–1012 (2006)
14. Sankoff, D., Lefebvre, J.-F., Tillier, E.R.M., Maler, A., El-Mabrouk, N.: The distribution of inversion lengths in bacteria. In: Lagergren, J. (ed.) RECOMB-WS 2004. LNCS (LNBI), vol. 3388, pp. 97–108. Springer, Heidelberg (2005)
15. Siepel, A.C.: An algorithm to find all sorting reversals. In: Proc. 6th Ann. Int'l. Conf. Comput. Mol. Biol. (RECOMB 2002). ACM Press, New York (2002)
16. Swenson, K.M., Rajan, V., Lin, Y., Moret, B.M.E.: Sorting signed permutations by inversions in $O(n \log n)$ time. In: Batzoglou, S. (ed.) RECOMB 2009. LNCS, vol. 5541, pp. 386–399. Springer, Heidelberg (2009)
17. Swenson, K.M., Lin, Y., Rajan, V., Moret, B.M.E.: Hurdles hardly have to be heeded. In: Nelson, C.E., Vialette, S. (eds.) RECOMB-CG 2008. LNCS (LNBI), vol. 5267, pp. 239–249. Springer, Heidelberg (2008)
18. Tannier, E., Bergeron, A., Sagot, M.-F.: Advances on sorting by reversals. Disc. Appl. Math. 155(6-7), 881–888 (2007)
19. Tannier, E., Sagot, M.: Sorting by reversals in subquadratic time. In: Sahinalp, S.C., Muthukrishnan, S.M., Dogrusoz, U. (eds.) CPM 2004. LNCS, vol. 3109, pp. 1–13. Springer, Heidelberg (2004)
20. Yang, Y., Szkely, L.A.: On the expectation and variance of reversal distance. Acta Univ. Sapientiae, Mathematica 1(1), 5–20 (2009)

Discovering Kinship through Small Subsets

Daniel G. Brown[1] and Tanya Berger-Wolf[2]

[1] Cheriton School of Computer Science, University of Waterloo
Waterloo, ON N2L 3G1, Canada
browndg@uwaterloo.ca
[2] Department of Computer Science, University of Illinois at Chicago
Chicago, IL 60607, USA
tanyabw@cs.uic.edu

Abstract. In kinship inference, we identify genealogical relationships among organisms. One such problem is sibgroup reconstruction: given a population of same-generation individuals, partition it into sibgroups resulting from mating events. Minimizing the number of matings is *NP*-hard to approximate, yet a simple heuristic, based on identifying population triplets that can be from the same sibgroup, performs comparably to better than integer programming algorithms in a fraction of the running time. With high probability if we study many loci in the genome, and large populations, our polynomial-time heuristic finds the true sibgroups, assuming a standard probabilistic inheritance model.

1 Introduction

A natural field biology question is to characterize genealogical relationships in wild populations, such as which organisms are siblings. Scientists use genetic markers from highly variable genomic loci to detect parentage: if three diploid organisms are father, mother and offspring, then at every position in the genome, the offspring has one copy of a paternal chromosome and one copy of a maternal chromosome. Assuming no mutations, if neither copy of a putative parent's DNA is found in the offspring, they cannot be parent and child. Similar arguments identify diploid siblings when many loci are considered. At each locus, siblings inherit one maternal and one paternal chromosome, from a choice of two in each case, so any overlap can be used to evaluate possible relationships. Still, any two individuals can be siblings, making studies of pairs of individuals problematic.

However, it is not possible for any *three* individuals to be siblings. This observation underlies our methods here in finding *sibgroups* of siblings in wild populations of diploid organisms. We identify all triplets of individuals that could be found in the same sibgroup. If two population members are found in *many* compatible triplets, they are likely siblings. We develop an extremely simple polynomial-time heuristic for finding sibgroups that gives comparable or superior results to the state of the art in much smaller runtimes.

Finally, we give our work a probabilistic basis. In a simple population generation model, a variant of our heuristic has failure probability shrinking exponentially with the number of loci sampled and the population size.

V. Moulton and M. Singh (Eds.): WABI 2010, LNBI 6293, pp. 111–123, 2010.

Kinship inference consists of a wide range of problems: identifying sibgroups is one of the simplest. However, our methods offer the suggestion of a new procedure for solving this entire class of problems heuristically and accurately.

2 Related Work

Statistical methods for this problem, as in COLONY [15], compute the likelihoods of putative sibgroups in a probabilistic model. More recently, KINALYZER [2] has introduced a combinatorial approach: it identifies subsets that could be complete sibgroups, and partitions the population in some optimal way (largely, minimizing the number of sibgroups). Minimizing this objective is *NP*-hard [1]; integer programming or other brute-force methods have worked for small data sets. This approach gives reasonable results [5], though much remains to be done for populations with small numbers of large families.

However, the approach of computing the minimum total number of sibgroups has limitations. First, the integer programs can be huge: the number of maximal potential sibgroups can be exponential in the size of the problem instance. Second, the problem often shows multiple optima: there may be many distinct partitions of the population sample into the same minimum number of sibgroups, and it is not obvious how to choose a preferred answer from this set of optima; Sheikh *et al.* develop a consensus approach, but it is still challenging to find a universally good answer [13]. Finally, it does not always solve the core problem, as evidenced by its moderate accuracy on some instances.

3 Preliminaries and Notation

We have a population of n *offspring* from the same generation, whom we wish to divide into sibgroups from individual mating events between unknown parents.

Individuals are diploid: they have two chromosomes of each type (we do not consider the sex chromosomes). One chromosome is inherited from each parent. At a single *locus* in the genome, an offspring inherits one copy of the local DNA from its mother, and one from its father. In both cases, the inherited copy is one choice from the two possibilities corresponding to the *parent*'s two chromosomes. At some loci in the genome, there is much variability in the DNA found at that locus in the population, so offspring may inherit different DNA at those loci from each parent. For example, microsatellites, ("TATATA...", for example), which are also known as Short Tandem Repeats (STRs) or Simple Sequence Repeates (SSRs), can be repeated a different number of times. An individual might have 5 copies on one chromosome, and 12 on the other, indicated by the *genotype* "(5,12)", denoting the pair of *alleles.*

We assume such marker DNA is available at m different genetically unlinked sites for each population member. (This requirement of no linkage is verified in microsatellite marker development.) We cannot identify which chromosome is derived from which parent of a population member. However, the genotype "(5,12)" indicates that one parent of the individual has allele 5 on one of *its* two chromosomes, and the other parent has allele 12 on one of *its* two chromosomes.

Some genotypes can be *homozygotic*, such as "(8,8)", where the offspring inherits the same allele from both parents, so each parent must have that allele on one of its chromosomes. Also, at different loci, the meaning of the alleles differs: at one locus, the allele 4 may indicate the presence of the sequence "TATATATA", while at another, it might indicate the presence of "GGAGGAGGAGGA".

Our input data is an $n \times m$ matrix of genotypes: each row corresponds to a single population member p_i and each column to a single locus ℓ_j. Our goal is to partition the set $P = \{p_1 \dots p_n\}$ into families $\{F_1 \dots F_f\}$ where each member of family F_i consists of full siblings from the same mating event.

For parsimony reasons, we minimize the total number of families, which is *NP*-hard to approximate to within a 153/152 ratio [1]. We give a heuristic which does well for this goal, and also at returning the true set of families.

4 Forbidden Sets

Previous work has focused on sets of population members that *can* be in the same sibgroup. We focus on sets that *cannot* be part of the same sibgroup. These are based on 3-element subsets of the population, a polynomial-sized group. This gives a small integer program, and some successful simple heuristics.

4.1 Forbidden Sets Are Triplets

Considering only the rules of genetic inheritance, any two offspring x and y can, in principle, be full siblings: if at a single locus x has genotype (a, b) and y has genotype (c, d), their parents may have genotypes (a, c) and (b, d).

However, it is not always possible that three offspring $\{x, y, z\}$ can be full siblings. If at the same locus x, y and z have genotypes $(1, 2), (3, 4)$ and $(5, 2)$, they cannot all be siblings: in the two genotypes of their parents, we must place five distinct alleles into four slots. Even with only three or four alleles in a set of offspring at a locus, a triplet of individuals may be incompatible: for example, the genotypes $(1, 1), (2, 3), (3, 4)$ cannot come from full siblings: the first indicates that both parents have the 1 allele, leaving three remaining alleles and only two slots. This leads to the 2-allele rule for a general set of offspring who can feasibly be siblings due to Berger-Wolf *et al.* [4]. Let a be the number of distinct alleles at a locus in a set of individuals T, and let R be the number of alleles that are found either homozygously or paired with at least three other alleles. The members of T can all be siblings iff at all loci, $a + R \leq 4$.

In fact, if a set of offspring cannot all be siblings, they must have an incompatible triplet as a subset.

Theorem 1. *If $T = \{p_1, \dots p_k\}$ cannot form a valid sibgroup, then there must exist a subset $\{x, y, z\} \subset T$ that is incompatible at at least one locus ℓ.*

Proof. There must be at least one locus ℓ where T does not satisfy the 2-allele rule: at that locus, $a + R > 4$. Assume for contradiction that $|T| > 3$ and all 3-element subsets of T do satisfy the 2-allele rule. We will have three cases.

1. If $R = 0$, then $a \geq 5$. Pick one member $x \in T$ with genotype (a_1, a_2). At most two members of T share an allele with x, so since $|T| \geq 4$, there is a member

of T with genotype (a_3, a_4) with which x shares no alleles. Finally, pick a member of T with the fifth allele, with genotype $(a_5, *)$ for some choice for $*$. These three members form the forbidden triplet $\{(a_1, a_2), (a_3, a_4), (a_5, *)\}$.
2. If $R = 1$, then $a \geq 4$. Either some allele a_1 appears with three other alleles, giving the forbidden triplet $\{(a_1, a_2), (a_1, a_3), (a_1, a_4)\}$, or there is a homozygous member with genotype (a_1, a_1). Pick that member, a member with genotype (a_2, a_3) that shares no allele with it and a member with the fourth allele, $(a_4, *)$. They form the forbidden triplet $\{(a_1, a_1), (a_2, a_3), (a_4, *)\}$.
3. If $R = 2$, then $a \geq 3$. Either an allele is found with three others (a forbidden triplet), or there are two homozygotes, (a_1, a_1) and (a_2, a_2). Joined to $(a_3, *)$, this gives the forbidden triplet $\{(a_1, a_1), (a_2, a_2), (a_3, *)\}$. □

4.2 Forbidden Triplets as a Means of Finding a Partition

Three incompatible members $\{x, y, z\}$ cannot be in the same sibgroup. We can verify in $O(m)$ time if a triplet is compatible: at each locus it is a constant-time test. In $O(n^3 m)$ time, we can consider each such triplet.

An Integer Programming Formulation.
This observation gives a polynomial-sized integer programming formulation of the problem of sibgroup reconstruction to minimize the number of sibgroups. Let $x_{i,j} = 1$ iff p_i and p_j belong to the same family (0 otherwise), and let $q_i = 1$ iff p_i is the numerically first member of the population in its sibgroup. We want to optimize this integer program:

$$\begin{aligned} \text{minimize} \sum_i q_i \quad & \text{subject to} \\ x_{i,j} + x_{j,k} + x_{i,k} \leq 1 \quad & \text{if } \{p_i, p_j, p_k\} \text{ are incompatible} \\ x_{i,j} + x_{j,k} - x_{i,k} \leq 1 \quad & \text{for all } i, j.k \text{ distinct} \\ q_i \geq 1 - \sum_{j=1 \ldots i-1} x_{j,i} \quad & \text{for all } i = 1, \ldots, n \\ q_i, x_{i,j} \in \{0, 1\} \quad & \text{for all } i, j \end{aligned}$$

The first set of constraints prevents incompatible triplets from joining the same family. The second set ensures the x relation is transitive. The rules for the q_i variables ensure we count each family. The program has $\Theta(n^2)$ variables, $\Theta(n^3)$ constraints, and is impractical in early experiments: for the 59-member shrimp population discussed below, the IP took 52 minutes to solve; our heuristic took 97 ms, and KINALYZER took a few minutes.

A Heuristic for Finding Likely Pairings.
An alternative approach uses *valid* triplets. If F_i is a true sibgroup, then for any $\{x, y\} \subset F_i$ and $z \in F_i$, $\{x, y, z\}$ is a compatible triplet; $\{x, y, z\}$ may often form compatible triplets even if $z \notin F_i$: at the loci studied, x and y may be similar, or the parents of z may be similar to the parents of x and y. If x and y are *not* from the same family, then the probability that they form a compatible triplet with an unrelated individual z is small, particularly if we study many loci.

This suggests a heuristic: if a pair $\{x, y\}$ is part of many compatible triplets, they likely belong to the same sibgroup. We build a weighted graph G with nodes for population members: the weight $w(e)$ of edge $e = \{x, y\}$ is the number of population members z that do not make $\{x, y, z\}$ a forbidden triplet. The graph G_t obtained when we filter out all edges with weight below t is highly clustered: its connected components often correspond to parts of true sibgroups.

For example, consider the graph shown in Figure 1: here, nodes represent population members, and the groups indicated are known population sibgroups. Nodes x and y share an edge if they are part of the at least three compatible triplets; it is red if they participate in at least six. The graph is highly clustered, with three obvious sibgroups. Further, three quarters of the edges in this graph are within sibgroups, and *all* edges of weight greater than 3 are within sibgroups.

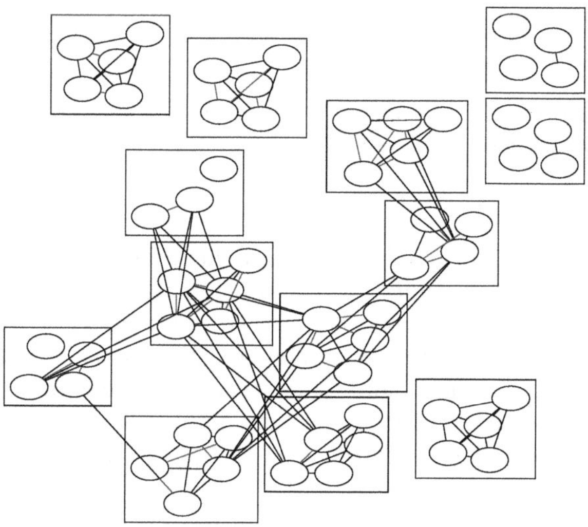

Fig. 1. Graph of compatibility found in a population of 59 shrimp. Nodes correspond to edges, clusters in the graph to true sibgroups. An edge exists between two nodes if they are found in at least 3 compatible triplets; it is red if they are found in at least 6. The highly clustered nature of this graph supports our simple heuristic.

We join clusters discovered together (if they are compatible) into putative sibgroups. Our algorithm has the following steps:

1. For each $e = \{x, y\}$, let w_e be the number of z where $\{x, y, z\}$ is compatible.
2. Let threshold $T = 0$ and X be the entire population. Set $C = \emptyset$ be our set of discovered clusters.
3. While X is non-empty:
   (a) Compute the connected components of the graph G_T with all edges $e = \{x, y\}$ such that $w_e \geq T$.

   (b) For each connected component c, if all elements of c are mutually compatible, add c to C and remove c from X.
   (c) Raise the threshold T by 1.
4. While there are still two clusters in C that can be joined together:
   (a) Join together the two compatible clusters $c_1, c_2 \in C$ that maximize the number of additional compatible triplets in a single cluster.

This simple procedure runs in $O(n^3m)$ time: the dominant step is the first one. We have explored alternative rules for picking which clusters to join, such as maximizing the average number of compatible triplets per node newly joined into the same cluster; all had similar performance on simulated data. Alternatively, instead of removing edges below a fixed threshold, we can start with singletons and keep adding heavy edges until there are clusters, which we join as before. We end up with a similar solution in a similar amount of time.

5 Experimental Results

We have validated the use of our heuristic for kinship inference on simulated and real data sets previously used by Chaovalitwongse *et al.* [5].

5.1 Accuracy Measure

To evaluate the performance of heuristics for this problem, the standard measure is 1 minus the partition distance [9]. If the true partition of the original set into families is P, and we computed partition C, the distance between the two is the minimum fraction of population members that must be removed from the population in order to make P and C identical on the remaining members. This is computable in polynomial time using maximum matching algorithms [9], and gives an accuracy score ranging from $\min(1/|P|, 1/|C|)$ to 1.

5.2 Simulation Results

The simulation data sets explore a wide variety of parameters: the number of adult females (and the equal number of males), p; the number of loci, m; the number of distinct alleles present at each locus, a; the number of families, f; and the size of each family (sibgroup), k. The model has the parents chosen at random from a pool of ten males and ten females, for each family. As such, two sibgroups may be half-sibs (sharing a parent), with both genders being promiscuous, which makes the problem harder. The simulations include data where $a = 2$, so any trio of population members are compatible. Our approach fails here (as does KINALYZER): both return a single family with all fk members.

We report statistics for the simulated data set, and compare our results to those for the IMCS method [5] used in KINALYZER. We implemented our approach in Python 2.6 on a 3.06 GHz Macintosh with 2 GB of memory.

Our new method, whose results are shown in Table 1, is approximately 5% more accurate than the IMCS procedure. The one exception is that for $k = 10$, our method gives results comparable to IMCS. (It is also 1000 times faster.)

Table 1. Simulation results for our new method, and those reported for the IMCS method. Our method is approximately 5% more accurate, and 700 times faster.

Fixed parameter	Parameter settings	IMCS runtime	IMCS accuracy	New method runtime	New method accuracy
m: number of loci	2	2.28 s	57.6%	66.9 ms	62.4%
	4	8.28 s	66.5%	71.9 ms	69.6%
	6	29.0 s	71.4%	75.3 ms	75.6%
	10	239 s	71.9%	86.7 ms	79.2%
a: number of alleles	2	0.21 s	26.7%	97.2 ms	26.7%
	5	30.5 s	72.2%	62.5 ms	75.6%
	10	225 s	81.8%	70.6 ms	90.4%
	20	22.8 s	86.8%	70.5 ms	94.2%
f: number of sibgroups	2	0.72 s	78.1%	1.9 ms	82.1%
	5	3.65 s	64.6%	27.4 ms	69.4%
	10	205 s	57.9%	196 ms	63.7%
k: size of sibgroup	2	2.67 s	54.4%	2.6 ms	65.1%
	5	14.4 s	69.8%	28.5 ms	74.1%
	10	192 s	76.4%	195 ms	75.9%

Where $k = 10$, if we remove the results from the instances where $a = 2$, our method gives 92.3% success, while IMCS gives 92.9%.

Our method is also much faster: 700 times faster for the entire experiment, and consistently much faster except for the degenerate cases like $a = 2$, where our algorithm must discover each triplet is compatible. The runtime is stable with some parameters that slow IMCS down, such as a (number of alleles per locus) and m (number of loci). We discover most incompatible triplets quickly, so increasing the number of loci or alleles has minimal effect. Increasing the population size slows both methods, though this effect is smaller for the new heuristic than for IMCS.

We can also use the heuristic to study far larger families: using the same breadth of choices for the other parameters, we find that if $k = 30$, our procedure requires an average of 10.9 s, and gives overal accuracy 76.7% (93.3% if $a \neq 2$), while for $k = 50$, our procedure takes 50.4 s and gives overall accuracy 77.3% (94.1% if $a \neq 2$). Note that for $k = 50, f = 10$, we are computing tests on populations of size 500, enumerating all 20.8 million triplets and testing them for compatibility; these runs averaged 132 s (102 s when $a \neq 2$), of which 95% or more of the runtime was spent enumerating and testing triplets.

5.3 Real Data

Benchmark data for kinship inference with known ground truth and no genotyping errors are uncommon. Here, we present results for a handful of data sets for which the data collection method allows some certainty in knowing which population members are from the same sibgroup.

These data originate from wild populations of shrimp, ants, salmon, turtles, flies and radishes. Despite their simplicity, our methods perform comparably to

Table 2. Results for biological data. Our procedure continues to be much faster than the IMCS procedure, though the enumeration of all triplets is slow for large populations like the radishes. Accuacy is comparable for the two systems.

Species	Sibgroups	Loci	Population size	IMCS runtime	IMCS accuracy	New method runtime	New method accuracy
Shrimp [12]	13	7	59	185 s	100%	97 ms	100%
Flies [3]	6	2	190	22.8 s	47.4%	12.9 s	46.3%
Salmon [11]	6	4	351	149 s	98.3%	20.8s	98.3%
Radishes [6]	2	5	531	26.3 s	52.5%	316 s	52.3%
Turtles [7]	26	3	175	N/A	N/A	9.35 s	41.3%
Ants [10]	10	6	377	N/A	N/A	62.6 s	98.4%

that of the IMCS method on the four sets for which the results of that method are available, while being substantially faster, as shown in Table 2. In particular, our algorithm is 2000 times faster on the small shrimp population.

One exception is the radish population: here, the population is properly divided into two very large subgroups, and only three loci offer any kinship information at all; most pairs of population members from the same family or different families are compatible with members of both groups, and there are many genotyping errors. Our algorithm enumerates all triplets, and then must move to a very high threshold in the clustering algorithm, and then essentially returns a random partition (and, indeed, the result is similar for IMCS, which simply takes less time producing this result). Similarly, for the turtle population, since there are only three loci, we cannot easily separate sibgroups.

6 Probabilistic Arguments

Why does such a simple heuristic work so well for a problem known not only to be *NP*-hard, but *MAXSNP*-hard? We give a partial answer to the question: assuming that families are fairly large, we separate between the number of compatible triplets $\{x, y, z\}$ of which members x and y of the *same* family are part and the number of compatible triplets of which members x and y of *different* families are part. In particular, if we sample populations from a natural probabilistic model at enough loci, then with high probability, the weighted graph we described in our heuristic will, at the correct threshold, divide cleanly into cliques exactly corresponding to the true families we seek.

Our bounds are fairly coarse; still, this argument goes far toward justifying the heuristic in the previous section.

6.1 A Probabilistic Model

We assume we are studying a population that arises from the following probabilistic model. There is a pool of adults who are not kin (first degree relatives). There are $f \geq 5$ matings of the adult pairs, with monogamy of both sexes, each mating producing k juvenile offspring. The juveniles are sampled at m loci. We assume that the parents have $a \geq 5$ alleles in each locus and each parent has two of these alleles chosen with replacement uniformly at random from among those a (thus, each allele has probability $1/a$). The offspring then have one of the two alleles of each parent (with probability 1/2) at each locus. The resulting population of juvenile offspring has kf members.

6.2 A Simplified Algorithm

Next, we assume k is known. For each pair of offspring population members $e = \{x, y\}$, let w_e be the number of compatible triplets $\{x, y, z\}$ containing x and y. Now, let G be a graph whose nodes are population members, and for which $e = \{x, y\}$ is an edge when $w_e \geq k - 2$. G contains as a subgraph f clique of size k corresponding to the families: for any pair $\{x, y\}$ in a single family F_i, it is compatible with all $k - 2$ other members of F_i, so $e = \{x, y\}$ is in G.

If G contains other edges, then our clustering idea may fail: we may need to raise the threshold T in our algorithm too high for the connected components to correspond to a true sibgroup. Thus, we consider whether *any* pair e of members of different families will have $w_e \geq k - 2$.

6.3 A Bound with Two Families

First, let $f = 2$, so we have $2k$ offspring, divided into families F_1 and F_2. We bound the failure probability by looking at a single pair $x \in F_1$ and $y \in F_2$:

Theorem 2. *Suppose we are given a 2-family population from our model. For an arbitrary $x \in F_1$ and $y \in F_2$, let $c_1(x, y) = |\{z \in F_1 \mid \{x, y, z\}$ are compatible$\}|$. The c_1 variables are identically distributed, and the failure probability of the clustering algorithm is bounded above by $2k^2 p$, where $p = \Pr[c_1(x, y) \geq \frac{k-2}{2}]$.*

Proof. All population members are drawn from the same distribution, so by symmetry, the c_1 random variables share the same distribution, though they are not independent. We know that the algorithm only fails if there exist $x \in F_1$ and $y \in F_2$ such that $w_{\{x,y\}} \geq k - 2$; for this to happen, one of the two families must contain $\frac{k-2}{2}$ valid third members of a compatible triplet. But this is equally likely for each family, so we can bound the probability by $2p$. Since there are k choices for each of i and j, the $2k^2 p$ bound holds. □

The Probability that k_1 is Very Large.
Now, what is the distribution of k_1? Once we consider only $y \in F_2$, we no longer care about the parents for F_2; at each locus, the genotype for y is the conflation of the toss of two a-sided dice. We also choose the genotype for x by tossing two a-sided dice. The final choice we must make is the choice of the two alleles (one paternal, one maternal) of the parents of F_1 *not* found in the genotype of x at that locus; these choices are *also* independent of the choices of genotypes for x and y.

For any $z \in F_1$, let q_z be the event that $\{x, y, z\}$ are compatible. These events are independent, conditioned on x, y and the hidden alleles in the parents of F_1: we make independent choices at each locus to pick the alleles at that site in z. At each locus, the probability that z is compatible with x and y is a multiple of

$1/4$, and is at least $1/4$, since we might choose both alleles for z that were found in x. The overall probability that $\{x, y, z\}$ are compatible is the product of these locus probabilities, and so is a multiple of 4^{-m}. Let this probability be q. Then k_1 is the sum of $f - 1$ independent Bernoulli trials, each with probability q.

Consider two kinds of bad events. First, if q is large, there will likely be many members of F_1 that make compatible triplets with x and y: this is when y appears closely related to F_1. The other bad event is that q is small, and yet many members of F_1 are compatible with x and y.

We first bound the second bad event. Since f is at least 5, we must have at least 2/5 of the members of $F_1 - \{x\}$ form a compatible triplet with x and y. Assume $q < \frac{1}{5e}$. By a standard application of the Chernoff bound [8], $\Pr[k_1 \geq .4(k-1)] \leq 2^{-.4(k-1)} < .76^{k+1}$.

Now, consider the other bad event, where $q \geq \frac{1}{5e}$. Recall that q is the product of m multiples of $1/4$, one for each locus. If, at two loci, the probability that x and y are compatible with a member of F_1 is $1/4$, then $q < 1/16$, so $q < \frac{1}{5e}$.

At any locus, we picked the six alleles (two in x, two in y and the two "hidden" alleles of F_1) uniformly from $\{1 \ldots a\}$. Let r be the probability that the only way $\{x, y, z\}$ can be compatible at a site is if z shares the alleles of x. This probability rises as a function of a, and is 0.08064 if $a = 5$, which can be shown by enumeration. So, at each locus, with at least 0.08 probability, the value of $1/4$ is multiplied into q. The probability that at most one such value of $1/4$ is multiplied in, then, is at most $.92^m + .08m(.92)^{m-1}$, as the loci are independent.

The overall probability, then, that $k_1 \geq \frac{k-2}{2}$ is bounded above by $.76^{k+1} + .92^m + .08m(.92)^{m-1}$, and the overall failure probability for our clustering algorithm is at most $2k^2(.76^{k+1} + .92^m + .08m(.92)^{m-1})$, which drops exponentially as both k, the family size, and m, the number of loci, grow.

Because of the quadratic dependency on k, however, it takes a while for this bound to become strong. For example, if $k = 40, m = 40, a = 10$, we have a bound of 0.05 on the failure probability; in practice, problems of size $k = 10, m = 10, a = 10$ never failed in 10,000 instances, despite the bound offering no guarantee for problems of this size.

6.4 Extending to Multiple Families

If $f > 2$, the problem becomes more difficult. As before, the $2f$ parents have their genotype at each locus chosen uniformly from $\{1 \ldots, a\}$. Now, the question is: are there pairs x, y from different families that have at least $k - 2$ compatible population members? If so, then there exists a family F_i such that $|\{z \in F_i | \{x, y, z\} \text{ compatible}\}| > \frac{k-2}{f}$. Let us call these events $A(x, y, i)$, where $x < y$ and x and y are in different families; we are interested in $\Pr[\bigcup_{x,y,i} A(x, y, i)] \leq \sum_{x,y,i} \Pr[A(x, y, i)]$. By symmetry, this bound equals $(fk)(f-1)\frac{k}{2} \sum \Pr[A(x, y, i)]$ for any $x \in F_1$ and $y \in F_2$; this, then equals $(fk)(f-1)\frac{k}{2}(2\Pr[A(x, y, 1)] + (f-2)\Pr[A(x, y, 3)])$, also by symmetry arguments.

Consider these two events $A(x, y, 1)$ and $A(x, y, 3)$ in turn. The first of these is parallel to what we had before: we need to bound the probability that x and y find many compatible third members in F_1, the family that contains x.

The genotype of $z \in F_1$ is chosen at each locus by two coin tosses, so the probability q that a member of $F_1 - \{x\}$ is compatible with x and y is a multiple of 1/4. If $q < \frac{2}{5ef}$, then by the same use of the Chernoff bound as before, $\Pr[A(x,y,1)|q < \frac{2}{5ef}] < 2^{-.8k/f} < .58^{k/f}$, dropping exponentially as k grows.

What is the probability that $q < \frac{2}{5ef}$? Consider how often we multiply 1/4 as one of the factors from the m loci that makes up q. If we include at least $2 + \log_4 f$ such factors of 1/4, then q will certainly be less than $\frac{1}{16f}$, so less than the minimum threshold needed. At each locus, we multiply a 1/4 into q with probability at least 0.08, and each locus is independent. Using straightforward Chernoff bounding techniques, we can bound the probability that $q < \frac{2}{5ef}$ by $(2 + \log_4 f)m^{2+\log_4 f}.92^{m-4-2\log_4 f}$, dropping exponentially in m.

The second bad event, $A(x,y,3)$, is easier to analyze: if we choose x and y randomly and the two parents for a third family F, at any given locus randomly, the probability that *no* child of F could be compatible with x and y is at least .16 if $a \geq 5$. As such, $\Pr[A(x,y,3)] \leq .84^m$.

So the overall probability that the algorithm fails is at most:

$$(fk)(f-1)\frac{k}{2}[2(.58^{k/f} + (2 + \log_4 f)m^{2+\log_4 f}.92^{m-4-\log_4 f}) + (n-2) \cdot .86^m)],$$

dropping exponentially as both k and m grow.

6.5 Extending to More Robust Models

This proof, which also has a weak bound, can be extended to models with non-identical allele frequencies; the only change is the constant corresponding to the probability that two members of one family and one of another will be compatible (found in both proofs) or that three members of different families will be compatible. The overall result, that the failure probability falls exponentially with the number of members per family and the number of loci, is still the case.

One situation that does not easily work is for variable family sizes: for small families, it is entirely possible that we will not compute the correct assignment.

We note finally that this heuristic algorithm experiences *better* performance as populations and loci expand. The runtime is cubic in the population size; moreover, as the overwhelming majority of the runtime is in the testing of triplets for compatibility, the heuristic is embarrassingly parallel.

7 Future Work

This simple heuristic approach to a single kinship discovery problem, that of full sibgroup detection, may be extended to other areas of this general problem domain. Recent work has moved to the discovery of half-sibling groups, with a different combinatorial goal [14], but it is likely that our simple procedures can discover good subgroups of true families here as well. It is not obvious how easily to move from the simple combinatorial goal of discovering sibgroups to more complex kinships, such as multi-generational families; however, this is a major challenge for current software in general.

Our current procedure is extremely fast, but still could easily be sped up: at the present time, the vast majority of work time is spent enumerating triplets, and comparing their values to triplets already known to be compatible or incompatible. However, with the three orders of magnitude of speedup that our procedure gives on small populations, it may make more sense to concentrate on solving instances more successfully. In particular, we might focus on joining the clusters found in the early stage of our algorithm using a more intelligent method than just looking at pairs of clusters.

Still, our current work gives an almost embarrassingly simple procedure for discovery of sibgroups, with extremely fast performance and success comparable to the state of the art. A probabilistic argument gives a justification for the argument: for large families and many loci, a generalization of our procedure has high probability of identifying exactly the correct families. This result is interesting, given the $MAXSNP$-hardness of finding the minimum number of sibgroups: can we give a complexity result, for example using parameterized complexity, that justifies this surprising dichotomy?

References

1. Ashley, M., Berger-Wolf, T., Berman, P., Chaovalitwongse, W., DasGupta, B., Kao, M.Y.: On approximating four covering and packing problems. J. Comp. and Sys. Sci. 75(5), 287–302 (2009)
2. Ashley, M.V., Caballero, I.C., Chaovalitwongse, W., DasGupta, B., Govindan, P., Sheikh, S., Berger-Wolf, T.Y.: Kinalyzer, a computer program for reconstructing sibling groups. Mol. Ecol. Res. 9, 1127–1131 (2009)
3. Wilson, A., Barker, J.: Isolation and characterization of 20 polymorphic microsatellite loci for *Scaptodrosophila hibisci*. Mol. Ecol. Notes 2, 242–244 (2002)
4. Berger-Wolf, T., Sheikh, S., DasGupta, B., Ashley, M., Caballero, I., Chaovalitwongse, W., Putrevu, S.L.: Reconstructing sibling relationships in wild populations. Bioinf. 23(13); Proceedings of ISMB 2007 (2007)
5. Chaovalitwongse, W., Chou, C., Berger-Wolf, T., DasGupta, B., Sheikh, S., Ashley, M.V., Caballero, I.C.: New optimization model and algorithm for sibling reconstruction from genetic markers. INFORMS J. Comp. 22(1), 180–194 (2010)
6. Conner, J.K.: Personal Communication (2006)
7. Crim, J., Spotila, L., Spotila, J., O'Connor, M., Reina, R., Williams, C., Paladino, F.: The leatherback turtle, *Dermochelys coriacea*. Mol. Ecol. 11(10), 2097–2106 (2002)
8. Devdatt, D., Panconesci, A.: Concentration of measure for the analysis of randomized algorithms. Cambridge Press, New York (2009)
9. Gusfield, D.: Partition-distance: A problem and class of perfect graphs arising in clustering. Info. Proc. Lett. 82(3), 159–164 (2002)
10. Hammond, R.L., Bourke, A.F.G., Bruford, M.W.: Mating frequency and mating system of the polygynous ant, *Leptothorax acervorum*. Mol. Ecol. 10(11), 2719–2728 (1999)
11. Herbinger, C.M., O'Reilly, P.T., Doyle, R.W., Wright, J.M., O'Flynn, F.: Early growth performance of atlantic salmon full-sib families reared in single family tanks versus in mixed family tanks. Aquaculture 173(1-4), 105–116 (1999)

12. Jerry, D., Evans, B., Kenway, M., Wilson, K.: Development of a microsatellite DNA parentage marker suite for black tiger shrimp *Penaeus monodon*. Aquaculture, 542–547 (2006)
13. Sheikh, S., Berger-Wolf, T., Khokhar, A., DasGupta, B.: Consensus methods for reconstruction of sibling relationships from genetic data. In: Proceedings of the 4th Workshop on Advances in Preference Handling (2008)
14. Sheikh, S., Berger-Wolf, T., Khokhar, A., Caballero, I., Ashley, M., Chaovalitwongse, W., DasGupta, B.: Parsimony-based reconstruction of half-sibling groups. J. Bioinf. and Comp. Biol. (to appear)
15. Wang, J.: Sibship reconstruction from genetic data with typing errors. Genetics 166, 1968–1979 (2004)

Fixed-Parameter Algorithm for Haplotype Inferences on General Pedigrees with Small Number of Sites

Duong D. Doan and Patricia A. Evans

Faculty of Computer Science, University of New Brunswick,
Fredericton, New Brunswick, Canada, E3B 5A3
{b89ct,pevans}@unb.ca

Abstract. The problem of computing the minimum number of recombination events for general pedigrees with a small number of sites is investigated. We show that this NP-hard problem can be parametrically reduced to the Bipartization by Edge Removal problem with additional parity constraints. The problem can be solved by an $O(2^k 2^{m^2} n^2 m^3)$ exact algorithm, where n is the number of members, m is the number of sites, and k is the number of recombination events.

1 Introduction

Human genomes contain two copies of each chromosome. Research shows that single chromosomes, called haplotypes, are useful to study complex genetic diseases. While genomic data, called genotypes, are abundant and easy to collect, haplotypes are rare and much more difficult to obtain by a biochemical method. Therefore, computationally inferring haplotypes from genotype data, called haplotyping, is necessary. Genotypes can be obtained from a population group where relationships between members are unknown or from a family pedigree with known relationships between members. We only consider pedigree data.

In the absence of recombination events, haplotypes of members in a pedigree follow the Mendelian law of inheritance, where the two haplotypes of a child are transferred from its parents, one haplotype from its father and the other from its mother. Various haplotyping algorithms exist for non-recombinant pedigree data [1,3], especially a linear algorithm for tree pedigrees [1] and a near-linear algorithm for general pedigrees [3]. Haplotype inference is complicated by recombination events and the complex structures of the data. In recombination events, complementary parts of both of a parent's haplotypes can be inherited as a single combined haplotype of a child. Structures of the pedigree can be complex, where there are multiple inheritance paths between some family members.

When recombination events are allowed, the problem of inferring haplotypes for pedigrees with the minimum number of recombination events is NP-hard, even for general pedigrees with only two sites or tree pedigrees with multiple sites [8]. For reconstructing haplotype configurations for pedigree data, Qian and Beckmann [11] proposed a rule-based algorithm with a time complexity

V. Moulton and M. Singh (Eds.): WABI 2010, LNBI 6293, pp. 124–135, 2010.

$O(2^d n^2 m^3)$, for n members, m sites, and family size $\leq d$. The main principle of their algorithm is that the best haplotype configuration for pedigree data is the one that minimizes the number of recombination events (the *MRHC problem*). Li and Jiang [7] proposed an integer linear programming (ILP) formulation for the MRHC problem. When the number of recombination events is strictly smaller than a positive number k, an $O(mn \cdot \log^{k+1} n)$ time probabilistic algorithm is given on tree pedigrees [12]. Doan and Evans [4] presented an $O(2^k \cdot n^2)$ time fixed-parameter algorithm for general pedigrees where each member has two sites, a special case of the problem that is still NP-complete.

We study the haplotype inference for general pedigrees with recombination events when the number of recombination events k and the number of sites m in an input pedigree are small. We also assume that there are no data missing and no data errors. We prove that our problem can be reduced to the problem of finding the *line index* of a *signed graph* [13] with additional parity constraints. We further show that finding the line index of a signed graph can also be reduced to the Graph Bipartization by Edge Removal (*GBER*) problem with parity constraints. The GBER problem is fixed-parameter tractable, but the existing solution [5] cannot satisfy the additional parity constraints. We present an algorithm that solves the problem while still satisfying the additional constraints, and thus show that the Recombinant Haplotype Configuration problem can be solved by a fixed-parameter algorithm with a running time of $O(2^k 2^{m^2} n^2 m^3)$, for n members, m sites, and k recombination events. This result extends our prior work for pedigrees with two sites to an arbitrary small number of sites.

2 Preliminaries

A member is an individual. A set of members is called a *family* if it includes only two parents and their children; it is a *parent-offspring trio* (hereafter a *trio*) if only two parents and one child are considered. A set of families connected through known family relationships is a *pedigree*.

In diploid organisms, a cell contains two copies of each chromosome. The description data of the two copies are called a *genotype* while those of a single copy are called a *haplotype*. A specific location in a chromosome is called a *site* and its state is called an *allele*. There are two main types of sites, *microsatellites* and *single nucleotide polymorphisms*. A microsatellite site has several different states while a single nucleotide polymorphism *(SNP)* site has exactly two possible states, denoted by 0 and 1. Only SNPs with two possible states are considered in this paper, as in other works on haplotype inference.

If the states at a specific site in two haplotypes are the same, then this site is a *homozygous* site (0-0 or 1-1); if they differ, it is *heterozygous* (0-1 or 1-0). Two haplotypes combine together to form one genotype. Each member u has two haplotypes, denoted by $h1_u$ and $h2_u$, which are vectors of 0 and 1's of length m, where m is the number of sites. The genotype of u, g_u, is a vector of 0's, 1's and 2's of length m, where $g_u[i] = 0$ means $h1_u[i] = 0 = h2_u[i]$, $g_u[i] = 1$ means $h1_u[i] = 1 = h2_u[i]$, and where $g_u[i] = 2$ means $\{h1_u[i], h2_u[i]\} = \{0, 1\}$. We say

$h1_u$ and $h2_u$ are consistent with g_u. The complement haplotype of a haplotype h at a heterozygous site is denoted by $\bar{h}$, where $\bar{h} = 1 - h$ so, $\bar{0} = 1$ and $\bar{1} = 0$.

The problem in this paper is to find the haplotypes $h1_u$ and $h2_u$ for all members u that minimize the number of recombination events, given their genotypes g_u. A set of haplotypes found for all members is called a *haplotype configuration.* When $g_u[i] = 0$ or 1, then $h1_u[i]$ and $h2_u[i]$ are known, but if $g_u[i] = 2$, we may not yet know the value of $h1_u[i]$ and $h2_u[i]$, in which case we give them the value "?", and say that the site is *unresolved.* Our problem is defined as follows.

RHC$_{opt}$: *Given the genotypes of a general pedigree P containing n members, where each member has m sites (m is small), find a haplotype configuration that minimizes the number of recombination events.*

This optimization problem, called *Recombination Haplotype Configuration* (RHC$_{opt}$) which is identical to MRHC, was proven NP-hard [8]. We investigate the corresponding decision version of RHC$_{opt}$.

RHC$_k$: *Given positive integers k and the genotypes of a general pedigree P containing n members, where each member has m sites (m is small), is there a haplotype configuration with at most k recombination events explaining P?*

3 Setting Up Graphs

Given a general pedigree with n members, where each member has m sites, we set up a pedigree graph $G = (V, E)$ and parity-constraint sets S_{pc}. A recombination event can only be detected if there is at least one heterozygous site on each side of a recombination breakpoint, e.g. we cannot detect if a recombination event happens between homozygous sites 1 and 3 of member u in Figure 1. We capture constraints between pairs of closest heterogynous sites and pairs of closest homozygous sites to detect possible recombination events in pedigrees.

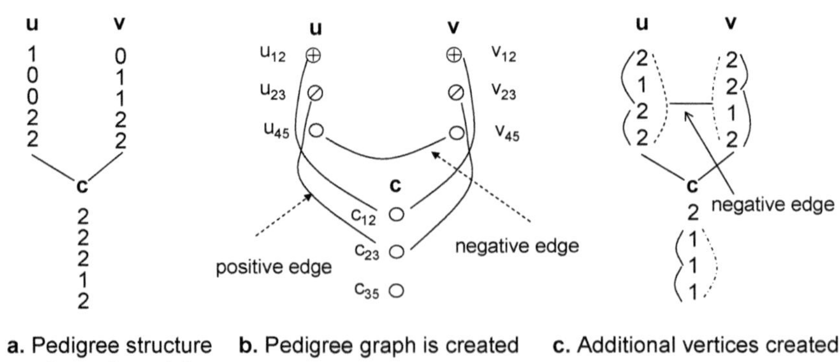

Fig. 1. Pedigree graph created from pedigree structure and genotype data. ⊕ denotes a red vertex, ⊘ denotes a green vertex, and O denotes a grey vertex.

3.1 Pedigree Graph

Create grey vertices. Let i be a heterozygous site in u ($i = 1, ..., m-1$). Let $j > i$ be the closest heterozygous site to i in u. We create a vertex u_{ij} from site i and site j and label this vertex *grey*. A grey vertex is an *unresolved vertex* and will later be resolved green if $h1_u[i] = h1_u[j] = 0$ or $h1_u[i] = h1_u[j] = 1$. It is resolved red otherwise. The resolution of a grey vertex depends on its adjacent vertices. Figure 1 shows a grey vertex u_{45} created from sites 4 and 5 of u.

Create red and green vertices. Let i be a homozygous site in u ($i = 1, ..., m-1$). Let $j > i$ be the closest homozygous site to i in u. We create a vertex u_{ij} from site i and site j, and label this vertex *red* if $g_u[i] \neq g_u[j]$ and *green* if $g_u[i] = g_u[j]$. A red or green vertex is a *resolved vertex*. Figure 1 shows a red vertex u_{12} created from sites 1 and 2, and a green vertex u_{23} from sites 2 and 3.

Insert positive edges. We insert positive edges between a parent u and its direct child v. For each vertex u_{ij} in u, if there is a vertex v_{ij} in v we insert a *positive edge* between u_{ij} and v_{ij}. If there is not a vertex v_{ij} in v and i and j are both homozygous sites or both heterozygous sites in v, we create a vertex v_{ij} in v and label this vertex properly, inserting a positive edge between u_{ij} and v_{ij}. We call v_{ij} a *supplementary vertex* as it is created by the need of member u.

Similarly, for each vertex v_{ij} in v, if there is not a vertex u_{ij} in u, and i and j are both homozygous sites or both heterozygous sites in u, we create a supplementary vertex u_{ij} in u and label this vertex properly, inserting a positive edge between u_{ij} and v_{ij}. Figure 1.b shows four positive edges linking u_{12} and c_{12}, u_{23} and c_{23}, v_{12} and c_{12}, v_{23} and c_{23}.

A positive edge between vertices u_{ij} and v_{ij} means vertex u_{ij} and v_{ij} should be resolved with the same color (both red or both green) unless a recombination event occurs in u. The reason for this is that if there is no recombination event in u, then v receives one full haplotype from u and another full haplotype from another parent. Therefore, the label of u_{ij} and the label of v_{ij} should be the same if there is no recombination event; otherwise, there is a recombination event in u. If u_{ij} is a resolved vertex and there is a positive edge between u_{ij} and a grey vertex v_{ij}, we color v_{ij} the same as the color of u_{ij}, since a recombination event at u_{ij} is not detectable and does not affect the color of v_{ij}.

Insert negative edges. We insert *negative edges* between two parents u and v of a common child c. If u_{ij} is a vertex in u but there is not a vertex c_{ij} in c (sites i and j are one homozygous and one heterozygous in c), two situations happen. If there is a vertex v_{ij} in v, we insert a negative edge between u_{ij} and v_{ij}. Otherwise, if there is not a vertex v_{ij} in v and i and j are both homozygous sites or both heterozygous sites, we create a supplementary vertex v_{ij} in v and label it properly. We insert a negative edge between u_{ij} and v_{ij}. Similarly, if v_{ij} is a vertex in v but there is not a vertex c_{ij} in c, there are two situations. If there is not a vertex u_{ij} in u, and i and j are both homozygous or both heterozygous, we create a supplementary vertex u_{ij} in u, and insert a negative edge between u_{ij} and v_{ij}. Figure 1.b shows a negative edge linking u_{45} and v_{45}.

A negative edge between u_{ij} and v_{ij} means vertices u_{ij} and v_{ij} should be resolved with different colors unless a recombination event occurs in one parent of c. This phenomenon can be explained as follows. If there is no recombination event and u_{ij} and v_{ij} have the same label (both red or both green), then sites i and j of c must be both homozygous or both heterozygous based on the Mendelian law of inheritance. Because sites i and j of c are one homozygous and one heterozygous, one recombination occurs if u_{ij} and v_{ij} have the same label when resolved, but no recombination event occurs if they are resolved differently.

Create additional vertices. Consider a grey vertex u_{ij} in u $(i < j)$. It is possible that u_{ij} has no incident edge but there is one recombination event occurring between site i and j. In this case none of the other two members in the trio has vertex created for site i and j. We delete vertex u_{ij} and create an additional vertex to capture the recombination event. Let j' be the closest heterozygous site from j in u $(j < j')$, where i and j' are both heterozygous sites or both homozygous sites in at least one member among the other two members, say v. If there is not a vertex $u_{ij'}$ in u, we create an *additional* grey vertex $u_{ij'}$ in u and create a supplementary vertex $c_{ij'}$ from sites i and j' in c if it does not exist. We color $c_{ij'}$ properly and insert a corresponding edge (positive or negative) between $u_{ij'}$ and $v_{ij'}$ depending on the relationship between u and v. Figure 1.c shows an additional vertex u_{14} created represented by a dashed edge between sites 1 and 4. A negative edge is inserted between u_{14} and v_{14}.

Pedigree graph. Pedigree graph $G = (V, E)$ created as described above is an undirected graph. Each vertex $y \in V$ has three possible labels, red, green, and grey. Each edge $e(y, z) \in E$ is either a positive edge, $e \in E_{pos}$, or a negative edge, $e \in E_{neg}$, with $E = E_{pos} \cup E_{neg}$. Graph G, set up this way, is a signed graph [13]. Let $N(y)$ be the set of adjacent vertices of y. Let $w(e)$ be the weight of edge e. If e is a positive edge, $w(e) = +1$. If e is a negative edge, $w(e) = -1$.

Observation 1. *There are at most $O(n \cdot m^2)$ vertices and $O(n \cdot m^2)$ edges in the pedigree graph.*

Each member has m sites. The total number of vertices created from pairs of sites for each member is $O(m^2)$. The whole pedigree graph with n members has $O(n \cdot m^2)$ vertices. A vertex has at most two positive edges linking it to two vertices in its parents. Therefore, the number of positive edges is linear in the number of vertices. The number of negative edges is also linear to the number of vertices. Thus the number of edges in the pedigree graph is $O(n \cdot m^2)$.

3.2 Parity-Constraint Sets

When a supplementary grey vertex u_{ij} is created in u by the need of an adjacent member, there must be more than one grey vertex already created from site i to site j in u. It is important to ensure that these grey vertices and u_{ij} when resolved will not result in an odd number of red vertices. Recall that a grey vertex is resolved red if $h1_u[i] \neq h1_u[j]$. In other words, the value of $h1_u$ flips

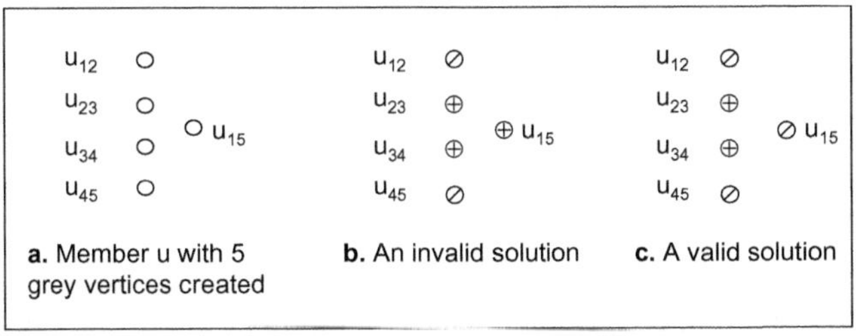

Fig. 2. Parity conflict between vertices within each member

from 0 to 1 and vice versa for a red vertex u_{ij}. Therefore there is a parity conflict if the number of red vertices from site i to site j including u_{ij} is odd.

In Figure 2.a, there are five grey vertices created for member u where vertices u_{12}, u_{23}, u_{34} and u_{45} are created from closest heterozygous sites, and a supplementary vertex u_{15} is created for a member adjacent to u. Figure 2.b shows an invalid solution with three resolved red vertices u_{23}, u_{34} and u_{15} in member u. A valid solution with an even number of red vertices is shown in Figure 2.c.

We create parity-constraint sets S_{pc} to capture parity constraints between each supplementary vertex and other vertices within each member. Let u_{ij} be a supplementary vertex and $u_{ip}, ..., u_{qj}$ be grey vertices from site i to site j. These vertices form a parity-constraint set, and its total number of red vertices must be even. There are $O(m^2)$ parity-constraint sets in each member and $O(nm^2)$ parity-constraint sets for the whole pedigree graph. A valid solution for RHC$_k$ must ensure that the number of red vertices in each parity-constraint set is even.

4 Signed Graph

A graph $G = (V, E)$ is a *signed graph* if it has both positive and negative edges ($E = E_{pos} \cup E_{neg}$) [13], where $w(e_{pos}) = 1$ and $w(e_{neg}) = -1$. Let (V_1, V_2) be a partition of V, and E^* be the set of edges between V_1 and V_2. The *line index* of the cut (V_1, V_2) is defined as:

$$l(V_1, V_2) = \sum_{e \in E^* \cap E_{pos}} w(e) + \sum_{e \in E_{neg} \setminus E^*} |w(e)| \quad (1)$$

The line index of graph G is defined as:

$$l(G) = \min_{V_1 \subseteq V} l(V_1, V_2) \quad (2)$$

The decision version of the line index of graph G is defined as follows.

LineIndex$_k$: *Given a signed graph G and a positive integer k, is there a line index of G at most k?*

Given a pedigree graph $G = (V, E)$, the RHC$_k$ problem can be solved by determining if we can label every grey vertex in G either red or green such that if we partition the set of vertices V into (V_{red}, V_{green}) and let E^* be the set of edges between V_{red} and V_{green} then

$$\sum_{e \in E^* \cap E_{pos}} w(e) + \sum_{e \in E_{neg} \setminus E^*} | w(e) | \leq k \tag{3}$$

and this partition (V_{red}, V_{green}) must satisfy parity-constraint sets S_{pc}.

Given a pedigree graph, any two adjacent members linked by a positive edge should be in the same partition, and any two adjacent members linked by a negative edge should be in different partitions. Any edge whose constraint is not satisfied represents a recombination event between the two adjacent members, or, in the case of a negative edge having endpoints in the same partition, between one parent and the child. Equation 3 thus counts the number of recombination events in the whole pedigree and ensures that it is at most k.

Clearly, the RHC$_k$ problem can be reduced to the LineIndex$_k$ problem with additional parity-constraint sets S_{pc} on its vertices. We will show that the LineIndex$_k$ problem can be reduced to the GBER problem, a classic NP-complete problem that is fixed-parameter tractable. The RHC$_k$ can therefore be solved through the GBER problem with additional parity-constraint sets S_{pc}.

Theorem 1. *A pedigree has at most k recombination events if and only if its corresponding signed graph has the line index of size at most k.*

Proof. We will show that one recombination event in the pedigree corresponds to exactly one negative edge within each partition or one positive edge crossing partitions in the signed graph.

$\Rightarrow$ Consider a recombination event in member u. To detect this recombination event there must be at least one heterozygous site on each side of the recombination breakpoint. Let i and j be the two closest heterozygous sites on the two sides of the recombination breakpoint. There are three possible types of vertices associated with this recombination event: a grey vertex u_{ij}, an additional vertex $u_{ij'}$, and supplementary vertices u_{pq} $(p \leq i, j \leq q)$.

If vertex u_{ij} has an incident positive edge to a vertex c_{ij}, the color u_{ij} should be different from the color of c_{ij} because of the recombination event and the positive edge between them would cross between partitions. On the other hand, if u_{ij} has an incident negative edge to a vertex v_{ij}, the color u_{ij} and v_{ij} should be the same because of the recombination event and the negative edge between them would be within the same partition. In both cases the line index increases by one. An additional vertex $u_{ij'}$ replaces u_{ij} when u_{ij} has no incident edge. The resolution of an additional vertex $u_{ij'}$ is similar to that of u_{ij}.

Consider a supplementary vertex u_{pq} constrained by a parity-constraint set S_{pc} where u_{pq} has an incident positive edge to a vertex c_{pq}. The color u_{pq} is determined by the swap of values in $h1_u$ by red vertices and recombination events from p to q, including the recombination from i to j. If no more recombinations happen, u_{pq} and c_{pq} must have the same color and the line index of the signed graph is the same. If u_{pq} and c_{pq} have different colors, there must be another recombination from sites p to q and the line index increases by one. A similar explanation follows for u_{pq} with an incident negative edge.

$\Leftarrow$ A negative edge links two vertices of two parents in a trio, and the two vertices are supposed to have different colors based on the Mendelian law of inheritance. Similarly, a positive edge links two vertices of a parent and a child and the two vertices are supposed to have the same color. Therefore, if a negative edge linking two vertices with the same color or a positive edge linking two vertices with different colors, one recombination event must happen.

5 Fixed-Parameter Algorithm

A NP-hard problem cannot be solved by a polynomial time algorithm unless P=NP. However, if we can restrict some parameters of the problem to small values, the running time of an algorithm for the problem can potentially be greatly reduced [10]. In this case, the problem is a *parameterized problem* and an algorithm that can solve the parameterized problem efficiently is a *fixed-parameter algorithm*, defined as follows [10].

Definition 1. *A parameterized problem is a language* $L \subseteq \Sigma^* \times \Sigma^*$, *where* Σ *is a finite alphabet. The second component is called the parameter of the problem.*

Practically, the parameter is a nonnegative integer or a set of nonnegative integers and therefore $L \subseteq \Sigma^* \times \mathbb{N}$. For $(x, k) \in L$, the size of the input is $n = |(x, k)|$, and the parameter is k.

Definition 2. *A parameterized problem* L *is fixed-parameter tractable (in class FPT) if it can be determined in* $f(k) \cdot n^{O(1)}$ *time whether or not* $(x, k) \in L$, *where* f *is a computable function only depending on* k.

5.1 Transforming to Bipartization by Edge Removal Problem

We review an important property of a signed graph given by [13].

Theorem 2. *Let* G *be a signed graph. If we replace each edge with weight* $w(e) > 0$ *by two consecutive edges with weight* $-w(e)$ *to get a graph* G' *then* $l(G) = l(G')$.

The pedigree graph is transformed into a new graph by replacing every positive edge by two consecutive negative edges and adding new intermediate vertices (*dum* vertices). We obtain a new weighted graph G' with all negative edges. This transformation does not affect the parity-constraint sets S_{pc}. The graph G' still has only $O(n \cdot m^2)$ vertices and $O(n \cdot m^2)$ edges. Equation 3 becomes

$$\sum_{e \in E_{neg} \setminus E^*} |w(e)| \leq k \tag{4}$$

This equation is to ensure that the total number of edges within V_1 and edges within V_2 is at most k. Removing these edges will make the graph bipartite.

To make the GBER algorithm [5] works on our partially colored graph, we merge all red vertices into one red vertex and all green vertices into one green vertex. We relabel the merged red vertex and the merged green vertex into two grey vertices, and insert $k+1$ negative edges between them. This transformation does not affect the parity-constraint set S_{pc}. We further transform our negative graph into a new graph with all positive edges by multiplying the weight of every edge by -1. Our problem becomes the GBER problem [5] with additional parity-constraint set S_{pc}. The k-Bipartization by Edge Removal problem is defined as follows.

Definition 3. *Given a graph G=(V,E) and a positive integer k, is there a set* $C \subseteq E$ *with* $|C| \leq k$ *whose removal produces a bipartite graph?*

GBER is a classical NP-hard problem [6] and is in FPT [5].

5.2 FPT Algorithm for Bipartization by Edge Removal

There are many techniques to solve an FPT problem such as kernelization, depth-bounded search trees, dynamic programming, crown reduction, greedy localization, and iterative compression. The iterative compression technique is used by Guo et al. [5] to solve the GBER problem with a running time of $O(2^k \cdot |E|^2)$, where $|E|$ is the number of edge in the graph and k is the number of edges to be deleted to make the graph bipartite. However, this algorithm does not enforce our parity constraints that require the number of red vertices in each set to be even. We thus need to modify this algorithm [5] to solve the RHC$_k$ problem while respecting the additional parity-constraint sets S_{pc}.

Given a graph $G = (V, E)$ where $E = \{e_1, ..., e_m\}$, let G_i be a graph induced by edges $\{e_1, ..., e_i\}$ of G ($1 \leq i \leq m$). If $i = 1$, the optimal edge bipartization set of G_1 is empty. If $i > 1$, let X be an optimal edge bipartization set of $G_i = G[e_1, ..., e_i]$ and $|X| = k'$. Consider graph $G_{i+1} = G[e_1, ..., e_{i+1}]$. If X is not an optimal edge bipartization set for G_{i+1} then $X' = X \cup \{e_{i+1}\}$ is clearly an optimal edge bipartization set for G_{i+1}. From the edge bipartization set X' of size $k' + 1$, we find an edge bipartization set of size at most k' or show that no such edge bipartization set of size at most k' exists. The algorithm assumes that an edge bipartization Y which is smaller than X' must be disjoint from X', $Y \cap X' = \varnothing$. This assumption can be made without loss of generality by a simple graph transformation, replacing each edge in X' by three consecutive edges and choosing the middle edge to be in the new X'. This graph transformation preserves the parities of lengths of all cycles and does not affect the parity constraint sets S_{pc}. Therefore the transformed graph has an edge bipartization set of size k' if and only if the original graph has an edge bipartization set of size k'. Let mapping Φ: $V(X') \rightarrow \{A, B\}$ be a valid partition of $V(X')$ if for each $\{y, z\} \in X$, we have $\Phi(y) \neq \Phi(z)$. Let A_Φ be $\Phi^{-1}(A)$ and B_Φ be $\Phi^{-1}(B)$. We enumerate all $2^{k'}$ valid partitions Φ of $V(X')$. For each valid partition Φ we find a minimum edge cut Y in $G \backslash X'$ between A_Φ and B_Φ. In other words, we use X' to partially color G and from the partially colored graph we compute a smaller bipartization set Y. This compression step is the core of the algorithm.

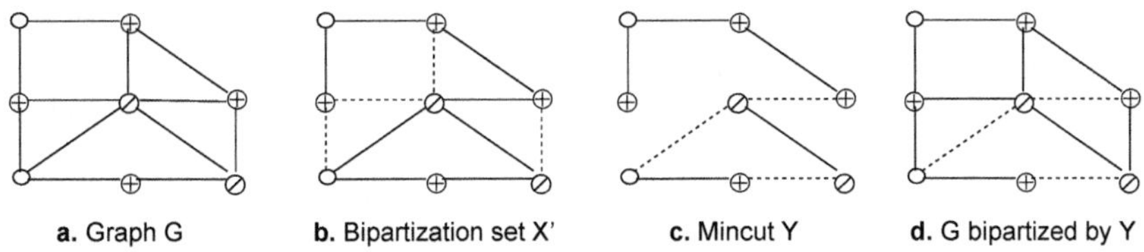

Fig. 3. Compression step

Theorem 3. *[5] Consider a graph $G = (V, E)$ and a minimal edge bipartization set X' for G. For a set of edges $Y \subseteq E$ with $X' \cap Y = \varnothing$, the following are equivalent:*

(1) Y is an edge bipartization set for G.
(2) There is a valid partition Φ for $V(X')$ such that Y is an edge cut in $G \backslash X'$ between $A_\Phi = \Phi^{-1}(A)$ and $B_\Phi = \Phi^{-1}(B)$.

Consider a graph G in Figure 3.a where $\oplus$ denotes a red vertex, $\oslash$ a green vertex, and O a grey vertex. A minimal edge bipartization set X' of size 4 illustrated by dashed lines is given in Figure 3.b. We compute a mincut Y for $G \backslash X'$ as in Figure 3.c. Set Y is the edge bipartization set of size 3 for G in Figure 3.d.

It remains to find a minimum edge cut Y between A_Φ and B_Φ that satisfies

(1) $|Y| \leq k'$ and
(2) graph G_i with set Y satisfies parity-constraint sets S_{pc}.

(s-t) Mincuts with parity constraints. A minimum edge cut Y between A_Φ and B_Φ can be computed in $O(k' \cdot |E|)$ time by the Edmonds-Karp algorithm [2] by finding at most k' augmenting paths; each path takes $O(|E|)$ time to find. If no min edge cut Y of size k' is found, we skip the current partition Φ and check a new valid partition. If a min edge cut Y of size k' is found, we need to check if G_i bipartized by Y satisfies the parity-constraint sets S_{pc}. Note that there can be many mincuts Y of size k' between A_Φ and B_Φ, and it is possible that the current mincut Y found does not make G_i satisfy S_{pc} while another mincut Y of size k' makes G_i satisfy S_{pc}. However, enumerating all mincuts in a graph is expensive. Consider a simple directed graph with n disjoint paths of length 2 from a source s to a sink t, where the weight of each edge is 1. Each (s-t) mincut has weight n and we have up to 2^n (s-t) mincuts. If a graph is an undirected graph, we replace each undirected edge by two directed edges with opposite directions and the number of (s-t) mincuts is still 2^n. Therefore enumerating all (s-t) mincuts in a graph in polynomial time, or in FPT, is impossible.

We do not enumerate all mincuts. Instead, we examine the structure of all mincuts in a graph by an algorithm in [9]. Given a graph $G = (V, E)$ including a source s and a sink t, where each directed edge $(i, j) \in E$ has a capacity c_{ij}, an (s-t) cut (S, S') is a cut where $S' = V - S$, $s \in S$ and $t \in S'$. If a graph is not directed, we replace every undirected edge by two oppositely directed edges. If a graph has multiple sources and sinks, we can transform the graph into a new graph with only a single source and a single sink by inserting edges of ∞ weights from a *super source* s to all sources, and from all sinks into a *super sink* t. Flows and mincuts in the new and old graphs correspond [2].

An (s-t) mincut is an (s-t) cut where the total capacity of all the edges between S and S' is minimum. We will call an (s-t) mincut a mincut hereafter. Ford and Fulkerson [2] show that the value of a minimum cut between s and t is equal the value of the maximum flow from s to t. Consider a binary relation R on V, a subset of vertices $V' \subseteq V$ is a *closure* for R if and only if for any two vertices i and j in V with iRj and $i \in V'$ we also have $j \in V'$. Given a relation iRj, we say that i is the *predecessor* of j and j is a *successor* and i. Picard and Queyranne [9] present the relationship between mincuts and closures as follows.

Theorem 4. *[9].*

Let f be a maximum flow in G. Define a relation R on the set of vertices V as follows:

iRj iff $(i,j) \in E$ and $f_{ij} < c_{ij}$, or $(j,i) \in E$ and $f_{ji} > 0$.

Then a cut (S,S') separating s from t is a minimum cut if an only if S is a closure for R containing s and not t.

Suppose we find a maximum flow in a graph by the Edmonds-Karp algorithm [2]. Clearly, the residual graph $G_r = (V, E_r)$ of G is defined by relation R where edge $(i,j) \in E_r$ iff iRj. We find strongly connected components in G_r and shrink each of them into a single vertex. Finding strongly connected components of a directed graph G_r can be done in $O(V+E)$ time using two depth first searches, one search on G_r and the other search on the transpose graph G_r^T of G_r [2].

Let V' be the reduced vertex set of V, we define a relation $\bar{R}$ on V' by $\bar{i}\bar{R}\bar{j}$ iff iRj for some $i \in \bar{i}$, $j \in \bar{j}$, and $\bar{i}, \bar{j} \in \bar{V}$. We eliminate component S containing source s and its successor components, and eliminate component T containing sink t and its predecessor components. Combining S and all successor components with any closure induced from the remaining components will produce a mincut. When the number of sites m is small, we can check if a member can satisfy its parity-constraint sets by a backtracking search on at most $O(m^2)$ components. Since the parity constraints involve vertices for an individual member, these searches can be done independently. Therefore we need to examine if a valid partition Φ satisfies S_{pc} on at most $2^{m^2} \cdot n$ cuts for the whole pedigree.

Theorem 5. *The RHC_k problem is solvable in $O(2^k 2^{m^2} n^2 m^3)$ time.*

Proof. Setting up the pedigree graph $G = (V, E)$ takes $O(|V|)$ time, where $|V| = |E| = O(nm^2)$. Generating parity-constraint sets S_{pc} takes $O(nm^3)$. Transforming the pedigree graph into a graph with all negative edges takes $O(|E|)$ time. The GBER problem can be solved by trying at most 2^k valid partitions Φ. For each partition, we can find the first mincut in $O(k \cdot |E|)$ time by finding at most k augmenting paths using Edmonds-Karp algorithm. We can find strongly connected components in $O(|E|)$ time. We do backtracking in at most 2^{m^2} cuts for each member to check if one can satisfy S_{pc}; each check takes $O(|E|)$ time. Therefore, each partition takes $O(k \cdot |E| + |E| + 2^{m^2} \cdot |E| \cdot n)$. The overall time complexity of the algorithm is $O(2^k 2^{m^2} n^2 m^3)$.

6 Conclusion

We have shown that given a general pedigree with n members, m sites, and k recombination events, where m and k are small, the haplotype inference can be done in $O(2^k 2^{m^2} n^2 m^3)$ time.

References

1. Chan, B.M.-Y., et al.: Linear-time haplotype inference on pedigrees without recombinations. In: Bücher, P., Moret, B.M.E. (eds.) WABI 2006. LNCS (LNBI), vol. 4175, pp. 56–67. Springer, Heidelberg (2006)
2. Cormen, T.H., Leiserson, C.E., Rivest, R.L., Stein, C.: Introduction to Algorithms, 2nd edn. MIT Press, McGraw-Hill (2001)
3. Doan, D.D., Evans, P.A., Horton, J.D.: A Near-Linear Time Algorithm for Haplotype Determination on General Pedigrees. Submitted to Journal of Computational Biology (2009) (Submission ID: JCB-2009-0133)
4. Doan, D.D., Evans, P.A.: Fixed-Parameter Algorithm for General Pedigrees with a Single Pair of Sites. In: Borodovsky, M., Gogarten, J.P., Przytycka, T.M., Rajasekaran, S. (eds.) ISBRA 2010. LNCS, vol. 6053, pp. 29–37. Springer, Heidelberg (2010)
5. Guo, J., et al.: Compression-Based Fixed-Parameter Algorithms for Feedback Vertex Set and Edge Bipartization. Journal of Computer and System Sciences 72(8), 1386–1396 (2006)
6. Karp, R.M.: Reducibility Among Combinatorial Problems. In: Miller, R.E., Thatcher, J.W. (eds.) Complexity of Computer Computations, pp. 85–103. Plenum, New York (1972)
7. Li, J., Jiang, T.: An exact solution for finding minimum recombinant haplotype configurations on pedigrees with missing data by integer linear programming. In: Proceedings of Research in Computational Molecular Biology, pp. 20–29 (2004)
8. Liu, L., Chen, X., Xiao, J., Jiang, T.: Complexity and approximation of the minimum recombinant haplotype configuration problem. Theoretical Computer Science 378, 316–330 (2007)
9. Picard, J., Queyranne, M.: On the structure of all minimum cuts in a network and applications. Mathematical Programming Study 13, 8–16 (1980)
10. Niedermeier, R.: Invitation to Fixed-Parameter Algorithms. Oxford University Press, Oxford (2006)
11. Qian, D., Beckmann, L.: Minimum-recombinant haplotyping in pedigrees. American Journal of Human Genetics 70(6), 1434–1445 (2002)
12. Xiao, J., Lou, T., Jiang, T.: An efficient algorithm for haplotype inference on pedigrees with a small number of recombinants. In: Fiat, A., Sanders, P. (eds.) ESA 2009. LNCS, vol. 5757, pp. 325–336. Springer, Heidelberg (2009)
13. Xu, S.: The line index and minimum cut of weighted graphs. European Journal of Operational Research 109, 672–682 (1998)

Haplotypes versus Genotypes on Pedigrees

Bonnie Kirkpatrick

Electrical Engineering and Computer Sciences, University of California Berkeley and International Computer Science Institute
bbkirk@eecs.berkeley.edu

Abstract. Genome sequencing will soon produce haplotype data for individuals. For pedigrees of related individuals, sequencing appears to be an attractive alternative to genotyping. However, methods for pedigree analysis with haplotype data have not yet been developed, and the computational complexity of such problems has been an open question. Furthermore, it is not clear in which scenarios haplotype data would provide better estimates than genotype data for quantities such as recombination rates.

To answer these questions, a reduction is given from genotype problem instances to haplotype problem instances, and it is shown that solving the haplotype problem yields the solution to the genotype problem, up to constant factors or coefficients. The pedigree analysis problems we will consider are the likelihood, maximum probability haplotype, and minimum recombination haplotype problems.

Two algorithms are introduced: an exponential-time hidden Markov model (HMM) for haplotype data where some individuals are untyped, and a linear-time algorithm for pedigrees having haplotype data for all individuals. Recombination estimates from the general haplotype HMM algorithm are compared to recombination estimates produced by a genotype HMM. Having haplotype data on all individuals produces better estimates. However, having several untyped individuals can drastically reduce the utility of haplotype data.

Pedigree analysis, both linkage and association studies, has a long history of important contributions to genetics, including disease-gene finding and some of the first genetic maps for humans. Recent contributions include fine-scale recombination maps in humans [4], regions linked to Schizophrenia that might be missed by genome-wide association studies [11], and insights into the relationship between cystic fibrosis and fertility [13]. Algorithms for pedigree problems are of great interest to the computer science community, in part because of connections to machine learning algorithms, optimization methods, and combinatorics [7,16,12,10,15].

Single-molecule sequencing is an attractive alternative to genotyping and would yield haplotypes for individuals in a pedigree [6]. Such technologies are being developed and may become commercial within five to ten years. Sequencing methods would apparently yield more information from the same set of sampled individuals, which is critical due to the limited availability of individuals for

V. Moulton and M. Singh (Eds.): WABI 2010, LNBI 6293, pp. 136–147, 2010.

sampling in multi-generational pedigrees (i.e. individuals usually must be living at the time of sampling). There is substantial evidence that haplotypes can be more useful than genotypes for both population and family based studies when using methods such as association studies [1,3] and pedigree analysis [2,8]. While it is intuitive that haplotypes provide more information than genotypes, there are instances with family data in which there are few enough typed individuals that there is little practical difference between haplotype and genotype data. Additionally, in order to exploit the information contained in haplotype data, we need to understand the instances where diploid inheritance is computationally tractable given haplotype data.

Pedigree analysis with genotype data is well studied in terms of complexity [12,10] and algorithms [5,9,14]. Less is known about haplotype data on pedigrees. This paper shows that, given haplotype data on a pedigree, finding both minimum recombination and maximum probability haplotypes is as tractable as computing the same quantities for pedigrees with genotype data (i.e., these problems are NP- and #P-hard, respectively). To obtain a reduction that applies equally well to several types of pedigree calculations, we will consider a modular polynomial-time mapping from the genotype problem to the haplotype problem. The reduction preserves the solutions to the analyses, meaning that the solution to the haplotype problem is the solution to the genotype problem after adjusting by constant factors or coefficients.

Since the reduction uses a biologically unlikely recombination scenario, we will investigate the accuracy and information of realistic examples with haplotypes and genotype data on the same pedigree. Pedigree data was simulated having a known number of recombinations. The recombination distributions were computed at a particular locus of interest and compared to the ground-truth. Since both the haplotypes and genotypes of a specific person contain the same alleles, the differences between the haplotype and genotype recombination distributions were determined by the extra information in the haplotype data. As expected, the haplotype data reliably yields greater accuracy when all the pedigree individuals are typed. However, as fewer pedigree individuals are typed, there is less practical difference between the utility of haplotype versus genotype data. The number of untyped generations that separate typed individuals influences whether haplotype data are actually more accurate than genotype data. For instance with two half-siblings, having two untyped parents results in estimates from genotype data that are nearly as accurate as the estimates computed from haplotype data.

Finally, there is an important instance where haplotype data is more computationally tractable than genotype data. When all individuals in the pedigree are typed, although unlikely from a practical perspective of obtaining genetic samples, the computational problem decomposes into conditionally independent sub-problems, and has a linear-time algorithm. This can be contrasted with the known hardness of the genotype problem even when all individuals are genotyped. The existence of this linear-time algorithm for haplotype data could facilitate useful approaches that combine population genetic and pedigree methods.

For instance, if the haplotypes of the founders are drawn from a coalescent and the pedigree individuals are all haplotyped, the probability of a combined model could easily be computed for certain coalescent models.

1 Introduction to Pedigree Analysis

A *pedigree* is a directed acyclic graph where the set of nodes, I, are individuals, and directed edges indicate genetic inheritance between parent and child. A diploid pedigree (i.e. for humans) necessarily has either zero or two incoming edges for each person. The set, F, of individuals without incoming edges are referred to as pedigree *founders*. An individual, i, with two parents is a *non-founder*, and we will refer to their two parents as $m(i)$ and $p(i)$.

As is commonly done to accommodate inheritance of genetic information, we will extend this model to include a representation of the alleles of each individual and of the inheritance origin of each allele. More formally, we represent a single chromosome as an ordered sequence of variables, x_j, where each variable takes on an *allele* value in $\{1, ..., k_j\}$. Each variable represents a *polymorphic site*, j, in the genome, where there are k_j possible sequence variants. Since diploid individuals have two copies of each chromosome, one copy inherited from each parent, we will use a superscript m and p to indicate the maternal and paternal chromosomes respectively. For a particular individual i, the information on both copies of a particular chromosome at site j is represented as $x^m_{i,j}$ and $x^p_{i,j}$.

Furthermore, we assume that inheritance in the pedigree proceeds with recombination and without mutation (i.e. Mendelian inheritance at each site). This imposes consistency rules on parents and children: the allele $x^m_{i,j}$ must appear in the mother $m(i)$'s genome as either the grand-maternal or grand-paternal allele, $x^m_{m(i),j}$ or $x^p_{m(i),j}$, and similarly for the paternal allele and the father $p(i)$'s genome.

Let x be a vector containing all the haplotypes x^m_i, x^p_i for all individuals $i \in I$, then we are interested in the probability

$$\mathbb{P}[x] = \prod_{f \in F} \mathbb{P}[x^p_f]\mathbb{P}[x^m_f] \prod_{i \in I \setminus F} \mathbb{P}[x^p_i | x^p_{p(i)}, x^m_{p(i)}]\mathbb{P}[x^m_i | x^p_{m(i)}, x^m_{m(i)}], \tag{1}$$

where the superscript m and p indicate maternal and paternal alleles, while the functions $m(i)$ and $p(i)$ indicate parents of i. The first product is over the independent founder individuals whose haplotypes are drawn from a uniform prior distribution, while the second product, over the non-founders, contains the probabilities for the children to inherit their haplotypes from their parents. The unobserved vector x is not immediately derived from observed haplotype data, since vector x contains haplotype alleles labeled with their parental origins for all the individuals. To compute this quantity, we need notation to represent the parental origins of each allele where differing origins for neighboring haplotype alleles will indicate recombination events.

For each non-founder, let us indicate the source of each maternal allele using the binary variable $s^m_{i,j} \in \{m, p\}$, where the value m indicates that $x^m_{i,j}$ allele has grand-maternal origin and p indicates grand-paternal origin. Similarly, we define $s^p_{i,j}$ for the origin of i's paternal allele. For a particular site, these indicators for all the individuals, s_j, is commonly referred to as the identity-by-descent (IBD) inheritance path. A recombination is observed at consecutive sites as a change in the binary value of a source vector, for instance, $s^m_{i,j} = p$ and $s^m_{i,j+1} = m$. To compute the inheritance portion of Equation 1, we will sum over the inheritance options $\mathbb{P}[x] = \sum_s \mathbb{P}[x|s]\mathbb{P}[s]$ where $\mathbb{P}[s] = 1/2^{2|I \setminus F|}$.

We can observe two kinds of data for pedigree individuals whose genetic material is available. The first, and most common, is genotype data, a tuple of alleles $(g^0_{i,j}, g^1_{i,j})$ that must appear in the variables $x^m_{i,j}$ and $x^p_{i,j}$ for each site j. Since these alleles are unlabeled for origin, we do not know which allele was inherited from which parent. The second type of data is haplotypes, where we observe two sequences of alleles h^0_i and h^1_i and each sequence represents alleles that were inherited together from the same parent. However, we do not know which sequence is maternal and which is paternal. For either type of data define a function $C_{i,j}$ for locus j which indicates compatibility of the assigned haplotype alleles with the data and requires inheritance consistency between generations. Specifically, for genotype data $C_{i,j} = 1$ if $x^m_{i,j} = x^{s^m_{i,j}}_{m(i),j}$, $x^p_{i,j} = x^{s^p_{i,j}}_{f(i),j}$, and $\{x^m_{i,j}, x^p_{i,j}\} = \{g^0_{i,j}, g^1_{i,j}\}$. Under haplotype data, the $C_{i,j} = 1$ when the first two equalities, above, hold and $\{x^m_{i,j}, x^p_{i,j}\} = \{h^0_{i,j}, h^1_{i,j}\}$, which are the haplotype alleles at locus j.

Now, we write Equation 1 as a function of the per-site recombination probability $\theta \leq 0.5$. For particular values of all the haplotype alleles $x^m_{i,j}$ and $x^p_{i,j}$, the haplotype probability conditional on the inheritance options and the observed data through $C_{i,j}$ is

$$\mathbb{P}[x|s] = \prod_{f \in F} \prod_{j=1}^{l} C_{f,j}\mathbb{P}[x^p_{f,j}]\mathbb{P}[x^m_{f,j}] \prod_{i \in I \setminus F} C_{i,1} \prod_{j=2}^{l} C_{i,j} \cdot \theta^{(R^m_{i,j} + R^p_{i,j})} \cdot (1-\theta)^{(2 - R^m_{i,j} - R^p_{i,j})}$$

where $R^m_{i,j} = \mathbb{I}[s^m_{i,j-1} \neq s^m_{i,j}]$ and $R^p_{i,j} = \mathbb{I}[s^p_{i,j-1} \neq s^p_{i,j}]$.

1.1 Pedigree Problem Formulations

Given a pedigree and some observed genotype or haplotype data, there are three problem formulations that we might be interested in. The first is to compute the probability of some observed data, while the last two problems find values for the unobserved haplotypes of individuals in the pedigree.

Likelihood. Find the probability of the observed data by summing over all the possible unobserved haplotypes, i.e. $\sum_x \sum_s \mathbb{P}[x|s]\mathbb{P}[s]$.

Maximum Probability. Find the values of $x^m_{i,j}$ and $x^p_{i,j}$ that maximize the probability of the data, i.e. $\max_x \sum_s \mathbb{P}[x|s]\mathbb{P}[s]$.

Minimum Recombination. Find the values of $x^m_{i,j}$ and $x^p_{i,j}$ that minimize the number of required recombinations, i.e. $\min_{x,s} \sum_i \sum_{j \geq 2} \mathbb{I}[s^p_{i,j-1} \neq s^p_{i,j}] + \mathbb{I}[s^m_{i,j-1} \neq s^m_{i,j}]$.

The likelihood is commonly used for estimating site-specific recombination rates, relationship testing, computing p-values for association tests, and performing linkage analysis. Haplotype and/or IBD inferences, obtained by maximizing the probability or minimizing the recombinations, are useful for non-parametric association tests, tests on haplotypes, and tests where there is disease information for unobserved genomes.

2 Hardness Results

With genotype data, the likelihood and minimum recombination problems are NP-hard, while the maximum probability problem is #P-hard. Piccolboni and Gusfield [12] proved the hardness of the likelihood and maximum probability computations by relying on a single locus sub-pedigree with half-siblings. Although their paper discussed a more elaborate setting involving a phenotype, their proof, however, applies to this setting. Li and Jiang proved the minimum recombination problem to be hard by using a two-locus sub-pedigree with half-siblings [10]. In all these proofs, half-siblings were pivotal to establishing reductions from well known NP and #P problems.

In this paper, we introduce a simple and powerful reduction that converts any genotype problem on a pedigree of n individuals into a haplotype problem on a pedigree of at most $6n$ individuals. This reduction is simple, because it merely introduces four full-siblings and an extra parent for each genotyped individual. We do not need complicated structures involving inbreeding or half-siblings. The reduction works equally well for all three problem formulations.

Mapping. Given a pedigree with genotype data, for any of the three pedigree problems, we define a polynomial mapping to a corresponding haplotype problem with exactly $5|G|$ individuals haplotyped. First we create the pedigree graph for the new haplotype instance, and later we construct the required haplotype observations from the genotype data.

Let $G \subset I$ represent the set of genotyped individuals in a pedigree having individuals I and edges E. We will create a haplotype instance of the problem, with individuals $H \cup I$ and edges $R \cup E$. To obtain the set H, we add five individuals, i_0, i_1, i_2, i_3, i_4, to H for every individual $i \in G$. The set of new relationship edges, R, will connect individuals in sets H and G. Specifically, the edges stipulate that i and i_0 are the parents of full-siblings i_1, i_2, i_3, and i_4 by including the edges: $i_0 \to i_1$, $i_0 \to i_2$, $i_0 \to i_3$, $i_0 \to i_4$, $i \to i_1$, $i \to i_2$, $i \to i_3$, and $i \to i_4$. We will refer to these five individuals, i_0, i_1, i_2, i_3, and i_4, and their relationships with i as the *proxy family* for individual i. For example in Figure 1, the 6-individual genotype pedigree in becomes a 21-individual haplotype pedigree. This produces a pedigree graph with exactly $5|G| + |I|$ individuals and $8|G| + |E|$ edges.

To obtain the new haplotype data from the genotype data, we type only individuals in $|H|$ such that the corresponding genotyped individual in G is required, by the rules of inheritance, to have the observed genotypes. Without loss of generality, assume that the genotype alleles are sorted such that $g^0_{i,j} < g^1_{i,j}$.

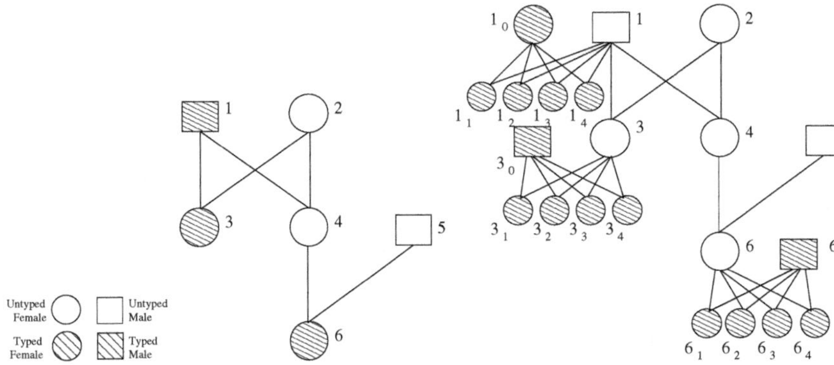

Fig. 1. Genotype and Haplotype Pedigrees. Values are averages computed from 500 simulations. (Left) Genotyped individuals are shaded, and all the individuals are labeled. Individuals 1, 2, and 5 are the founders, and individual 6 is the grandchild of 1 and 2. (Right) Haplotyped individuals are shaded, and individuals have the same labels. For each of the genotyped individuals, i, from the previous figure, the mapping adds a nuclear family containing five new individuals labeled i_0, i_1, i_2, i_3, i_4.

Now we can easily constrain the parental genotype for individual $i \in G$ by giving the spouse, i_0, homozygous haplotypes of all ones while giving child i_1 the haplotypes $\{\mathbf{1}, g_i^0\}$, child i_2 haplotypes $\{\mathbf{1}, g_i^1\}$. This guarantees the correct genotype, but does not ensure that the haplotypes of that genotype have the same probability or number of recombinations.

Since there is an arbitrary assorting of genotype alleles at neighboring loci into the parent haplotypes x_i^p and x_i^m, we will use the remaining two children to represent possible re-assortments of the genotyped parent's T_i heterozygous loci, indexed by t_j where $1 \leq j \leq T_i$. In addition to the haplotype $\mathbf{1}$, child i_3, will have haplotype consisting of $h_{i_3,t_j} := g_{i,t_j}^{1-j \bmod 2}$ while child i_4 has the genotyped parent's complementary alleles $h_{i_4,t_j} := g_{i,t_j}^{j \bmod 2}$. This results in child i_3 and i_4 alternating in having the smaller allele at every other heterozygous locus.

This reduction preserves the solutions to the three problems up to constant factors or constant coefficients. Specifically, the solution to the haplotype version of the problem is the solution to the genotype version with the values of the functions being related by constant factors or coefficients, depending on whether the function is a recombination count or a probability.

Lemma 1. *Let r_g be the minimum number of recombinations in the genotype problem instance. The mapping yields a haplotype problem instance having $r_h = r_g + \sum_{i \in G} 2(T_i - 1)$ for the minimum number of recombinations, where T_i is the number of heterozygous sites in genotype i.*

To prove this result, we exploit the alternating pattern of alleles assigned to the four children. This pattern forces there to be two recombinations, among the four children, between consecutive heterozygous loci.

After applying the mapping, the haplotype probability turns out to have a coefficient that is independent of the haplotype assignment to the non-founding parent of the proxy family. This coefficient can be computed in linear time from the genotype data using a Markov chain. The Markov chain has 16 states and has a transition step between each pair of neighboring loci. This small Markov model can be thought of in the sum-product algorithm as an elimination of the typed individuals in the proxy family; alternatively, it is also equivalent to peeling-off the typed proxy individuals in the Elston-Stewart algorithm [5]. Once we have this coefficient, independent of the haplotype assignment, it is clear that the likelihood and maximum probability haplotype problems also have haplotype solutions related proportionally to the genotype solution.

Lemma 2. *The mapping yields a haplotype problem instance having haplotype probabilities proportional to the haplotype probabilities of the genotype instance. Specifically, for all* x*,*

$$\mathbb{P}_h[x] = \left(\mathbb{P}_g\left[\{x_i | i \in I\}\right]\right) \prod_{i \in G} p_t(i) \prod_{j} \mathbb{P}[x^p_{i_0,j} = 1]\mathbb{P}[x^m_{i_0,j} = 1]$$

where the proxy family transmission probability is a function of genotype g_i*, the recombination rate* $\theta \leq 0.5$*, and of the transition matrices* P*,* Q_{0110}*, and* Q_{1001}*,*

$$p_t(i) = \left(\frac{1}{16}\right) \mathbf{1} \cdot P^{h_0} \prod_{j=0}^{T_i} \left(O_j Q_{0110} + (1 - O_j) Q_{1001}\right) \cdot P^{h_j} \cdot \mathbf{1}^T$$

and O_j *indicates whether index* j *is odd,* h_0 *is the number of homozygous loci that begin proxy parent's genotype, and* h_j *is the number of consecutive homozygous loci after the* j*'th heterozygous locus where there are* T_i *heterozygous loci for proxy parent* i*. The transition probabilities are given by* $P_{ij} = \theta^{H(i,j)}(1-\theta)^{4-H(i,j)}$ *where* $H(i,j)$ *is the Hamming distance between inheritance states* i *and* j*. Let* Q_{0110} *be a transition matrix having non-zero recombination probabilities only in column* 0110 *(i.e.* $Q_{0110,i,j} = P_{ij}$ *when* $j = 0110$*). Similarly, let* Q_{1001} *be a transition matrix with non-zero recombination probabilities only in column* 1001*.*

Although this reduction establishes the hardness of these haplotype pedigree problems, it does so by constructing children whose haplotypes require many recombinations and would be extremely unlikely to occur naturally. Accordingly, we suspect that realistic instances of these haplotyping problems may provide more information about the locations of recombinations than genotype instances.

3 Algorithms and Accuracy of Estimates

One indication that the haplotype problem might be practically more tractable is the amount of information in the haplotype data relative to the genotype data. To understand this, we can consider a pedigree with a fixed set of sampled individuals. Assume that there are two input data sets available, either the haplotype or the genotype data, for all the sampled individuals. Note that the alleles observed will be identical in both the haplotype and genotype data, so we are interested in the distribution that these data impose on the inheritance probabilities. By comparing the accuracy of the recombination estimates under these two data sets, we can get an idea for how useful the respective probability distributions are.

Let R_j be a random variable representing the number of recombinations in the whole pedigree that occur between loci $j-1$ and j. Similar to our notation before, $R_j = \sum_{i \in I} R_{i,j}^p + R_{i,j}^m$. We want to compute the distribution of R_j under both the genotype and haplotype inheritance probability distributions. These two inheritance distributions are different precisely because there are haplotypes and inheritance paths that are consistent with the genotype constraints but disallowed by the haplotype constraints.

These distributions are obtained by constructing a hidden Markov model for the linkage dependencies along the genome. At each locus, the HMM considers the constraints given by either the haplotype or genotype data (i.e. the haplotype data HMM is a variation on the Lander-Green algorithm [9]). We first use the forward-backward algorithm to compute the marginal inheritance probabilities for each locus using a hidden Markov model. Once we have the marginal probabilities, we can easily obtain the distribution for R_j.

3.1 General Haplotype and Genotype HMMs

The likelihood can be modeled using a hidden Markov model along the genome with inheritance paths as hidden states. An *inheritance path* is a graph with nodes being the alleles of individuals and directed edges between alleles that are inherited from parent to child. The transition probabilities are functions of θ and the number of recombinations between a given pair of inheritance graphs.

Given the data, we compute the marginal inheritance path probabilities at each site by using the forward-backward algorithm for HMMs. Sobel and Lange described a method for enumerating the inheritance paths compatible with the allele data observed at each locus [14]. There are at most $k = 2^{2|I \setminus F|}$ inheritance paths when $I \setminus F$ is the set of non-founder individuals, and both the forward and backward recursions do an $O(k^2)$ calculation at each site.

To compute the analogous probability for haplotype data, we use a similar HMM. For haplotypes, the hidden states must consider the *haplotype orientations*, which specify the parental origins of all the observed haplotypes. Notice that these orientations are not equivalent to inheritance paths, since they only specify inheritance edges between haplotyped individuals and their parents. For each of the $2^{2|H|}$ haplotype orientations, where H is the set of haplotyped individuals, we enumerate the inheritance paths compatible with the haplotype alleles, their orientations, and the pedigree relationships. Alternatively, each of the inheritance paths enumerated for the genotype algorithm induces a particular orientation on the haplotypes heterozygous for that locus (i.e. parental origin of the entire haplotype). Thus, the hidden states for the haplotype HMM are the cross-product of the orientations and the inheritance paths.

The haplotype HMM has transition probabilities that are nearly identical to the genotype HMM with the exception that transitions between inheritance paths with different haplotype orientations have probability zero. Recombinations are only allowed when they do not occur between typed haplotypes.

The forward-backward algorithm is also used on the haplotype HMM. However, there are $2^{2(|I|+|H|-|F|)}$ hidden states, yielding a slightly slower calculation. Fortunately, the haplotype recursions can be run simultaneous with the genotype recursions, meaning that the inheritance paths need only be enumerated once.

3.2 Haplotype Likelihoods in Linear Time

There is one obvious instance of the haplotyping problems where there are polynomial-time algorithms. Even though it is impractical to assume that we can sample genetic material from deceased individuals in a multi-generational pedigree, for a moment, let us consider the case where all the individuals in the pedigree are haplotyped.

The Elston-Stewart algorithm [5] for genotype data has a direct analogue for haplotype data. This algorithm calculates the likelihood via the belief propagation algorithm by eliminating individuals recursively from the bottom up. Each individual is "peeled off", after their descendants have been peeled off, by using a forward-backward algorithm on the HMM for the mother-father-child trio.

The haplotype version of this algorithm is linear when all the individuals are haplotyped, since each elimination step is conditionally independent of all the others. Given the parents' haplotypes, regardless of which was inherited from which grand-parent, the probability of the child's haplotype is independent of all other trios. Therefore, we can take a product over the likelihoods for all the trios, and compute each trio likelihood using a 4-state HMM. Then for k non-founding individuals, and l loci, this algorithm has $O(kl)$ running time.

This same intuition carries through to the minimum recombination problem, and each trio can be considered independent of the others. This contrasts with the genotype minimum recombination problem which is known to be hard, even when all the individuals are genotyped [10].

3.3 Results

To simulate realistic pedigree data, SNPs were selected from HapMap that span 100mb on both sides of a loosely-linked pair of sites. There are 40 SNPs total, with 20 tightly linked SNPs on each side of a strong recombination breakpoint having $\theta = 0.25$. The haplotypes for these SNPs were selected randomly from HapMap. Pedigree haplotype and genotype data were simulated for each child by uniformly selecting one of the parental alleles for the first locus, and subsequent loci were selected on the same parental haplotype with probability θ_j for each locus j. Inheritance was simulated for 500 simulation replicates.

The simulation yielded completely typed pedigrees. For each pedigree, we removed the genotype and haplotype information for increasing numbers of untyped individuals. For each instance of a specific number of untyped individuals, two values were computed on the estimated number of recombinations between the central pair of loci: the haplotype and genotype accuracies. Accuracy was computed as a function of the l_1 distance between the deterministic number of recombinations and the calculated distribution. Specifically, accuracy was $2 - \sum_{i \geq 0} |x_i - a_i|$, where x_i was the estimated probability for i recombinations and a_i was the deterministic indicator of whether there were i recombinations in the data simulated on the pedigree.

In all the instances we observed a trend where the best accuracy was obtained with haplotype data where everyone in the pedigree was haplotyped. For example, a five-individual pedigree with two half-siblings is shown in Figure 2, left panel. With the three founders untyped, the haplotype data yielded similar accuracy as the genotype data. Consider a three-generation pedigree having two parents, their two children, an in-law, and a grandchild for a total of six individuals, three of them founders. This pedigree has a similar trend in accuracy as the number of untyped founders increases, Figure 2, right panel. As the number of untyped individuals increases, the accuracies of genotype and haplotype estimates appear to converge.

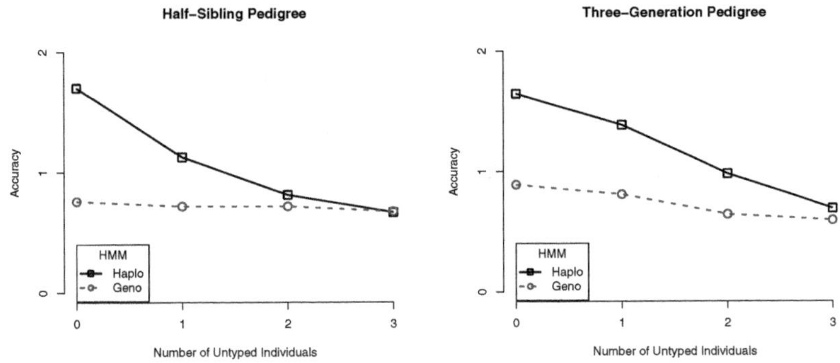

Fig. 2. Predicting Recombinations. The left panel is the average accuracy for predictions from a pedigree with two half-siblings and three parents. The right panel shows results from a six-individual, three-generation pedigree. In both cases, 500 simulation replicates were performed, and the average accuracy of estimates from the haplotype data is superior to those from genotype data. However, as the number of untyped founders increases, in both cases, the accuracy of estimates from haplotype data drop relative to the accuracy from genotype data. The accuracies of genotype and haplotype estimates appear to converge.

4 Discussion

Sequencing technologies would seem to solve the phasing problem by yielding haplotype data. However, if we wish to consider diploid inheritance with recombination, the phasing problem remains, even when we are given chromosome-length haplotype data. This is demonstrated by reduction of the phasing problem for genotypes to the phased version of the same problem for three common pedigree problems. This theoretical result is due largely to the unavailability of genetic material for deceased individuals.

Three pedigree calculations were discussed: likelihood, maximum probability, and minimum recombination. Each of these calculations on haplotype data have the same computational complexity as the same computation on genotype data. In the worst case, it takes only a single generation to remove the correlation between sites in the haplotype. This worst case provided the reduction that proves the the complexity results for the haplotype computations, and it worked equally well for all three pedigree computations.

The worst-case is not biologically realistic, since it requires roughly $2(m-1)$ recombinations for m sites in 4 meioses. This is very unlikely to occur under typical models for inheritance. To investigate more likely scenarios, sequences were simulated in a region of the genome surrounding a recombination breakpoint. From haplotype and genotype data, we estimated the distribution of the number of recombinations at the breakpoint and compared the estimates to the ground-truth for accuracy.

When typing everyone in the pedigree, the estimates from haplotype data were very accurate, because the haplotype data provides enough constraints to determine where the recombinations must have occurred. With decreasing numbers of typed individuals, the accuracy of haplotype-based estimates dropped until it seemed to converge to the genotype accuracy due to a lack of constraints. From the structure of the calculations, we observed that with fewer typed individuals there were more haplotype orientations to consider, and the haplotype calculation more closely resembled the genotype calculation. However, the haplotype calculation had more constraints and lost accuracy at a slower rate.

Several interesting open problems remain. First, approximation algorithms might be a useful approach for haplotypes on pedigrees. The existence of a linear-time algorithm when all individuals are haplotyped may suggest that the general haplotype problem instance could be amenable to approximation algorithms. Second, these proofs apply when there is no missing data in a genotyped individual (i.e. a proxy parent). The proof requires knowing whether the proxy parent is heterozygous or homozygous at each locus, and this is unknown when there is missing data. Third, there is an interesting case of mixed haplotypes and genotypes. For this case to be interesting, the ends of haplotypes must occur at different locations in different individuals in the pedigree. Otherwise, the haplotypes that start and end at the same positions in all individuals can easily be converted into multi-allelic genotypes, with an allele for each haplotype. The mixed haplotype-genotype problem is not amenable to the proof techniques used here. However, the haplotype HMM in Section 3.1 can easily be revised to handle the mixed case. This is important because the data produced by single polymer sequencing is more likely to resemble the mixed case than either the haplotype or the genotype cases.

Acknowledgments. I want to thank Richard Karp for reviewing a draft of the manuscript and the National Science Foundation for support through the Graduate Research Fellowship.

References

1. Barrett, J.C., Hansoul, S., Nicolae, D.L., Cho, J.H., Duerr, R.H., Rioux, J.D., Brant, S.R., Silverberg, M.S., Taylor, K.D., Barmada, M.M., et al.: Genome-wide association defines more than 30 distinct susceptibility loci for crohn's disease. Nature Genetics 40, 955–962 (2008)
2. Burdick, J.T., Chen, W., Abecasis, G.R., Cheung, V.G.: In silico method for inferring genotyeps in pedigrees. Nature Genetics 38, 1002–1004 (2006)
3. Chen, W.-M., Abecasis, G.R.: Family-based association tests for genomewide association scans. American Journal of Human Genetics 81, 913–926 (2007)
4. Coop, G., Wen, X., Ober, C., Pritchard, J.K., Przeworski, M.: High-Resolution Mapping of Crossovers Reveals Extensive Variation in Fine-Scale Recombination Patterns Among Humans. Science 319(5868), 1395–1398 (2008)
5. Elston, R.C., Stewart, J.: A general model for the analysis of pedigree data. Human Heredity 21, 523–542 (1971)
6. Eid, J., et al.: Real-Time DNA Sequencing from Single Polymerase Molecules. Science 323(5910), 133–138 (2009)
7. Geiger, D., Meek, C., Wexler, Y.: Speeding up HMM algorithms for genetic linkage analysis via chain reductions of the state space. Bioinformatics 25(12), i196 (2009)
8. Kirkpatrick, B., Halperin, E., Karp, R.M.: Haplotype inference in complex pedigrees. Journal of Computational Biology (2010) (in press)
9. Lander, E.S., Green, P.: Construction of multilocus genetic linkage maps in humans. Proceedings of the National Academy of Science 84(5), 2363–2367 (1987)
10. Li, J., Jiang, T.: An exact solution for finding minimum recombinant haplotype configurations on pedigrees with missing data by integer linear programming. In: Proceedings of the 7th Annual International Conference on Research in Computational Molecular Biology, pp. 101–110 (2003)
11. Ng, M.Y., Levinson, D.F., et al.: Meta-analysis of 32 genome-wide linkage studies of schizophrenia. Mol. Psychiatry 14, 774–785 (2009)
12. Piccolboni, A., Gusfield, D.: On the complexity of fundamental computational problems in pedigree analysis. Journal of Computational Biology 10(5), 763–773 (2003)
13. Romero, I.G., Ober, C.: CFTR mutations and reproductive outcomes in a population isolate. Human Genet. 122, 583–588 (2008)
14. Sobel, E., Lange, K.: Descent graphs in pedigree analysis: Applications to haplotyping, location scores, and marker-sharing statistics. American Journal of Human Genetics 58(6), 1323–1337 (1996)
15. Thatte, B.D.: Combinatorics of pedigrees (2006)
16. Xiao, J., Liu, L., Xia, L., Jiang, T.: Efficient Algorithms for Reconstructing Zero-Recombinant Haplotypes on a Pedigree Based on Fast Elimination of Redundant Linear Equations. SIAM Journal on Computing 38, 2198 (2009)

Haplotype Inference on Pedigrees with Recombinations and Mutations

Yuri Pirola[1], Paola Bonizzoni[1], and Tao Jiang[2]

[1] DISCo, Univ. degli Studi di Milano-Bicocca, Milan, Italy
{pirola,bonizzoni}@disco.unimib.it
[2] Department of Computer Science and Engineering,
University of California, Riverside, CA
jiang@cs.ucr.edu

Abstract. Haplotype Inference (HI) is a computational challenge of crucial importance in a range of genetic studies, such as functional genomics, pharmacogenetics and population genetics. Pedigrees have been shown a valuable data that allows us to infer haplotypes from genotypes more accurately than population data, since Mendelian inheritance restricts the set of possible solutions. In order to overcome the limitations of classic statistical haplotyping methods, a combinatorial formulation of the HI problem on pedigrees has been proposed in the literature, called Minimum-Recombinant Haplotype Configuration (MRHC) problem, that allows a single type of genetic variation events, namely recombinations. In this work, we define a new problem, called Minimum-Change Haplotype Configuration (MCHC), that extends the MRHC formulation by allowing also a second type of natural variation events: *mutations*. We propose an efficient and accurate heuristic algorithm for MCHC based on an L-reduction to a well-known coding problem. Our heuristic can also be used to solve the original MRHC problem and it can take advantage of additional knowledge about the input genotypes, such as the presence of *recombination hotspots* and different rates of recombinations and mutations. Finally, we present an extensive experimental evaluation and comparison of our heuristic algorithm with several other state-of-the-art methods for HI on pedigrees under several simulated scenarios.

1 Motivations

After the first draft of the human genome was published in 2000, a lot of research efforts have been devoted to the discovery of genetic differences among same-species individuals and to the characterization of their impact to the expression of different phenotypic traits such as disease susceptibility or drug resistance. Most of these efforts are driven by the *International HapMap Project* [14], which discovered, investigated and characterized millions of genomic positions (called *loci* or *sites*) where different individuals carry different genetic subsequences (called *alleles*). In practice, unordered pairs of alleles coming from both parents of each individual studied are routinely collected, since determining the parental source

V. Moulton and M. Singh (Eds.): WABI 2010, LNBI 6293, pp. 148–161, 2010.

of each allele is too much time-consuming and expensive to be performed on large studies [3]. The pairs of alleles located at a given set of loci of an individual are called the (multi-locus) *genotype* of the individual, while the sequence of alleles that were inherited from a single parent is called a *haplotype*. The advance of high-throughput and high-density genotyping technologies, combined with a consistent reduction of genotyping costs, had led to a great abundance of genotypic data. Such genotypes (also called SNP genotypes) are generally biallelic (*i.e.*, at each locus only two distinct alleles are observed in the population) and they will be the focus of this work. A number of association studies based on SNP genotypes have been carried out but, since haplotypes substantially increase the power of genetic variation studies [15], accurate and efficient computational prediction of haplotypes from genotypes is highly desirable. Mendelian inheritance laws, which govern the transmission of genetic material from parents to children, have been effectively used to improve the accuracy of haplotyping methods. However, the increasing density and length of SNP genotypes challenge classic statistics-based methods (such as Lander-Green [7] and Elson-Stewart [4] methods) because they do not scale well on large datasets and they do not take directly into account the presence of Linkage Disequilibrium among loci. Combinatorial formulations have been proposed to overcome such limitations. Among them, the most popular formulation is represented by the MINIMUM-RECOMBINANT HAPLOTYPE CONFIGURATION (MRHC) problem [12,9]. The aim of this formulation is the computation of a haplotype configuration which is consistent with an input genotyped pedigree and induces the minimum number of recombinations. The formulation naturally arises since recombinations are the most common form of variation events. However, with the progressive increase of the size of genetic variation studies, the incidence of other types of variation events (such as *mutations*) will inevitably become noticeable.

The above observation motivates the work in this paper, where the Haplotype Inference (HI) problem on pedigrees admitting recombination and mutation events, called MINIMUM-CHANGE HAPLOTYPE CONFIGURATION (MCHC), is studied. Polynomial-time exact algorithms for MCHC are unlikely to exist since it is possible to prove that MCHC is APX-hard even on simple instances. The main contribution of this paper is an efficient and accurate heuristic algorithm for MCHC. Our algorithm is based on an L-reduction [2] of MCHC to a fundamental problem of coding theory: the NEAREST CODEWORD PROBLEM (NCP) [2, probl. MS3]. Although NCP is theoretically hard to approximate [1], there exists several heuristics that compute near-optimal solutions of NCP in practice [6]. Our idea is to transform the instance of MCHC to an instance of NCP, to solve it with a custom-tailored version of a heuristic for NCP, and, finally, to reconstruct a solution of the original instance of MCHC from the solution of NCP. Our L-reduction guarantees that the transformation of the instance and the reconstruction of the solution are performed in polynomial-time while preserving the solution cost.

The work is structured as follows. First, in Section 2, we formalize the MINIMUM-CHANGE HAPLOTYPE CONFIGURATION problem and define the related basic terminology. Then, in Section 3, we present a heuristic algorithm based on an L-reduction from MCHC to NCP. Finally, in Section 4, we discuss the results of an experimental evaluation of our algorithm under several simulated scenarios and compare the accuracy and efficiency of our algorithm with those of several state-of-the-art methods for HI on pedigrees in the literature.

2 The Computational Problem

In this section we define the basic concepts and formalize the computational problem that will be studied in the rest of the work.

A *pedigree graph* is an oriented acyclic graph $P = (V, E)$ such that (i) vertices correspond to individuals and are partitioned into *male* and *female* vertices (*i.e.*, $V = M \cup F$, with M and F disjoint), (ii) each vertex has indegree 0 or 2, and (iii) if a vertex has indegree 2, then one edge must come from a male node and the other from a female node. For each edge $(p, c) \in E$, we say that p is a parent of c and c is an offspring (or child) of p. More precisely, we say that p is the father (mother, resp.) of c if p is male (female, resp.). A *trio* is a triplet (f, c, m) where f is the father and m is the mother of c. Individual f and individual m are said to be *mates* in such a trio. A pedigree graph contains a *mating loop* if there exists two nodes a and d such that they are connected by two distinct paths. A pedigree graph is a *tree pedigree* if it does not contain mating loops.

Let Σ be an ordered set $\langle l_1, \ldots, l_m \rangle$ of m loci and c an individual of the pedigree P. A *haplotype* of individual c is an m-dimensional vector over the set $\{0, 1\}$. The *genotype* g_c of individual c is an m-dimensional vector over the set $\{0, 1, 2\}$, where the i-th element (denoted with $g_c[i]$) represents the pair of alleles that individual c possesses at locus l_i. We follow the convention of encoding pair $\{0, 0\}$ as 0, $\{1, 1\}$ as 1, and $\{0, 1\}$ as 2.

A *genotyped* (*haplotyped*, respectively) *pedigree* is a pedigree such that every individual has been associated with a genotype (an ordered pair of haplotypes, respectively). We use g_c to denote the genotype associated with an individual c of a genotyped pedigree and $\langle h_c^0, h_c^1 \rangle$ the haplotypes associated with an individual c of a haplotyped pedigree. Moreover, we say that h_c^0 is the paternal haplotype of c and h_c^1 is the maternal haplotype of c. A haplotyped pedigree P_h is *consistent* with a genotyped pedigree P_g of the same set of individuals if for each individual c, the genotype g_c is resolved by the pair of haplotypes $\langle h_c^0, h_c^1 \rangle$. An individual is called a *founder* if its indegree is 0. Otherwise it is called a *non-founder*. The *grandparental source vector* of a non-founder individual c w.r.t. one of its parents p, is an m-long binary vector $s_{p,c}$ defined as follows. Let l_i be a locus of Σ. If p is the father (mother, resp.) of c, then $s_{p,c}[i] = 0$ if the allele of the paternal (maternal, resp.) haplotype of c at locus l_i has been inherited from the paternal haplotype of p. On the other hand, $s_{p,c}[i] = 1$ if the allele has been inherited from the maternal haplotype of p. Given a genotyped pedigree P_g, a (consistent) *haplotype configuration* of P_g is a pair (P_h, S) where P_h is a (consistent) haplotyped pedigree of P_g and S an assignment of two grandparental source vectors to each individual of P.

The Haplotype Inference (HI) problem on pedigrees asks for a haplotype configuration (or the set of haplotype configurations) consistent with a given genotyped pedigree. However, since there can exist an exponential number of consistent haplotype configurations, additional constraints are generally imposed. A particularly successful approach is the formulation that attempts to minimize the number of genetic variation events that are induced in the resulting haplotyped pedigree [12,9]. Two types of variation events will be considered, *recombinations* and *mutations*, defined as follows. Let (P_h, S) be a consistent haplotype configuration of a genotyped pedigree P_g. The haplotype configuration induces (or contains) a *recombination* at locus l_i between an individual c and one of its parents p if $s_{p,c}[i] \neq s_{p,c}[i+1]$. The haplotype configuration induces (or contains) a *mutation* at locus l_i between c and its parent p if $h_c^j[i] \neq h_p^s[i]$ where $s = s_{p,c}[i]$ and $j = 0$ ($j = 1$, resp.) if p is the father (mother, resp.) of c.

In this work we are interested in the computational problem of computing a haplotype configuration that is consistent with a given genotyped pedigree and that induces the minimum number of variation events. We call such a problem MINIMUM-CHANGE HAPLOTYPE CONFIGURATION (MCHC) problem. The MCHC problem is a generalization of two problems proposed in the literature: the MINIMUM-RECOMBINANT HAPLOTYPE CONFIGURATION (MRHC) problem (where only recombinations are allowed [9]), and the MINIMUM-MUTATION HAPLOTYPE CONFIGURATION (MMHC) problem (where only mutations are allowed [16]). Differently from [16], in the following we do not restrict the number of mutations at each locus (among all individuals) to be at most one. It is possible to prove that MCHC is APX-hard even on instances where genotypes are defined on only 2 loci or where each individual has at most one mate and one child. Due to the page limit, the proof is deferred to the full version of this paper.

3 A Heuristic Algorithm for MCHC

The presentation of the heuristic algorithm that we propose is divided into three parts. First, we give an extension of the system of linear equations over the field $\mathbb{Z}_2$ proposed by Xiao *et al.* [17] for representing the set of haplotype configurations that are consistent with the input genotyped pedigree. In the extended system that we propose, recombinations and mutations are explicitly modeled as variables of the equations. In the second part, we establish an L-reduction from MCHC to the well-known NEAREST CODEWORD PROBLEM (NCP) by splitting the system into two parts where one part contains only variables needed for the haplotype reconstruction and the other contains only recombination and mutation variables. Finally, we present a tailored version of a well-known heuristic algorithm for NCP. Using this heuristic, we can guarantee that a feasible solution for NCP (and hence for MCHC) is found.

3.1 A System of Linear Equations for MCHC

In this part, we first illustrate the linear system over $\mathbb{Z}_2$ proposed in [17] for the HI problem where no recombinations or mutations are permitted (*i.e.*, the zero-recombinant haplotype configuration problem or ZRHC), and then we describe how it can be extended to accommodate recombinations and mutations events. For simplicity of presentation, we denote with the symbol + the addition over $\mathbb{Z}_2$ instead of using $\oplus$.

3.2 A Linear System for ZRHC

Computing the paternal haplotypes of all individuals is sufficient to fully describe the haplotyped pedigree because the maternal haplotype can be reconstructed from the paternal haplotype and the genotype of the individual. Therefore, we introduce a variable $h_i[l]$ for each individual i and locus l which represents the allele present at locus l of the paternal haplotype of i. Secondly, we need to represent the grandparental source. Let i be an individual and p one of its parents. Since no recombinations are admitted, the grandparental source is denoted as a single variable $s_{p,i}$. Variable $s_{p,i}$ is equal to 0 if i has inherited from p the paternal haplotype of p, or 1 otherwise. To express concisely the linear equations, we need two additional sets of constants: the w- and the d-constants. For each locus l and individual i, constant $w_i[l]$ is equal to 0 if i is homozygous at locus l, and 1 otherwise. For each locus l and pair of individuals p and i such that p is a parent of i, constant $d_{p,i}[l]$ is equal to 0 if p is the father of i and equal to $w_i[l]$ if p is the mother of i. Finally, since the paternal haplotype (and hence the maternal haplotype) is known at homozygous loci, we set $h_i[l] = g_i[l]$ for every individual i and locus l such that $g_i[l] \neq 2$.

A case-by-case analysis shows that any solution of the following linear system over $\mathbb{Z}_2$ is a zero-recombinant haplotype configuration consistent with the genotyped pedigree (and vice versa) [17]. For all loci l and individuals i,

$$\begin{cases} h_p[l] + s_{p,i} \cdot w_p[l] = h_i[l] + d_{p,i}[l] & \text{for each parent } p \text{ of } i \\ h_i[l] = g_i[l] & \text{if } g_i[l] \neq 2 \\ w_i[l] = 0 & \text{if } g_i[l] \neq 2 \\ w_i[l] = 1 & \text{if } g_i[l] = 2 \\ d_{p,i}[l] = 0 & \text{if } p \text{ is the father of } i \\ d_{p,i}[l] = w_i[l] & \text{if } p \text{ is the mother of } i \end{cases} \tag{1}$$

Notice that, if the pedigree has n members and the genotypes are defined over a set of m loci, then we have nm h-variables, at most $2n$ s-variables, and at most $2nm$ equations.

3.3 A Linear System for MCHC

We now show how the previous linear system can be modified for representing all the consistent haplotype configurations that may contain recombinations and mutations.

To accommodate recombinations, we introduce a set of δ-variables defined as follows. For each locus l, variable $\delta_{p,i}[l]$ is equal to 1 if a recombination has occurred at locus l between an individual p and one of its children i, and 0 otherwise. The grandparental source vector of a consistent haplotype configuration can be expressed as a (linear) function of an s-variable and a subset of

δ-variables. In particular, by induction on l, it is easy to prove that the grand-parental source of i w.r.t. p at locus l, $s_{p,i}[l]$, is equal to $s_{p,i}+\sum_{j=1}^{l}\delta_{p,i}[j]$. Denote as $\Delta_{p,i}[l]$ the sum $\sum_{j=1}^{l}\delta_{p,i}[j]$. By replacing $s_{p,i}$ with $(s_{p,i}+\Delta_{p,i}[l])$ in Eq. 1, we obtain a linear system that represents all the haplotype configurations consistent with the genotyped pedigree and allows recombination events. Since mutations are point events that replace an allele inherited from the parent with another allele, it suffices to add a term in the first equation of the original linear system to model mutation events. We denote this term as $\mu_{p,i}[l]$ and set $\mu_{p,i}[l]=1$ if a mutation at locus l between p and i has occurred, and $\mu_{p,i}[l]=0$ otherwise. The following lemma is straightforward.

Lemma 1. *Let P_g be a genotyped pedigree. Then each solution of the system*

$$\begin{cases} h_p[l] + (s_{p,i} + \Delta_{p,i}[l]) \cdot w_p[l] = h_i[l] + d_{p,i}[l] + \mu_{p,i}[l] & \text{for each parent } p \text{ of } i \\ h_i[l] = g_i[l] & \text{if } g_i[l] \neq 2 \\ w_i[l] = 0 & \text{if } g_i[l] \neq 2 \\ w_i[l] = 1 & \text{if } g_i[l] = 2 \\ d_{p,i}[l] = 0 & \text{if } p \text{ is the father of } i \\ d_{p,i}[l] = w_i[l] & \text{if } p \text{ is the mother of } i \end{cases} \tag{2}$$

concerning all loci l and individuals i represents a haplotype configuration consistent with P_g that admits recombination and mutation events. Conversely, a haplotype configuration consistent with P_g that admits recombination and mutation events is represented by a solution of the linear system.

By construction, a haplotype configuration that induces k variation events is represented by a solution of the linear system where exactly k δ- and μ-variables are non-zero.

3.4 Reducing MCHC to NCP

The Nearest Codeword Problem is the problem of coding theory that reconstructs the original codeword of a given received message by minimizing the Hamming distance between them. More formally, given an $r \times n$ matrix H over $\mathbb{Z}_2$, and a column vector $q \in \mathbb{Z}_2^r$, the Nearest Codeword Problem [2, probl. MS3] asks for a vector $e \in \mathbb{Z}_2^n$ with the minimum number of non-zero entries such that $H \cdot e = q$. The number of non-zero entries of a vector v is called the *weight* of the vector and is denoted as $\|v\|$.

The basic idea of our reduction is to split the linear system of Lemma 1 into two linear systems: one containing only h- and s-variables, and the other one containing only δ- and μ-variables. The second part of the system is, directly, an instance of NCP.

Since all $w_i[l]$ and $d_{p,i}[l]$ assume constant (predetermined) values, we can write the linear system of Eq. 2 as the following matricial equation:

$$A_{h,s} \cdot x_{h,s} + A_{\delta,\mu} \cdot x_{\delta,\mu} = b \tag{3}$$

where: $x_{h,s}$ is the column vector of the h- and s-variables, $x_{\delta,\mu}$ the column vector of the δ- and μ-variables, $A_{h,s}$ the $n \times m_1$ matrix of the coefficients of the h- and s-variables, $A_{\delta,\mu}$ the $n \times m_2$ matrix of the coefficients of the δ- and μ-variables, and b a column vector composed by constant entries.

Let k be the rank of the matrix $A_{h,s}$ and $A_{h,s}^T$ be its transpose. Suppose w.l.o.g. that the first k rows of $A_{h,s}$ are linearly independent. Now, we construct the instance of NCP associated to an instance of MCHC as follows. Let $B = \{v_1, \ldots, v_r \mid v_i \in \mathbb{Z}_2^n\}$ be a basis of the vector space $\ker(A_{h,s}^T) = \{y \in \mathbb{Z}_2^n \mid A_{h,s}^T \cdot y = \mathbf{0}\}$, where $\mathbf{0}$ denotes the all-zero column vector. Collate vectors v_i to form a $r \times n$ matrix V such that the i-th row is equal to v_i^T. Then, the instance I' of NCP associated with an instance $I = (A_{h,s}, A_{\delta,\mu}, x_{h,s}, x_{\delta,\mu}, b)$ of MCHC is the pair $I' = (H, q)$ where $H = VA_{\delta,\mu}$ and $q = Vb$. Clearly, the transformation of I into I' can be computed in polynomial-time via Gaussian elimination (to compute V) and two matrix multiplications (to compute H and q).

We complete the L-reduction from MCHC to NCP by proving the following two lemmas. Lemma 2 illustrates how to reconstruct in polynomial-time a solution of an MCHC instance given a solution for the associated NCP instance, and Lemma 3 shows how to compute (in polynomial-time) a solution for an instance I' of NCP associated with an instance I of MCHC given a solution for I. Since both above transformations preserve the cost of solutions, the reduction is an L-reduction with parameters $\beta = \gamma = 1$. See [2, Def. 8.4] for the formal definition of L-reduction and an explanation of these parameters. Due to the page limit, the proofs of the lemmas are deferred to the full version of the paper.

Lemma 2. *Let $I = (A_{h,s}, A_{\delta,\mu}, x_{h,s}, x_{\delta,\mu}, b)$ be an instance of* MCHC *and $I' = (H, \tilde{y})$ the* NCP *instance associated with I. Then, given a solution e of* NCP *on I', it is possible to compute in polynomial-time a haplotype configuration $(\widehat{x}_{h,s}, \widehat{x}_{\delta,\mu})$ of I that induces $\|e\|$ variation events.*

Lemma 3. *Let $S = (\widehat{x}_{h,s}, \widehat{x}_{\delta,\mu})$ be a solution of* MCHC *on the instance $I = (A_{h,s}, A_{\delta,\mu}, x_{h,s}, x_{\delta,\mu}, b)$ and $I' = (H, q)$ the* NCP *instance associated with I. Then, vector $e := \widehat{x}_{\delta,\mu}$ is a solution of* NCP *on I'.*

The following corollary easily follows from Lemma 2 and Lemma 3.

Corollary 1. MCHC *is L-reducible to* NCP *with parameters $\beta = \gamma = 1$.*

3.5 The Heuristic Algorithm

In this section, we present an efficient heuristic algorithm that solves the MCHC problem. In addition, this heuristic can be also used to solve the MRHC and MMHC problems by restricting the types of variation events that are allowed. An implementation of the heuristic described below can be freely downloaded from the web page `http://www.algolab.eu/Heu-MCHC/`.

The algorithm is based on the above L-reduction from MCHC to NCP. Since NCP $\notin$ APX [1], there do not exist algorithms that can guarantee a good (*i.e.*, constant) approximation ratio unless P = NP. Nevertheless, it has been shown that the *sum-product* (SP) algorithm [6] (independently proposed in Artificial Intelligence as the *belief-propagation* algorithm [11]) is an effective and efficient heuristic for NCP. The SP algorithm computes an approximation of the likelihood that each bit of the received message has been "flipped" during the transmission of the message. Such an approximation is computed by employing the set of parity constraints of the linear code and a vector q (called *syndrome*) representing the constraints that are not satisfied by the received message.

Our idea is to consider the variation events (recombinations and mutations) as the "errors" that we have to reconstruct and, once the "errors" (variation events) have been determined, it is easy to reconstruct the haplotyped pedigree (by Gaussian elimination). The L-reduction in Corollary 1 formalizes this idea. The set of parity constraints and the syndrome q are obtained from the genotyped pedigree (represented by the matrices $A_{h,s}$ and $A_{\delta,\mu}$) as illustrated in the previous section. The likelihoods computed by the SP algorithm on this instance represents the likelihoods that each δ- or μ-variable is equal to 1. In other words, for each possible variation event, it computes the likelihood that the event has occurred on the pedigree.

Our heuristic iteratively adds the most likely variation event (as computed by the SP algorithm) to a set E of *imputed* variation events until a haplotype configuration that induces exactly the imputed events can be found. Given a set of variation events E, the reconstruction of the haplotype configuration that induces E can be performed efficiently. Indeed, it suffices to solve the linear system of Lemma 1 with the δ- and μ-variables assigned to 1 if the corresponding events (the mutations or the recombinations they represent) belong to E, or 0 otherwise. Initially, no variation events are imputed (thus $E = \varnothing$) while a set N contains all the possible variation events (represented by the corresponding δ- and μ-variables). For each binary linear code, the set of parity constraints is represented by a particular binary matrix H, called the *check matrix*, such that $H \cdot y = \mathbf{0}$ if and only if y is a valid codeword. In our L-reduction, the check matrix associated with the MCHC instance is computed as $H = V \cdot A_{\delta,\mu}$ for some matrix V. As a consequence, matrix H has the same number of columns as $A_{\delta,\mu}$, each of which is associated with a δ- or μ-variable. We associate each column i of H with the δ- or μ-variable that is associated with the i-th column of $A_{\delta,\mu}$. For simplicity, we identify each column i of H with the associated variable.

The haplotype configuration is computed in two steps: first the set of variation events E that makes the reconstruction of a haplotype configuration possible is computed, then the haplotype configuration is recovered using the imputed events E. The first step iteratively constructs the set of variation events. Using the SP algorithm, it computes an event e^* that most likely is induced in a haplotype configuration consistent with the pedigree. If more than one event have the maximum likelihood, one of them is chosen at random. Once e^* has been determined, the corresponding δ- or μ-variable is set to 1, and the syndrome is updated according to the check matrix H. Then, the column of H associated with event e^* can be removed, and e^* can be moved from the set of possible events N to the set of imputed events E. Based on the remaining parity constraints, we check if the presence (or absence) of other variation events is implied by e^* and the other events contained in E. This check can be performed by the Gaussian elimination algorithm. This step ends if all the remaining parity constraints are satisfied. The second step reconstructs the haplotype configuration consistent with the input genotyped pedigree by solving the linear system of Eq. 2 using, as a partial solution, the set E of imputed events.

An important remark is in order. The SP algorithm considers as an additional input the prior probability that each variation event e has occurred. Although we have not incorporated this feature into the current algorithm, it could be extremely useful to model *recombination hotspots* (by increasing the prior probability of recombination events in regions where recombinations occurs more frequently), to differentiate the rates of recombinations and mutations (by increasing the prior probability of a recombination event with respect to a mutation event), and/or to model additional knowledge about the input genotypes. This feature of the SP algorithm could also allow us to combine the combinatorial formulation of the problem presented here with some elements of statistics-based methods.

The time complexity of the heuristic depends on several parameters. Let n be the number of individuals of the genotyped pedigree and m the number of loci. The NCP instance I' is calculated by the Gaussian elimination algorithm on $A^T_{h,s}$ and by two matrix multiplications, requiring $O(n^3m^3)$ time. The check-matrix H has $O(nm)$ rows and at most $4nm$ columns (one for each variation event). Therefore the reduction from the pedigree to the NCP instance is computed in $O(n^3m^3)$ time. The time required by each iteration is bounded by $O(n^3m^3)$ since the check of the existence of predetermined events (by Gaussian elimination) requires $O(n^3m^3)$ time, the SP algorithm requires linear time in the number of one-entries of matrix H, and the other operations that update parity constraints and the syndrome can be accomplished in $O(n^2m^2)$ time. The resolution of the final linear system can be performed in $O(n^3m^3)$ time by the Gaussian elimination algorithm. Let k be the number of events that are imputed, then the overall time complexity of the heuristic is $O(kn^3m^3)$.

4 Experimental Results

Our approach has been experimentally analyzed under several simulated scenarios. The experimental analysis is divided into two parts. In the first part, we evaluate the accuracy and efficiency of our heuristic on randomly generated MCHC instances. In the second part, we compare the performance of our heuristic with that of three state-of-the-art approaches for MRHC and MMHC: PedPhase v2.1 [10], SimWalk2 [13], and MMPhase [16].

4.1 Evaluation on Random Instances

The first part of our experimentation involves randomly generated instances under several choices of 4 parameters: pedigree size (n), the number of loci (m), recombination probability (θ_r), and mutation probability (μ_r). For each choice of the parameters, we generated 30 haplotype configurations on 6 different random pedigree graphs. We applied a variation event at each locus with probability θ_r for recombinations and μ_r for mutations. For each instance, we ran our heuristic 10 times and we picked the solution with the minimum number of induced events.

We evaluated the quality of the results considering *phase error* (the ratio between the number of incorrectly predicted haplotype alleles and twice the number of heterozygous loci) and *approximation ratio* (the ratio between the number of predicted events and the number of generated events). The approximation ratio can be less than 1.0 because the generated haplotype configuration might be suboptimal. Finally, we also evaluated the total running time required by the heuristic on all 10 executions.

We chose a base set of values for the parameters n, m, θ_r, and μ_r and we conducted three series of tests. In each series, we modified the value of one of these parameters: pedigree size in the first, genotype length in the second, and the two probabilities θ_r and μ_r in the third. The base values were: pedigree size $n = 40$, number of loci $m = 40$, recombination probability $\theta_r = 0.02$, and mutation probability $\mu_r = 0.004$. The detailed results of the three series of tests are summarized in Table 1. In the first series of tests, we varied the pedigree size n and analysed the cases $n = 40$, $n = 60$, and $n = 100$ on both tree pedigrees and "general" pedigrees (*i.e.*, pedigree with mating loops). In all cases, the heuristic never required more than 6 minutes (169 seconds on average) on a standard PC with a 1.66GHz CPU and 2GB of main memory and it always found a haplotype configuration that induces fewer variation events than the generated one (*i.e.*, the approximation ratio is always smaller than 1.0). Although this fact does not imply that the heuristic computed the optimal solution, it increases our confidence in the soundness of the approach. The values of the quality measures are similar in all cases and, on average, are equal to 0.02 and 0.97 for phase error and approximation ratio, respectively. In the second series of tests, we varied the number of loci m and considered the following cases: $m = 40$, $m = 60$, and $m = 100$. Similarly to the previous series, we obtained 0.039 as the average phase error and 0.96 as the average approximation ratio, with an average running time of 182 seconds. In the third series of tests, we varied the probabilities of recombinations and mutations in the range of $(0.02, 0.004)$ to $(0.10, 0.02)$. In this case, the quality of the results decreases with the increase of the number of generated events. The worst results were obtained when recombination and mutation probabilities were at the maximum. Note that when this happens, the generated haplotype configuration significantly deviates from the parsimony principle that MCHC assumes. In fact, our heuristic reconstructs a solution with much fewer events than the generated haplotype configuration (in such a situation the average approximation ratio is 0.85).

4.2 Comparison with State-of-the-Art Methods

In the second part of the experimental evaluation, we compare the accuracy and efficiency of our heuristic with those of some state-of-the-art approaches

Table 1. Summary of the results obtained by our heuristic on randomly generated instances. Each table reports the quality and performance measures of the heuristic on a subset of instances where a parameter has been varied. The default settings of the parameters are: pedigree size $n = 40$, number of loci $m = 40$, recombination probability $\theta_r = 0.02$, and mutation probability $\mu_r = 0.004$.

(a) Increasing pedigree size (n)

	Tree pedigrees			General pedigrees			
Pedigree size $n =$	40	60	100	40	60	100	Mean
Avg. no. of generated events	22.0	30.4	55.2	25.5	35.8	63.6	38.7
Avg. no. of predicted events	21.3	29.5	53.2	24.5	34.9	61.8	37.5
Avg. phase error	0.027	0.029	0.028	0.022	0.024	0.024	0.026
Avg. approximation ratio	0.968	0.975	0.965	0.963	0.975	0.972	0.970
Avg. time (in seconds)	36	73	265	62	118	460	169

(b) Increasing the number of loci (m)

	Tree pedigrees			General pedigrees			
Number of loci $m =$	40	60	100	40	60	100	Mean
Avg. no. of generated events	24.0	34.7	53.2	26.4	38.0	61.0	39.6
Avg. no. of predicted events	23.1	33.0	51.2	25.7	36.9	59.6	38.3
Avg. phase error	0.035	0.057	0.042	0.026	0.026	0.044	0.039
Avg. approximation ratio	0.964	0.956	0.964	0.975	0.972	0.976	0.968
Avg. time (in seconds)	41	95	247	76	148	485	182

(c) Increasing recombination and mutation probabilities (θ_r and μ_r)

	Tree pedigrees			General pedigrees			
Recombination prob. $\theta_r =$	0.02	0.04	0.10	0.02	0.04	0.10	
Mutation probability $\mu_r =$	0.004	0.01	0.02	0.004	0.01	0.02	Mean
Avg. no. of generated events	24.5	48.8	111.5	24.9	48.9	121.8	63.4
Avg. no. of predicted events	23.8	45.7	94.8	24.0	45.8	105.3	56.6
Avg. phase error	0.035	0.061	0.114	0.020	0.053	0.099	0.064
Avg. approximation ratio	0.973	0.937	0.848	0.963	0.939	0.866	0.921
Avg. time (in seconds)	45	74	164	63	86	248	113

for HI on pedigrees. Popular approaches to HI on pedigrees do not allow for both recombinations and mutations at the same time. Therefore, we separately considered two classes of algorithms. The first one consists of algorithms for MRHC (*i.e.*, only recombinations are allowed) and the second class consists of algorithms for MMHC (*i.e.*, only mutations are allowed). We adapted our heuristic algorithm to the two problems by keeping only the variables associated with the type of events that are allowed (δ-variables for MRHC and μ-variables for MMHC).

Comparison with MRHC *Algorithms.* Several algorithms for MRHC have been proposed in the literature. For our comparison, we chose two popular approaches with different computational characteristics: PedPhase v2.1 [10] (an exact ILP-based algorithm) and SimWalk2 [13] (a popular statistical approach for HI). We generated 750 instances using SimPed [8], a simulation program for the generation of haplotyped pedigrees based on user-supplied biological information

Table 2. Summary of the comparison with other methods for MRHC on 750 instances whose sizes vary from pedigrees with 8 members and 10 loci to pedigrees with 100 members and 100 loci. For each method, we report the number of instances that have been solved within an hour of computation (*i.e.*, completed instances), the average number of predicted recombinations, the average phase error, and the average running time. To facilitate the comparison, for each row, we report in brackets the same performance measure values obtained by our heuristic on the instances completed by the other two methods.

	Heuristic	PedPhase 2.1 [10]		SimWalk2 [13]	
Completed instances	750	565		495	
Avg. no. of recombinations	26.42	14.25	(14.27)	17.37	(7.21)
Avg. phase error	0.030	0.030	(0.031)	0.037	(0.029)
Avg. running time (s)	4.7	35.8	(1.0)	787.7	(0.2)

Table 3. Summary of the comparison with another method for MMHC on 300 instance whose sizes vary from pedigrees with 50 members and 50 loci to pedigrees with 150 members and 150 loci. For each method, we report the number of instances that have been solved within an hour (*i.e.*, completed instances), the average number of computed mutations, the average phase error, and the average running time.

	Heuristic	MMPhase [16]
Completed instances	298	297
Avg. no. of mutations	38.83	37.77
Avg. phase error	0.0030	0.0030
Avg. running time (s)	483.58	167.54

(such as intramarker distances and allele frequencies). The same biological information have then been used to correctly initialize the input parameters of SimWalk2. The instance sizes ranged from pedigrees with 8 members and 10 loci to pedigrees with 100 members and 100 loci. We limited the running time on each instance to 1 hour. Our heuristic was the only method that completed all the 750 instances within this time limit. PedPhase completed 565 instances and SimWalk2 only 495 of them. PedPhase took over 5 hours to solve the 565 instances that it was able to tackle, while our heuristic on the same instances took only 575 seconds. Our heuristic was able to compute a solution with the same number of recombinations as PedPhase (*i.e.*, an optimal solution) in 560 of the 565 cases. SimWalk2 is a much slower approach; it took nearly 108 hours of computation while our heuristic completed the same set of instances in less than 90 seconds. Moreover, SimWalk2 never computed a solution with fewer recombinations than our heuristic. On the other hand, the average phase errors of the three approaches are almost identical (PedPhase 0.030, SimWalk2 0.037, and our heuristic 0.030), implying that the sets of recombinations computed by the three approaches, albeit different, are similar. A summary of the results is presented in Table 2.

Comparison with MMPhase. We compared our heuristic with MMPhase [16], the only other algorithm that has been explicitly proposed for MMHC in the literature (to the best of our knowledge). MMPhase is an ILP-based approach for MMHC with two restrictions: the model explicitly forbids two mutations at the same locus in different individuals (called the *infinite-site assumption*) and the current implementation is only able to handle tree pedigrees. Therefore, we generated 300 random instances of different sizes according to these restrictions. In particular we considered 4 different pedigree sizes (50, 75, 100, 150) and 3 different numbers of loci (50, 100, 150) and we generated 25 instances for each possible combination of the two parameters. The comparison revealed that MMPhase is noticeably faster than our heuristic (on average MMPhase required 167 seconds per instance vs. 483 seconds for our heuristic). However we observe that MMPhase exploits the infinite-site assumption in order to reduce the solution space, while our method allows more than one mutation on the same locus. Moreover, while MMPhase was able to solve 297 of the 300 instances within an hour, our method was able to solve 298 instances in the same time limit. Although our heuristic obtained solutions with slightly more mutations than MMPhase on 38 of the 298 instances, the average phase errors of the two methods are identical (0.0030). A summary of the comparison results is presented in Table 3.

5 Conclusion

In this paper, we presented a heuristic method for the haplotype inference problem on pedigrees allowing two types of variation events: recombinations and mutations. The experimental evaluation under several simulated scenarios showed that the heuristic is both accurate and efficient. The heuristic also compares favorably with several other state-of-the-art methods. It is faster than (but as accurate as) the other methods that consider only recombinations. Moreover, the only method considered in this study that is faster than our heuristic (MMPhase, which allows only point mutations) requires and exploits more restrictive assumptions about the input data than our method. The heuristic algorithm could handle moderately large pedigrees very well (in some of our tests, it was able to process tree pedigrees with 50 individuals and 1000 loci in approximately 2.5 hours of computation time on a standard PC). However, it cannot be applied directly to genome-scale data with millions of loci. Fortunately, the haplotype block structure observed in the human genome [5] provides a straightforward way of partitioning long genotypes into short blocks which can be readily handled by our method.

Acknowledgments. This research was supported in part by FAR MIUR 60% grant "Algorithmic methods and combinatorial structures in Bioinformatics" (Univ. di Milano-Bicocca) to YP and PB, and NIH grant 2R01LM008991 and NSF grant IIS-0711129 to TJ. Part of the work was done when YP was visiting at University of California, Riverside. We would like to thank Wei-Bung Wang for valuable discussions and sharing the MMPhase code with us.

References

1. Arora, S., Babai, L., Stern, J., Sweedyk, Z.: The hardness of approximate optima in lattices, codes, and systems of linear equations. J. of Computer and System Sciences 54(2), 317–331 (1997)
2. Ausiello, G., Crescenzi, P., Gambosi, G., Kann, V., Marchetti-Spaccamela, A., Protasi, M.: Complexity and Approximation: Combinatorial optimization problems and their approximability properties. Springer, Heidelberg (1999)
3. Bonizzoni, P., Della Vedova, G., Dondi, R., Li, J.: The haplotyping problem: An overview of computational models and solutions. J. of Computer Science and Technology 18(6), 675–688 (2003)
4. Elson, R.C., Stewart, J.: A general model for the analysis of pedigree data. Human Heredity 21, 523–542 (1971)
5. Gabriel, S.B., Schaffner, S.F., Nguyen, H., Moore, J.M., et al.: The structure of haplotype blocks in the human genome. Science 296(5576), 2225–2229 (2002)
6. Gallager, R.G.: Low-Density Parity-Check Codes. MIT Press, Cambridge (1963)
7. Lander, E., Green, P.: Construction of multilocus genetic linkage maps in human. Proceedings of the National Academy of Sciences USA 84, 2363–2367 (1987)
8. Leal, S.M., Yan, K., Müller-Myhsok, B.: SimPed: a simulation program to generate haplotype and genotype data for pedigree structures. Human heredity 60(2), 119–122 (2005)
9. Li, J., Jiang, T.: Efficient inference of haplotypes from genotypes on a pedigree. J. of Bioinformatics and Computational Biology 1(1), 41–69 (2003)
10. Li, J., Jiang, T.: Computing the minimum recombinant haplotype configuration from incomplete genotype data on a pedigree by integer linear programming. J. of Computational Biology 12(6), 719–739 (2005)
11. Pearl, J.: Reverend Bayes on inference engines: A distributed hierarchical approach. In: Proc. of the American Ass. of Artificial Intelligence National Conference on AI, Pittsburgh, PA, pp. 133–136 (1982)
12. Qian, D., Beckmann, L.: Minimum-recombinant haplotyping in pedigrees. American J. of Human Genetics 70(6), 1434–1445 (2002)
13. Sobel, E., Lange, K.: Descent graphs in pedigree analysis: applications to haplotyping, location scores, and marker-sharing statistics. American J. of Human Genetics 58(6), 1323–1337 (1996)
14. The International HapMap Consortium: A second generation human haplotype map of over 3.1 million SNPs. Nature 449(7164), 851–861 (2007)
15. Trégouët, D.A., König, I.R., Erdmann, J., et al.: Genome-wide haplotype association study identifies the SLC22A3-LPAL2-LPA gene cluster as a risk locus for coronary artery disease. Nature genetics 41(3), 283–285 (2009)
16. Wang, W.B., Jiang, T.: Efficient inference of haplotypes from genotypes on a pedigree with mutations and missing alleles (extented abstract). In: Kucherov, G., Ukkonen, E. (eds.) CPM 2009. LNCS, vol. 5577, pp. 353–367. Springer, Heidelberg (2009)
17. Xiao, J., Liu, L., Xia, L., Jiang, T.: Efficient algorithms for reconstructing zero-recombinant haplotypes on a pedigree based on fast elimination of redundant linear equations. SIAM J. on Computing 38(6), 2198–2219 (2009)

Identifying Rare Cell Populations in Comparative Flow Cytometry

Ariful Azad[1], Johannes Langguth[2], Youhan Fang[1], Alan Qi[1], and Alex Pothen[1]

[1] Dept. Computer Science, Purdue University, West Lafayette, IN 47907, USA
[2] Department of Informatics, University of Bergen, Bergen, Norway
{aazad,yfang,alanqi,apothen}@cs.purdue.edu, johannes.langguth@ii.uib.no

Abstract. Multi-channel, high throughput experimental methodologies for flow cytometry are transforming clinical immunology and hematology, and require the development of algorithms to analyze the high-dimensional, large-scale data. We describe the development of two combinatorial algorithms to identify rare cell populations in data from mice with acute promyelocytic leukemia. The flow cytometry data is clustered, and then samples from the leukemic, pre-leukemic, and Wild Type mice are compared to identify clusters belonging to the diseased state. We describe three metrics on the clustered data that help in identifying rare populations. We formulate a generalized edge cover approach in a bipartite graph model to directly compare clusters in two samples to identify clusters belonging to one but not the other sample. For detecting rare populations common to many diseased samples but not to the Wild Type, we describe a clique-based branch and bound algorithm. We provide statistical justification of the significance of the rare populations.

Keywords: flow cytometry, edge cover, clique, mixture modeling, KL divergence, acute promyelocytic leukemia (APL).

1 Introduction

We describe two algorithms to identify rare cell populations characteristic of diseases such as leukemia by analyzing flow cytometric data obtained from diseased and healthy samples. The recent development of high-throughput, multi-channel flow cytometry creates high-dimensional and large-scale data that requires the concomitant development of algorithms for comparative analyses of data from diseased and healthy samples, and from diseased samples at various stages of disease. Specifically, we need algorithms that can match cell populations among diseased samples, and differentiate between cell populations that belong to diseased and healthy states. Such studies could distinguish cancer cells from healthy cells, and identify cancer stem cells that are responsible for generating new cancerous cells, which could lead to therapies targeting such cells.

In flow cytometry, fluorescently labeled antibodies are bound to antigens on the cell, and on excitation with a laser as cells flow in a fluid stream, the fluorochrome emits light of a specific wavelength, thus identifying the cell

V. Moulton and M. Singh (Eds.): WABI 2010, LNBI 6293, pp. 162–175, 2010.

populations that express the antigen. Flow cytometry is routinely used in the diagnosis of diseases and has many applications in clinical practice and research. Initially flow cytometry permitted the investigation of only one fluorophore, but recent advances allow close to twenty parallel channels to be monitored [5,12]. Various techniques have been developed in the past [7,10] to analyze this high dimensional data. A recent survey of data analysis methods in flow cytometry is provided in [2].

Early work on analyzing this high dimensional data has relied on projecting the data to lower dimensions and manual gating, which is labor intensive and influenced by analyst bias. Hence the development of efficient and accurate algorithms for analyzing the large-scale, high dimensional data is a critical need.

Performing comparative analysis of samples at the cell level is computationally expensive, and hence a more practical approach is to cluster cells in each sample first, and then perform the comparative analyses across the samples. Various techniques have been proposed to cluster flow cytometry data and form groups of cells [3,4,7], but there has been little work on the post-processing of the clustered data to identify common and distinct cell populations among diseased and healthy states.

A recent approach for downstream analysis of clustered data, flow analysis with automated multivariate estimation (FLAME), was proposed by Pyne et al. [10]. The fluorescense intensity matrix with rows corresponding to cells and columns corresponding to antibodies is first clustered into cell populations using the skew t distribution. The clusters across all samples are then pooled and a set of *global metaclusters* are obtained from them using an approach called Partitioning across Medoids. Each sample is then compared with the set of global metaclusters using an integer programming formulation of a weighted b-matching in a bipartite graph with additional constraints.

Our work is closest to the FLAME approach, while differing from it in significant ways. First, we use a non-parametric infinite mixture model in clustering phase, whereas FLAME used the skew t mixture model. Second, we compare clusters in two or more samples directly without creating metaclusters from the clusters in all samples. We propose a generalized edge cover formulation in a bipartite graph as a model for discovering outlying clusters using pairwise comparisons of samples. Third, we propose a weighted clique approach to compare multiple samples to identify outlying clusters and classify them further into distinctive and common outliers.

2 Problem Formulation

2.1 Description of Data

We analyze two different flow cytometry datasets on mouse bone marrow cells from Brigham and Women's Hospital in Boston [15]. In this work, an oncogene PML-RARα, was expressed in mice leading to acute promyelocytic leukemia (APL) in a course of weeks. Each dataset consists of flow cytometry data of cells from three leukemic mice (P^i), one pre-leukemic mouse (H) that has the oncogene expressed but has not developed APL yet, and a Wild Type (WT) sample that does not have the oncogene expressed. Each sample consists of multidimensional (6- or 7-dimensional) flow cytometry data of cells from a single mouse with each dimension representing a specific characteristic of the cell. A sample is represented as a matrix of size $N \times d$, where N is the number of cells, and d is the dimension of data. The data is shown in Table 1. We normalize each column of the matrix by converting it to a vector with mean equal to zero and standard deviation equal to one, and then apply a clustering method to be described in Sec. 3.1.

2.2 Model of the Data

Let each dataset consist of samples from N patients, labeled P^1 P^2, ..., P^N, and one WT. The i-th patient P^i has n_{P^i} clusters $P^i = \{u_1^i, u_2^i, \ldots, u_{n_{Pi}}^i\}$, where u_j^i is the j-th cluster in the i-th patient data. Similarly, WT has n_{WT} clusters $WT = \{w_1, w_2, \ldots, w_{n_{WT}}\}$. If multiple WT samples are available they can be combined beforehand to construct a unique WT model.

We use the Kullback-Leibler divergence as the measure of distance between two clusters. The KL-divergence [6], also known as the relative entropy, between two probability density functions $p(x)$ and $q(x)$ is:

$$KL(p\|q) = -\int p(x)ln\left\{\frac{q(x)}{p(x)}\right\}dx. \tag{1}$$

For two d-dimensional Gaussian distributions N_0 and N_1 with means μ_0, μ_1 and covariance matrices Σ_0, Σ_1, respectively, the KL divergence has a closed-form expression:

$$\begin{aligned} KL(N_0\|N_1) = \frac{1}{2}\Bigg[\ln\left(\frac{\det \Sigma_1}{\det \Sigma_0}\right) + tr(\Sigma_1^{-1}\Sigma_0) \\ + (\mu_1 - \mu_0)^T \Sigma_1^{-1}(\mu_1 - \mu_0) - \frac{d}{2}\Bigg]. \end{aligned} \tag{2}$$

We make the distance measure symmetric by setting the average of $KL(p\|q)$ and $KL(q\|p)$ as the distance $d(p,q)$ between two clusters p and q. A few additional terms are needed to discuss our objective function.

2.3 Basic Definitions

Definition 1. Cohesion Index(CI): Given a set of clusters from N patients, $S = \{u_1, u_2 ... u_N\}$ such that $u_i \in P^i$, and $d(u_i, u_j)$ is the distance between clusters u_i and u_j, the Cohesion Index of the set S is the average distance between pairs of clusters (u_i, u_j) in the set S:

$$CI(S) = \frac{2}{N(N-1)} \sum_{\substack{u_i, u_j \in S \\ i<j}} d(u_i, u_j). \tag{3}$$

A small value of CI means that the clusters in S are similar, while large values indicate that they are dissimilar. The Cohesion Index CI for set S consisting of clusters represented by the large filled circles (within the circles denoting the P^i vertices) in Fig. 1 is the sum of the weights of the edges joining these clusters divided by ten.

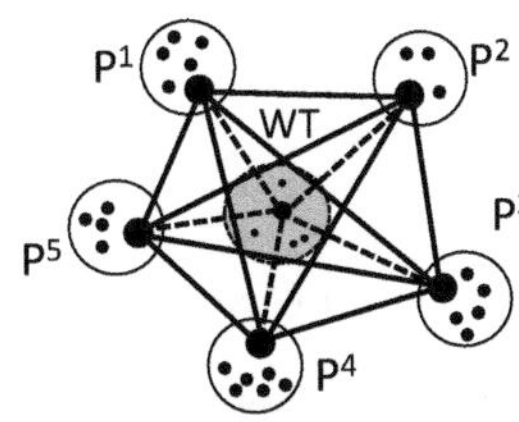

Fig. 1. Graph model of data with 5 patients and 1 Wild Type. The data from each individual has been clustered; vertices in the graph are the clusters, and the edge weights are derived from the Kullback-Leibler divergences between clusters.

Definition 2. Divergence Index(DI): Given a set of clusters from N patients, $S = \{u_1, u_2...u_N\}$ such that $u_i \in P^i$, and $d(w, u_i)$ is the KL-divergence between clusters $w \in WT$ and $u_i \in S$, the Divergence Index (DI) is the minimum sum of distances between each pair (w, u_i) in the set S:

$$DI(S) = \frac{1}{N} \min_{w \in WT} \left\{ \sum_{u_i \in S} d(w, u_i) \right\}. \tag{4}$$

A large value of DI means clusters in S are dissimilar from any WT cluster, while a small value of DI means the clusters in S are similar to some cluster in WT. In Fig. 1, the central grey circle represents the WT sample, the large filled circle within it corresponds to a cluster in WT with the least sum of distances from a set S of patient clusters denoted by the filled circles, and the average length of the edges between the WT and patient clusters yields DI for the set S.

To identify groups of similar outliers we look for sets of clusters S with low values of $CI(S)$ and high values of $DI(S)$. However, maximizing $(DI(S) - CI(S))$ does not suffice to guarantee both a low value of CI and a high value of DI. This observation leads to the next definition.

Definition 3. Coherence Confidence (CC): Given a set of clusters $S = \{u_1, u_2...u_N\}$ such that $u_i \in P^i$, the Coherence Confidence (CC) is the product of the normalized difference between $DI(S)$ and $CI(S)$ and a damping factor:

$$CC(S) = \frac{DI(S) - CI(S)}{DI(S) + CI(S)} \left[1 - a^{-(CI(S) + DI(S))} \right], \tag{5}$$

where a is a constant greater than one. The damping factor prevents the ratio from becoming unstable for small values of the sum $CI(S) + DI(S)$. If this sum is small, then the factor is small enough to keep the value of $CC(S)$ low. As the sum increases, the factor increases to its maximum value of one, and does not significantly influence the CC value. The range of values of CC is $[-1, 1]$. The constant a in the damping factor is chosen so that it should not grow too quickly to its maximum value. We tried various values for the constant and $a \sim 1.2$ worked reasonably well with the data here. We will identify groups S consisting of similar outlying clusters from sets with large positive values of $CC(S)$.

2.4 Objectives

We now state the objectives of our analysis.

1. **Pairwise Outliers:** Identify dissimilar clusters in a diseased sample by pairwise comparison with WT. These are pairwise outliers which contain both distinctive and common outliers described below.
2. **Distinctive Outliers:** Identify the clusters in a diseased sample that are dissimilar to any WT clusters as well as clusters from other diseased samples. These are *distinctive outliers* that fail to form groups with low values of CI.
3. **Common Outliers:** Identify group of similar outliers, i.e., groups with members similar to each other in diseased samples but dissimilar to any WT cluster. These *common outliers* have high values for CC.

3 Methods

3.1 Clustering

We denote the flow cytometry data from a sample as $X = [x_1^T, x_2^T, \ldots, x_N^T]$, where x_i^T corresponds to the data from the i-th cell. We assume the data are generated from a hierarchical Bayesian model. First, the observation x_i is sampled from a likelihood function $f(\theta_i)$ where θ_i is the likelihood parameter for the i-th observation. Second, the parameter θ_i follows a distribution G, which is sampled from a Dirichlet process $DP(\alpha, G_0)$ with a concentration parameter α and a base distribution G_0. Note that the use of a Dirichlet prior will make many $\{\theta_i\}$'s share the same value, naturally inducing clustering of data. The model is known as the *Dirichlet Process Mixture* (DPM) model [1,9] and can be summarized as follows:

$$x_i|\theta_i \sim F(\theta_i), \quad \theta_i|G \sim G, \quad G \sim DP(\alpha, G_0), \tag{6}$$

where $X \sim S$ means that X follows the distribution S. Since G is a distribution, $G \sim DP(\alpha, G_0)$ suggests that the Dirichlet Process $DP(\alpha, G_0)$ is a distribution over distributions.

We used a publicly available Matlab implementation of DPM clustering by Teh [14], which is based on a Chinese Restaurant Process representation of the DPM model and uses simple Gibbs sampling. The computational cost per iteration is $O(Nd^2k)$, where N is the number of rows in a data sample matrix

X, d is the number of columns of X, and k is the number of clusters in the current iteration (k changes with iterations). The value of N is large (see Table 1 in Sec. 4), which makes each iteration computationally expensive even for a small number of clusters. Moreover, the quality of clustering improves with the number of iterations. In our experiments, a hundred iterations work reasonably well for the data as the clustering changes relatively little after that.

Although the DPM inference is expensive for large samples, we prefer the nonparametric model, DPMs, over classical parametric cluster methods, e.g, K-means. The reason is that the computational cost of selecting the number of clusters for a parametric model is prohibitively expensive for flow cytometry data analysis. DPMs circumvents the model selection problem by automatically determining the number of clusters for each sample, making it well suited as a clustering tool before the application of our outlier detection algorithms.

The DPM model is an infinite model in the sense that it contains a mixture of countably infinite components. For example, if $F(\cdot)$ is a Gaussian distribution, the DPM model can be viewed as a mixture of infinite Gaussians [11]. Given a finite number N data points, however, we compute the posterior distribution of the DPM model using Bayes theorem; the expected number of components in the posterior distribution is always finite and, often, much smaller than the number of data points.

3.2 Pairwise Comparison: Generalized Edge Cover

One method we used to identify outliers in the clustered data is pairwise comparison between samples. We model a pair of samples, say A_1 and A_2, using a complete bipartite graph with each cluster represented by a vertex, and edges joining pairs of clusters in different samples. Formally $G = (V_1, V_2, E)$ is a complete bipartite graph, where V_1 contains all clusters from A_1, V_2 contains all the clusters from A_2, and the edge weight function is $c : E \to \mathbb{R}$ where c_{ij} is the weight of edge $\{u_i, u_j\}$, with $u_i \in V_1$ and $u_j \in V_2$. The weight of an edge is the average KL divergence of its endpoint clusters. In this bipartite graph we seek to identify clusters that are common to samples A_1 and A_2, and also those that belong to only one sample.

Since low edge weight implies high similarity among clusters we could find a minimum-weight matching among all maximum cardinality matchings in the graph G and declare unmatched vertices to be outliers. However, this attempt at a model for outlier detection has a significant drawback. Since the number of clusters in the two samples is generally not the same, some clusters will remain unmatched even if they are highly similar to another cluster, and should not be identified as outliers. We address this issue by formulating the problem as a minimum-weight edge cover on a complete bipartite graph. An *edge cover* is a subset of edges such that each vertex in the graph has *at least* one edge incident on it, whereas a *matching* is a subset of edges such that each vertex in the graph has *at most* one edge incident on it. However, even an edge cover fails to accurately model the problem since clusters that represent outliers should not be covered in an edge cover. Hence we find a *generalized edge cover* that permits some vertices not to be covered, at the cost of a penalty, by adding a weight λ for each uncovered vertex to the weight of an edge cover. Thus λ acts as a cut-off value for long edges that would not be included in a generalized edge cover. This leads to a generalized edge cover formulation of the problem, where the cover EC leaves some uncovered vertices $V_{uc} \subseteq V_1 \cup V_2$, while minimizing the objective function:

$$min\left(\sum_{(v_i,v_j)\in EC} c_{ij} + \lambda * |V_{uc}|\right). \tag{7}$$

A generalized edge cover in G can be computed from an edge cover in a transformed graph G'. Let the graph G' be obtained from G by introducing two new distinguished vertices $v_1 \in V_1$ and $v_2 \in V_2$, and adding an edge $\{v_1, v_2\}$ with $c(\{v_1, v_2\}) = 0$, and edges $\{v_1, u_2\}$ for each $u_2 \in V_2$, $\{v_2, u_1\}$ for each $u_1 \in V_1$, with $c(\{u_1, v_2\}) = c(\{u_2, v_1\}) = \lambda$. If a minimum-weight edge cover includes added edges with weight λ, for each such edge, we leave the original vertex in G incident on this edge uncovered in a generalized edge cover of the original graph, thus paying a price of λ for the vertex, without changing the weight or structure of the remaining edge cover.

A minimum-weight edge cover in a graph can be computed in polynomial time by making a copy of the graph and connecting each vertex to its twin in the copy by an edge with weight equal to twice the minimum weight among original edges incident on it. A minimum-weight perfect matching in this graph can be used to compute a minimum-weight edge cover in the original graph [13].

Following the above discussion, our pairwise comparison algorithm for outlier detection can be formulated in the following stages:

1. **Pre-processing:** Add distinguished vertices $v_1 \in V_1$ and $v_2 \in V_2$, and an edge $\{v_1, v_2\}$ with $c(\{v_1, v_2\}) = 0$. Given a cut-off value λ, add edges $\{v_1, u_2\}$ for each $u_2 \in V_2$, and $\{v_2, u_1\}$ for each $u_1 \in V_1$, all with a weight of λ. Let $G' = (V', E')$ be the resulting graph.
2. **Duplicate Graph:** Let $\tilde{G}' = (\tilde{V}', \tilde{E}')$ be a disjoint copy of G'. Let $\bar{G}$ be the the graph formed by taking the union of G' and $\tilde{G}'$ and adding an edge $\{v, \tilde{v}\}$ connecting every $v \in V'$ with its twin $\tilde{v} \in \tilde{V}'$. Let $c(\{v, \tilde{v}\}) = 2\mu(v)$ for each $v \in V'$, where $\mu(v)$ is the minimum weight of the edges of G' incident on v.
3. **Matching:** Compute a minimum-weight perfect matching M in $\bar{G}$.
4. **Edgecover:** Obtain a minimum-weight edge cover EC' of G' by replacing every edge $\{v, \tilde{v}\} \in M$ by an edge of weight $\mu(v)$ in G' incident on v.
5. **Post-processing:** Remove all edges $\{v_1, o\}$, $\{v_2, o\}$ from EC', where o denotes an original vertex in $V_1 \cup V_2$; add each vertex o to the set of outliers O. Remove the distinguished vertices v_1 and v_2 from EC'. The resulting edge cover EC^* together with the set of uncovered vertices O is a solution to the generalized edge cover problem in G.

Lemma: The Algorithm described above computes an optimal generalized edge cover in G.

Proof: The correctness of the algorithm for computing the edge cover EC' in the graph G' was shown in [13]. We obtain a generalized edge cover in G by deleting, the vertices v_1 and v_2, the edges incident on these vertices in EC', and all vertices adjacent to these vertices in EC'. Let O be the set of the deleted vertices adjacent to v_1 and v_2, which will be identified as outliers. Let EC'' be the edges remaining from the edge cover EC' in the modified graph $G'' = G \setminus O$.

We claim that EC'' together with O is an optimal solution for the generalized edge cover problem in G. Assume that there is an optimal solution in G consisting of a set of outliers O and an edge cover EC^* in $G \setminus O$. Clearly $c(EC'') = c(EC^*)$, for otherwise one of the solutions could be improved upon, thereby contradicting their minimality in G'' or $G \setminus O$ respectively. It remains to prove that there is no solution in G with a different outlier set and smaller cost. Let $EC^{*'}$ together with O' be such a solution for G having a smaller cost $c' < c(EC')$. Then $EC^{*'}$ together with an edge $\{o, v_1\}$ or $\{o, v_2\}$ for every $o \in O'$ and the edge $\{v_1, v_2\}$ is an edge cover in G' with cost $c' < c(EC')$, contradicting the optimality of EC'. □

Thus we obtain a generalized edge cover. Note that a vertex $u \in V_k$, where $k = 1$ or 2, and $\mu(u)$ is the minimum weight among the edges of G incident on u, will always be an outlier if $\mu(u) \geq 2\lambda$, and can never be an outlier if $\mu(u) < \lambda$. Otherwise, it will be an outlier if and only if it is not matched to a vertex in G' during step 3 of the algorithm.

For a graph with n vertices and m edges, an edge cover of minimum weight can be computed in time $O(n(n + m \log n))$ [13]. In this context, since there are at most K clusters in each patient, $n \leq 2K$, and $m \leq K(K-1)/2$, and thus the time complexity of pairwise comparison to identify outliers is $O(K^3 \log K)$.

3.3 Comparing Multiple Clusters

Formation of Coherent Groups. We now consider an approach that compares multiple diseased samples to identify clusters common to them but not belonging to the Wild Type. A *group* of clusters S is a set of distinct clusters from each patient, $S = \{u_1, u_2 ... u_N\}$, with $u_i \in P^i$. In Sec. 4.5 we relax this to form groups that do not cover all patients. To identify common outliers we find such groups that exhibit high similarity among their clusters while being dissimilar to the Wild Type.

A graph representation of a group S of clusters is a clique consisting of one cluster from each patient. The cost of a group S is the average weight of all the edges of the corresponding clique, which is the Cohesion Index $CI(S)$. It is easy to show that finding a group with minimum CI score is NP-hard via a reduction from MAX-CLIQUE. To identify groups with low CI score we use a branch and bound technique, which provides good performance for a reasonable number of clusters and patients. We omit the details here due to space considerations.

The branch and bound procedure is called once with each cluster as seed, and it finds a group with minimum cost $CI(S)$ containing the seed cluster, resulting in NK groups in total. The method works very well in practice, although it has a worst-case running time exponential in N. Since the distance measure is not metric, no obvious approximation guarantee exists.

Once we have obtained coherent groups of clusters with small CI scores, we calculate the DI and CC scores for those sets. Since we have a single Wild Type sample with K clusters we can find the minimum Divergence Index for a group S in $O(NK)$ time. The decision about distinctive and common outliers is based on the following rules:

Distinctive Outliers: If a group has high value for CI, then declare the seed cluster of that set as a distinctive outlier, since it fails to form a close group with clusters from other patients.

Common Outliers: Among the remaining groups with small CI scores, find those groups having large CC values. These sets are close to one another while being distinct from any WT cluster.

4 Results

4.1 Clustering Results

We cluster each normalized sample using a DPM clustering algorithm and the results are shown in Table 1. All subsequent downstream analysis is built upon these clustering results.

Table 1. Clustering the flow cytometry data for two datasets. WT represents the Wild Type, P^i denotes a leukemic mouse, and H is a pre-leukemic mouse with an oncogene expressed.

Dataset-1				Dataset-2			
Sample	Dimension	#Cells	#Clusters	Sample	Dimension	#Cells	#Clusters
WT	6	115,407	18	WT	7	49,316	21
H	6	131,850	23	H	7	68,886	22
P^1	6	107,299	22	P^1	7	78,406	21
P^2	6	131,575	28	P^2	7	6,050	12
P^3	6	236,392	31	P^3	7	48,998	21

The computational cost of the clustering step using the Matlab DPM code is about six to ten hours depending on the dataset size, while the edge cover and branch and bound computations run under a minute on a 3 GHz PC. The DPM clustering code should be much faster when it is implemented efficiently in a non-interpreted environment, but it would still be the dominant cost of the current computation. Improving its performance was not the scope of this work.

4.2 Pairwise Comparison Results

The generalized edge cover approach compares leukemic mouse samples with WT and identifies outliers depending on a cut-off value λ. The optimal cut-off value is dependent on the Kullback-Leibler divergences of the clusters involved. The number of outliers is inversely related to λ in that a large value of λ yields very few outliers, and vice versa. A plot is shown in Figure 2.

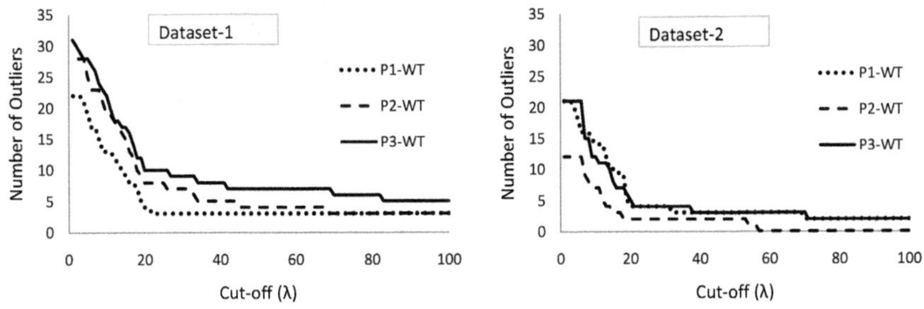

Fig. 2. Outliers from pairwise comparison between WT and leukemic samples with different cut-off (λ) values for two datasets.

The outlier profile in both datasets shows a sharp change approximately at $\lambda = 20$, which we choose to be a good cut-off value for the detection of extreme outliers. Table 2 shows all the outliers obtained for two different values of λ. Note that pairwise comparison of a leukemic sample with the WT sample cannot distinguish between distinctive and common outliers.

Table 2. Outlying clusters in leukemic samples (Dataset 1) for two cut-off λ values

Sample	Outliers at $\lambda = 20$	Outliers at $\lambda = 10$
P^1	2,10,13,14	2,3,10,12,13,14,16,17,18,19,21
P^2	4,12,17,18,20,24,25,28	2,3,4,11,12,16,17,18,19,20,21,22,24,25,26,27,28
P^3	11,12,13,16,17,18,19,20,21,29	8,9,11,12,13,15,16,17,18,19,20,21,23,24,26,28,29,31

4.3 Coherent Groups

Every cluster in each sample is used as a seed cluster to construct a group S with a minimum value of the Cohesion Index. The significance of the CI scores of the identified groups can be assessed using the permutation test [8]. We randomly select one cluster from each leukemic mouse to form a group and construct $N_{perm} = 100,000$ random groups in total. For any (non-randomly constructed) group S, let N_S be the number of random groups (S_{rand}) having $CI(S_{rand}) \leq CI(S)$. The significance measure, the p-value of S, can then be calculated as $p(CI(S)) = (N_S + 1)/(N_{perm} + 1)$. Groups with small $p(CI(S))$ values are significant since the chance of finding them at random is small. The histogram of the N_{perm} permutations is shown in the left subfigure in Figure 3 with the broken vertical line indicating 5% confidence level. We observe that most of the non-random groups fall within the 5% confidence interval.

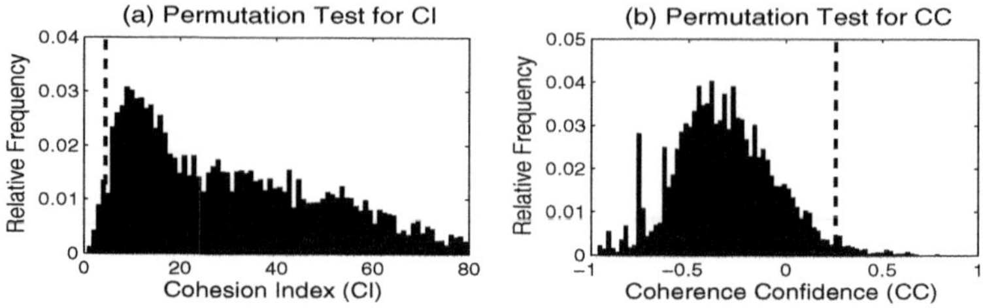

Fig. 3. Histogram of the permutation tests for CI (left) and CC (right) scores from Dataset 1. Groups at a 5% confidence level are to the left of the broken vertical line for CI scores, and to the right of the broken vertical line for CC scores.

Table 3. Groups with CI scores and p-values of CI scores. Seed clusters are shown in grey and distinctive outliers are shown in boxed squares.

Dataset-1					Dataset-2				
P1	P2	P3	CI	$p(CI)$	P1	P2	P3	CI	$p(CI)$
1	7	8	0.62	0.00024	3	6	4	1.232	0.00064
2	4	18	296.7	-	6	8	9	24.495	-
4	7	8	1.404	0.0002	8	10	14	1.204	0.00043
10	26	18	154.658	-	10	10	7	0.627	0.00012
11	13	14	1.27	0.00013	11	4	12	44.63	-
14	1	30	74.604	-	12	5	15	3.397	0.00815
15	11	28	3.031	0.00335	13	5	15	3.918	0.01196
15	11	21	49.91	-	14	4	10	107.167	-
17	21	17	1.675	0.00046	18	5	15	3.326	0.0076
18	26	31	3.054	0.00345	21	12	21	7.234	0.053099

Several representative groups with their p-values are presented in the Table 3, where seed clusters are highlighted in grey. Notice that multiple seeds may construct the same group (e.g., $\{4, 7, 8\}$ in dataset-1). Such groups are usually tight with low p-values. Also notice the three groups in $\{12, 5, 15\}$, $\{13, 5, 15\}$, $\{18, 5, 15\}$ in Dataset-2, where the same clusters from P^2 and P^3 are grouped with different clusters from P^1 with similar CI scores. If clusters 12, 13, 18 from P^1 all have small KL divergence from each other, then merging these three clusters produces a unified cluster. Thus we can use the group formation approach to refine clusters obtained from the clustering algorithm.

4.4 Distinctive and Common Outliers

Seed clusters that fail to form a group with significantly low CI scores are declared as *distinctive outliers* and are shown in boxed squares in Table 3. The p-values for such groups will be large, bearing little significance in this context.

Common outliers are groups of clusters with high Cohesion Confidence (CC) values, and do not have a distinctive outlier as a member. We performed the permutation test on the CC value to assess its significance in the same way as we did for the CI score. However, we are now interested in the confidence limit to the right side of the broken vertical line in the histogram in the right subfigure of Figure 3. Again, we find that groups with high CC values are significant since the chance of finding them at random is small. We report distinctive and common outliers discovered by the clique approach in Table 4. All distinctive and common outliers identified by this approach were also identified by the pairwise comparison approach (Table 2), but the converse is generally not true. Hence the clique approach is more powerful in detecting and classifying outliers than the edge cover approach.

Table 4. Distinctive and Common Outlying clusters identified by the weighted clique approach. Here u_i is a cluster belonging to a leukemic mouse P^i.

Distinctive Outliers				Common Outliers					
Dataset-1		Dataset-2		Dataset-1			Dataset-2		
Sample	Clusters	Sample	Clusters	$\{u_1, u_2, u_3\}$	CC	$p(CC)$	$\{u_1, u_2, u_3\}$	CC	$p(CC)$
P^1	2,10,14,16	P^1	6,11	17,21,17	0.64	0.00007	10,10,7	0.78	0.00013
P^2	4,20	P^2	4	1,7,8	0.63	0.00011	4,9,1	0.64	0.00078
P^3	12,16,19,20,21	P^3	5,10	9,5,3	0.58	0.00015	17,2,6	0.612	0.0016

4.5 Probabilistic Model for Groups

We describe a probabilistic model to refine the groups by relaxing our initial requirement that a group must necessarily include one cluster from each patient. Let f_{ij} be the number of times two clusters u_i and u_j are grouped together, and let $f_i = \sum_{j \neq i} f_{ij}$ be the number of times cluster u_i appears in any group. Then for a group $S = \{u_1, u_2 ... u_N\}$ the probability of u_i being a member of S, $P(u_i|S)$, and the probability of the whole group, $P(S)$ can be calculated by

$$P(u_i|S) = \frac{\sum_{\substack{u_j \in S \\ i \neq j}} f_{ij}}{f_i}, \quad \text{and} \quad P(S) = \prod_{\substack{u_i \in S \\ 1 \leq i \leq N}} P(u_i|S). \tag{8}$$

Within a group, a low $P(u_i|S)$ value and high $P(u_j|S)$ value for all $j \neq i$, suggests that u_i is a member with weak cohesion to S. We can refine the group S by deleting the weak member u_i, and inserting a gap in that position. A high $P(u_i|S)$ and low $P(u_j|S)$, for all $j \neq i$, also indicates a weakly formed group. In this case, we allow u_i to form a group by itself deleting all other members of S, making u_i a distinctive outlier. We present several representative groups from Dataset-1 in Table 5, where clusters with high support are marked in grey.

Consider the groups in rows 2 and 3 of the Table. Each has one weak member(u_1 and u_3, respectively) that can be safely removed from the corresponding groups. However, rows 4 and 5 show groups with only one strong member (u_2 and u_1, respectively), which can be declared as a distinctive outlier by removing all other members from the groups. The group in the sixth row has a low probability.

Table 5. Group-uniqueness probabilities for Dataset-1. Clusters in grey have the highest probabilities of belonging to the group.

u_1	$P(u_1\|S)$	u_2	$P(u_2\|S)$	u_3	$P(u_3\|S)$	$P(S)$
11	1	13	1	14	1	1
17	.29	16	1	6	1	.29
19	.83	22	.83	28	.33	.23
2	.33	4	1	18	.33	.11
21	1	6	.20	27	.25	.05
18	.17	19	.23	27	.25	.01

4.6 Effect of APL on Bone Marrow Cells

Wojiski et al. [15] compared the populations of a number of cell types in the bone marrow of WT, leukemic and pre-leukemic (with oncogene PML-RARα expressed in the latter two groups) mice. They reported that WT and pre-leukemic (H) mice had similar cell populations of hematopoietic stem cells (LSKs), common myeloid progenitor cells (CMPs), granulocyte-monocyte progenitor cells (GMPs), and megakaryocyte erythrocyte progenitor cells (MEPs); but in leukemic mice (P) cell populations of LSKs, CMPs and MEPs are reduced and GMPs are increased, relative to the WT and pre-leukemic mice. They also found that mature granulocytes were increased in pre-leukemic mice relative to WT.

In a pairwise comparison of flow cytometry data from WT and H using the edge cover approach, we found that of the 18 clusters in WT and 23 clusters in H in the first data set, only 3 clusters from each set were left uncovered when a value $\lambda = 20$ was used. Similar results were obtained for the second data set also, confirming the general correspondence of populations of various cell types in these two kinds of mice. Generally, a group of clusters from leukemic mice that has a high value of CC (hence it is distant from any cluster in the WT) also has a high value of CC when clusters from a pre-leukemic mouse are used in the place of WT. However, we found some clusters in the pre-leukemic mouse that were closer to the leukemic mice rather than the WT. We performed this experiment by treating the pre-leukemic sample as an additional leukemic sample, and using the branch and bound algorithm to identify sets with high CC values. In Dataset-1, we found the clusters $\{8, 14, 15, 5\}$ and $\{17, 21, 17, 18\}$; and in Dataset-2, we found the clusters $\{4, 9, 1, 4\}$ and $\{10, 10, 7, 8\}$; here in each set the first three clusters are from leukemic mice and the last is from the pre-leukemic mouse, and these clusters are all distant from any cluster in the WT. Identifying these specific cell types through further experimental work could shed light on disease progression in the murine model of APL.

Acknowledgments. This research was supported through a PRF grant from the College of Science at Purdue, NSF grant CCF-0830645, and Department of Energy grant DE-FC02-08ER25864 (CSCAPES Institute).

References

1. Antoniak, C.E.: Mixtures of Dirichlet processes with applications to Bayesian nonparametric problems. Annals of Statistics 2(6), 1152–1174 (1974)
2. Bashashati, A., Brinkman, R.: A survey of flow cytometry data analysis methods. In: Advances in Bioinformatics, pp. 1–19 (December 2009)
3. Boedigheimer, M., Ferbas, J.: Mixture modeling approach to flow cytometry data. Cytometry A 73, 421–429 (2008)
4. Chan, C., Feng, F., Ottinger, J., et al.: Statistical mixture modeling for cell subtype identification in flow cytometry. Cytometry A 73(A), 693–701 (2008)
5. Herzenberg, L., Tung, J., Moore, W., et al.: Interpreting flow cytometry data: A guide for the perplexed. Nature Immunology 7(7), 681–685 (2006)
6. Kullback, S.: Information Theory and Statistics. Dover Publications Inc., Mineola (1968)
7. Meur, N., Rossini, A., Gasparetto, M., Smith, C., Brinkman, R., Gentleman, R.: Data quality assessment of ungated flow cytometry data in high throughput experiments. Cytometry A 71A, 393–403 (2007)
8. Moore, D., McCabe, G.: Introduction to the Practice of Statistics. W. H. Freeman & Co., New York (2006)
9. Neal, R.: Markov chain sampling methods for Dirichlet process mixture models. Journal of Computational and Graphical Statistics 9, 249–265 (2000)
10. Pyne, S., Hu, X., Wang, K., et al.: Automated high-dimensional flow cytometric data analysis. PNAS 106(21), 8519–8524 (2009)
11. Rasmussen, C.E.: The infinite Gaussian mixture model. In: Solla, S., Leen, T., Muller, K.R. (eds.) Advances in Neural Information Processing Systems, vol. 12. MIT Press, Cambridge (2000)
12. De Rosa, S., Brenchley, J., Roederer, M.: Beyond six colors: A new era in flow cytometry. Nature Medicine 9(1), 112–117 (2003)
13. Schrijver, A.: Combinatorial Optimization — Polyhedra and Efficiency, Volume A: Paths, Flows, Matchings. Algorithms and Combinatorics, vol. 24. Springer, New York (2003)
14. Teh, Y.W.: DPM Software (2010), `http://www.gatsby.ucl.ac.uk/~ywteh/research/software.html`
15. Wojiski, S., Gubal, F.C., Kindler, T., et al.: PML-RARα initiates leukemia by conferring properties of self-renewal to committed promyelocytic progenitors. Leukemia 23, 1462–1471 (2009)

Fast Mapping and Precise Alignment of AB SOLiD Color Reads to Reference DNA

Miklós Csűrös[1], Szilveszter Juhos[2], and Attila Bérces[2]

[1] Department of Computer Science and Operations Research, University of Montréal, Canada
csuros@iro.umontreal.ca
[2] Omixon, Chemistry Logic Kft, Budapest, Hungary
www.omixon.com

Abstract. Applied Biosystems' SOLiD system offers a low-cost alternative to the traditional Sanger method of DNA sequencing. We introduce two main algorithms of mapping SOLiD's color reads onto a reference genome. The first method performs mapping by adapting a greedy alignment framework. In such an alignment, reads are mapped to approximate genome positions, allowing for a pre-specified bound on sequence difference that combines nucleotide mismatches, gaps, and sequencing errors. The second method for precise alignment relies on a pair hidden Markov model framework, combining a DNA sequence evolution model and sequencing errors (from read quality files).

1 Introduction

Next-generation sequencing (NGS) methods [1] provide economical alternatives to the traditional Sanger method of DNA sequencing. Various commercially available platforms can generate large amounts of information which enable important biological and medical applications [2], including, perhaps most notably, the sequencing of personal and somatic genomes [3,4], or even entire ecosystems [5]. In a typical genome analysis pipeline, NGS reads are mapped to reference sequences, and the alignments are further examined to detect variations within the target DNA sample, and with respect to the reference.

Currently available software for large-scale NGS mapping [6] use indexing techniques in order to speed up the search for similarities. The underlying algorithms rely either on hashtable-based indexes (*seed-and-extend*), or on compressed indexes exploiting the Burrows-Wheeler Transformation (BWT). BWT-based methods use little memory, and have an impressive computing speed [7,8]. Seed-and-extend has an increasing advantage with higher sequence divergences, due to flexible tailoring choices for seeding methods [9].

The AB SOLiD sequencing platform from Applied Biosystems, Inc. (Foster City, Cal.) poses even greater challenges for bioinformatics than other widely used NGS technologies, due to the sheer size of the produced data (up to about a billion 35bp or 50bp reads in one production run), and the employed dinucleotide encoding by "colors." We introduce algorithmic solutions to different problems encountered when mapping AB SOLiD reads to a reference genome. First, we

V. Moulton and M. Singh (Eds.): WABI 2010, LNBI 6293, pp. 176–188, 2010.

propose a seed-and-extend framework for mapping color reads to locations along a reference DNA. The novelty of the framework is a greedy extension procedure employed in filtering the hits, which combines sequencing errors and DNA sequence differences. The seeding and the extension use the same "phase" representation of the color sequences, in order to minimize the number of executed arithmetic operations. The mappings are immediately useful for inferring structural variations [10] or phylogenetic classifications [11] (when multiple reference genomes are considered). Our second algorithmic solution addresses fine-scale alignments in a statistical framework. A notable feature of the approach is that color read quality values (sequencing error probabilities) are incorporated into a pair hidden Markov model. The statistical framework helps inferring the alignment with maximum expected accuracy or alignment metric accuracy (AMAP). The model assigns posterior probabilities to all target sequence variations, which can be used directly to deduce the consensus between overlapping reads without a multiple alignment.

2 Methods

2.1 Sequences and Numerical Encoding

The AB SOLiD system relies on the ligase-driven synthesis of PCR-amplified target DNA fragments. The sequencing read is produced in "color" encoding, where colors correspond to the dinucleotides sampled by fluorescently labeled probes in iterated synthesis cycles, arranged in their physical order along the target fragment. In the rest of the paper, we use a convenient numerical encoding for nucleotides and colors (or fluorescent dies):

$$\mathtt{A} = 0, \mathtt{C} = 1, \mathtt{G} = 2, \mathtt{T} = 3$$
$$\text{FAM/blue} = 0, \text{Cy3/green} = 1, \text{TXR/orange} = 2, \text{Cy5/red} = 3.$$

With this encoding, the mapping between colors and dinucleotides is simply the bitwise exclusive OR operation, denoted by $\oplus$: dinucleotide xy is encoded by the color $c = x \oplus y$.

The error-free color encoding for a DNA sequence $\mathbf{t} = t_{0..m}$ is the sequence $\mathbf{s} = c_{1..m}$ where $c_i = t_i \oplus t_{i-1}$. Notice that the same $\mathbf{c}$ translates into four possible $\mathbf{t}$ determined by t_0. The *read alignment* problem is that of aligning an unknown *target sequence* $\mathbf{t}$ to a known reference DNA sequence $\mathbf{s} = s_{1..n}$, using a color sequence $\mathbf{c}$ that encodes $\mathbf{t}$ but may contain sequencing errors. The alignment is evaluated with respect to the implied nucleotide mismatches and gaps, as well as the implied sequencing errors. Figure 1 illustrates this concept. An alignment is composed of column types M1–M4, D and I1–I2, where each column contains three cells: a reference cell s, a color cell c and a target cell t. For all three, $s, c, t \in \{0, 1, 2, 3, \square\}$, where $\square$ is the indel character. Concatenated non-indel characters in the color cells give the complete sequence $c_{1..m}$, and those in the reference cells yield a reference region $s_{i..i'}$. Indel characters may not occupy all three cells, and indels appear together in the color and target cells.

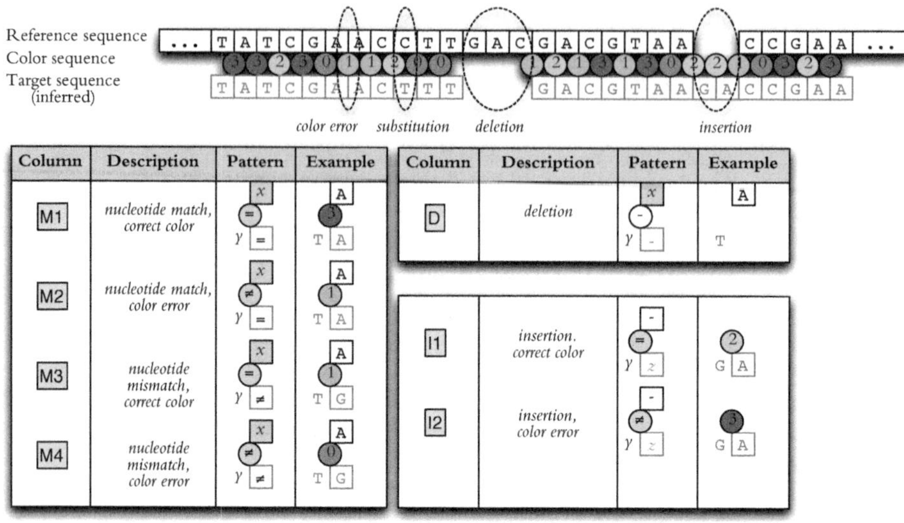

Fig. 1. Alignment between reference DNA, color sequence, and target sequence. The tables on the **bottom** enumerate possible alignment columns. Every column is annotated by the preceding nucleotide y in the target, defining sequencing errors.

Here we consider the simplest alignment scoring system, called the *edit distance*, which is computed by penalizing columns of type M2, M3, D and I1 with 1, and columns of type M4 and I2 with 2. Columns of type M1 are not penalized. The classic Smith-Waterman-Gotoh alignment [12,13] is readily adaptable to find an optimal alignment [14,15]. In order to track sequencing errors, it is necessary to include dinucleotide information in the formulas. Formally, there is a color error in a non-D column that is not the leftmost such column, if $t \oplus t' \neq c_j$ where t' is in the target cell, c_j is in the color cell, and t is the target cell content in the closest preceding non-D column.

The following lemma (proof omitted) shows that there is an optimal alignment that contains no columns with penalty 2.

Lemma 1. *There is an alignment with minimum edit distance that contains neither* M4 *nor* I2 *columns.*

In a run of ℓ consecutive perfect matches (M1) between $s_{i..i+\ell-1}$ and $c_{j..j+\ell-1}$,

$$c_j = y \oplus s_i \tag{1a}$$

$$c_{j+k} = s_{i+k-1} \oplus s_{i+k} \qquad \{1 \leq k < \ell\}, \tag{1b}$$

where y denotes the last aligned target nucleotide preceding the run. For convenience, we introduce the *phase representation* $\phi_{0..m}$ of the color sequence: $\phi_0 = 0$, and $\phi_k = c_1 \oplus c_2 \oplus c_3 \oplus \cdots \oplus c_k = \phi_{k-1} \oplus c_k$ for $k > 0$. From (1),

$$s_{i+k} = y \oplus c_j \oplus c_{j+1} \oplus \cdots \oplus c_{j+k} = y \oplus \phi_{j-1} \oplus \phi_{j+k} \tag{2}$$

for all $k = 0, \ldots, \ell - 1$. In other words, there exists an u (in particular, $u = y \oplus \phi_{j-1}$) with which $s_{i+k} = u \oplus \phi_{j+k}$ holds for all $k < \ell$. The phase representation

$\phi_{0..m} = t_{0..m}$ is thus the translation of the color read into DNA, assuming that the target sequence starts with $t_0 = \phi_0 = 0$ (A).

2.2 Color Read Indexing

In a *seed-and-extend* framework [9], local alignments between two DNA sequences R, T are found by using a seeding function $h\colon \{\mathtt{A}, \mathtt{C}, \mathtt{G}, \mathtt{T}\}^\ell \mapsto \mathcal{H}$, which filters the (i, j) position pairs where local alignments are worth being looked for. Specifically, an index table is built for R which gives the set of positions $h_R^{-1}(x) = \{i\colon h(R_{i..i+\ell-1}) = x\}$ for all $x \in \mathcal{H}$. A pair (i, j) is *hit* when $h(T_{j..j+\ell-1}) = h(R_{i..i+\ell-1})$, or $i \in h_R^{-1}(h(T_{j..j+\ell-1}))$. Hits are found by sliding a window along T and consulting the index table for $h(T_{j..j+\ell-1})$ in each position j. Hits are *extended* by performing a local alignment in a region around (i, j).

In the simplest case, h is the identity function, and hits correspond to matching ℓ-mers. Other widely used seeding functions rely on so-called *spaced seeds.* An (ℓ, w) spaced seed is defined by a set $\{\delta_1, \delta_2, \ldots, \delta_w\} \subseteq \{1, 2, \ldots, \ell\}$ of sampled positions, corresponding to the seeding function $h(x_{1..\ell}) = x_{\delta_1} \cdots x_{\delta_w}$. Accordingly, (i, j) pairs are hit when $R_{i+\delta_k-1} = T_{j+\delta_k-1}$ for all $k = 1, \ldots, w$. Spaced seeds perform theoretically and practically better [9] than ℓ-mers as seeding functions.

Seeding is not straightforward with color reads, because $\mathbf{s}$ and $\mathbf{c}$ do not encode DNA in the same way. Equation (2) suggests a possible way of adapting spaced seeds to indexing color reads. For a hit, $s_{i+\delta_k-1} = t_{j+\delta_k-1}$ holds in all sample positions $k = 1, \ldots, w$. Assuming no sequencing errors in $c_{j..j+\ell-1}$, Eq. (2) implies that $s_{i+\delta_1-1} \oplus s_{i+\delta_k-1} = \phi_{j+\delta_1-1} \oplus \phi_{j+\delta_k-1}$ for all $k = 2, \ldots, w$. Consequently, the hits can be found by indexing the reads in the phase representation: the seeding function is $h(x_{1..\ell}) = y_{1..w-1}$ with $y_k = x_{\delta_1} \oplus x_{\delta_{k+1}}$. For a corresponding hit, $h(s_{i..i+\ell-1}) = h(\phi_{j..j+\ell-1})$. (Existing tools like [14] translate instead the reference sequence into color space, so that for an (i, j)-hit, $s_{i+\delta_k-2} \oplus s_{i+\delta_k-1} = c_{j+\delta_k-1} = \phi_{j+\delta_k-2} \oplus \phi_{j+\delta_k-1}$ at all k, which corresponds to a seeding function $h(x_{0..\ell}) = y_{1..w}$ with $y_k = x_{\delta_k-1} \oplus x_{\delta_k}$ in our notation.)

2.3 Greedy Alignment between Color Read and DNA Sequence

Hits are extended by adapting the classic greedy procedure of Wu et al. [16]. An (i, j) hit between the reference DNA $s_{1..n}$ and color read $c_{1..m}$ is extended by computing the longest prefix of the color sequence that can be aligned starting at reference position $(i - j + 1)$ within prespecified bounds on the edit distance. Specifically, the procedure uses an argument $d_{\max}$ bounding the number of allowed indels between the reference and the inferred target sequence, and an argument $e_{\max}$ that bounds the edit distance. The procedure is explained best in terms of the *edit graph.* The edit graph's vertices are $\{(i, j, t)\colon 0 \le i \le n; 0 \le j \le m; 0 \le t \le 3\}$. The edges are weighted, and correspond to alignment columns of Fig. 1. By Lemma 1, it suffices to consider column types M1–M3, D and I1. The t component of the vertex triple contains phase information on the color sequence. An edge of type M1 has weight 0, and by (2), connects (i, j, t) to $(i + 1, j + 1, t)$ where $t = s_{i+1} \oplus \phi_{j+1}$. All other edge types have weight 1: M2

$(i, j, t) \to (i+1, j+1, s_{i+1} \oplus \phi_{j+1})$ with $t \neq s_{i+1} \oplus \phi_{j+1}$, M3 $(i, j, t) \to (i+1, j+1, t)$ with $t \neq s_{i+1} \oplus \phi_{j+1}$, I1 $(i, j, t) \to (i, j+1, t)$, and D $(i, j, t) \to (i+1, j, t)$.

A path in the edit graph corresponds to an alignment. Given a bound $e_{\max}$, we restrict our attention to paths from any of the $(i, 0, t)$ vertices reach some (i', j, t') with maximum $j \leq m$, and have at most $e_{\max}$ non-M1 edges. In other words, we are searching for the longest alignable prefix within the bound. Define the *diagonal* $d = 0, 1, \ldots, n$ as the vertex set $\{(i, i-d, t)\}$. Our greedy algorithm considers paths along diagonals $0, \ldots, 2d_{\max}$ only. Let $R_t^d(e) = j$ if $(j+d, j, t)$ is the farthest reachable vertex from any $(i, i-d, t')$ on a path with vertices on diagonals $d \leq 2d_{\max}$, and with edge weight sum at most $e \leq e_{\max}$. Algorithm Greedy computes all $R_t^d(e)$.

Algorithm Greedy$(s_{1..n}, \phi_{0..m}, d_{\max})$
Output: longest prefix of ϕ alignable within $e_{\max}$ errors on diagonals $0, \ldots 2d_{\max}$.
G1 **for** $t \leftarrow 0, \ldots, 3$ and $d = 0, \ldots, 2d_{\max}$ **do** $R_t^d(0) \leftarrow 0; \forall e > 0\colon R_t^d(e) \leftarrow -\infty$
G2 **for** $e \leftarrow 0, \ldots, e_{\max}$ **do**
G3   **for** $t \leftarrow 0, \ldots, 3$ and $d = 0, \ldots, 2d_{\max}$ **do**
G4 $j \leftarrow R_t^d(e); i \leftarrow j + d$
G5   **if** $j \neq -\infty$ **then**
G6   **while** $i + 1 < n$ and $j + 1 < m$ and $s_{i+1} \oplus \phi_{j+1} = t$ **do**
G7 $i \leftarrow i + 1; j \leftarrow j + 1$ ▷ *run of* M1 *edges*
G8   **if** $j \geq m - (e_{\max} - e)$ **then return** m **else** $R_t^d(e) \leftarrow j$
G9   **if** $e \neq e_{\max}$ **then**
G10   **for** $t \leftarrow 0, \ldots, 3$ and $d = 0, \ldots, 2d_{\max}$ **do**
G11 $j \leftarrow R_t^d(e); i \leftarrow j + d$
G12   **if** $j \neq -\infty$ **then**
G13   **if** $d \neq 2d_{\max}$ **then** Update$(d+1, t, e+1, j)$ ▷ D *edge*
G14   **if** $d \neq 0$ **then** Update$(d-1, t, e+1, j+1)$ ▷ I1 *edge*
G15   **if** $i < n$ **then**
G16   Update$(d, t, e+1, j+1)$ ▷ M3 *edge*
G17   Update$(d, s_{i+1} \oplus \phi_{j+1}, e+1, j+1)$ ▷ M2 *edge*
G18 **return** $\max_{t=0,\ldots,3; d=0,\ldots,2d_{\max}}\{R_t^d(e_{\max})\}$

Algorithm Update(d, t, e, j)
U1 **if** $R_t^d(e) < j$ **then** $R_t^d(e) \leftarrow j$

When extending a hit at (i, j) for the reference sequence $\mathbf{s}$ and the phase sequence $\phi_{0..m}$, Algorithm Greedy$(s_{i'..i'+m+2d_{\max}-1}, \phi_{0..m}, d_{\max})$ is called, where $i' = i - j + 1 - d_{\max}$ is the starting position of the region within which the extension is performed. By an analogous argument to [16], the running time is $O(m + d_{\max} e_{\max})$ on average (for random sequences), and $O(m d_{\max})$ in the worst case. The greedy framework can be adapted to slightly more general scoring systems (match/mismatch penalties), but it is unclear whether it could accommodate symbol-dependent scoring and affine gap penalization [17]. Therefore, Greedy is more useful for filtering hits than for retrieving optimal alignments.

2.4 Statistical Alignment for Color Reads

We perform statistical alignment by using a pair hidden Markov model [18], or pair-HMM. A pair-HMM defines a probability distribution over alignments.

The advantages of having a well-defined probabilistic model are manifold [19]. Likelihoods can be used to recognize unrelated sequence pairs, or to optimize model parameters. Posterior probabilities quantify discrepancies between the two sequences in a statistically principled manner.

For the alignment of color reads to a reference DNA, we introduce a pair-HMM with state set $\mathcal{Q} = \{\mathsf{S}, \mathsf{E}\} \cup \big(\{\mathsf{M}, \mathsf{I}, \mathsf{D}\} \times \{0, 1, 2, 3\}\big)$. The HMM generates a state sequence $q_0, \ldots, q_\ell \in \mathcal{Q}^\ell$ as a random Markov chain determined by transition probabilities between the states. A transition is followed by the random emission of a pair $w = (s, c)$ where $s \in \{0, 1, 2, 3, \Box\}$ is a numerically encoded nucleotide and $c \in \{0, 1, 2, 3, \Box\}$ is a numerically encoded color. A *run* of the hidden Markov model [20] consists of a random state sequence $q_0, \ldots, q_\ell$ coupled with the random emitted pairs $w_1, \ldots, w_\ell$. States S and E emit unaligned prefixes and suffixes of the reference sequence. States (M, t), (D, t), (I, t) encode the rightmost inferred target nucleotide t, and correspond to match, deletion, and insertion. A transition from (x, t) to (x', t') with $x' \in \{\mathsf{M}, \mathsf{I}\}$ entails the emission of a color character c: the color is correct if $t \oplus t' = c$. The SOLiD sequencing system provides error estimates in so-called quality files that encode the error probability ν on an integer scale using a formula originally introduced for Sanger sequencing in the phred program [21]: $\mathsf{qual} = \lfloor -10 \cdot \log_{10} \nu \rfloor$. We thus assume that a sequence of error probabilities $\nu_{1..m}$ is available with the color sequence $c_{1..m}$. Subsequently to a state transition $(x, t) \to (x', t')$, the emission of the color character c_j occurs with probability $\gamma_j(t \oplus t')$, where

$$\gamma_j(c_j) = 1 - \nu_j \quad \text{and} \quad c \neq c_j : \gamma_j(c) = \nu_j / 3. \tag{3}$$

The emission of reference nucleotides is dictated by an assumed Markov model of DNA sequence evolution [22], like the F84 model [23]. In general, we assume that the nucleotide substitutions between reference and target happen according to a Markov model that specifies the stationary distribution π and the substitution probabilities $p(s \to t)$, and that the model is reversible ($\pi_s p(s \to t) = \pi_t p(t \to s)$). Transitions to states (x, t) with different $t \in \{\mathtt{A}, \mathtt{C}, \mathtt{G}, \mathtt{T}\}$ thus happen by probabilities proportional to π_t. The emission of a reference nucleotide $s \neq \Box$ occurs with probability $p(t \to s)$ on arrival to state (M, t).

Transition probabilities determine the expected lengths of unaligned prefixes and suffixes, as well as the frequency and length of gaps. In particular, we assume that the prefix and suffix regions have a geometric prior length distribution with mean $1/\eta$, that insertions and deletions start with a probability δ, and that gaps have a geometric prior length distribution with mean $1/(1 - \epsilon)$. When aligning a color sequence of length m, we are interested in state sequences with exactly m states emitting color characters (M and I). For that reason, we impose the non-emitting state transition $M \to E$ and $I \to E$ after emitting m color characters. The transition out of state S to (M, t) or (I, t), which sets the first target nucleotide $t_0 = t$, is also non-emitting. Figure 2 summarizes the state transitions and the emissions.

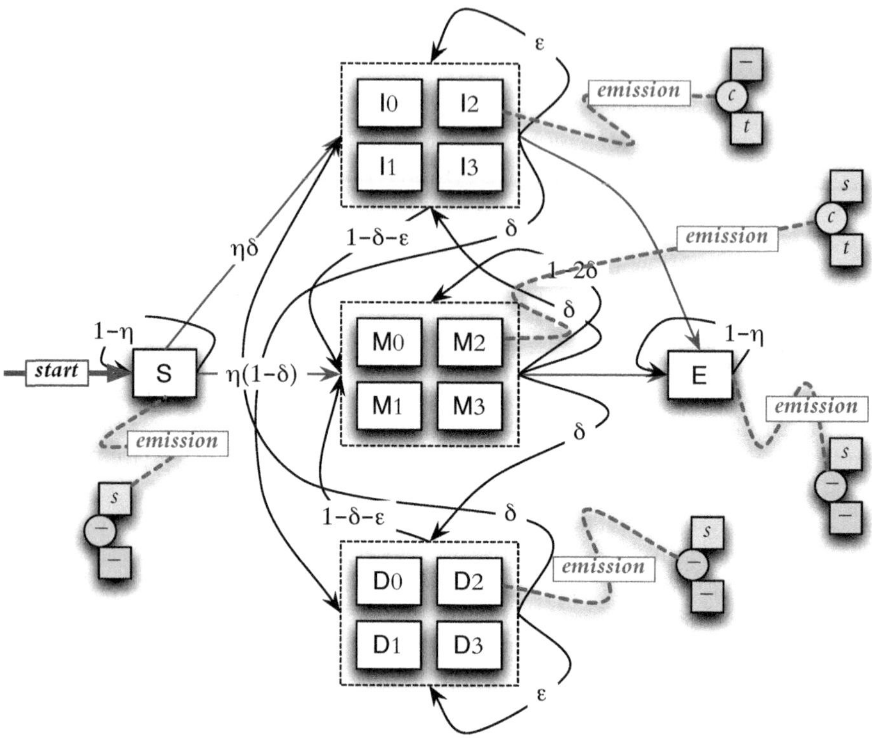

Transition	probability	Emission	probability
$\mathsf{S} \to \mathsf{S}$	$1-\eta$	$(s, \square)$	π_s
$\mathsf{S} \to (\mathsf{M}, t)$	$\eta(1-\delta)\pi_t$	$(\square, \square)$	1
$\mathsf{S} \to (\mathsf{I}, t)$	$\eta\delta\pi_t$	$(\square, \square)$	1
$(\mathsf{M}, t) \to (\mathsf{M}, t')$	$(1-2\delta)\pi_{t'}$	$(s, t \oplus t')$	$p(t' \to s)$
$(\mathsf{M}, t) \to (\mathsf{D}, t)$	δ	$(s, \square)$	π_s
$(\mathsf{M}, t) \to (\mathsf{I}, t')$	$\delta\pi_{t'}$	$(\square, t \oplus t')$	1
$(\mathsf{M}, t) \to \mathsf{E}$	$1[\text{after } m \text{ colors}]$	$(\square, \square)$	1
$(\mathsf{D}, t) \to (\mathsf{M}, t')$	$(1-\epsilon-\delta)\pi_{t'}$	$(s, t \oplus t')$	$p(t' \to s)$
$(\mathsf{D}, t) \to (\mathsf{D}, t)$	ϵ	$(s, \square)$	π_s
$(\mathsf{D}, t) \to (\mathsf{I}, t')$	$\delta\pi_{t'}$	$(\square, t \oplus t')$	1
$(\mathsf{I}, t) \to (\mathsf{M}, t')$	$(1-\delta-\epsilon)\pi_{t'}$	$(s, t \oplus t')$	$p(t' \to s)$
$(\mathsf{I}, t) \to (\mathsf{D}, t)$	δ	$(s, \square)$	π_s
$(\mathsf{I}, t) \to (\mathsf{I}, t')$	$\epsilon\pi_{t'}$	$(\square, t \oplus t')$	1
$(\mathsf{I}, t) \to \mathsf{E}$	$1[\text{after } m \text{ colors}]$	$(\square, \square)$	1
$\mathsf{E} \to \mathsf{E}$	$1-\eta$	$(s, \square)$	π_s

Fig. 2. Pair HMM for alignment of color reads and the reference DNA. π and p are the parameters of the nucleotide subsitution model (stationary distribution, and Markov-chain transition probabilities, respectively); δ, ϵ and η are the HMM's state transition parameters (gap open, gap extend, and overhang, respectively). Only the correct colors are shown in the **Emission** column, i.e., the sequencing error probability ν is 0 in this table.

2.5 Likelihood and Posterior Probabilities

A run of the pair-HMM in Fig. 2 produces an alignment, but the indels cannot be observed, only the produced sequences. Given a reference $s_{1..n}$ and a color sequence $c_{1..m}$, we can compute the likelihood that such a pair is generated by the model, while admitting color errors by known probabilities $\nu_{1..m}$. In order to compute the likelihood (and various posterior probabilities later), we use *forward* and *backward* probabilities [18,20]. The forward probabilities are denoted by $S[i] = S[i,0], E[i] = E[i,m], M_t[i,j], I_t[i,j], D_t[i,j]$ with $i = 0,\ldots,n$ and $j = 0,\ldots m$. The quantity $q[i,j]$ denotes the probability that the pair-HMM generates the prefixes $s_{1..i}$ and $c_{1..j}$ in a run that ends with state q. Forward probabilities can be computed in a recursive manner, as shown in Table 1.

Table 1. Recursions for forward probabilities

$$\begin{aligned}
S[0] &= 1; \qquad E[0] = 0 \\
S[i] &= \pi_{s_i} \cdot (1-\eta) \cdot S[i-1] && \{i > 0\} \\
M_t[i,0] &= \pi_t \cdot \eta(1-\delta) \cdot S[i]; \quad I_t[i,0] = \pi_t \cdot \eta\delta \cdot S[i]; \quad D_t[i,0] = 0 && \{i \geq 0\} \\
M_t[i,j] &= \pi_t p(t \to s_i) \sum_{t'} \Big(\gamma_j(t' \oplus t) && \{i,j > 0\} \\
&\quad \times \Big((1-2\delta) \cdot M_{t'}[i-1,j-1] \\
&\quad +(1-\delta-\epsilon) \cdot \big(I_{t'}[i-1,j-1] + D_{t'}[i-1,j-1]\big)\Big)\Big) \\
I_t[i,j] &= \pi_t \sum_{t'} \gamma_j(t' \oplus t) \Big(\epsilon \cdot I_{t'}[i,j-1] && \{i \geq 0, j > 0\} \\
&\quad +\delta \cdot \big(M_{t'}[i,j-1] + D_{t'}[i,j-1]\big)\Big) \\
D_t[i,j] &= \pi_{s_i} \Big(\epsilon \cdot D_t[i-1,j] + \delta \cdot \big(M_t[i-1,j] + I_t[i-1,j]\big)\Big) && \{i,j > 0\} \\
E[i] &= \pi_{s_i}(1-\eta) \cdot E[i-1] + \sum_t \Big(M_t[i,m] + I_t[i,m] \Big) && \{i > 0\}
\end{aligned}$$

The backward probabilities $S'[i] = S'[i,0], E'[i] = E'[i,m], M'_t[i,j], I'_t[i,j], D'_t[i,j]$ capture a symmetric concept. The quantity $q'[i,j]$ is the probability that the pair-HMM produces the suffixes $s_{i+1..n}$ and $c_{j+1..n}$ in a run starting with state q. The backward probabilities are calculated by analogous recursions to those in Table 1.

Now, given the color error probabilities $\nu_{1..m}$, the likelihood for the observed sequences is $L(s_{1..n}, c_{1..m}) = E[n] = S'[0]$. The forward and backward probabilities are combined to calculate posterior probabilities for visiting various states. The posteriors are $\psi(q)[i,j] = \frac{q[i,j] \cdot q'[i,j]}{L(s_{1..n}, c_{1..m})}$ for $q = M_t, I_t, D_t$ and $\psi(q)[i] = \frac{q[i] \cdot q'[i]}{L(s_{1..n}, c_{1..m})}$ for $q = S, E$. The posterior probabilities can be used to assign confidence to a triple alignment column. A state transition to $q = (\mathsf{M}, t)$, followed by the emission of (s, c) corresponds to an alignment column (s, c, t) of type M1–M4. Hence, $p(i \diamond j, t) = \psi(M_t)[i, j]$ is the probability that such a column aligning $s = s_i$ and $c = c_j$ is correct. The probability that s_i is deleted in the target sequence is $p(i \diamond \cdot) = \sum_{j,t} \psi(D_t)[i, j] = 1 - \sum_{j,t} p_t(i \diamond j, t)$. The probability that a column of type I1 or I2 containing $(\square, c_j, t)$ should appear in the alignment is $p(\cdot \diamond j, t) = \sum_i \psi(I_t)[i, j]$. Finally, the probability that reference nucleotide i is part of the skipped prefix or suffix is $\alpha(i) = \psi(S)[i] + \psi(E)[i] - \sum_t \bigl(\psi(M_t)[i, m] + \psi(I_t)[i, m]\bigr)$, where the non-emitting transitions into E are taken into account.

With the posterior probabilities at hand, we can find the so-called *AMAP alignment* that maximizes metric accuracy [24]. Consider an alignment with ℓ columns $\bigl((s_k, c_k, t_k) \colon k = 1, \dots, \ell\bigr)$ Let $T(k)$ be the type of column k, and let $s_k^\#$, $c_k^\#$ denote the number of non-indel reference and color characters emitted in columns $1, \dots, k$. Using a gap-factor $G \in [0, 1]$, the alignment maximizes the score $(1 - G) \cdot \sum_{k \colon T(k) \in \mathcal{M}} p(s_k^\# \diamond c_k^\#, t_k) + G \cdot \Bigl(\sum_{k \colon T(k) = \mathsf{D}} p(s_k^\# \diamond \cdot) + \sum_{k \colon T(k) \in \mathcal{I}} p(\cdot \diamond c_k^\#, t_k) + \sum_{k \colon T(k) \in \mathcal{S}} \alpha(s_k^\#) \Bigr)$, where $\mathcal{M} = \{\mathsf{M1}, \mathsf{M2}, \mathsf{M3}, \mathsf{M4}\}$, $\mathcal{I} = \{\mathsf{I1}, \mathsf{I2}\}$ and $\mathcal{S} = \{\mathsf{S}, \mathsf{E}\}$. The gap-factor sets a tradeoff between specificity and sensitivity: $G = 0$ corresponds to the alignment with maximum expected accuracy [18], and $G = 1/3$ provides a neutral setting. Computing the AMAP alignment is straightforward by dynamic programming after the posterior probabilities are calculated.

Small-scale variations such as nucleotide substitutions and short gaps can be readily identified with statistical confidence. The probability that s_i is aligned with a target nucleotide $t \in \{0, 1, 2, 3\}$ is $p(i \sim t) = \sum_j p(i \diamond j, t)$. The probability that the reference nucleotide s_i is aligned with a gap is $p(i \diamond \cdot)$. Figure 3 illustrates AMAP alignments and sequence variants. The probabilities of the

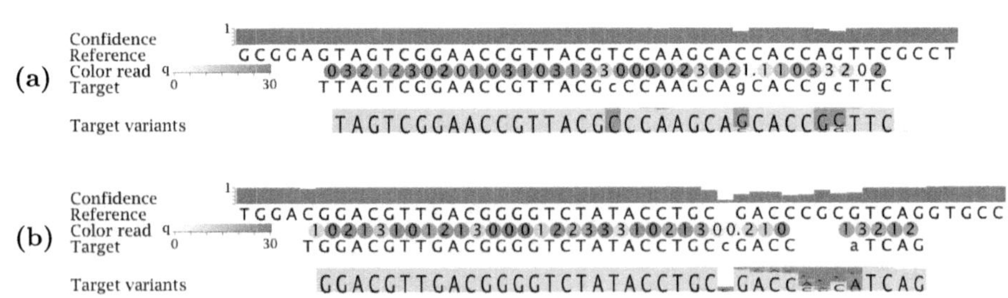

Fig. 3. AMAP alignments and sequence variations. "Confidence" is the probability of the column being correct. Shading indicates the quality values along the color sequence; a dot '.' denotes a color error. Sequence variants are shown by the logos. The height of each logo box is proportional to the probability $1 - \alpha(i)$ that the nucleotide is covered by the alignment; posterior probabilities for homology statements are shown by the relative symbol height. **(a)** Mismatches with different credibility. **(b)** Homology statements may be stronger than alignment confidence (see `GACC` before the deletion).

homology statements can be combined across different reads that align to the same reference region, in order to infer sequence variations in the target DNA.

3 Experiments

We implemented the algorithms in a Java software package called Crema, and used it on sequencing reads for *Escherichia coli* DH10B. The reads (35bp long reads, no mate pairs) were downloaded from the Applied Biosystems website (`http://download.solidsoftwaretools.com/frag/R1a007_20080307_2_EG017_F3.csfasta.zip`), with the accompanying quality file. We selected 1 million reads randomly, and mapped them against the genome of *Shigella flexneri* 2a str. 301 (Genbank accession number NC_004337.1). In the experiments, we compared our implementation with Bowtie [8] version 0.12.5, and SHRiMP [14] version 1.3.2. All programs were tested on an ordinary Linux machine (Amazon Elastic Compute Cloud, Standard Instance).

Read mapping. Table 2 shows the mapping results. In the greedy extension, we mapped the reads by retaining hits where all 35 positions be aligned within an edit distance of $e_{\max} = 6$, along a band of ± 3 diagonals. At comparable sensitivities, the greedy extension (with a platform-independent implementation) is faster than Bowtie, or SHRiMP.

Table 2. Mapping DH10B sequencing reads to *S. flexneri*. Mappings with different seeds (numbers denote length and weight) are compared with other tools at parameter settings resulting in comparable sensitivities. "Unique" reads are mapped to a single locus with maximal alignment score.

Method	CPU time	Mapped reads	Unique
SHRiMP (`-M 35bp,fast`)	263 s	593785	561301
SHRiMP (`-M 35bp,sensitive`)	717 s	605348	572317
Bowtie (`--best`)	216 s	488137	
Crema (19,17)-seed + greedy	118 s	511263	492230
Crema (16,14)-seed + greedy	192 s	569865	546495
Crema (14,12)-seed + greedy	581 s	605938	576419

Read alignment. We computed the alignments for uniquely mapped reads by first optimizing the pair-HMM parameters using a random subset of 100 thousand reads. We employed the F84 model [23] of DNA sequence evolution, with equal base frequencies (GC-content of *E. coli* is close to 50%), and a transition/transversion ratio of 2. The sequence divergence, and the gap open/extend probabilities were set in an Expectation-Maximization procedure by computing the expected numbers of substitutions and indels: convergence was achieved after four iterations with a divergence of 0.0145, gap open probability $\delta = 0.00025$ and gap extension probability $\epsilon = 0.5$. Instead of directly using the Phred formula for transforming quality scores into probabilities, we used our own mapping based on the expected number of color errors at different scores, as computed by the pair-HMM model.

Table 3. Alignments of DH10B sequencing reads with *S. flexneri*. "Validated" reads and nucleotides appear in BLAST alignments to the DH10B reference. "Incorrect" nucleotides differ from the DH10B genome sequence.

	Reads		All inferred nucleotides		Substitutions		Insertions	
	unique	validated	validated	incorrect	validated	incorrect	validated	incorrect
Bowtie (best)	465 024	463 785	15 274 713	13 862 (0.09%)	70 110	2 783 (4.0%)	*(does not infer indels)*	
SHRiMP (sensitive)	572 317	570 107	19 569 714	42 863 (0.22%)	184 254	28 364 (15.4%)	290	10 (3.4%)
Crema (AMAP alignment)	576 419	574 490	19 962 920	40 308 (0.20%)	249 107	27 162 (10.9%)	1 108	72 (6.5%)

In order to validate alignment results, we used **blastn** [25] to align the inferred target sequences to the assembled DH10B genome (Genbank accession number NC_010473.1), with default parameters and an E-value cutoff of 10^{-6}. BLAST found an alignment for 99.7–99.6% of the reads. The alignments (as reported in SAM [`http://samtools.sourceforge.net/`] format's CIGAR strings) of uniquely mapped reads were scanned to validate the inferred target nucleotides. Table 3 shows the results. Bowtie, designed to map human sequence variants, captures only very similar sequences, with an overall error rate of 0.09%. SHRiMP and Crema are much more sensitive, but have a similar 0.2% overall error. Crema is, however, better than SHRiMP at finding actual sequence differences: about 35% more substitutions are predicted, with 30% fewer errors. The framework is especially useful in annotating the computed alignments. The posterior probabilities for the inferred nucleotides can be encoded in the QUAL field of the SAM format using the phred transformation [21]. Figure 4 illustrates that high-scoring

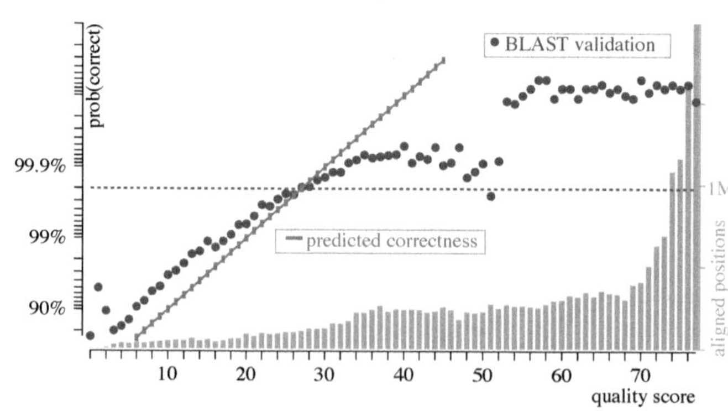

Fig. 4. Quality scores for inferred nucleotides and actual correctness ("BLAST validation") in validating BLAST hits. The horizontal dashed line shows the overall fraction of correctly inferred nucleotides. "Predicted correctness" uses the Phred formula with small bars denoting rounding errors. Vertical bars plot the frequency of quality scores with scaling shown on the right.

positions have a much lower error level. For instance, inferred nucleotides with a quality score at least 20 (96% of positions) are wrong only 0.045% of the time. The plot also shows that quality values under 30 are predicted fairly accurately (Bowtie quality values are underestimated by more than 20 on the same interval — data not shown).

4 Conclusion

We presented a seed-and-extend framework for efficient color read mapping, and a statistical alignment framework for precise alignments. The experiments demonstrate that they offer valuable options in the comparative sequencing of bacterial genomes.

References

1. Shendure, J., Li, H.: Next-generation DNA sequencing. Nat. Biotechnol. 26(10), 1135–1145 (2008)
2. Shendure, J., Mitra, R.D., Varma, C., Church, G.M.: Advanced sequencing technologies: Methods and goals. Nat. Rev. Genet. 5, 335–344 (2004)
3. Wheeler, D.A., et al.: The complete genome of an individual by massively parallel DNA sequencing. Nature 452, 872–876 (2008)
4. Pleasance, E.D., et al.: A comprehensive catalogue of somatic mutations from a human cancer genome. Nature 463, 191–196 (2010)
5. Venter, J.C., et al.: Environmental genome shotgun sequencing of the Sargasso Sea. Science 304, 66–74 (2004)
6. Flicek, P., Birney, E.: Sense from sequence reads: methods for alignment and assembly. Nat. Methods 6(11s), S6–S12 (2009)
7. Li, H., Durbin, R.: Fast and accurate short read alignment with Burrows-Wheeler transform. Bioinformatics 25(14), 1754–1760 (2009)
8. Langmead, B., Trapnell, C., Pop, M., Salzberg, S.L.: Ultrafast and memory-efficient alignment of short DNA sequences to the human genome. Genome Biol. 10 (2009)
9. Brown, D.G., Li, M., Ma, B.: A tutorial of recent developments in the seeding of local alignment. J. Bioinform. Comput. Biol. 2(4), 819–842 (2004)
10. Medvedev, P., Stanciu, M., Brudno, M.: Computational methods for discovering structural variation with next-generation sequencing. Nat. Methods 6(11s), S13–S20 (2009)
11. Huson, D.H., Auch, A.F., Qi, J., Schuster, S.C.: MEGAN analysis of metagenomic data. Genome Res. 17, 377–386 (2007)
12. Smith, T.F., Waterman, M.S.: Identification of common molecular subsequences. J. Mol. Biol. 147, 195–197 (1981)
13. Gotoh, O.: An improved algorithm for matching biological sequences. J. Mol. Biol. 162(3), 705–708 (1982)
14. Rumble, S.M., Lacroute, P., Dalca, A.V., Fiume, M., Sidow, A., Brudno, M.: SHRiMP: Accurate mapping of short color-space reads. PLoS Comput. Biol. 5(5), e1000386 (2009)
15. Homer, N., Merriman, B., Nelson, S.F.: Local alignment of two-base encoded DNA sequence. BMC Bioinformatics 10, 175 (2009)

16. Wu, S., Manber, U., Myers, G., Miller, W.: An $O(NP)$ sequence comparison algorithm. Inform. Process. Lett. 35(6), 317–323 (1990)
17. Zhang, Z., Schwartz, S., Wagner, L., Miller, W.: A greedy alignment for aligning DNA sequences. J. Comput. Biol. 7(1/2), 203–214 (2000)
18. Durbin, R., Eddy, S.R., Krogh, A., Mitchison, G.: Biological Sequence Analysis: Probabilistic Models of Proteins and Nucleic Acids. Cambridge University Press, UK (1998)
19. Lunter, G., Drummond, A.J., Miklós, I., Hein, J.: Statistical alignment: Recent progress, new applications, and challenges. In: Nielsen, R. (ed.) Statistical Methods in Molecular Evolution. Springer, Heidelberg (2005)
20. Rabiner, L.R.: A tutorial on Hidden Markov Models and selected applications in speech recognition. Proc. IEEE 77(2), 257–286 (1989)
21. Ewing, B., Green, P.: Base-calling of automated sequencer traces using phred: II. error probabilities. Genome Res. 8, 186–194 (1998)
22. Liò, P., Goldman, N.: Models of molecular evolution and phylogeny. Genome Res. 8, 1233–1244 (1998)
23. Felsenstein, J., Churchill, G.A.: A Hidden Markov Model approach to variation among sites in rate of evolution. Mol. Biol. Evol. 13(1), 93–104 (1996)
24. Schwartz, S.A., Pachter, L.: Multiple alignment by sequence annealing. Bioinformatics 23(2), 24–29 (2007)
25. Altschul, S.F., Gish, W., Miller, W., Myers, E.W., Lipman, D.J.: Basic local alignment search tool. J. Mol. Biol. 215(3), 403–410 (1990)

Design of an Efficient Out-of-Core Read Alignment Algorithm

Arun S. Konagurthu[1,2,⋆,⋆⋆], Lloyd Allison[1,⋆⋆], Thomas Conway[1,2], Bryan Beresford-Smith[1,2], and Justin Zobel[2,1]

[1] National ICT Australia (NICTA) Victoria Research Laboratory, Department of Electronics and Electrical Engineering The University of Melbourne, Parkville, Victoria 3010 Australia
[2] Department of Computer Science and Software Engineering, The University of Melbourne, Parkville, Victoria 3010 Australia
{firstname.lastname}@nicta.com.au

Abstract. New genome sequencing technologies are poised to enter the sequencing landscape with significantly higher throughput of read data produced at unprecedented speeds and lower costs per run. However, current *in-memory* methods to align a set of reads to one or more reference genomes are ill-equipped to handle the expected growth of read-throughput from newer technologies.

This paper reports the design of a new out-of-core read mapping algorithm, Syzygy, which can scale to large volumes of read and genome data. The algorithm is designed to run in a constant, user-stipulated amount of main memory – small enough to fit on standard desktops – irrespective of the sizes of read and genome data. Syzygy achieves a superior spatial locality-of-reference that allows all large data structures used in the algorithm to be maintained on disk. We compare our prototype implementation with several popular read alignment programs. Our results demonstrate clearly that Syzygy can scale to very large read volumes while using only a fraction of memory in comparison, without sacrificing performance.

1 Introduction

The landmark publications of Margulies *et al.* [1] and Shendure *et al.* [2] in 2005 heralded a new era of non-Sanger based, massively parallel genome sequencing technologies. Today's major commercial *next-generation sequencing* (NGS) systems include Roche's (454) Genome Sequencer FLX, Illumina-Solexa's Genome Analyzer (GA) II, and Applied Biosystem's SOLiD. The volume of data generated from these new sequencers is already staggering. (For example, Illumina's latest GA IIe sequencer produces about 1.75×10^9 bases of read data in a day's run.) More recently, several new sequencing systems, such as Helicos' Genetic

⋆ Corresponding author.
⋆⋆ These authors contributed equally to this work.

V. Moulton and M. Singh (Eds.): WABI 2010, LNBI 6293, pp. 189–201, 2010.

Analysis system, Pacific Biosciences' Single Molecule Real Time (SMRT) system, Oxford Nanopores' Nanopore sequencer, and Visigen's Genetic sequencer, have been announced promising higher read throughput at faster speeds and significantly reduced costs per run. Some of these technologies are already in business.

Mapping (or aligning) a set of reads to a reference genome is a fundamental task in genome resequencing studies. Massive volumes of read data (growing faster than Moore's law) and very large genome sizes make the read mapping problem computationally very challenging. Neither the classical methods for pattern matching on strings [3–5] nor the methods from traditional sequence bioinformatics [6, 7] can cope with the large volumes of data from modern sequencers.

Since 2007, several methods catering specifically to NGS were published [8–19].[1] These methods can be broadly classified into four groups:

1. Methods which rely on hashing the read set (*e.g.* see [8–12]);
2. Methods which rely on hashing the reference (*e.g.* see [13–15]);
3. Methods based on advances in Stringology (*e.g.* see [16–18]);
4. A method [19] based on a *sort-and-join* approach.

Methods in categories 1 and 2 maintain a large hash index in main memory when performing read alignment. The growth of data (read or genome, depending on which set is maintained as a hash index) translates to increasing demands on main memory for these methods. Methods in category 3, especially those that use the Burrows-Wheeler index [20] of the reference sequence are comparatively memory efficient when aligning reads to a single genome [16, 17]. However if reads were to be mapped on more than one large genome (for example, multiple human genomes simultaneously), even these methods begin to have impractical memory demands. Slider [19] is currently a solitary method in category 4 which relies on a simple sort-and-join strategy. The program is slow and requires both a large amount of memory as well as disk space.

A common problem with the current programs is that they do *not* scale elegantly to handle very large data volumes due to impractical memory requirements or very long run times (in some cases, both). Moreover, the random nature of data accesses to the index structures maintained by the methods in the first three categories pose a major hurdle for their implementation out of core.

This paper describes the design of a new method, `Syzygy`, to efficiently align massive numbers of reads simultaneously against multiple genomes. The design allows the program to run in a fixed, user-stipulated amount of memory, small enough to be deployed even on standard computers. Broadly, our method is based on a *sort-and-join* strategy similar to the one used by Slider [19]. However the details of our algorithm and its implementation is radically different, especially in the way it handles the *approximate* read mapping problem. `Syzygy` reorganizes the read mapping problem to achieve a superior *spatial locality* of accesses to various data structures maintained by the method. This reorganization results

[1] A comprehensive list of NGS read mapping tools is maintained by Heng Li at `http://lh3lh3.users.sourceforge.net/NGSalign.shtml`

in accesses to all data structures being predominantly *linear*, facilitating an *out-of-core* implementation among other performance optimizations; all large data structures in the algorithm are maintained on disk, requiring only a small *in-memory* working set to proceed with the alignment.

2 Algorithm

2.1 Definitions

DNA sequence: A DNA sequence of length n is a string $S = (s_1 \cdots s_n)$ containing n 'bases', where each base is from the alphabet of DNA nucleotides, $\aleph = \{A, C, G, T\}$.
Reference genome set: Let $\mathcal{G} = \{\mathcal{G}^1 \cdots \mathcal{G}^N\}$ be a set containing N reference genomes where any genome $\mathcal{G}^i = (g_1^i \cdots g_{n_i}^i)$ (assume) is a DNA sequence containing n_i bases.
Read set: Let $\mathcal{R} = \{r^1 \cdots r^m\}$ be a set containing m sequence reads, where any read $r^i = (r_1^i \cdots r_L^i) \in \mathcal{R}$ of length L is a short DNA sequence.
k*-mer:* Given any sequence $S = (s_1 \cdots s_l)$ of length l and a constant $k \leq l$, a k-mer of S defines another sequence $\mathcal{K} = (s_i \cdots s_{i+k-1}), 1 \leq i \leq l - k + 1$ which is a substring of S.
Reverse complement: A reverse complement of a DNA sequence S, denoted by $\overline{S}$, is a sequence of bases which reverses S and replaces each base with its Watson-Crick conjugate (A with T, G with C, and vice versa).
Key: A `key`(S) denotes an integral hash value of a sequence S using some key-generation function which transforms strings uniquely to integers. A straightforward key generation function of DNA sequences over the alphabet $\aleph$ is the integral value as a result of representing the sequence using a 2 bits-per-base encoding. (For example, $\{00, 01, 10, 11\}$ for $\{A, C, G, T\}$.)

2.2 The Basic Sort-and-Join Method

We build our exposition by introducing first the basic sort-and-join method to align reads simultaneously to a set of genomes. In the basic approach we use the ideal scenario where reads R are matched *exactly* (that is, without errors) with the genome set $\mathcal{G}$. (To make the exposition clearer, in the entire paper we assume that all reads in the set $\mathcal{R}$ are of a fixed length L. We note, however, that it is straightforward to generalize our algorithm to variable length reads.) The basic algorithm involves three simple steps:

Step 1: Reference list generation. A reference list defines a sorted list of records G corresponding to every L-mer in $\mathcal{G}$. Each L-mer, $\mathcal{L} = (g_j^i \cdots g_{j+L-1}^i), 1 \leq i \leq N, 1 \leq j \leq n_i - L + 1$, contributes the fields (h, p) to form a record in G, where $h =$ `key`$(\mathcal{L})$ is the key of $\mathcal{L}$, and $p = [i, j]$ is the positional coordinate (sequence number and offset in sequence) of $\mathcal{L}$ in $\mathcal{G}$. The records in list G are sorted on the field h.

Step 2: Read list generation. For each read $r = (r_1 \cdots r_L) \in \mathcal{R}$, construct a read list $R = \{(h, r)\}$ where $h =$`key(r)`. R is also sorted on the field h.
Step 3: Join list generation. A join list $J = G \bowtie R$ is derived by joining $G = \{(h, p)\}$ and $R = \{(h, r)\}$ on the field h, resulting in $J = \{(p, r)\}$. Each record in this list gives a position p of the *exact* occurrence of a read r.

Observe the predominantly linear nature of data accesses in this method. In Steps 1 and 2, genome and reference data are read sequentially while creating G and R respectively. Again, the generation of the join list J requires all sequential accesses through the sorted lists G and R, giving the list of matches of reads against the genome(s). We note here that sorting of small keys (of fixed size) is *near-linear* [21, 22].

Below, we use the framework of the basic sort-and-join strategy to address the problem of approximate matching of reads to multiple genomes.

2.3 Sort-and-Join Method for Mapping with Errors

In practice, a large number of reads will not map *exactly* to reference sequence(s) due to the presence of sequencing errors in the reads as well as other natural genomic variations between the sample and the reference draft assembly. Therefore it becomes necessary to map the reads to the reference genomes allowing a certain number of errors or mismatches. A common strategy to handle approximate matches is based on a lossless k-mer filtering technique. This technique relies on the observation that two sequences of length L which are at a Hamming distance of *at most* δ should share *at least* one k-mer of size $k = \lfloor \frac{L}{\delta+1} \rfloor$. Indeed this observation generalizes to Levenshtein distances between two strings. Our method for approximate matching uses this observation. Below we describe the extension of the basic sort-and-join strategy to map the reads in the presence of errors under a threshold of Hamming distance δ:

Step 1: Reference list generation. Build a sorted reference list G such that the tuples correspond to k-mers (instead of L-mers, previously) such that $k = \lfloor \frac{L}{(\delta+1)} \rfloor$.
Step 2: Read list generation. Each read is partitioned into fixed size (non-overlapping) tiles of length $k = \lfloor \frac{L}{\delta+1} \rfloor$. The pigeonhole principle suggests that if a read matches under a threshold of δ at some position in the reference, then there must be *at least* one of the $\delta + 1$ read tiles that must match *exactly* to a corresponding k-mer in the reference. Note that due to the chemistry involved in the sequencing process each read should also be examined for a match against the reference using its reverse complement.

Each read, therefore, contributes $2 \times (\delta + 1)$ non-overlapping (tiled) k-mers (in both forward and reverse complement directions), where $k = \lfloor \frac{L}{(\delta+1)} \rfloor$. Specifically, a read $r^i = (r^i_1 \cdots r^i_L)$ and its corresponding reverse complement $\overline{r^i} = (\overline{r^i}_1 \cdots \overline{r^i}_L)$ contributes to these non-overlapping k-mer tiles: $\{(r^i_1 \cdots r^i_k),\ (r^i_{k+1} \cdots r^i_{2k}),\ \cdots,\ (r^i_{\delta \times k+1} \cdots r^i_{(\delta+1)\times k})\}$ and $\Big\{(\overline{r^i}_1 \cdots \overline{r^i}_k),$ $(\overline{r^i}_{k+1} \cdots \overline{r^i}_{2k}),\ \cdots,\ (\overline{r^i}_{\delta \times k+1} \cdots \overline{r^i}_{(\delta+1)\times k})\Big\}$.

Hence each tile of a read now contributes to the following fields: (h, r, o, s), where h is the key of a tile, r the read from which it came, o the offset of the tile from the start of the read, and s the 'sense' (forward or reverse complement) of the tile. For example, the k-mer tile $\mathcal{K} = ({r^i}_{k+1} \cdots {r^i}_{2k})$ and its corresponding reverse complement $\overline{\mathcal{K}} = (\overline{{r^i}}_{k+1} \cdots \overline{{r^i}}_{2k})$ contribute to R: $\big(\texttt{key}(\mathcal{K}), r^i, (k+1), \rightarrow\big)$, and $\Big(\texttt{key}(\overline{\mathcal{K}}), \overline{r^i}, (k+1), \leftarrow\Big)$, where '$\rightarrow$' and '$\leftarrow$' indicate matching in the *sense* (forward) and *antisense* (reverse complement) directions of the k-mer tiles respectively. Finally, the list $R \equiv \{(h, r, o, s)\}$ is sorted on the field h.
Step 3: Join list generation. Join the sorted lists $G = (h, p)$ and $R = (h, r, o, s)$, to give a new list $J = G \bowtie R \equiv \{(p', r, s)\}$, where $p' = p - o$ is the adjusted positional coordinate on the genome set which allows different tiles of the same read to coalesce back together. (See Step 4.) For a given read if more than one tile matches exactly at some position in the genome, the adjustment p' ensures that they point to the same starting position on the reference. This provides the necessary efficiency in the post-processing step below.
Step 4: Verification and post-process. The list J is sorted on the fields p', r, and s in that order. Let a *context* in J define a set of items in J that have the same (adjusted) position, read, and sense tuples. Sorting J on (p', r, s) ensures that tiles which share the same context will group together. Each context, containing one or more tiles, identifies a unique position $p' \in \mathcal{G}$, read r, and the directionality of the match s. While traversing linearly in the newly sorted list J, for each unique context, just one tile is enough to *verify* the Hamming distance of the L-mer in $\mathcal{G}$ starting at position p' with respect to the read r, in the direction specified by s. The result of the verification is a list of mappings of the set $\mathcal{R}$ on $\mathcal{G}$ under the approximate Hamming distance threshold δ.

We note that the extension of the sort-and-join strategy to approximate matching still retains its sequential data access characteristic which is important for any out-of-core implementation. In step 4, each Hamming distance verification of the join record requires the extraction of a read-length sized substring from the reference (string) data set which is then compared with the corresponding read available in the join record. Since the join list is sorted primarily on the (adjusted) position in the reference, accesses to the reference string(s) are sequential, giving the crucial advantage of spatial locality of reference.

3 Implementation of Syzygy

3.1 Encoding, Key Generation and Other Bitwise Tricks

A nucleotide sequence from the alphabet containing four bases $\{A, T, G, C\}$ is *packed* into an array of unsigned 64-bit integer data types, where each base is represented using a *2 bits-per-base* encoding. Specifically, Syzygy uses the $\{00, 01, 10, 11\}$ binary encoding for $\{A, C, G, T\}$ respectively. Each unsigned 64-bit word (or, plainly, *word*) can encode information of up to 32 bases of a sequence. For example, a DNA sequence of length 100 is packed into an array of 4 encoded words. (This requires an implicit convention to align strings to a consistent word boundary, necessary to decode strings whose length are not an integral multiple of 32. Assuming a *little-endian* architecture, in Syzygy, the start of strings are aligned to the *most-significant* word boundary). In the implementation of Syzygy, the key of a k-mer is simply the integer defined by its encoded word(s).

In practice, both the genome and read sequences come from an *extended* alphabet to account for ambiguous or unknown bases. On the genome side, we store the strings in the encoded form by converting each ambiguous base to a random unambiguous base in $\aleph$, while ignoring the contributions to the reference list G by the k-mers in the regions containing ambiguous bases. For the read set we ignore all sequences containing more than two ambiguous bases.

```
#define MASK 0x5555555555555555UL

uint8_t hammingDistance(
 uint64_t *w1,
 uint64_t *w2,
 size_t n
) {
   uint64_t tmp ;
   uint8_t  d = 0 ;

   for( size_t i = 0 ; i <  n ; i++ ) {
     /* XOR ith words */
     tmp = w1[i] ^ w2[i];

     /* convert to a popcount problem */
     tmp = (tmp & MASK) |
                 (tmp>>1 & MASK) ;

     /* count set bits in tmp */
     d += popcount( tmp ) ;
   }
   return d ;
}
```

Fig. 1. Code for Hamming distance computation of two strings packed into an array of n words $w1$ and $w2$

This encoding has some convenient advantages. Hamming distance and reverse complement generation can be performed cheaply using a few bitwise operations. The code for determining Hamming distance is shown in Fig. 1. (There are several bitwise tricks to facilitate fast population count using `popcount()` in Fig. 1. See [23] for a comprehensive summary on accelerated population counting. Alternatively, a rapid way to perform this operation would be to invoke a `POPCNT` instruction which comes as a part of the instruction set on most modern microprocessors.) Coincidentally, the specific encoding used in Syzygy also allows a fast computation of reverse complements of DNA sequencing which relies on the fact that, in our encoding, Watson-Crick conjugates have their bits flipped. (See Fig. 2 below.)

3.2 Main List Data Structures

Reference list: Recall, the reference list G is a list of records of the form $\{(h, p)\}$. The field h, storing the key (the integer encoding) of a k-mer, can be denoted using an array of one or more words depending on the size of the k-mer. Field p on the other hand can be simply stored in a single word. In Syzygy, however, the reference list is generated using 32-mers from the reference genome set, requiring just one word each to store the fields h and p. (See also the special construction of the reference list, explained in section 3.3, to handle *'blowups'* in the join.) In addition, we observe that a 32-mer based reference list containing (h, p) tuples subsumes all reference lists corresponding to any $(k < 32)$-mers. This holds primarily due to the encoding Syzygy uses and the nature of its key generation function. A fully sorted 32-mer keys implies the sorted order of any of its prefixes. For example, a

16-mer reference list can be derived from a 32-mer list by trivially masking out the encoded bits corresponding to the trailing 16 bases. (Note however that some ($k < 32$)-mers in the end of each genome – or any discontinuous region – will be lost when sliding along the genome set with a window of size 32. But this is a minor issue which can be trivially remedied.) Using the above observation, a 32-mer reference list can be *preprocessed* and used for mapping under variable tile sizes, calculated based on the hamming distance threshold and the read length. This is especially meaningful because often it is the read set which changes while the set of reference genomes remains mostly static.

Read list: Reference list R is a list of records of the form $\{(h, r, o, s)\}$. To conserve space, we do not store the field h but, instead, it is calculated on the fly using the remaining fields: r, o and s. Field r is an array of words representing the encoded read. In addition to the encoded read data, it makes practical sense to carry along a read identifier. Read identifier (a number) and fields o and s are packed together into one word. We will call this word, *read metadata*.

Syzygy computes an appropriate tile size depending on the read length L and the parameters for approximate matching δ as $\min\{\lfloor L/(\delta+1)\rfloor, 32\}$.

```
#define MASK1 0x3333333333333333UL
#define MASK2 0x0F0F0F0F0F0F0F0FUL
#define MASK3 0x00FF00FF00FF00FFUL
#define MASK4 0x0000FFFF0000FFFFUL

void reverseComplement(
 uint64_t *w,
 size_t n
) {
   uint64_t tmp ;

   /* First, reverse complement one word at a time */
   for( size_t i = 0 ; i < n ; i++ ) {
     /* A base complement trick on whole word */
     w[i] = ~w[i] ;

     /* Reverse base (NOT bits) order in each word*/
     w[i] = ((w[i]>>2)&MASK1) | ((w[i]&MASK1)<<2) ;
     w[i] = ((w[i]>>4)&MASK2) | ((w[i]&MASK2)<<4) ;
     w[i] = ((w[i]>>8)&MASK3) | ((w[i]&MASK3)<<8) ;
     w[i] = ((w[i]>>16)&MASK4) | ((w[i]&MASK4)<<16) ;
     w[i] = ((w[i]>>32)) | (w[i]<<32) ;
   }

   /* Next, reverse order of words in array  */
   for( size_t i = 0, j = n-1 ; i < j ; i++, j-- ) {
     tmp  = w[i] ; w[i] = w[j] ; w[j] = tmp ;
   }
  /* additional shift of bases across words may be
     needed to align to consistent word boundary */
}
```

Fig. 2. Code of reverse complement computation of a sequence w encoded into an array of n words

Join list: The resultant join list J consists of, for each item in the list, a word corresponding to a k-mer's (adjusted) position in the genome set, a word of the read metadata and remaining words corresponding to the encoded read.

Memory and disk usage: Syzygy runs within the user-defined amount of memory($\geq$ 32 MB). The program operates under the stipulated memory limit by maintaining all main data structures on disk rather than in memory. Each of the large data structures in the algorithm, G, R, and J is simply represented as one large linear array of words.

3.3 Managing "Blowups" in the Join List

A major challenge in this approach is to manage the size of the join list J. Consider a range of records in $R = \{(h, r, o, s)\}$ which share the same key h with another range of records in $G = \{(h,p)\}$. During the join operation, the join list will be populated with the *product*—Cartesian product of two sets containing items from the two ranges—of the items in these two ranges. For preponderant k-mers (such as poly-As in the Human genome which occur a very large number of times), this product would be staggering causing an unmanageable "blowup" in the size of the join.

Syzygy uses the following strategies to handle blowups. The program initially computes and stores (in memory) all k-mers (for a given tile size in the run) and their corresponding preponderances when they exceed a predefined threshold (100, 000). Now, when generating the read list, tiles which can potentially cause a blowup are excluded from the read list at the time of its generation. A read is blacklisted (and written out to a blacklist file) if one or more of its tiles cause a blowup. For a given read it is however possible that only a few tiles are excluded while the remainder enter the read list, in which case the program does *not* guarantee to find all matches under the chosen distance parameters for that read. The program generates a 'greylist' of reads to inform the user of such cases.

We observe that the join list usually contains many *false-positives* which are removed at a later verification stage (in step 4). A vast portion of records in the join can be filtered out using a fast *early-verification* step. Recall (from section 3.1) that each record in the reference list has 32 bases. For a given tile size T, the join is performed by masking out unrelated $32 - T$ trailing bases. For a tile size of (say) 20, the bits corresponding to the 12-mer suffixes (that is, 24 bits) in all 32-mer keys from G are masked out. Since we are carrying in R the read data corresponding to all k-mer tiles, an early verification can be performed during the join on the remaining $32 - k$ bases between the read and the reference. If a tile fails the chosen distance threshold so would the read it belongs to. Hence such tiles are barred from entering into the join list.

While this procedure prunes the number of false positives drastically, it is possible, under some parameters, that the last tiles of the reads do not benefit from the early-verification. Take for example a read of length 50 mapped under an Hamming distance of 2. Each read will contain 3 tiles of length 16. While the first two tiles can be verified on 32-bases, the last tile can only be verified on the remaining two bases(that is 49th and 50th base in the read). Since we are running under a Hamming distance of 2 and there are only two additional bases to check, it is easy to see that all the last tiles of the read in such a run will pass unfiltered into the join list.

An alternative construction of Reference list used in Syzygy to aid early-verification of all tiles: A modification in the reference list generation will guarantee even the last tiles to benefit from early-verification. Recall that the reference list is constructed on 32-mers (and their positions) lexicographically sorted on the entire 32-mer keys. Modify the sort procedure by sorting the keys *only* on their 28-mer suffixes, instead of the complete 32-mers. To illustrate why this helps, we use the example in the above paragraph. The first 16-mer tile can now be verified on 28-bases (of which we know that 16-bases match exactly).

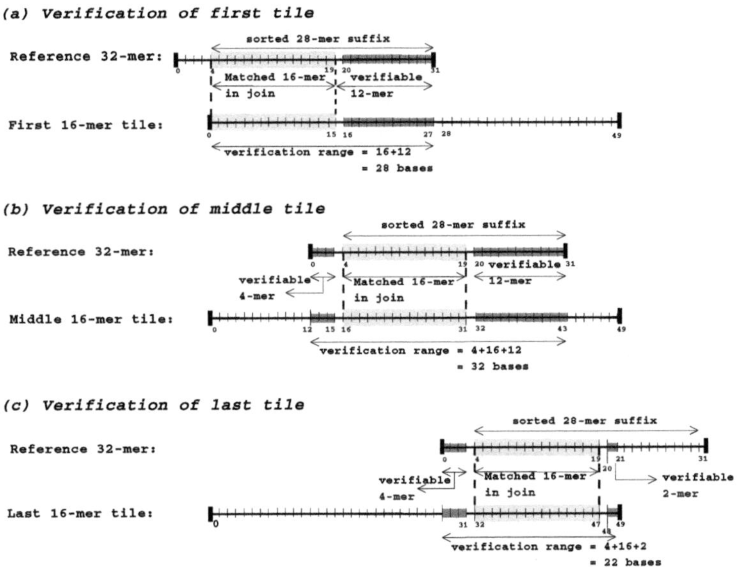

Fig. 3. An illustration of early-verification. (Assume: read length = 50; Hamming distance = 2; Tile size = 16. The reference list G is constructed on 32-mers sorted on their 28-mer suffixes.) **(a) Verification of all first tiles:** When the key of a first tile in a read matches a 16-mer key from the reference, bases at the positions [0-15] of the read are equivalent to those at the positions [4-19] of the reference. Early-verification can be performed on bases at the positions [16-27] of the read with those in positions [20-31] of the reference. **(b) Verification of all middle tiles:** When the key of a middle tile in a read matches a 16-mer key from the reference, bases at the positions [16-31] of the read are equivalent to those at the positions [4-19] in the reference. Early-verification can be performed on bases at the positions [12-15] and [32-43] of the read with those at the positions [0-3] and [20-31] of the reference respectively. **(c) Verification of all last tiles:** When the key of a last tile in some read matches with a 16-mer key from the reference, bases at the positions [32-47] in the read are equivalent to bases at the positions [4-19] in the reference. Early-verification can be performed on bases at the positions [28-31] and [48-49] of the read with those in the positions [0-3] and [20-21] of the reference respectively.

(See Fig. 3(a).) The middle tile can be verified on all 32-positions including the 4 base prefix that was left out from sorting. (See Fig. 3(b).) Importantly, the last tile can now be verified on the remaining 2 bases (49th and 50th) in the read, plus on the 4 bases preceding the last tile. (See Fig. 3(c).) In summary, early-verification filters out a significant portion of false-positives from entering into the join list, thereby severely constraining the final size of the join list. This naturally translates to a significant boost in the program's run time since it is implemented out of core.

3.4 Overlapped I/O

`Syzygy` uses a system of memory-mapped files and buffered writing for all I/O involving its data structures that are maintained on disk.

Memory mapping is an efficient alternative to standard file I/O on Unix-based systems where the kernel provides an interface to seamlessly map portions of very large files into memory. POSIX.1 standardizes the system call `mmap()` which most current Unix-based distributions implement. In addition, the user can advise the kernel in advance on how a memory-mapped block can be utilized. Linux provides, for example, `madvise()` system call for this purpose. Where the kernel recognizes (or is advised) that the accesses to memory mapped blocks are sequential, it automatically optimizes the read-performance by *caching* asynchronously pre-fetched pages of data through aggressive read-ahead from the point where data is being processed in the memory-mapped block.

`Syzygy` uses `write()` defined in POSIX.1 to write data to disk. The behaviour of write() is optimized by Linux. Write calls return immediately because the kernel copies the data into its buffers, batching many writes together and deferring the *writeback* to disk at a later time to be written asynchronously.

3.5 Sorting

`Syzygy` implements an efficient *external* sorting that is capable of sorting a very large number of records (and their payloads) on disk. Broadly, the sorting is performed in two steps. In the first step the collection is partitioned into several consecutive blocks or *runs*, where each run is fully sorted in memory. The size of each partition depends on the user-defined memory limit. `Syzygy` implements a variant of *least significant digit* (LSD) radix sort to sort the runs in memory. We note that the radix sort in general has a complexity of $O(kn)$, where k is the number of radixes in the sort-key while n is the size of the run. Our implementation uses byte-size radices. Sorting a sort-key of size one 64-bit word requires 8 linear passes (using byte-size radices) on the run. Our implementation additionally benefits from several algorithmic and hardware-directed optimizations derived from the works of [21] and [22]. The second step involves a merge step which merges partially sorted runs of a collection on disk into a fully sorted collection.

4 Results and Discussion

One tool from each of the three categories of read mapping programs was chosen to compare against our prototype implementation of `Syzygy`. Specifically, `maq` [10] from category 1, `soap(v.1)` [13] from category 2 and `bwa` [16] from category 3 were chosen for comparisons.

Reads used here belong to sample NA19240, sequenced using *Illumina* technology, downloaded from the *1000 Genome Project*.[2] Human genome (draft 18)

[2] `http://www.1000genomes.org/`

Table 1. Results of comparison of `Syzygy` with `maq`, `soap(v.1)` and `bwa` when mapping varying sizes of reads on the Human genome under a Hamming distance threshold of 2. All programs were run on a single core as a sequential program. '–' in various columns indicate that the program ran exceeded 48 hours and hence were aborted. Runs on `Syzygy` were performed with a 1GB memory limit. (`Syzygy` runs faster with a larger memory limit.)

Program	nReads	Time (wall)	Memory	Scratch
maq	1 million	1.9 hrs	1.1 GB	0
soap(v.1)	1 million	7.4 hrs	14.1 GB	0
bwa	1 million	0.4 hrs	2.3 GB	0
Syzygy	1 million	0.7 hrs	1 GB	5.7 GB
maq	10 million	21.6 hrs	5.7 GB	0
soap(v.1)	10 million	—	—	0
bwa	10 million	3.7 hrs	2.3 GB	0
Syzygy	10 million	3.4 hrs	1 GB	27 GB
maq	50 million	—	—	0
soap(v.1)	50 million	—	—	0
bwa	50 million	20.5 hrs	2.3 GB	0
Syzygy	50 million	8.2 hrs	1 GB	65.4 GB
maq	100 million	—	—	0
soap(v.1)	100 million	—	—	0
bwa	100 million	40.0 hrs	2.3 GB	0
Syzygy	100 million	13.9 hrs	1 GB	103 GB

was used as the reference genome. All experiments were carried out on a single node of a AMD Quad-core server with 32GB of main memory. The server is connected to a large array of Serially Connected SCSI (SAS) disks accessible via fast Ethernet (giving a upper limit of 125 MB/s on the disk bandwidth).[3]

Table 1 summarizes the results of comparison between various programs. From the table we can see that as the volume of reads increases, the run times of other programs grow drastically. `soap(v.1)` has the worst run time of all the methods, becoming impractical even at a read volume of 10 million. At a read volume of 50 million, `maq` becomes impractical. Only `bwa` scales to 50 million reads and beyond. `Syzygy` runs significantly faster than all other program, especially on very large read volumes. At the read volume of 100 million, `Syzygy` is $\sim$ 2.8x faster than `bwa`. The results clearly show that `Syzygy` scales elegantly compared to other methods.

Notice in Table 1 that the memory usages of various programs. `Syzygy` runs in the user-stipulated amount of memory (in this case 1GB) while the rest of the programs have a variable memory footprint depending on the size of various indexes they maintain. `soap(v.1)` uses an inverted hash table on the reference genome. For a human genome the size of this data structure is roughly 14GB

[3] The runs of `Syzygy` in Table 1 uses two independent disks to alternate reading the writing operations of the algorithm to reduce the I/O latency.

which has to be maintained in memory. maq uses an inverted index composed of pairs of k-mer tiles (that is, spaced tiles). However, since such an index prepared on large reference genomes is huge, maq chooses to prepare the inverted index on the read set. To further conserve space, maq indexes the read set in batches compromising on program speed. bwa which uses a Burrow-Wheeler index [20] has the most concise index structure among the tools. However, for the program to work efficiently, the entire index has to be maintained in memory. This comes in the way of mapping a set of reads on multiple (instead of one) genome.

Syzygy uses a large amount of scratch space on disk to allow the program to run in a fixed, user-defined amount of main memory. The last column of Table 1 gives the amount of scratch space used by the program for various runs. Note that, compared to main memory, disks are inexpensive and vastly bigger in sizes. (A standard 1TB disk costs only a couple of hundred Dollars.)

5 Conclusion

The design of an efficient read mapping algorithm, Syzygy, has been described in this paper. The program reorganizes the read mapping problem to maximize spatial locality of reference of data accesses in the algorithm, a crucial ingredient for any performance optimization. This facilitates the program to maintain all its data structures on disk, and hence allowing Syzygy to run in any user-defined amount of memory. Syzygy scales elegantly to volumes of read and genome data unachievable by current read-mapping programs. Our future work will extend Syzygy to handle paired-end read mapping while generalizing the alorithm for mapping under an edit distance threshold. We are also working towards parallelizing Syzygy for symmetric as well as distributed memory multiprocessors. An academic version of the program will be available shortly from http://www.csse.unimelb.edu.au/~arun/syzygy.

Acknowledgement

National ICT Australia (NICTA) is funded by the Australian Government's Department of Communications, Information Technology and the Arts, the Australian Research Council through Backing Australia's Ability, and the ICT Centre of Excellence programs.

References

1. Margulies, M., Egholm, M., Altman, W., et al.: Genome sequencing in microfabricated high-density picolitre reactors. Nature 437, 376–380 (2005)
2. Shendure, J., Porreca, G.J., Reppas, N.B., Lin, X., Mccutcheon, J.P., Rosenbaum, A.M., Wang, M.D., Zhang, K., Mitra, R.D., Church, G.M.: Accurate multiplex polony sequencing of an evolved bacterial genome. Science 309, 1728–1732 (2005)
3. Boyer, R.S., Moore, J.S.: A fast string searching algorithm. Commun. ACM 20(10), 762–772 (1977)

4. Knuth Jr., D.E., Pratt, V.R.: Fast pattern matching in strings. SIAM Journal on Computing 6(2), 323–350 (1977)
5. Karp, R.M., Rabin, M.O.: Efficient randomized pattern-matching algorithms. IBM Journal of Research and Development 31(2), 249–260 (1987)
6. Altschul, S.F., Gish, W., Miller, W., Myers, E.W., Lipman, D.J.: Basic local alignment search tool. Journal of Molecular Biology 215, 403–410 (1990)
7. Kent, W.J.: BLAT–the blast-like alignment tool 12, 656–664 (April 2002)
8. Cox, A.J.: Ultra-high throughput alignment of short sequence tags (2007) (unpublished)
9. Rumble, S.M., Lacroute, P., Dalca, A.V., Fiume, M., Sidow, A., Brudno, M.: SHRiMP: accurate mapping of short color-space reads. PLoS Computational Biology 5 (May 2009)
10. Li, H., Ruan, J., Durbin, R.: Mapping short dna sequencing reads and calling variants using mapping quality scores. Genome Research (August 2008)
11. Lin, H., Zhang, Z., Zhang, M.Q., Ma, B., Li, M.: ZOOM! zillions of oligos mapped. Bioinformatics 24, 2431–2437 (2008)
12. Jiang, H., Wong, W.H.: SeqMap: mapping massive amount of oligonucleotides to the genome. Bioinformatics 24, 2395–2396 (2008)
13. Li, R., Li, Y., Kristiansen, K., Wang, J.: SOAP: short oligonucleotide alignment program. Bioinformatics 24, 713–714 (2008)
14. Eaves, H.L., Gao, Y.: MOM: maximum oligonucleotide mapping. Bioinformatics 25, 969–970 (2009)
15. Campagna, D., Albiero, A., Bilardi, A., Caniato, E., Forcato, C., Manavski, S., Vitulo, N., Valle, G.: PASS: a program to align short sequences. Bioinformatics 25, 967–968 (2009)
16. Li, H., Durbin, R.: Fast and accurate short read alignment with Burrows-Wheeler transform. Bioinformatics 25, 1754–1760 (2009)
17. Langmead, B., Trapnell, C., Pop, M., Salzberg, S.L.: Ultrafast and memory-efficient alignment of short dna sequences to the human genome. Genome Research 10 (March 2009)
18. `http://www.vmatch.de/`
19. Malhis, N., Butterfield, Y.S., Ester, M., Jones, S.J.: Slider–maximum use of probability information for alignment of short sequence reads and snp detection. Bioinformatics 25, 6–13 (2009)
20. Ferragina, P., Manzini, G.: Opportunistic data structures with applications. In: Proceedings of Foundations of Computer Science, pp. 390–398 (2000)
21. McIlroy, P.K., Bostic, K., Mcilroy, M.D.: Engineering radix sort. Computing Systems 6, 5–27 (1993)
22. Kärkkäinen, J., Rantala, T.: Engineering radix sort for strings. In: Amir, A., Turpin, A., Moffat, A. (eds.) SPIRE 2008. LNCS, vol. 5280, pp. 3–14. Springer, Heidelberg (2008)
23. The quest for an accelerated population count. In: Oram, A., Wilson, G. (eds.) Beautiful code, pp. 147–160. O‘ Reilly, Sebastopol (2007)

Estimation of Alternative Splicing isoform Frequencies from RNA-Seq Data*

Marius Nicolae[1], Serghei Mangul[2], Ion Măndoiu[1], and Alex Zelikovsky[2]

[1] Computer Science & Engineering Department, University of Connecticut
371 Fairfield Way, Storrs, CT 06269
{man09004,ion}@engr.uconn.edu

[2] Computer Science Department, Georgia State University
University Plaza, Atlanta, Georgia 30303
{serghei,alexz}@cs.gsu.edu

Abstract. In this paper we present a novel expectation-maximization algorithm for inference of alternative splicing isoform frequencies from high-throughput transcriptome sequencing (RNA-Seq) data. Our algorithm exploits disambiguation information provided by the distribution of insert sizes generated during sequencing library preparation, and takes advantage of base quality scores, strand and read pairing information if available. Empirical experiments on synthetic datasets show that the algorithm significantly outperforms existing methods of isoform and gene expression level estimation from RNA-Seq data. The Java implementation of IsoEM is available at http://dna.engr.uconn.edu/software/IsoEM/.

1 Introduction

Ubiquitous regulatory mechanisms such as the use of alternative transcription start and polyadenylation sites, alternative splicing, and RNA editing result in multiple messenger RNA (mRNA) isoforms being generated from a single genomic locus. Most prevalently, alternative splicing is estimated to take place for over 90% of the multi-exon human genes [19], and thought to play critical roles in early stages of development and normal function of cells from diverse tissue types. Thus, the ability to reconstruct full length isoform sequences and accurately estimate their frequencies is critical for understanding gene functions and transcription regulation mechanisms.

Three key interrelated computational problems arise in the context of transcriptome analysis: *gene expression level estimation (GE)*, *isoform discovery (ID)*, and *isoform expression level estimation (IE)*. Targeted GE has long been a staple of genetic studies, and the completion of the human genome has enabled genome-wide GE performed using expression microarrays. Since expression microarrays have limited capability of detecting alternative splicing events, specialized splicing arrays have been developed to interrogate genome-wide both (annotated) exons and exon-exon junctions. However, despite sophisticated deconvolution algorithms [1,15], the fragmentary information provided by splicing

* Work supported in part by NSF awards IIS-0546457, IIS-0916401, and IIS-0916948.

V. Moulton and M. Singh (Eds.): WABI 2010, LNBI 6293, pp. 202–214, 2010.

arrays is often insufficient for unambiguous identification of transcribed isoforms [6,9]. High-throughput transcriptome sequencing, commonly referred to as RNA-Seq, is quickly replacing microarrays as the technology of choice for performing GE due to the far wider dynamic range and more accurate quantitation capabilities [20]. Unfortunately, most RNA-Seq studies to date either ignore alternative splicing or, similar to splicing array studies, restrict themselves to surveying the presence/expression levels of exons and exon-exon junctions. The main difficulty lies in the fact that current technologies used to perform RNA-Seq generate short reads (from few tens to hundreds of bases), many of which cannot be unambiguously assigned to individual isoforms.

1.1 Related Work

RNA-Seq analyses typically start by mapping sequencing reads onto the reference genome, transcript libraries, exon-exon junction libraries, or combinations thereof. Early RNA-Seq studies have recognized that short read lengths result in a significant percentage of so called *multireads*, i.e., reads that map equally well at multiple locations in the genome. A simple (and still commonly used) approach is to discard multireads, and estimate expression levels using only the so called *unique* reads. Mortazavi et al. [12] proposed a multiread "rescue" method whereby initial gene expression levels are estimated from unique reads and used to fractionally allocate multireads, with final expression levels re-estimated from total counts obtained after multiread allocation. An expectation-maximization (EM) algorithm that extends this scheme by repeatedly alternating between fractional read allocation and re-estimation of gene expression levels was recently proposed in [13].

A number of recent works have addressed the IE problem, namely isoform expression level estimation from RNA-Seq reads. Under a simplified "exact information" model, [9] showed that neither single nor paired read RNA-Seq data can theoretically guarantee unambiguous inference of isoform expression levels, but paired reads may be sufficient to deconvolute expression levels for the majority of annotated isoforms. The key challenge in IE is accurate assignment of ambiguous reads to isoforms. Compared to the GE context, read ambiguity is much more significant, since it affects not only multireads, but also reads that map at a unique genome location expressed in multiple isoforms. To overcome this difficulty, [8] proposed a Poisson model of single-read RNA-Seq data explicitly modeling isoform frequencies. Under this model, maximum likelihood estimates are obtained by solving a convex optimization problem, and uncertainty of estimates are obtained by importance sampling from the posterior distribution. Li et al. [11] introduced an expectation-maximization (EM) algorithm similar to that of [13] but apply it to isoforms instead of genes. Unlike the method of [8], which estimates isoform frequencies only from reads that map to a unique location in the genome, the algorithm of [11] incorporates multireads as well. The IE problem for single reads is also tackled in [14], who propose an EM algorithm for inferring isoform expression levels from read coverage of exons (reads spanning exon junctions are ignored).

The related isoform discovery (ID) problem has also received much interest in the literature. De novo transcriptome assembly algorithms have been proposed in [2,7]. Very recently, [4] and [18] proposed methods for simultaneously solving ID and IE based on paired RNA-Seq reads. Assuming known genomic positions for alternative transcription start and polyadenylation sites as well as exon boundaries, [4] formulate IE as a convex quadratic program (QP) that can be efficiently solved for each gene locus after discarding multireads. ID is solved by iteratively generating isoform candidates from the splicing graph derived from annotations and reads spanning exon-exon junctions. The process is continued until the p-value of the objective value of the QP corresponding to the set of selected isoforms, assumed to follow a χ^2 distribution, exceeds an empirically selected threshold of 5%. However, pair read information is not directly used in isoform frequency estimation, contributing only as secondary data to filter out false positives in the process of isoform selection. Trapnell et al. [18] also describe a method, referred to as Cufflinks, for simultaneously solving ID and IE. Unlike the method of [4], Cufflinks requires no genome annotations (but can use them if available). After performing spliced alignment of (paired) reads onto the genome using TopHat [17], Cufflinks constructs a read overlap graph and generates candidate isoforms by finding a minimal size path cover via a reduction to maximum matching in a weighted bipartite graph. Reads that match equally well multiple locations in the genome are fractionally allocated to these locations, and estimation is then performed independently at different transcriptional loci, using an extension to paired reads of the methods in [8].

1.2 Our Contributions

In this paper we focus on the IE problem, namely estimating isoform expression levels (interchangeably referred to as frequencies) from RNA-Seq reads, under the assumption that a complete list of candidate isoforms is available. Projects such as [3] and [16] have already assembled large libraries of full-length cDNA sequences for humans and other model organisms, and the coverage of these libraries is expected to continue to increase rapidly. Although an incomplete isoform library may lead to estimation biases [18], statistical tests such as the one in [4] can be used to detect the presence of isoforms not represented in the library. Inferring expression at isoform level provides information for finer-resolution biological studies, and also leads to more accurate estimates of expression at the gene level by allowing rigorous length normalization. Indeed, as shown in Section 3, genome-wide gene expression level estimates derived from isoform level estimates are significantly more accurate than those obtained directly from RNA-Seq data using isoform-oblivious GE methods such as the widely used counting of unique reads, the rescue method of [12], or the EM algorithm of [13].

Our main contribution is a novel expectation-maximization algorithm for isoform frequency estimation from (any mixture of) single and paired RNA-Seq reads. A key feature of our algorithm, referred to as IsoEM, is that it exploits the information provided by the distribution of insert sizes, which is tightly controlled during sequencing library preparation under current RNA-Seq protocols.

The recently published [18] is the only other work we are aware of that exploits this information (that is not captured by the "exact" information models of [6,9]) in conjunction with paired read data. We show that modeling insert sizes is also highly beneficial in conjunction with single RNA-Seq reads. Insert sizes contribute to increased estimation accuracy in two different ways. On one hand, insert sizes help disambiguating the isoform of origin for the reads. In IsoEM, insert lengths are combined with base quality scores, and, if available, read pairing and strand information to probabilistically allocate reads to isoforms during the expectation step of the algorithm. As in [11], the genomic locations of multireads are also resolved probabilistically in this step, further contributing to overall accuracy compared to methods that ignore or fractionally pre-allocate multireads. On the other hand, insert size distribution is used to accurately adjust isoform lengths during frequency re-estimation in the M step of the IsoEM algorithm; an equivalent adjustment was independently employed in [18].

We also present preliminary experimental results on synthetic datasets generated with various sequencing parameters and distribution assumptions. The results show that IsoEM algorithm significantly outperforms existing methods of isoform and gene expression level estimation from RNA-Seq data. Furthermore, we empirically evaluate the effect of sequencing parameters such as read length, read pairing, and strand information on estimation accuracy. Our experiments confirm the finding of [11] that, for a fixed total number of sequenced bases, longer reads do not necessarily lead to better accuracy for estimation of isoform and gene expression levels.

2 Methods

2.1 Read Mapping

As with most RNA-Seq analyses, the first step of IsoEM is to map the reads. Our approach is to map them onto the library of known isoforms using any one of the many available aligners (we used Bowtie [10] with default parameters in our experiments). An alternative strategy is to map the reads onto the genome using a spliced alignment tool such as TopHat [17], as done in [18]. However, preliminary experiments with TopHat resulted in fewer mapped reads and increased mapping uncertainty. Since further increases in read length coupled with improvements in spliced alignment algorithms could make genome mapping more attractive in the future, we made our IsoEM implementation compatible with both mapping approaches by converting read alignments to genome coordinates and performing all operations in genome space.

2.2 Finding Read-Isoform Compatibilities

The candidate set of isoforms for each read is obtained by putting together all genome coordinates for reads and isoforms, sorting them and using a line sweep technique to detect read-isoform compatibilities. During the line sweep, reads are grouped into equivalence classes defined by their isoform compatibility sets; this speeds up the E-steps of the IsoEM algorithm by allowing the processing of an entire read class at once.

Some of the reads match multiple positions in the genome, which we refer to as *alignments* (for paired end reads, an alignment consists of the positions where the two reads in the pair align with the genome). Each alignment a can in turn be compatible with multiple isoforms that overlap at that position of the genome. During the line sweep, we compute the relative "weight" of assigning a given read/pair r to isoform j as $w_{r,j} = \sum_a Q_a F_a O_a$, where the sum is over all alignments of r compatible with j, and the factors of the summed products are defined as follows.

- Q_a represents the probability of observing the read from the genome locations described by the alignment. This is computed from the base quality scores as $Q_a = \prod_{k=1}^{|r|}[(1-\varepsilon_k)M_{a_k} + \varepsilon_k(1-M_{a_k})]$, where $M_{a_k} = 1$ if position k of alignment a matches the genome and 0 otherwise, while ε_k denotes the error probability of kth base of r.
- F_a represents the probability of the fragment length needed to produce alignment a from isoform j. For paired end reads, the length of the fragment can be inferred from the positions of the two reads. For single reads, we can only estimate a maximum fragment length: if the alignment is on the same strand as the isoform, we use the distance from the start of the alignment to the end of the isoform, otherwise we use the distance from the end of the alignment to the start of the isoform.
- O_a is 1 if alignment a of r is consistent with the orientation of isoform j, and 0 otherwise. Consistency between the orientations of r and j depends on whether or not the library preparation protocol preserves the strand information. For single reads $O_a = 1$ when reads are generated from fragment ends randomly or, for directional RNA-Seq, when they match the known isoform orientation. For pairs, $O_a = 1$ if the two reads come from different strands, point to each other, and, in the case of directional RNA-Seq, the orientation of first read matches the known isoform orientation.

Weigths $w_{r,j}$ can be further adjusted to account for biases introduced by sequencing library preparation or the sequencing process once a model of this biases, such as the one in [5], is available.

2.3 The IsoEM Algorithm

The IsoEM algorithm starts with the set of N known isoforms. For each isoform we denote by $l(j)$ its length and by $f(j)$ its (unknown) frequency. If we ignore library preparation and amplification biases, the probability that a read is sampled from isoform j is proportional with $(l(j) - \mu + 1)f(j)$ where μ is the mean fragment length from the sample preparation. To see why this is true, we write the expected number of reads coming from an isoform by summing over all possible fragment lengths. For each fragment length k we expect the number

Algorithm 1. IsoEM algorithm

```
assign random values to all f(i)
while not converged do
  initialize all n(j) to 0
  for each read r do
    sum = Σ_{j:w_{r,j}>0} w_{r,j} f(j)
    for each isoform j with w_{r,j} > 0 do
      n(j)+ = w_{r,j} f(j)/sum
    end for
  end for
  s = Σ_j n(j)/(l(j) − μ + 1)
  for each isoform j do
    f(j) = (n(j)/(l(j)−μ+1)) / s
  end for
end while
```

of fragments of that length to be proportional to the number of valid starting positions for a fragment of that length in the isoform. If $p(k)$ denotes the probability of a fragment of length k and $n(j)$ denotes the number of reads coming from isoform j then $E[n(j)] \propto \sum_k p(k)(l(j) - k + 1) = l(j) - \mu + 1$. Thus, if the isoform of origin is known for each read, the maximum likelihood estimator for $f(j)$ is given by $c(j)/(c(1) + \ldots + c(N))$, where $c(j) = n(j)/(l(j) - \mu + 1)$ denotes the length-normalized fragment coverage.

Unfortunately, some reads match multiple isoforms, so their isoform of origin cannot be established unambiguously. The IsoEM algorithm (see Algorithm 1) overcomes this difficulty by simultaneously estimating the frequencies and imputing the missing read origin within an iterative framework. After initializing frequencies $f(j)$ at random, the algorithm repeatedly performs the next two steps until convergence:

- E-step: Compute the expected number $n(j)$ of reads that come from isoform j under the assumption that isoform frequencies $f(j)$ are correct, based on weights $w_{r,j}$
- M-step: For each j, set the new value of $f(j)$ to $c(j)/(c(1) + \ldots + c(N))$, where normalized coverages $c(j)$ are based on expected counts computed in previous step

3 Experimental Results

3.1 Simulation Setup

We tested IsoEM on simulated human RNA-Seq data. The human genome sequence (hg18, NCBI build 36) was downloaded from UCSC together with the coordinates of the isoforms in the KnownGenes table. Genes were defined as clusters of known isoforms defined by the GNFAtlas2 table. The dataset contains a total of 66803 isoforms pertaining to 19372 genes. The isoform length distribution and the number of isoforms per genes are shown in Figure 1.

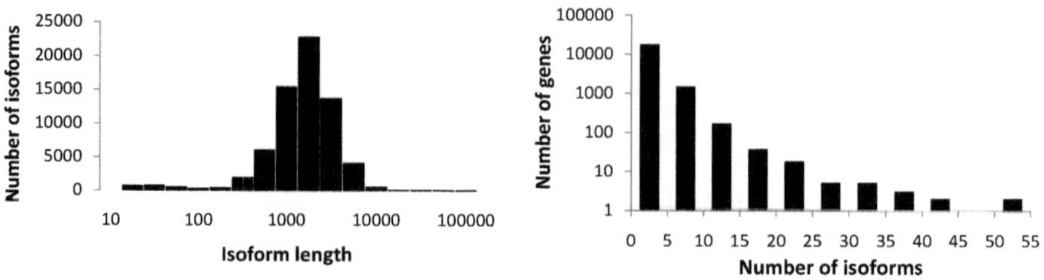

Fig. 1. Distribution of isoform lengths (left panel) and gene cluster sizes (right panel) for the UCSC KnownGenes dataset

Single and paired-end reads were randomly generated by sampling fragments from the known isoforms. Each isoform was assigned a *true frequency* based on the abundance reported for the corresponding gene in the first human tissue of the GNFAtlas2 table, and a probability distribution over the isoforms inside a gene cluster. Thus, the true frequency of isoform j is $a(g)p(j)$, where $a(g)$ is the abundance of the gene g for which j is an isoform and $p(j)$ is the probability of isoform j among all the isoforms of g. We simulated datasets with uniform and geometric ($p = 0.5$) distributions for the isoforms of each gene. Fragment lengths were simulated from a normal probability distribution with mean 250 and standard deviation 25. We simulated between 1 and 60 million single and paired reads of lengths ranging from 25 to 100 base pairs, with or without strand information.

We compared IsoEM to several existing IE and GE algorithms. For IE we included in the comparison the isoform analogs of the Uniq and Rescue methods used for GE [12], an improved version of Uniq (UniqLN) that estimates isoform frequencies from unique read counts but normalizes them using adjusted isoform lengths that exclude ambiguous positions, the Cufflinks algorithm of [18], and the RSEM algorithm of [11]. For the GE problem, the comparison included the Uniq and Rescue methods, our implementation of the EM algorithm described in [13] (GeneEM), and estimates obtained by summing isoform expression levels inferred by Cufflinks, RSEM, and IsoEM. All methods except Cufflinks use alignments obtained by mapping reads onto the library of isoforms with Bowtie [10] and then converting them to genome coordinates. As suggested in [18], Cufflinks uses alignments obtained by mapping the reads onto the genome with TopHat [17], which was provided with a complete set of annotated junctions.

Frequency estimation accuracy was assessed using the coefficient of determination, r^2, along with the *error fraction (EF)* and *median percent error (MPE)* measures used in [11]. However, accuracy was computed against true frequencies, not against estimates derived from true counts as in [11]. If $\hat{f}_i$ is the frequency estimate for an isoform with true frequency f_i, the *relative error* is defined as $|\hat{f}_i - f_i|/f_i$ if $f_i \neq 0$, 0 if $\hat{f}_i = f_i = 0$, and ∞ if $\hat{f}_i > f_i = 0$. The error fraction with threshold τ, denoted EF_τ is defined as the percentage of isoforms with relative error greater or equal to τ. The median percent error, denoted MPE, is defined as the threshold τ for which $EF_\tau = 50\%$.

Table 1. r^2 for isoform and gene expression levels inferred from 30M reads of length 25 from reads simulated assuming uniform, respectively geometric expression of gene isoforms

Isoform Expression			Gene Expression		
Algorithm	Uniform	Geometric	Algorithm	Uniform	Geometric
Uniq	0.466	0.447	Uniq	0.579	0.586
Rescue	0.693	0.675	Rescue	0.724	0.724
UniqLN	0.856	0.838	GeneEM	0.636	0.637
Cufflinks	0.661	0.618	Cufflinks	0.778	0.757
RSEM	0.919	0.911	RSEM	0.939	0.934
IsoEM	**0.979**	**0.964**	IsoEM	**0.988**	**0.978**

3.2 Comparison between Methods

Table 1 gives r^2 values for isoform, respectively gene expression levels inferred from 30M reads of length 25, simulated assuming both uniform and geometric isoform expression. IsoEM significantly outperforms the other methods, achieving an r^2 values of over .96 for all datasets. For all methods the accuracy difference between datasets generated assuming uniform and geometric distribution of isoform expression levels is small, with the latter one typically having a slightly worse accuracy. Thus, in the interest of space we present remaining results only for datasets generated using geometric isoform expression.

For a more detailed view of the relative performance of compared IE and GE algorithms, Figure 2 gives the error fraction at different thresholds ranging between 0 and 1. The variety of methods included in the comparison allows us to tease out the contribution of various algorithmic ideas to overall estimation accuracy. The importance of rigorous length normalization is demonstrated by the IE accuracy gain of UniqLN over Uniq – clearly larger than that achieved by ambiguous read reallocation as implemented in the IE version of Rescue. Proper length normalization is also the main reason for the accuracy gain of isoform-aware GE methods (Cufflinks, RSEM, and IsoEM) over isoform oblivious GE methods. Similarly, the importance of modeling insert sizes even for single read data is underscored by the IE and GE accuracy gains of IsoEM over RSEM.

For yet another view, Tables 2 and 3 report the MSE and $EF_{.15}$ measures for isoform, respectively gene expression levels inferred from 30M reads of length 25, computed over groups of isoforms with various expression levels. IsoEM consistently outperforms the other IE and GE methods at all expression levels except for isoforms with zero true frequency, where it is dominated by the more conservative Uniq algorithm and its UniqLN variant.

3.3 Influence of Sequencing Parameters

Although high-throughput technologies allow users to make tradeoffs between read length and the number of generated reads, very little has been done to determine optimal parameters even for common applications such as RNA-Seq.

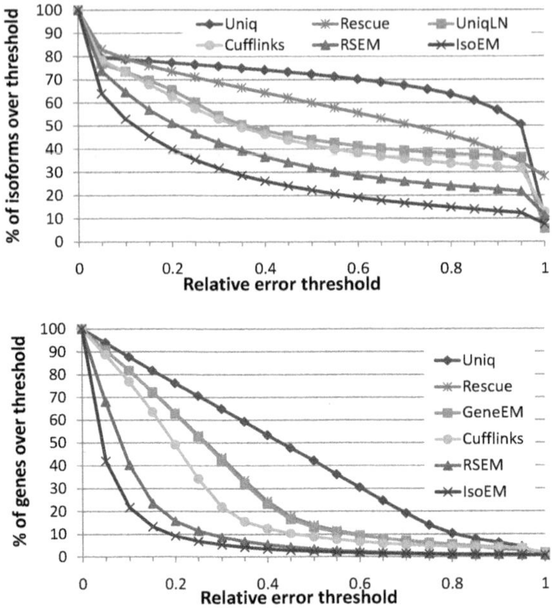

Fig. 2. Error fraction at different thresholds for isoform (top panel) and gene (bottom panel) expression levels inferred from 30M reads of length 25 simulated assuming geometric isoform expression.

Table 2. Median percent error (MPE) and 15% error fraction ($EF_{.15}$) for isoform expression levels inferred from 30M reads of length 25

	Expression range	0	$(0, 10^{-6}]$	$(10^{-6}, 10^{-5}]$	$(10^{-5}, 10^{-4}]$	$(10^{-4}, 10^{-3}]$	$(10^{-3}, 10^{-2}]$	All
	# isoforms	13290	10024	23882	18359	1182	66	66803
MPE	Uniq	**0.0**	**100.0**	98.4	97.1	98.5	96.6	95.4
	Rescue	**0.0**	294.7	75.5	49.2	30.4	28.3	71.9
	UniqLN	**0.0**	**100.0**	80.8	30.3	26.4	24.8	36.0
	Cufflinks	**0.0**	**100.0**	49.7	25.5	27.2	44.6	34.1
	RSEM	**0.0**	**100.0**	31.9	13.5	11.4	13.0	21.2
	IsoEM	**0.0**	**100.0**	**22.7**	**7.3**	**3.5**	**2.5**	**11.8**
$EF_{.15}$	Uniq	**0.2**	98.4	97.2	96.9	97.0	95.5	78.0
	Rescue	48.4	95.5	86.2	73.1	61.5	56.1	76.0
	UniqLN	**0.2**	97.2	86.2	82.8	83.3	77.3	69.8
	Cufflinks	17.6	96.4	81.3	71.0	74.7	80.3	67.9
	RSEM	19.9	93.7	71.1	46.4	39.8	47.0	56.9
	IsoEM	5.1	**91.2**	**62.8**	**29.3**	**15.8**	**7.6**	**45.5**

The intuition that longer reads are better certainly holds true for many applications such as de novo assembly. Surprisingly, [11] found that *shorter* reads are better for IE when the total number of sequenced bases is fixed. Figure 3 plots

Table 3. Median percent error (MPE) and 15% error fraction ($EF_{.15}$) for gene expression levels inferred from 30M reads of length 25

Expression range		$(0, 10^{-6}]$	$(10^{-6}, 10^{-5}]$	$(10^{-5}, 10^{-4}]$	$(10^{-4}, 10^{-3}]$	$(10^{-3}, 10^{-2}]$	All
# genes		120	5610	11907	1632	102	19372
MPE	Uniq	37.4	43.6	42.7	43.0	48.2	43.0
	Rescue	32.8	28.7	26.0	25.1	28.8	26.7
	GeneEM	30.6	28.2	25.7	25.1	28.0	26.3
	Cufflinks	33.0	21.1	19.0	20.2	40.2	19.7
	RSEM	23.6	11.0	7.2	7.9	11.4	8.1
	IsoEM	**18.3**	**8.4**	**3.3**	**2.2**	**2.1**	**4.0**
$EF_{.15}$	Uniq	77.5	82.4	81.7	79.7	82.4	81.7
	Rescue	74.2	74.0	71.6	72.8	76.5	72.4
	GeneEM	72.5	73.8	71.5	73.0	74.5	72.3
	Cufflinks	73.3	64.7	62.3	66.2	82.3	63.5
	RSEM	64.2	37.3	17.4	16.3	41.2	23.5
	IsoEM	**57.5**	**28.3**	**6.8**	**6.5**	**4.9**	**13.3**

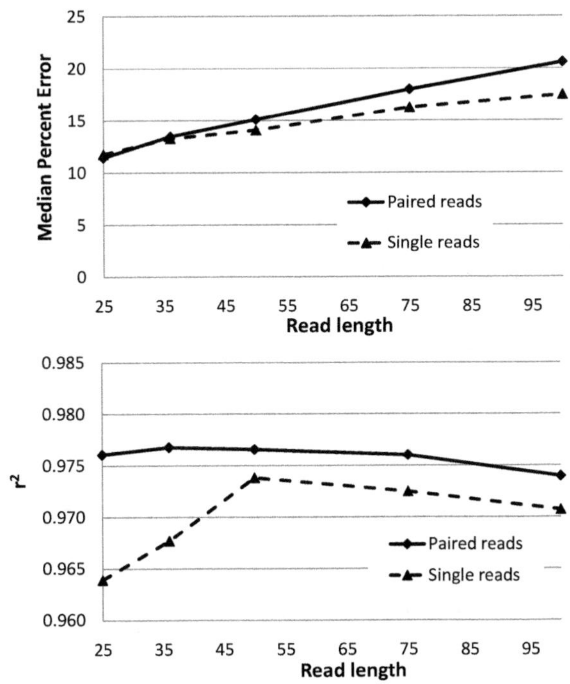

Fig. 3. IsoEM MPE (top panel) and r^2 values (bottom panel) for 750Mb of data generated using single and paired-end sequencing with read length between 25 and 100

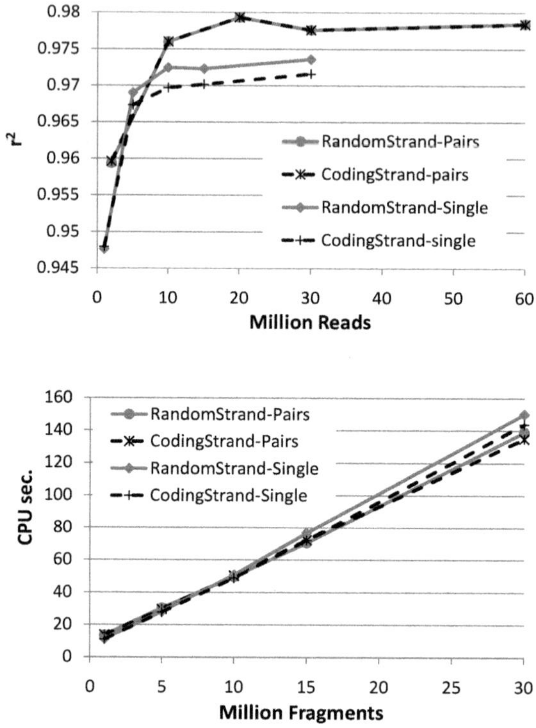

Fig. 4. IsoEM r^2 (top panel) and CPU time (bottom panel) for 1-60 million single/paired reads of length 75, with or without strand information

IE estimation accuracy for reads of length between 25 and 100 when the total amount of sequence data is kept constant at 750M bases. Our results confirm the finding of [11], although the optimal read length is somewhat sensitive to the accuracy measure used and to the availability of pairing information. While 25bp reads optimize the MPE measure regardless of the availability of paired reads, the read length that maximizes r^2 is 36 for paired reads and 50 for single reads. While more experiments are needed to determine how the optimum length depends on the amount of sequence data and transcriptome complexity, this does suggest that, for isoform and gene expression estimation accuracy, increasing the number of reads may be more useful than increasing read length beyond a certain limit.

The top panel of Figure 4 shows, for reads of length 75, the effects of paired reads and strand information on estimation accuracy as measured by r^2. Not surprisingly, for a fixed number of reads, paired reads yield better accuracy than single reads. Also not very surprisingly, adding strand information to paired sequencing yields no benefits to genome-wide IE accuracy (although it may be helpful, e.g., in identification of novel transcripts). Quite surprisingly, performing strand-specific single read sequencing is actually *detrimental* to IsoEM IE (and hence GE) accuracy under the simulated scenario, most likely due to the reduction in sampled transcript length.

As shown in the bottom panel of Figure 4, the runtime of our Java implementation of IsoEM scales roughly linearly with the number of *fragments*, and is largely insensitive to the type of sequencing data (single or paired reads, directional or non-directional). IsoEM was tested on a DELL PowerEdge R900 server with 4 Six Core E7450Xeon Processors at 2.4Ghz (64 bits) and 128Gb of internal memory. None of the datasets require more than 16GB of memory to complete, however, increasing the amount of memory made available to the Java virtual machine significantly decreases runtime by reducing the time needed for garbage collection. The runtimes in Figure 4 were obtained by allowing IsoEM to use up to 32GB of memory, in which case none of the datasets took more than 3 minutes to solve.

4 Conclusions and Ongoing Work

In this paper we have introduced an expectation-maximization algorithm for isoform frequency estimation assuming a known set of isoforms. Our algorithm, called IsoEM, explicitly models base quality scores, insert size distribution, strand and read pairing information. Experiments on synthetic data sets generated using two different assumptions on the isoform distribution show that IsoEM consistently outperforms existing algorithms for isoform and gene expression level estimation with respect to a variety of quality metrics.

The open source Java implementation of IsoEM is freely available for download at http://dna.engr.uconn.edu/software/IsoEM/. In ongoing work we are extending IsoEM to perform allelic specific isoform expression and exploring integration of isoform frequency estimation with identification of novel transcripts using the iterative refinement framework proposed in [4].

References

1. Anton, M., Gorostiaga, D., Guruceaga, E., Segura, V., Carmona-Saez, P., Pascual-Montano, A., Pio, R., Montuenga, L., Rubio, A.: SPACE: an algorithm to predict and quantify alternatively spliced isoforms using microarrays. Genome Biology 9(2), R46 (2008)
2. Birol, I., Jackman, S.D., Nielsen, C.B., Qian, J.Q., Varhol, R., Stazyk, G., Morin, R.D., Zhao, Y., Hirst, M., Schein, J.E., Horsman, D.E., Connors, J.M., Gascoyne, R.D., Marra, M.A., Jones, S.J.M.: De novo transcriptome assembly with ABySS. Bioinformatics 25(21), 2872–2877 (2009)
3. Carninci, P., et al.: The Transcriptional Landscape of the Mammalian Genome. Science 309(5740), 1559–1563 (2005)
4. Feng, J., Li, W., Jiang, T.: Inference of isoforms from short sequence reads. In: Berger, B. (ed.) RECOMB 2010. LNCS, vol. 6044, pp. 138–157. Springer, Heidelberg (2010)

5. Hansen, K.D., Brenner, S.E., Dudoit, S.: Biases in Illumina transcriptome sequencing caused by random hexamer priming. Nucl. Acids Res. p. gkq224 (2010) (advance access)
6. Hiller, D., Jiang, H., Xu, W., Wong, W.H.: Identifiability of isoform deconvolution from junction arrays and RNA-Seq. Bioinformatics 25(23), 3056–3059 (2009)
7. Jackson, B., Schnable, P., Aluru, S.: Parallel short sequence assembly of transcriptomes. BMC Bioinformatics 10(suppl. 1), S14+ (2009)
8. Jiang, H., Wong, W.H.: Statistical inferences for isoform expression in RNA-Seq. Bioinformatics 25(8), 1026–1032 (2009)
9. Lacroix, V., Sammeth, M., Guigo, R., Bergeron, A.: Exact transcriptome reconstruction from short sequence reads. In: Crandall, K.A., Lagergren, J. (eds.) WABI 2008. LNCS (LNBI), vol. 5251, pp. 50–63. Springer, Heidelberg (2008)
10. Langmead, B., Trapnell, C., Pop, M., Salzberg, S.: Ultrafast and memory-efficient alignment of short DNA sequences to the human genome. Genome Biology 10(3), R25 (2009)
11. Li, B., Ruotti, V., Stewart, R.M., Thomson, J.A., Dewey, C.N.: RNA-Seq gene expression estimation with read mapping uncertainty. Bioinformatics 26(4), 493–500 (2010)
12. Mortazavi, A., Williams, B.A.A., McCue, K., Schaeffer, L., Wold, B.: Mapping and quantifying mammalian transcriptomes by RNA-Seq. Nature methods (2008)
13. Paşaniuc, B., Zaitlen, N., Halperin, E.: Accurate estimation of expression levels of homologous genes in RNA-seq experiments. In: Berger, B. (ed.) RECOMB 2010. LNCS, vol. 6044, pp. 397–409. Springer, Heidelberg (2010)
14. Richard, H., Schulz, M.H., Sultan, M., Nurnberger, A., Schrinner, S., Balzereit, D., Dagand, E., Rasche, A., Lehrach, H., Vingron, M., Haas, S.A., Yaspo, M.-L.: Prediction of alternative isoforms from exon expression levels in RNA-Seq experiments. Nucl. Acids Res. 38(10), e112+ (2010)
15. She, Y., Hubbell, E., Wang, H.: Resolving deconvolution ambiguity in gene alternative splicing. BMC Bioinformatics 10(1), 237 (2009)
16. Temple, G., et al.: The completion of the Mammalian Gene Collection (MGC). Genome Research 19(12), 2324–2333 (2009)
17. Trapnell, C., Pachter, L., Salzberg, S.L.: TopHat: discovering splice junctions with RNA-Seq. Bioinformatics 25(9), 1105–1111 (2009)
18. Trapnell, C., Williams, B.A., Pertea, G., Mortazavi, A., Kwan, G., van Baren, M.J., Salzberg, S.L., Wold, B.J., Pachter, L.: Transcript assembly and quantification by RNA-Seq reveals unannotated transcripts and isoform switching during cell differentiation. Nature biotechnology 28(5), 511–515 (2010)
19. Wang, E.T., Sandberg, R., Luo, S., Khrebtukova, I., Zhang, L., Mayr, C., Kingsmore, S.F., Schroth, G.P., Burge, C.B.: Alternative isoform regulation in human tissue transcriptomes. Nature 456(7221), 470–476 (2008)
20. Wang, Z., Gerstein, M., Snyder, M.: RNA-Seq: a revolutionary tool for transcriptomics. Nat. Rev. Genet. 10(1), 57–63 (2009)

Improved Orientations of Physical Networks

Iftah Gamzu[1], Danny Segev[2], and Roded Sharan[1]

[1] Blavatnik School of Computer Science, Tel-Aviv University, Tel-Aviv 69978, Israel
{iftgam,roded}@tau.ac.il
[2] Department of Statistics, University of Haifa, Haifa 31905, Israel
segevd@stat.haifa.ac.il

Abstract. The orientation of physical networks is a prime task in deciphering the signaling-regulatory circuitry of the cell. One manifestation of this computational task is as a maximum graph orientation problem, where given an undirected graph on n vertices and a collection of vertex pairs, the goal is to orient the edges of the graph so that a maximum number of pairs are connected by a directed path. We develop a novel approximation algorithm for this problem with a performance guarantee of $O(\log n/\log\log n)$, improving on the current logarithmic approximation. In addition, motivated by interactions whose direction is pre-set, such as protein-DNA interactions, we extend our algorithm to handle mixed graphs, a major open problem posed by earlier work. In this setting, we show that a polylogarithmic approximation ratio is achievable under biologically-motivated assumptions on the sought paths.

1 Introduction

A fundamental problem in the study of biological networks is the inference of causal relations that are often not covered by current experimental techniques. One prime example for such deficiency concerns protein-protein interaction (PPI) networks. While PPIs have been measured at large scale across tens of organisms for over a decade now, current technologies do not provide information on the direction in which signal flows. Such information may be indirectly obtained from perturbation experiments in which a gene is perturbed and as a result other genes change their expression levels. Assuming that the expression changes imply directed pathways from the perturbed, or causal, gene to the affected genes, several authors have successfully inferred interaction directions [15].

The inference of interaction directions that best fit the perturbation experiments can be cast as a *maximum graph orientation* (MGO) problem. An instance of this problem consists of an undirected graph on n vertices, representing the PPI network, and a collection C of requests. Each request is given as an ordered pair of source-target vertices, representing a causal gene and an affected gene. The goal is to *orient* the graph, i.e., choose a single direction for each of its edges, such that a maximal number of requests admit a directed path from the source to the target. We note that any instance of MGO can be reduced to one

V. Moulton and M. Singh (Eds.): WABI 2010, LNBI 6293, pp. 215–225, 2010.

where the underlying graph is a tree. Indeed, if the input graph contains cycles, one can sequentially contract them, one after the other. In each step, the edges of an arbitrary cycle are all oriented in the same direction (either clockwise or counter-clockwise). As a result, every pair of vertices on this cycle admit a directed path between them and, thus, the cycle can be contracted into a single vertex.

Medvedovsky et al. [15] were the first to study MGO. They demonstrated that the seemingly simple setting when the underlying graph is a star is equivalent to the *maximum directed cut* problem. The latter problem admits a semidefinite programming based 0.874-approximation algorithm [4,13], while approximating it within factors of $11/12 \approx 0.916$ and $\alpha_{\mathrm{GW}} \approx 0.878$ is NP-hard [10] and Unique Games-hard [12], respectively. These hardness bounds clearly follow to MGO. On the positive side, they devised an $O(\log n)$ approximation algorithm for arbitrary trees, and proposed an exact dynamic-programming algorithm for the special case of path graphs.

Further research along these lines focused on variants of the maximum graph orientation problem. For instance, Hakimi, Schmeichel, and Young [8] studied the special setting in which the set of requests contains all vertex pairs, and developed an exact polynomial time algorithm. Arkin and Hassin [2] established hardness results for the problem of deciding whether one can orient a *mixed* graph, i.e., a graph in which the orientation of some edges is predetermined, to satisfy a given set of requests.

Our contribution in this paper is two-fold: (i) We propose a deterministic algorithm for the maximum graph orientation problem whose approximation ratio is $O(\log n/\log\log n)$. This result improves on the current $O(\log n)$ approximation due to Medvedovsky et al. [15]. Our algorithm optimizes with respect to the optimal solution, which is crucial to our sublogarithmic performance guarantee. In contrast, previous results made use of the request set cardinality, $|C|$, as a reference point, and were therefore limited by the observation that there are certain trees in which any orientation cannot satisfy more than a logarithmic fraction of the entire set of requests [14]. (ii) We devise an approximation algorithm for the generalized scenario of mixed graphs. This scenario is motivated by the need to include in the network protein-DNA interactions (PDIs), which are both fundamental to signal transduction and key mediators in the observed expression changes. The approximation guarantee of our algorithm depends on the number of PDI segments in the underlying pathways. For typical cases, in which the pathways contain at most one segment, our algorithm achieves an approximation ratio of $O(\log n)$.

The rest of the paper is organized as follows: In Section 2, we describe and analyze the algorithm for the maximum graph orientation problem, while in Section 3, we study the mixed-graphs scenario.

2 Orienting Undirected Graphs

In this section, we devise a deterministic algorithm that achieves an approximation guarantee of $O(\log n/\log\log n)$ for MGO. The algorithm employs the classify-and-select paradigm, that is, it exploits various structural properties of the input graph to partition the collection of requests into $O(\log n/\log\log n)$ pairwise-disjoint classes. For each such class, given the additional structure imposed, we separately compute a graph orientation that satisfies a constant fraction of the optimal number of satisfiable requests in this class. Consequently, the above-mentioned approximation ratio follows by picking, out of the set of all the computed orientations, the one that satisfies a maximum number of requests. Below we describe the classification and orientation steps in detail.

2.1 The Classification Process

To specify the process by which requests are partitioned into classes, we begin by presenting the notion of an almost-balanced decomposition, which can be viewed as a generalization of the well-known centroid decomposition [5]. Note that structural properties in this spirit have been explored and exploited in various settings (see, e.g., [3,7,9]).

Definition 1. *Let $T = (V, E)$ be a tree. An* almost balanced k-decomposition *of T is a partition of T into k edge-disjoint subtrees $T_1, \ldots, T_k$ such that each subtree contains between $|E|/(3k)$ and $3|E|/k$ edges.*

Lemma 1. ([6]) *Let $T = (V, E)$ be a tree with $|E| \geq k$. An almost balanced k-decomposition of T exists and can be found in polynomial time. In addition, the number of vertices that are shared by at least two subtrees is less than k.*

The classification process corresponds to a recursive decomposition of the input tree T. Let $\mathcal{T}_1 = \{T_1, \ldots, T_k\}$ be an almost balanced k-decomposition of T into k edge-disjoint subtrees. We say that a decomposition *separates* a request i when its endpoints s_i and t_i reside in different subtrees of the decomposition (see Figure 1 for an example). The first class of requests, C_1, consists of all requests separated by $\mathcal{T}_1$. To classify the remaining set of requests, $C \setminus C_1$, we recursively apply the previously-described procedure with respect to the collection of subtrees in $\mathcal{T}_1$. Specifically, in the second level of the recursion, an almost balanced k-decomposition is computed in each of the subtrees $T_1, \ldots, T_k$, to obtain a set $\mathcal{T}_2$, comprising of k^2 subtrees. The second class of requests, C_2, consists of all yet-unclassified requests separated by $\mathcal{T}_2$. In other words, the endpoints of each request $i \in C_2$ reside in different subtrees of $\mathcal{T}_2$, but in the same subtree of $\mathcal{T}_1$. The remaining classes $C_3, C_4, \ldots$ are defined in a similar manner. It is important to note that the recursive process ends as soon as we arrive at a subtree with strictly less than k edges. In this case, we make use of the trivial decomposition, where the given subtree is broken into its individual edges.

In the above description, k was treated as a parameter whose value has not been determined yet. To obtain the desired approximation ratio, we set $k = \lceil \log n \rceil$. It follows that the overall number of levels in the recursion, or equivalently, the number of request classes is $O(\log_k n) = O(\log n/\log\log n)$. This claim is immediately implied by observing that the maximum size of a subtree in level ℓ of the recursion is at most $(3/k)^{\ell} \cdot |E|$.

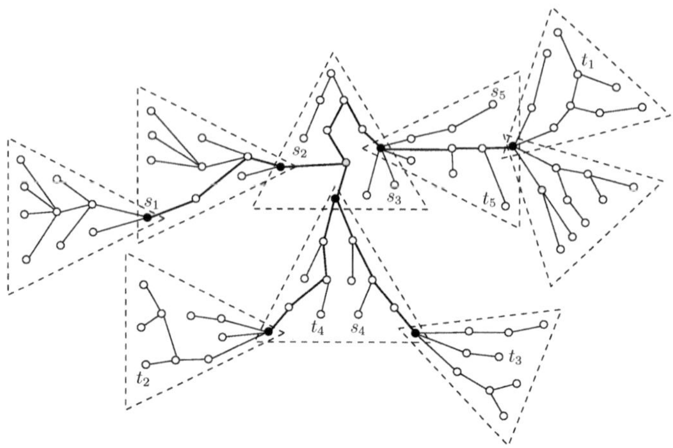

Fig. 1. An almost balanced 9-decomposition. Here, requests 1, 2, and 3 are separated, whereas requests 4 and 5 are not.

2.2 An Orientation Algorithm for a Single Decomposition

Notice that a class of requests, say C_ℓ, generally consists of several subsets of requests, each created when different subtrees in $\mathcal{T}_{\ell-1}$ are partitioned by the decomposition $\mathcal{T}_\ell$. More specifically, assuming that the subtrees in $\mathcal{T}_{\ell-1}$ are $T_1, T_2, \ldots$, the class C_ℓ can be written as the disjoint union of $C_\ell^1, C_\ell^2, \ldots$, where C_ℓ^j is the set of requests that are first separated when T_j is partitioned. Recall that the path of any request separated by some subtree decomposition must be contained in that subtree (otherwise, this request would have been separated in previous recursion steps). This observation implies that it is sufficient to compute an orientation for a single subtree decomposition and its induced set of separated requests. Given a polynomial-time algorithm that computes such an orientation, one can *sequentially* apply it to each of the subtree decompositions in the same recursion level. The resulting orientations (in edge-disjoint subtrees) can then be "glued" to form a single orientation, defined for the entire edge set, satisfying at least as many requests as the overall number of requests satisfied in all individual subtrees.

In what follows, we focus our attention on a single decomposition, and devise a randomized algorithm that computes an orientation which satisfies, in expectation, a constant fraction of the optimal number of satisfiable requests for this decomposition. Formally, an instance of the problem in question consists of a tree $T = (V, E)$, and a partition $\mathcal{T} = \{T_1, \ldots, T_k\}$ of this tree into k edge-disjoint subtrees, where $k \leq \lceil \log n \rceil$, and the number of vertices shared by at least two subtrees is less than k. In addition, we are given a collection C of requests, where each request path is separated by $\mathcal{T}$, meaning that s_i and t_i reside in different subtrees of the decomposition $\mathcal{T}$.

We need the following notation (exemplified in Figure 2). Let OPT denote the number of satisfied requests in some fixed optimal orientation of T. Let $V_B \subseteq V$

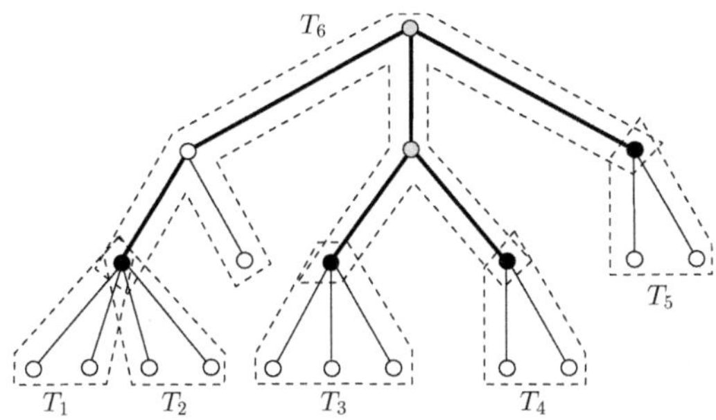

Fig. 2. An almost balanced 6-decomposition. Note that black vertices are border vertices, gray vertices are junction vertices, and bold edges make up the skeleton of the decomposition.

be the set of *border vertices* of $\mathcal{T}$, that is, the set of vertices that are shared by at least two subtrees in $\mathcal{T}$. Moreover, let $S \subseteq T$ be the *skeleton* of $\mathcal{T}$, namely, the minimal subtree spanned by all border vertices. Note that this subtree consists of the union of paths connecting any two vertices in V_B. Finally, let $V_J \subseteq V$ the set of *junction vertices*, defined as non-border skeleton vertices with degree at least 3 (counting only skeleton edges).

The algorithm. We are now ready to present the orientation algorithm. Our algorithm consists of two phases: *segment guessing*, where the optimal direction state of disjoint subpaths of the skeleton is attained, followed by *randomized assignment*, in which individual edges are assigned a direction.

PHASE I: SEGMENT GUESSING. Let us name the vertex set $V_B \cup V_J$ the *core* of the skeleton S. One can easily verify that $|V_B \cup V_J| < 2k$ as $|V_J| < |V_B| < k$, by Lemma 1. We now partition the skeleton into a collection $\Sigma(S)$ of edge-disjoint paths, which are referred to as *segments*. Each such segment is a subpath of S whose endpoints are core vertices, but its interior traverses only non-core vertices. Clearly, $|\Sigma(S)| = |V_B \cup V_J| - 1 < 2k$. We now argue that one could obtain in polynomial time the direction state that the optimal orientation induces on each segment $\sigma \in \Sigma(S)$, *simultaneously* for all segments. To this end, notice that any skeleton segment $\sigma = \langle v_1, v_2, \ldots, v_\ell \rangle$ may be in one of three possible direction states:

1. *Right direction*: all edges are consistently directed from v_1 towards v_ℓ, i.e., $v_1 \rightarrow v_2, v_2 \rightarrow v_3, \ldots, v_{\ell-1} \rightarrow v_\ell$.
2. *Left direction*: all edges are consistently directed from v_ℓ towards v_1, namely, $v_1 \leftarrow v_2, v_2 \leftarrow v_3, \ldots, v_{\ell-1} \leftarrow v_\ell$.
3. *Mixed direction*: the direction of segment edges is non-consistent.

These definitions imply that the total number of segment direction states to be examined is of polynomial size since $3^{|\Sigma(S)|} = 3^{O(k)} = 3^{O(\log n)} = n^{O(1)}$. As a consequence, we may assume without loss of generality that the set of direction states induced by the optimal orientation on all the segments of $\Sigma(S)$ is known in advance. This assumption can be easily enforced by enumerating over all $n^{O(1)}$ possible segment direction states.

PHASE II: RANDOMIZED ASSIGNMENT. The goal of this phase is to orient the graph while making sure that the edge directions respect the outcome of the segment guessing phase. For this purpose, we begin by considering skeleton segments that have a consistent direction, namely, segments in either right or left direction states, and assign all the edges in these segments their implied direction. The assignment procedure proceeds with two randomized assignment steps:

1. Each segment in a mixed direction state is assigned, independently and uniformly at random, a right or left direction. All segment edges are oriented according to the chosen direction.
2. Each of the decomposition subtrees $T_1, \ldots, T_k$ is assigned, independently and uniformly at random, the role of a *sender* or a *receiver*. All the edges of each sender subtree are oriented towards the skeleton (in its simplest form, when the subtree contains a single border vertex, all edges are oriented toward that vertex). In contrast, all the edges of each receiver subtree are oriented away from the skeleton. We refer the reader to an example in Figure 3(a).

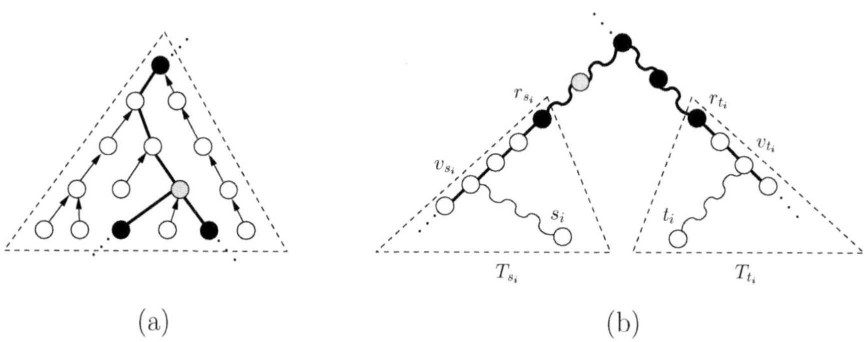

Fig. 3. (a) An orientation of a sender subtree, where the bold edges are part of the skeleton. (b) A partition of a request path into five parts.

We turn to prove that the expected number of satisfied requests is within a constant factor of optimal, as formally stated in the following theorem.

Theorem 1. *The resulting orientation satisfies at least* OPT/16 *requests in expectation.*

Proof. Recall that we have previously assumed the endpoints of each request to reside in different subtrees of the decomposition $\mathcal{T}$. In particular, this implies that each request path must traverse at least one border (core) vertex. For this reason, as shown in Figure 3(b), we can divide each request path, with endpoints s_i and t_i, into five (some possibly empty) parts:

1. A subpath between s_i and its closest skeleton vertex v_{s_i}.
2. A subpath, along a partial skeleton segment, between v_{s_i} and its closest core vertex r_{s_i}.

3. A subpath between t_i and its closest skeleton vertex v_{t_i}.
4. A subpath, along a partial skeleton segment, between v_{t_i} and its closest core vertex r_{t_i}.
5. A subpath between r_{s_i} and r_{t_i}, along a sequence of complete skeleton segments.

With these definitions in mind, let us focus on some request i that is satisfied in the optimal orientation. We now argue that, with probability at least 1/16, this request is satisfied in the random orientation constructed by the algorithm. Consequently, by linearity of expectation, the overall expected number of satisfied requests is OPT/16. A key observation one should make to establish this argument is that all the segments along the subpath between r_{s_i} and r_{t_i} must have a consistent direction in the optimal orientation; otherwise, this request would not have been satisfied. Accordingly, we may assume that our algorithm assigned the same direction to all the edges in these segments. Now, notice that the request under consideration is satisfied if the following four probabilistic events occur: (1) the edges in the subpath between s_i and v_{s_i} are oriented towards v_{s_i}; (2) the edges in the subpath between v_{s_i} and r_{s_i} are oriented towards r_{s_i}; (3) the edges in the subpath between v_{t_i} and r_{t_i} are oriented towards v_{t_i}; and (4) the edges in the subpath between t_i and v_{t_i} are oriented towards t_i. One can easily validate that these four events are independent, and that each one of them occurs with probability of at least 1/2. For example, the edges in the subpath between s_i and v_{s_i} are oriented towards v_{s_i} if the underlying subtree T_{s_i} is selected as a sender. As a result, the probability that request i is satisfied in the random orientation is at least 1/16.

Derandomization. The avid reader may have already noticed that the extent to which we utilize randomization is rather limited, and that its foremost purpose is to make the presentation of our algorithm simpler. Specifically, each segment in a mixed direction state is randomly assigned one of two possible directions, while each decomposition subtree is randomly assigned one of two possible roles. In other words, all we need to obtain a deterministic algorithm is a uniform sample space, with two possible values for $O(\log n)$ independent random variables. This can be constructed in polynomial time either explicitly, as there are only $n^{O(1)}$ possible outcomes, or in a more compact way, by observing that fourwise-independence is sufficient for the preceding analysis (see, for instance, [1, Chap. 15]).

A semi-oblivious property. In view of the derandomization procedure, we may reinterpret our algorithm as the following two-stage process: initially, we generate a set of polynomially-many potential orientations, determined by all possible outcomes of both the segment guessing and randomized assignment phases, and then, we select an orientation that maximizes the number of satisfied requests. Now, notice that the first stage of this process is semi-oblivious. Specifically, the set of generated orientations is independent of the collection of requests, and only builds on the structure of the underlying network. This property allows us to employ the algorithm in generalized requests settings. For instance, a natural generalization of maximum graph orientation is when each request is characterized by a collection of ordered source-target pairs, rather than a single pair. In this setting, a request is regarded as satisfied if at least one of its source-target pairs admits a directed path in the oriented graph. One can easily verify that our algorithm attains the same performance guarantees for this setting by applying nearly identical analysis.

One interesting scenario that is captured by the above-mentioned generalization is of maximum graph orientation *with groups*. In this scenario, a request i is satisfied if there is a directed path from some vertex $s_i \in S_i$ to some vertex $t_i \in T_i$ in the oriented graph. Here, S_i and T_i are vertex sets that characterize the request.

3 Orienting Mixed Graphs

Thus far, we have restricted our attention to undirected graphs. In practice, signaling pathways contain various types of interactions whose direction is specified in advance, most notably protein-DNA interactions. This implies that the input to the graph orientation problem is, in its utmost general setting, a mixed graph. A key difficulty in this setting is that, unlike the seemingly easier-to-handle scenario of unoriented edges, there is no trivial reduction to tree instances. What prevents us from contracting cycles is the possible existence of cycles with oriented edges pointing in opposite directions, for which there does not seem to be an easy way to decide in advance on the orientation of remaining edges.

Despite the inherent difficulty in a mixed graph input, the biological setting provides us with several constraints on the input graph, which we exploit in our approximation algorithm. The first biologically-motivated constraint relates to the occurrence of PDI edges along pathways. Reviewing real pathways, we observed that signaling pathways do not jump back and forth between PPIs and PDIs, rather in the vast majority of the cases there is a single switch from PPIs to PDIs. Precisely, define a PDI *segment* in a linear path as a series of consecutive PDIs along the path that is flanked by PPI edges or by the start/end of the path. To gather statistics on the number of PDI segments in real pathways, we downloaded 116 human pathways from KEGG [11]. For each pathway, we counted the number of PDI segments in its longest linear path. Only 35 of the 116 pathways contained PDIs, and 18 of which had at least one PDI segment in their longest path. Notably, 17 of the 18 contained a single PDI segment; the remaining pathway contained two segments.

A second constraint is a refinement of the first one: In two thirds of the 18 KEGG pathways with at least one PDI segment, the segment occurred at the end of the pathway. Interestingly, our algorithm can be directly applied to this latter scenario (of a single PDI segment that occurs at the end of the pathway) as each cause-effect pair can then be translated into a group request where the cause should connect to any of the genes that have a directed PDIs path to the effect. Below, we consider the general case and show that if the sought pathways contain at most ℓ segments then an $O(\log^{\ell} n)$ approximation is possible.

3.1 The Approximation Algorithm

Let $G = (V, E)$ be a mixed graph whose edge set can be described as $E = E_{\mathcal{O}} \cup E_{\mathcal{U}}$, where $E_{\mathcal{O}}$ consists of edges with predefined directions, and $E_{\mathcal{U}}$ are unoriented edges. Even though the input graph may contain cycles with oriented edges pointing in opposite directions, we can still contract unoriented cycles, and more generally, cycles where all oriented edges are consistently pointing in the same direction. Therefore, from this point on we assume that such cycles have already been contracted. With this setting in mind, an *unoriented component* (or, $\mathcal{U}$-component, for short) is defined as a maximal connected component of the unoriented subgraph $(V, E_{\mathcal{U}})$. It is worth noting that any $\mathcal{U}$-component is necessarily a tree, or otherwise, there must be unoriented cycles, which should have been contracted earlier on. Also note that there are no oriented edges with both head and tail residing in the same $\mathcal{U}$-component since any such edge induces a cycle that should have been contracted before. As a consequence of the preprocessing steps described above, the input graph can be represented as a directed acyclic graph on the $\mathcal{U}$-components, as illustrated in Figure 4(a). That is, the collection of $\mathcal{U}$-components can be topologically sorted such that all oriented edges between them are pointing from left to right.

Let $T = (V_T, E_T)$ be some $\mathcal{U}$-component. We first show how to compute a random orientation of E_T such that *any* pair of vertices in T is connected with probability $\Omega(1/\log n)$. For this purpose, suppose we execute the classification process suggested in Section 2.1 where the collection of requests consists of all vertex pairs in T. However, rather than using almost-balanced k-decompositions with $k = \lceil \log n \rceil$, we will simplify the process by picking $k = 2$. Even though the number of resulting request classes slightly blows up to $O(\log n)$, each time a tree is being decomposed, we obtain only two almost-balanced edge-disjoint subtrees which intersect in a common vertex. On top of picking an alternative value of k, instead of testing all $O(\log n)$ request classes as potential candidates for the class that separates the maximal number of pairs, we will pick one such class uniformly at random. Given this class, the orientation of each decomposed tree (two edge-disjoint subtrees) is determined, independently and uniformly at random, from one of the following two alternatives, as shown in Figure 4(b):

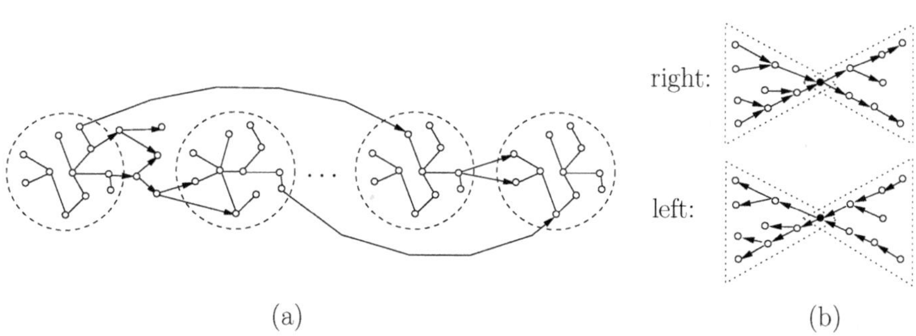

Fig. 4. (a) A directed acyclic graph on $\mathcal{U}$-components. (b) The two possible orientations of a decomposed tree.

- *Right orientation*: All edges of the first subtree are oriented towards the common root and all edges of the second subtree are oriented away from that root.
- *Left orientation*: All edges of the second subtree are oriented towards the common root and all edges of the first subtree are oriented away from that root.

Following Section 2.1, it is not difficult to verify that any pair of vertices $(s,t) \in V_T \times V_T$ is connected with probability $\Omega(1/\log n)$ since the particular class that separates s and t is picked with probability $\Omega(1/\log n)$, and given that this class has been picked, there is a directed path from s to t with probability 1/2.

We handle an arbitrary mixed graph $G = (V, E_{\mathcal{O}} \cup E_{\mathcal{U}})$, which has already been brought to the structural form of a directed acyclic graph on the collection of its $\mathcal{U}$-components, by independently running the randomized single-tree procedure in each $\mathcal{U}$-component. As a result, we obtain a random orientation of the graph, whose performance guarantee depends on the minimal number of $\mathcal{U}$-components that must be traversed in order to satisfy any request. Specifically, let (s_i, t_i) be the i-th request pair, and suppose ℓ_i stands for the minimal number of components that have to be traversed in an orientation that connects s_i to t_i. Then the following result can be stated:

Theorem 2. *The random orientation algorithm constructs an orientation that satisfies $\Omega(\mathrm{OPT}/\log^{\ell} n)$ requests in expectation, where $\ell = \max \ell_i$.*

Proof. Let us focus on request i, and let P_i be a directed s_i-t_i path that traverses ℓ_i components in some orientation of G; we denote these $\mathcal{U}$-components by $\mathcal{U}^1, \ldots, \mathcal{U}^{\ell_i}$, indexed in topological order. Furthermore, let x_j and y_j be the entry vertex and exit vertex of P_i in $\mathcal{U}^j$, respectively. Notice that every (x_j, y_j) pair is connected with probability $\Omega(1/\log n)$ since the randomized single-tree procedure is independently run in each $\mathcal{U}$-component. This implies that s_i and t_i are connected with probability $\Omega(1/\log^{\ell_i} n)$. Consequently, the expected number of satisfied requests is $\Omega(\mathrm{OPT}/\log^{\ell} n)$ by linearity of expectation.

4 Conclusions

We have designed a novel approximation algorithm for maximum graph orientation that achieves an $O(\log n/ \log\log n)$ ratio. We have further shown an extension of the algorithm that handles mixed graphs and provides a polylogarithmic approximation ratio under biologically-motivated assumptions. On the theoretical side, we believe that the techniques presented here are of independent interest, and may be applicable in other settings as well. On the practical side, the algorithmic extension to mixed graphs tackles a major open problem posed in [14] and is expected to yield much more realistic network orientations by integrating knowledge on PDIs into the orientation process.

Acknowledgments

We thank Yael Silberberg for her help in gathering statistics on the KEGG pathways. I.G. was supported by the Israel Science Foundation, by the European

Commission under the Integrated Project QAP funded by the IST directorate as Contract Number 015848, by a European Research Council (ERC) Starting Grant, and by the Wolfson Family Charitable Trust. R.S. was supported by a research grant from the Israel Science Foundation (grant no. 385/06).

References

1. Alon, N., Spencer, J.H.: The Probabilistic Method, 2nd edn. Wiley, Chichester (2000)
2. Arkin, E.M., Hassin, R.: A note on orientations of mixed graphs. Discrete Applied Mathematics 116(3), 271–278 (2002)
3. Even, G., Garg, N., Könemann, J., Ravi, R., Sinha, A.: Covering graphs using trees and stars. In: Arora, S., Jansen, K., Rolim, J.D.P., Sahai, A. (eds.) RANDOM 2003 and APPROX 2003. LNCS, vol. 2764, pp. 24–35. Springer, Heidelberg (2003)
4. Feige, U., Goemans, M.X.: Aproximating the value of two prover proof systems, with applications to MAX 2SAT and MAX DICUT. In: Proceedings 3rd Israel Symposium on Theory and Computing Systems, pp. 182–189 (1995)
5. Frederickson, G.N., Johnson, D.B.: Generating and searching sets induced by networks. In: de Bakker, J.W., van Leeuwen, J. (eds.) ICALP 1980. LNCS, vol. 85, pp. 221–233. Springer, Heidelberg (1980)
6. Gamzu, I., Segev, D.: A sublogarithmic approximation for highway and tollbooth pricing. In: Proceedings of the 37th International Colloquium on Automata, Languages and Programming (to appear, 2010), `http://arxiv.org/abs/1002.2084`
7. Goldschmidt, O., Hochbaum, D.S., Levin, A., Olinick, E.V.: The SONET edge-partition problem. Networks 41(1), 13–23 (2003)
8. Hakimi, S.L., Schmeichel, E.F., Young, N.E.: Orienting graphs to optimize reachability. Information Processing Letters 63(5), 229–235 (1997)
9. Hassin, R., Segev, D.: Robust subgraphs for trees and paths. ACM Transactions on Algorithms 2(2), 263–281 (2006)
10. Håstad, J.: Some optimal inapproximability results. Journal of the ACM 48(4), 798–859 (2001)
11. Kanehisa, M., Goto, S., Furumichi, M., Tanabe, M., Hirakawa, M.: KEGG for representation and analysis of molecular networks involving diseases and drugs. Nucleic Acids Research 38(2), D355–D360 (2010)
12. Khot, S., Kindler, G., Mossel, E., O'Donnell, R.: Optimal inapproximability results for MAX-CUT and other 2-variable CSPs? SIAM Journal on Computing 37(1), 319–357 (2007)
13. Lewin, M., Livnat, D., Zwick, U.: Improved rounding techniques for the MAX 2-SAT and MAX DI-CUT problems. In: Proceedings 9th International Conference on Integer Programming and Combinatorial Optimization, pp. 67–82 (2002)
14. Medvedovsky, A.: An algorithm for orienting graphs based on cause-effect pairs and its applications to orienting protein networks. Master's thesis, Tel-Aviv University, Israel (2009)
15. Medvedovsky, A., Bafna, V., Zwick, U., Sharan, R.: An algorithm for orienting graphs based on cause-effect pairs and its applications to orienting protein networks. In: Crandall, K.A., Lagergren, J. (eds.) WABI 2008. LNCS (LNBI), vol. 5251, pp. 222–232. Springer, Heidelberg (2008)

Enumerating Chemical Organisations in Consistent Metabolic Networks: Complexity and Algorithms

Paulo Vieira Milreu[1,2], Vicente Acuña[1,2], Etienne Birmelé[3], Pierluigi Crescenzi[4], Alberto Marchetti-Spaccamela[5], Marie-France Sagot[1,2], Leen Stougie[6,7], and Vincent Lacroix[1,2]

[1] Université de Lyon, F-69000 Lyon, Université Lyon 1, CNRS, UMR5558, Laboratoire de Biométrie et Biologie Evolutive, F-69622 Villeurbanne, France
[2] INRIA Rhône-Alpes, 38330 Montbonnot Saint-Martin, France
{milreu,viacuna,lacroix}@biomserv.univ-lyon1.fr,
marie-france.sagot@inria.fr
[3] Lab. Statistique et Génome, CNRS UMR8071 INRA1152, Université d'Évry, France
etienne.birmele@genopole.cnrs.fr
[4] Università di Firenze, Dipartimento di Sistemi e Informatica, I-50134 Firenze, Italy
pierluigi.crescenzi@unifi.it
[5] Sapienza University of Rome, Italy
alberto.marchetti@dis.uniroma1.it
[6] VU University and CWI, Amsterdam, The Netherlands
lstougie@feweb.vu.nl

Abstract. The structural analysis of metabolic networks aims both at understanding the function and the evolution of metabolism. While it is commonly admitted that metabolism is modular, the identification of metabolic modules remains an open topic. Several definitions of what is a module have been proposed. We focus here on the notion of chemical organisations, i.e. sets of molecules which are closed and self-maintaining. We show that finding a reactive organisation is NP-hard even if the network is flux-consistent and that the hardness comes from blocking cycles. We then propose new algorithms for enumerating chemical organisations that are theoretically more efficient than existing approaches.

1 Introduction

Until recently, metabolism was analysed via the pathways composing it, which were traditionally established in a non automatic way by experts interested in some specific function (glycolysis for instance, or anaerobic respiration). The pathways were studied independently from each other, even though molecules could be shared. The advent of full genome sequences now enables to infer genome-scale metabolic networks (see [8] for an overview) and the study of these networks has revealed extensive cross-talk between traditionally defined pathways, as well as the use by different organisms of alternative pathways, that is, different metabolic routes to a same final overall product goal. While the notion

V. Moulton and M. Singh (Eds.): WABI 2010, LNBI 6293, pp. 226–237, 2010.

of metabolic pathway remains useful as a reference definition of functional modules, the exact frontiers between pathways can now be questioned, and other definitions of the concept of metabolic module can be proposed. While several formal definitions have already been suggested, none of them is able to capture all the expert knowledge that led to the first pathways while at the same time providing an insight into alternative functional ones. It is actually an open question in the field whether such definition exists. Most likely, not a single model will suffice. The motivation of this paper is to explore one model for metabolic modules called chemical organisations, both in terms of the complexity of enumerating such modules and of exact algorithms for performing the enumeration. The results obtained constitute a solid algorithmic ground for the study of chemical organisations, a necessary first step to widen the use of this notion for the computational analysis of metabolic networks.

Several formal definitions of pathways and modules can be found in the literature on metabolism, the best known of which may be elementary modes [12] or any of its close cousins (see [8,9] for a survey). Elementary modes may be informally described as metabolic subnetworks that can function at steady state, meaning that all internal metabolites are produced and consumed in equal rates (that is, nothing accumulates internally). This is a fine definition, but has at least one drawback: it is restricted to the analysis of the system at steady state and does not allow to describe states of the system where metabolites can accumulate. However, such states are relevant as they could correspond to intermediary steps in the evolution of metabolism, or temporary states in the dynamics of metabolism.

As far as we know, two models in the literature enable to study such states. One is Petri nets and the other is a more recent model called *chemical organisations*. Because it is algebraically easier to manipulate chemical organisations as their formulation follows closely that of (hyper)graphs and matrices, we focus our attention in this paper on chemical organisations. The concept was introduced in 2005 by Peter Dittrich and his group [4], building on earlier work by Fontana and Buss [6] and can be used not only for metabolism, but also for any kind of reaction system, including regulatory networks. In this paper, however, we focus exclusively on metabolism.

Chemical organisations are sets of molecules that are self-maintaining and closed (in this paper, we use the terms metabolite and molecule with no distinction). Informally, a self-maintaining set is a set where molecules can accumulate – the feature we were seeking – provided no molecule vanishes. A set is closed if all metabolites produced from reactions for which all the inputs are present in the set will also be present and thus part of the set. By convention, this includes all reactions that take their input from the environment, *i.e.* are external. All external inputs are therefore considered as being available *and used.* This introduces a second contrast with elementary modes (EMs). Indeed, EMs may use only part of the externally available inputs. More generally, EMs are not closed.

Finally, as we have already said, the theory of chemical organisations has been proposed for general reaction systems. Its application to metabolic networks raises new specific questions, as the networks have specific properties. They are indeed expected to be flux-consistent (each reaction belongs to at least one elementary mode).

The objectives of this paper are thus twofold. The first is to revisit chemical organisations in the context of flux-consistent networks. In particular, finding a chemical organisations was shown to be hard in [2]. A legitimate and non trivial question is whether this remains true in biologically more realistic flux-consistent networks. Section 2 presents the main definitions on chemical organisations and consistency of networks. Section 3 shows that even for consistent networks the enumeration problem is hard. We go however further by identifying the specific structural properties of the network that account for this hardness. Those are discussed in Section 4, while Section 5 fulfills the second objective of this paper. This is to describe a new algorithm that takes advantage of such properties to obtain an exact method that is in all cases theoretically more efficient for consistent networks than the enumeration algorithms presented in [2] because, at best, a smaller part of the solution space needs to be explored. Due to space limitations some of the proofs are omitted here and will be presented in the journal version.

2 Preliminaries

A metabolic network, like any reaction system, can be modelled as a *weighted directed hypergraph* $G = (M, R)$ with M the set of *vertices* corresponding to the metabolites and R the set of *hyperarcs* corresponding to the reactions. A directed hyperarc (*i.e.* a reaction) $r \in R$ is an ordered pair of sets of vertices (*i.e.* metabolites) $r = (subs(r), prod(r))$ where $subs(r)$ is the set of substrates of r and $prod(r)$ is the set of products of r. For each x in $subs(r)$ (in $prod(r)$) the weight of x with respect to r denotes the stoichiometric coefficient of x in r, that is, the number of units of x consumed (or produced) when r fires. Note that x can belong to both $subs(r)$ and $prod(r)$; in this case there are two weights associated to x w.r.t. r. Note also that, according to the above definitions, the set of substrates of a reaction r can be empty: in this case, we say that the metabolites in $prod(r)$ are *inputs* of the network.

Metabolic networks have also been often modelled using matrices [11]. The *stoichiometric matrix* S has $|M|$ rows and $|R|$ columns where $S_{i,j}$ is the stoichiometric coefficients of molecule i in reaction j. $S_{i,j}$ is negative if i is consumed and it is positive if i is produced. We notice here that while the stoichiometric matrix can always be derived from the weighted hypergraph, the reverse is not true. Indeed, metabolites involved as substrates and products of the same reaction cannot be handled in the matrix representation.

For some of the results presented, we also use the concept of the *underlying graph* of G, which is a directed multigraph with the same set of vertices of G and arcs $x \rightarrow y$ for every pair of vertices x, y for which there is an hyperarc r such that $x \in subs(r)$ and $y \in prod(r)$. A reaction is said to be on a path/cycle of the underlying directed graph if any of its (substrate,product)-pairs is an arc of the path/cycle.

In the context of metabolic networks, we say that a *flux* over the network is the rate at which each reaction occurs. A flux can be represented as a *flux vector* $v \in \mathbb{R}^{|R|}$ with $v[i]$ denoting the rate of reaction i.

A metabolic network is **flux-consistent** if there exists a flux vector $v > 0$,i.e $\forall i \in R$ the flux $v[i] > 0$, such that $Sv = 0$ [1]. This is the same as saying that every reaction of the network belongs to at least one elementary mode, thus checking for the usefulness of each reaction. For more information on elementary modes, see [12] and [11].

We denote by $\mathcal{R}_A \subseteq R$ the subset of reactions that can be fired when the metabolites in set $A \subseteq M$ are present, *i.e.*, $\mathcal{R}_A = \{r \in R | subs(r) \subseteq A\}$. We now introduce the basic definitions that will be used throughout the paper.

Definition 1. *A set $C \subseteq M$ is* ***closed*** *if, for all reactions $r \in \mathcal{R}_C$, $prod(r) \subseteq C$. Moreover, given a set $C \subseteq M$, the* ***closure*** *of C, denoted by Cl_C, is the smallest closed set H that contains C.*

Note that if C is a closed set of molecules, then C must contain all inputs of the network (since the empty set is a subset of C and input reactions therefore belong to $\mathcal{R}_C$). In particular, the closure of the empty set will contain all inputs and whatever can be produced from them.

Definition 2. *A set of molecules $C \subseteq M$ is* ***self-maintaining*** *if there is a flux vector v such that:*

1. *for all reactions $r \in \mathcal{R}_C$, $v[r] > 0$;*
2. *for all reactions $r \notin \mathcal{R}_C$, $v[r] = 0$;*
3. *for all molecules $i \in C$, the production rate $(Sv)[i] \geq 0$.*

A set of molecules is self-maintaining if there exists a flux vector such that the molecules present in the set can accumulate ($(Sv)[i] > 0$) or be consumed and produced at the same rate ($(Sv)[i] = 0$) but none of them may disappear (3rd condition). Conditions 1 and 2 basically specify that all reactions that can fire with molecules from the set will fire. In particular, reactions which produce molecules outside the set will also fire. A self-maintaining set is therefore really self-maintaining, even in the presence of "leaks".

Definition 3. *A set of molecules $O \subseteq M$ is an* ***organisation*** *if it is closed and self-maintaining. O is said to be* ***reactive connected*** *if:*

- *(reactive) each metabolite in O takes part as substrate or product in at least one reaction inside $\mathcal{R}_O$;*
- *(connected) for any two molecules x and y in O, there is a path from x to y in the underlying undirected graph.*

We present a network in Figure 1 that has 3 connected components and 8 organisations but only 2 of them are reactive connected organisations. The others are just combinations of organisations which cannot directly interact among them. Notice also that some sets are not organisations, such as for instance the sets

$\{s, p\}$ and $\{b, c, d\}$, since they are not closed, or the sets $\{s, q, r\}$ and $\{c, d\}$ that are not closed nor self-maintaining. On the other hand, the closure of the empty set is $\{s\}$ and it is an organisation. This organisation must be present in all organisations even if there is no possible interaction between their molecules as in the case of organisation $\{s, a\}$. Clearly, any set of disconnected nodes will form an organisation, as long as we consider its union with the closure of the empty set. In the following, we shall ignore such organisations and focus on organisations where the molecules interact among them. This is our motivation to find only reactive connected organisations.

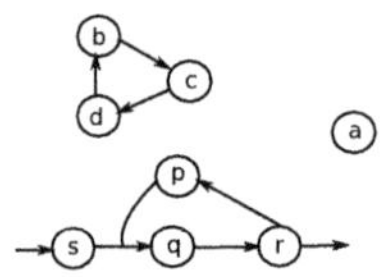

Fig. 1. A metabolic network with (a) 3 connected components, (b) 8 organisations: $\{\{s\}, \{s, p, q, r\}, \{s, a\}, \{s, b, c, d\}, \{s, a, b, c, d\}, \{s, a, p, q, r\}, \{s, b, c, d, p, q, r\}, \{s, a, b, c, d, p, q, r\}\}$, and (c) 2 reactive connected organisations: $\{\{s\}, \{s, p, q, r\}\}$

For this reason, from now on we restrict our networks to have only one connected component and some input and output metabolites. For general networks, indeed, we can without loss of generality work on each connected component separately and then combine the results.

Since throughout the paper we need to compute closures of sets, we recall here the forward propagation procedure [10] that in an iterative process enables to obtain the closure of a given set C. Informally, this consists in starting from C itself, adding $prod(r)$ for every $r \in \mathcal{R}_C$, and repeating this procedure until no new metabolites are added to C.

As already mentioned, all inputs of the network need to be considered together in order to compute organisations. This is a modelling choice that implies that if one wished to compute organisations for different subsets of the inputs, then it would be necessary to edit the network and recompute the organisations for the subsets of interest.

3 Chemical Organisations in Consistent Networks

It was shown that deciding whether a network contains at least one organisation is NP-complete [2]. However the proof was based on a network that was not flux-consistent. We now characterise organisations in consistent networks. First of all, we observe that it is easy to check whether a set C is an organisation by inspecting the reaction rules to check closure and self-maintenance using linear programming.

The following theorem shows how to compute two possible organisations.

Theorem 4. *If a network is flux-consistent then the whole network and the closure of the empty set are organisations.*

Proof. The whole network is always closed. By definition of flux-consistency, we have a flux vector v that covers the whole network, satisfying the condition of self-maintenance. Therefore the whole network is an organisation. Analogously, if the closure of the empty set produces the whole network then it is an organisation. Otherwise, since every metabolite is produced from the empty set, we can easily obtain a valid flux vector v satisfying the condition of self-maintenance. □

Notice that the closure of the empty set is the smallest possible organisation since it has to be contained in all other organisations.In the following, we say that the whole network and the closure of the empty set are **trivial organisations**.

Observe that the closure of the empty set may not always produce the whole network. An example is given in Figure 2 since the closure of the empty set for that network is $\{a\}$.

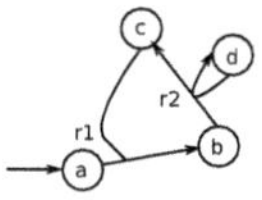

Fig. 2. Network in which the closure of the empty set does not produce the whole network

Theorem 5. *If the network is flux-consistent and acyclic, i.e. the underlying directed graph of the hypergraph is acyclic, then the whole network is the only organisation.*

Proof. The smallest organisation is given by the closure of the empty set, which can be obtained by applying the forward propagation algorithm to the empty set. As the network is flux-consistent and acyclic, from the inputs any metabolite can be reached, *i.e.*, produced. Hence, the closure of the empty set is the entire network. From the flux-consistency of the network and from Theorem 4, it follows that the smallest organisation is the whole network. □

The next result shows that the problem of finding a non trivial organisation in a flux-consistent network is NP-hard. The proof is based on a reduction from the 3-SAT problem, which is an appropriate modification of the original reduction given in [2], that showed that finding a reactive organisation in a general reaction system is NP-hard.

Theorem 6. *Deciding if a flux-consistent network contains a non trivial organisation is NP-hard.*

4 Enumerating Chemical Organisations

Theorem 6 immediately implies that it is not possible to enumerate all organisations in a flux-consistent network in polynomial-time-delay in the size of the network.

We now observe that Theorem 5 indicates that for flux-consistent networks, the difficulty of finding non trivial organisations comes from the presence of cycles in the network. Indeed, as shown in Figure 3(b), cycles may interrupt the forward propagation if there exists a reaction that can produce a new metabolite a but needs for this a metabolite b which is not available and therefore blocks the reaction.

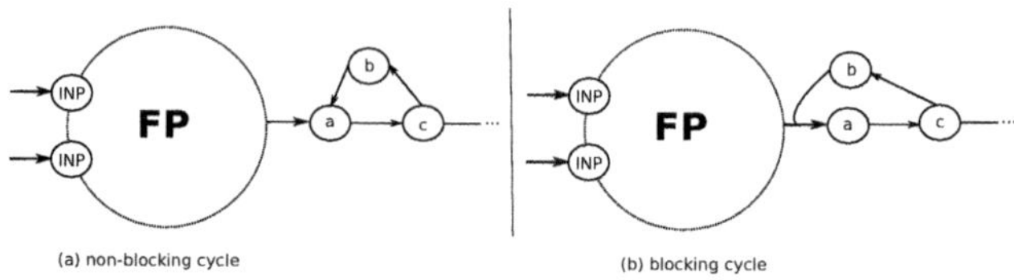

Fig. 3. (a) Non blocking cycle that will be traversed by the forward propagation procedure. (b) Forward propagation blocked by a unreached metabolite b.

In order to find the reactive connected organisations, we need to process cycles every time the forward propagation procedure stops. This simple observation gives an upper bound of 2^k on the number of reactive connected organisations in flux-consistent networks, where k is the number of cycles in the network. In order to proceed we first define cycles formally. By a **cycle** in the metabolic network we mean a simple *directed* cycle in the underlying graph. Self-loops are also considered as cycles.

Definition 7. *A **hitting set** of a set of cycles is a set of metabolites such that each cycle contains at least one element of the hitting set.*

Theorem 8. *Let H be a hitting set of all the cycles of a directed hypergraph. The set of all reactive connected organisations, denoted as $\mathcal{O}$, is such that*

$$\mathcal{O} \subseteq \bigcup_{C \subseteq H} \{Cl_C\}$$

Proof. It is sufficient to show that if A is a reactive connected organisation then $A = Cl_C$, where $C = A \cap H$.

First observe that, since A is closed and C is a subset of its metabolites, it follows that $Cl_C \subseteq A$.

Let us suppose that A contains vertices which are not in Cl_C. We colour these vertices white and the vertices of Cl_C black. Consider any white metabolite a_1. Since A is an organisation, a_1 cannot be vanishing. Moreover, it is not an input of the network as otherwise it would be black. Therefore, there exists a reaction r fired by A that has a_1 as product and has a white substrate a_2, $a_2 \neq a_1$, otherwise a_1 would again be black by closure.

By iterating the above reasoning, it follows that the subgraph of the underlying directed graph induced by white vertices has minimum in-degree at least 1 and contains a directed cycle. This contradicts the fact that H hits all the cycles. The set of white vertices is therefore empty and $A = Cl_C$. □

The bound of the previous theorem is tight. Indeed, an example where the number of organisations reaches $2^{|H|}$ is given in Figure 4 where from the input, k metabolites are produced by k independent reactions and all of them are blocked by cycles. Any combination of these independent paths can be de-blocked and produce a new organisation, and therefore we have 2^k organisations.

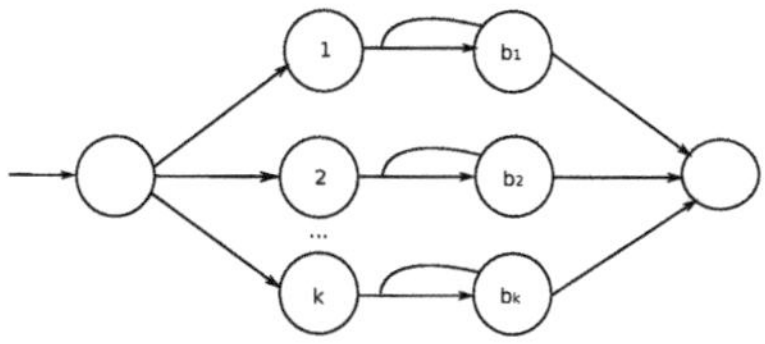

Fig. 4. Example with k parallel blocking cycles and 2^k organisations

However, some cycles never interrupt the forward propagation procedure as illustrated in Figure 3(a). Other cycles, on the other hand, exhibit structural properties that may lead to a blocking situation. Such cycles are called **potentially blocking cycles**. A basic solution to find all organisations is to know how to unblock all cycles of the network independently of whether they are potentially blocking cycles or not. Therefore, instead of finding a hitting set for all cycles of the network, it is enough to break all potentially blocking cycles to compute all reactive connected organisations. In order to prove this, we first introduce a more formal definition of potentially blocking cycle.

Definition 9. *A* ***potentially blocking cycle*** *is a cycle such that there exists a reaction* $r = (\{s_1, \ldots, s_h\}, \{p_1, \ldots, p_k\})$ *in the network satisfying the following two conditions: (1) there exists* i *and* j *such that* (s_i, p_j) *is an edge of the cycle, and (2) there exists* ℓ *such that* s_ℓ *is not in the cycle.*

A potentially blocking cycle may or may not interrupt the forward propagation depending on the metabolites that were produced by the procedure once the cycle is reached. Figure 5(a) shows an example in which the cycle will be traversed, while Figure 5(b) shows an example in which it will block the forward propagation algorithm.

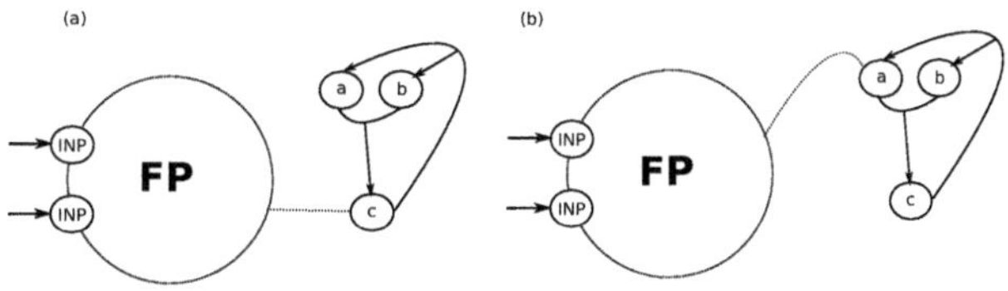

Fig. 5. Example of a potentially blocking cycle formed by the vertices a and c and reactions $a + b \rightarrow c$ and $c \rightarrow a + b$ (note that vertices b and c also form a symmetric cycle). If the forward propagation procedure (FP) reaches the cycle through c, the cycle is traversed, but if it reaches it through a, it is blocked.

Theorem 10. *Let H be a hitting set of all the potentially blocking cycles of a directed hypergraph. The set of all reactive connected organisations, denoted as $\mathcal{O}$, is such that*

$$\mathcal{O} \subseteq \bigcup_{C \subseteq H} \{Cl_C\}$$

Proof. As in the proof of Theorem 8, it is sufficient to show that, given a reactive connected organisation A, $A = Cl_C$, where $C = A \cap H$ (in the following, all paths and cycles are meant directed in the underlying directed graph). Once again, it is easy to see that $Cl_C \subseteq A$. Let us then suppose that A contains vertices which are not in Cl_C and let us colour them white and those of Cl_C black.

Let a_i ($i = 1, 2, \ldots, k$) be the set of white vertices such that there exists a reaction r_i having a_i as product and at least one black substrate. Then r_i has also at least one white substrate, which we denote by w_i, as otherwise a_i would be black by closure. Note that a white vertex does not have to belong to the set of the a_i's, but that this set is not empty as the organisation is connected and the set of white vertices is not empty.

If $w_i = a_i$, the selfloop induced by r_i is a potentially blocking cycle that contains no vertex of H, leading to a contradiction. Thus we may assume that w_i is distinct from a_i.

For $1 \leq i \leq k$, define T_i as follows: a vertex w is in T_i if either $w = w_i$ or there exists a white path starting from w and ending in w_i. Up to a reordering, we may assume that $|T_1| \leq |T_i|$ for $2 \leq i \leq k$. As T_1 is not empty, it has to contain at least one vertex that is a product of a reaction having a black substrate. If that vertex is a_1, there exists a path from a_1 to w_1, which yields a white cycle with the edge (w_1, a_1). That cycle is a potentially blocking cycle (because of the reaction r_1) which contains no vertex in H, leading to a contradiction.

In other words, T_1 contains a vertex among $(a_2, \ldots, a_k)$, say a_2. This implies that every path ending in w_2 can be extended to a path ending in w_1 and thus $T_2 \subseteq T_1$. Therefore, by minimality of T_1, $T_1 = T_2$.

This implies that $w_1 \in T_2$: hence, there exists a white path from w_1 to w_2. As $a_2 \in T_1$, there also exists a white path from a_2 to w_1. Thus, we can construct a white path from a_2 to w_2. Considering the edge (w_2, a_2), we again obtain a white cycle, which is potentially blocking (because of the reaction r_2) and contains no vertex in H, leading to a contradiction.

Thus, the set of white vertices is empty and $A \subseteq Cl_C$. □

Even in the case of the previous theorem, the bound is tight. Indeed, an example where the number of organisations is $2^{|H|}$ is, once again, given in Figure 4, since all cycles presented in the example are potentially blocking.

5 Hitting Set Approach to Enumerate Organisations

Two exact algorithms were proposed in [2] to enumerate organisations. The first one consisted in enumerating all closed sets and then checking for their self-maintenance, while the second one consisted in enumerating all self-maintaining sets and then checking linear combinations of them in order to obtain closed sets. A third approach was also proposed that was based on the second one but avoided enumerating all self-maintaining sets. This algorithm however was a heuristic not guaranteed to find all self-maintaining sets (and thus organisations). Finally, a variation of the first algorithm was proposed in order to enumerate only reactive connected organisations. This algorithm comes closer to the one we describe later and works as follows. First, the forward propagation of the empty set is computed. Once the procedure is blocked, all possible combinations of metabolites that are connected to the produced set X are considered for addition, in order to obtain further closed sets that include X. At this point, the algorithm recursively continues.

Note that in the above procedures, no concept of blocking cycles has been formally identified and used. Now that we know that the hardness comes from such cycles, two different approaches can be applied in flux-consistent networks. One is to find a **global hitting set** for all cycles of the network and then, following Theorem 8, to apply the forward propagation procedure on each subset of the hitting set to produce closed sets which together form all candidate organisations and, finally, to check through LP if the candidates are self-maintaining. However, the problem of finding a minimum hitting set for all cycles of a directed graph is NP-hard as indeed it corresponds to the **feedback vertex sex (FVS)** problem [7]. Nevertheless approximation algorithms such as the one described in [13] can be used in order to perform this step.

A second possibility is to find a **local hitting set**. According to Theorem 10 only potentially blocking cycles need to be considered. This is a superset of the blocking cycles that can be identified when the forward propagation procedure stops because it is at this moment that we know we are dealing with actually blocking cycles. A more efficient algorithm to enumerate reactive connected organisations is thus the following one: apply the forward propagation algorithm and once blocked, identify the set B of metabolites that are blocking the closure and find a hitting set that unblocks only the cycles which directly or indirectly involve these blocking metabolites.

In [5], the authors presented an approximation algorithm to a generalisation of the FVS problem, called SUBSET-FVS, in which only a subset of the directed cycles in the graph is considered interesting, more specifically the ones that intersects a set of special vertices. In our case, the set of special vertices would be the blocking metabolites locally identified as described in the previous paragraph. The authors in [5] gave two approximation algorithms for the SUBSET-FVS problem. The first algorithm achieves an approximation factor of $O\left(\log^2 |B|\right)$. The second achieves an approximation factor of $O\left(\min\{\log T \log\log T, \log n \log\log n\}\right)$, where T is the value of the optimum fractional solution of the problem at hand, and n is the number of vertices in the graph.

Before proving that this idea can be correctly used to exactly solve our problem, we need to define the concept of a blocking cycle in relation to a given set C of metabolites.

Definition 11. *Let C be a set of metabolites. A C-**blocking cycle** is a cycle of vertices which are not in C such that there exists a reaction r in the cycle whose set of substrates contains at least one metabolite in C.*

C-blocking cycles correspond to those which actually stop the forward propagation procedure.

Theorem 12. *Let C be a closed set and H a hitting set of the C-blocking cycles of a metabolic network. Let A be a reactive connected organisation whose metabolites contain C. Then either $C = A$ or there exists a non empty subset B of H such that the closure of $C \cup B$ is still a subset of A.*

Proof. Let A be a reactive connected organisation containing C as a subset of its metabolites and let $B = A \cap H$. Let us suppose that $C \neq A$. To prove the theorem, it is then sufficient to prove that $B = A \cap H$ is not empty.

We colour the vertices of $A \setminus C$ in white and those of C in black. Since A is a reactive connected organisation, there exist edges between the white and the black metabolites and some of them go from a black to a white vertex, as otherwise, white vertices would be vanishing. Let $(a_1, \ldots, a_k)$ be the set of white vertices reached by at least one edge coming from a black vertex.

The same argument of the proof of Theorem 10 can now be applied, showing that, as the set of white vertices is not empty, it contains a white C-blocking cycle. Therefore, B is not empty. □

Corollary 13. *Every reactive connected organisation is included in the set $\mathcal{CO}$ returned by the procedure given in Algorithm CCO.*

Proof. Let A be a reactive connected organisation. It has to contain C_0 as every organisation contains the closure of the empty set. Let C be maximum among the elements of $\mathcal{CO}$ which are subsets of A. Then Theorem 12 implies, by maximality of C, that $A = C$. □

```
Algorithm CCO(G)
Require: a metabolic network represented as a hypergraph G = (M, R);
Ensure: the set CO of all candidates for being organisations.
  CO ← {C0} where C0 is the closure of the empty set (Cl_{})
  for all elements C in CO which have not been treated before do
     Compute a hitting set H of the C-blocking cycles
     for every B ⊂ H do
        Compute Cl_{C∪B} and add it to CO if it was not present already
  return CO
```

Notice that the size of the hitting set computed by the algorithm is never greater than the number of blocking metabolites. Thus we can guarantee that our algorithm is theoretically better than existing algorithms which consider all blocking metabolites and then test all subsets.

6 Conclusion

All the results presented in this paper correspond to enumerating closed sets as potential organisations. The problem of enumerating self-maintaining sets is still open. Such an approach could enable us to design a method that enumerates stoichiometrically valid precursor sets, which would be an important follow-up of the work on minimal precursors sets presented in [3]. Finally, the algorithms introduced in this paper do not take any specific advantage of the fact that the network should be mass-consistent and exploiting this might lead to better algorithms for enumerating self-maintaining sets.

References

1. Acuña, V., Chierichetti, F., Lacroix, V., Marchetti-Spaccamela, A., Sagot, M.-F., Stougie, L.: Modes and cuts in metabolic networks: Complexity and algorithms. Biosystems 95(1), 51–60 (2009)
2. Centler, F., Kaleta, C., di Fenizio, P.S., Dittrich, P.: Computing chemical organizations in biological networks. Bioinformatics 24(14), 1611–1618 (2008)
3. Cottret, L., Milreu, P.V., Acuña, V., Marchetti-Spaccamela, A., Martinez, F.V., Sagot, M.F., Stougie, L.: Enumerating Precursor Sets of Target Metabolites in a Metabolic Network. In: Crandall, K.A., Lagergren, J. (eds.) WABI 2008. LNCS (LNBI), vol. 5251, pp. 233–244. Springer, Heidelberg (2008)
4. Dittrich, P., di Fenizio, P.S.: Chemical organisation theory. Bull. Math. Biol. 69(4), 1199–1231 (2007)
5. Even, G., Naor, J., Schieber, B., Sudan, M.: Approximating minimum feedback sets and multicuts in directed graphs. Algorithmica 20, 151–174 (1998)
6. Fontana, W., Buss, L.: The Arrival of the fittest: towards a theory of biological organization. Bull. Math. Biol. 56, 1–64 (1994)
7. Karp, R.M.: Reducibility among combinatorial problems. In: Complexity of Computer Computations, pp. 85–103 (1972)
8. Lacroix, V., Cottret, L., Thbault, P., Sagot, M.F.: An Introduction to Metabolic Networks and Their Structural Analysis. TCBB 5(4), 594–617 (2008)
9. Papin, J., Stelling, J., Price, N., Klamt, S., Schuster, S., Palsson, B.: Comparison of network-based pathway analysis methods. Trends Biotechnol. 22, 400–405 (2004)
10. Romero, P., Karp, P.D.: Nutrient-related analysis of pathway/genome databases. In: PSB 2001, pp. 470–482 (2001)
11. Schuster, S., Dandekar, T., Fell, D.A.: Detection of elementary flux modes in biochemical networks: a promising tool for pathway analysis and metabolic engineering. Trends in Biotechnology 17(2), 53–60 (1999)
12. Schuster, S., Hilgetag, C.: On elementary flux modes in biochemical reaction systems at steady state. J. Biol. Syst. (2), 165–182 (1994)
13. Seymour, P.D.: Packing directed circuits fractionally. Combinatorica 15(2), 281–288 (1995)

Efficient Subgraph Frequency Estimation with G-Tries

Pedro Ribeiro and Fernando Silva

CRACS & INESC-Porto LA
Faculdade de Ciências, Universidade do Porto, Portugal
{pribeiro,fds}@dcc.fc.up.pt

Abstract. Many biological networks contain recurring overrepresented elements, called network motifs. Finding these substructures is a computationally hard task related to graph isomorphism. G-Tries are an efficient data structure, based on multiway trees, capable of efficiently identifying common substructures in a set of subgraphs. They are highly successful in constraining the search space when finding the occurrences of those subgraphs in a larger original graph. This leads to speedups up to 100 times faster than previous methods that aim for exact and complete results. In this paper we present a new efficient sampling algorithm for subgraph frequency estimation based on g-tries. It is able to uniformly traverse a fraction of the search space, providing an accurate unbiased estimation of subgraph frequencies. Our results show that in the same amount of time our algorithm achieves better precision than previous methods, as it is able to sustain higher sampling speeds.

Keywords: complex networks, network motifs, subgraph frequency, sampling, g-tries.

1 Introduction

A wide variety of real-life systems can be modeled and analyzed with complex networks [4]. It has been found that many of these networks contain recurring elements, called network motifs [15]. These are overrepresented subnetworks, i.e., subgraphs that appear in higher frequency than it would be expected in randomized networks with similar topological characteristics.

Network motif analysis has a broad multidisciplinary applicability. Just to name a few domains, it has been applied on biological systems (like in brain networks [20], protein-protein interactions [1] or gene regulation [5]), social networks [9]), engineering systems like electronic circuits [8] and even on software architecture [21]. Discovering these motifs is a computationally hard task closely related to the graph isomorphism problem. Currently, this is done by computing the frequency of subgraph classes of a determined size both in the original network and in a randomized ensemble of networks sharing similar topological features, namely the degree sequence.

Discovering subgraph frequencies is the main bottleneck of the whole computation, with an explosive combinatorial effect as the subgraph size increases.

V. Moulton and M. Singh (Eds.): WABI 2010, LNBI 6293, pp. 238–249, 2010.

This is typically tackled using one of two approaches: either we compute the frequency of each possible individual subgraph class, one at the time (*subgraph-centric* approach) [7], or we enumerate all subgraphs and then we compute which ones are isomorphic (*network-centric* approach) [15,23].

Recently we have proposed a new specialized data-structure, g-tries [17]. It takes advantage of common subgraph substructures in order to avoid redundant computations, matching an entire set of the subgraph classes at the same time in a given network. This leads to significant performance gains when compared to previous methods, up to one hundred times faster for some networks.

In the network-centric approach, approximation techniques have been developed in order to improve execution time at the cost of reducing the accuracy [11,23,16]. This is done by sampling a fraction of the subgraph occurrences, instead of exhaustively enumerating all of them.

Our main contribution is an efficient heuristic sampling algorithm for discovering network motifs using g-tries. We take the already existing g-trie exhaustive and complete algorithm and extend it in order to obtain an unbiased sample that can be used to estimate the desired subgraph frequencies. This leads to a new algorithm that, by taking advantage of g-tries, achieves higher sampling rates and thus is able to reach more accurate predictions than previous algorithms for the same computing time. To substantiate this claim, we empirically evaluate the sampling speed, accuracy and total execution time of the algorithm in a set of representative networks. Our results show that in the same amount of time our algorithm can potentially reach higher subgraph and graph sizes. It can also only sample subgraphs from a predefined set.

The remainder of this paper is organized as follows. Section 2 establishes a network terminology and gives an overview of related work. Section 3 overviews the used g-trie data structure and details our sampling algorithm. Section 4 discusses the obtained results on a set of representative networks. Section 5 concludes the paper, with comments on the results and possible future work.

2 Preliminaries

To ensure a coherent network terminology, we briefly review the main concepts and notation that will be used throughout the paper, and discuss related work.

2.1 Graph Terminology

A *graph* G is composed by the set of vertices $V(G)$ and the set of edges $E(G)$. The *size* of a graph is $|V(G)|$, the number of vertices. A k-graph has size k. An edge is a pair $(a, b) : a, b \in V(G)$. If the graph is *directed* the order of the pair expresses direction, while in *undirected* graphs there is no direction in edges. The *neighborhood* of a vertex u is defined as $N(u) = \{v : (v, u) \vee (u, v) \in E(G)\}$. All vertices are assigned consecutive integer numbers starting from 0, and the comparison $v < u$ means that the index of v is lower than that of u. The adjacency matrix of a graph G is denoted as G_{Adj}, and $G_{Adj}[a][b]$ represents a possible edge between vertices with index a and b.

A *k-subgraph* G_k of a graph G is a k-graph such that $V(G_k) \subseteq V(G)$ and $E(G_k) \subseteq E(G)$. This subgraph is said to be *induced* if $u, v \in V(G_k)$ and $(u, v) \in E(G)$ implies $(u, v) \in E(G_k)$. Two graphs G and H are said to be *isomorphic* ($G \sim H$) if there is a one-to-one mapping between the vertices of both graphs where two vertices of G share an edge if and only if their corresponding vertices in H also share an edge.

2.2 Network Motifs and Frequency Count

In network motif discovery, frequency count is the central subproblem being addressed, and thus, we define it more precisely:

Definition 1 (Subgraph Counting Problem). *Given a set of subgraphs S_G and a graph G, count the number of all induced occurrences of subgraphs of S_G in G. Two occurrences are considered different if they have at least one node or edge that they do not share. Other nodes and edges can overlap.*

Note especially that we only count induced occurrences and how we distinguish occurrences. Although other frequency concepts exist [19], we resort to the standard definition for the network motif discovery problem [18]. It has direct implications on the number of occurrences and on the tractability of the problem, with no downward closure property [12] on the frequencies, i.e., a subgraph may appear more times than a subgraph contained in it.

2.3 Related Work

A general and informal survey on algorithms for network motifs discovery can be seen in [3], and [18] provides a more technical comparison of the algorithms. The overall most efficient exhaustive network-centric algorithms are ESU [23] and Kavosh [10]. MODA [16] and Grochow and Kellis [7] provide efficient subgraph-centric algorithms and we provided the g-trie data-structure for an efficient intermediate appproach [17].

Regarding heuristic approximate algorithms, there are three different approaches that we are aware of. Kashtan et al [11] propose to sample one subgraph at a time, following a random graph walk, which results in a biased estimator. RAND-ESU [23] algorithm provides unbiased sampling by associating probabilities with each recursive search tree branch of the ESU algorithm. MODA [16] chooses nodes with a probability proportional to their degree. Our sampling algorithm differs from all previous approaches since we use a different underlying data structure and its associated methodology.

3 Sampling Algorithm

3.1 G-Tries Data Structure

A g-trie is a data structure designed to store a set of graphs. It is conceptually inspired in prefix trees (trie) in the sense that it tries to identify common graph substructures in the same way a trie identifies common prefixes of sequences.

A g-trie is a tree where each tree node contains information about a single graph vertex and its correspondent connection to the vertices of ancestor tree nodes. Every node can have an arbitrary number of children and the path from the root to a node (possible a leaf) defines a single subgraph. Note that all descendants of a node share the same initial g-trie substructure and therefore have a common subtopology in graph terms. Figure 1 gives an example of a g-trie with 6 undirected subgraphs.

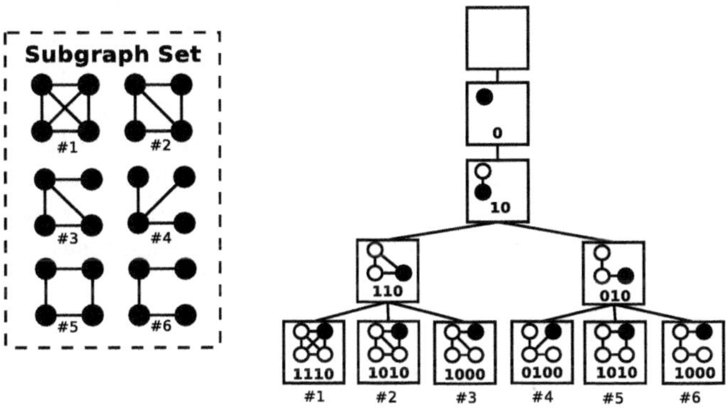

Fig. 1. A g-trie representing a set of 6 undirected subgraphs. Each g-trie node adds a new vertex (in black) to the already existing vertices in the ancestor nodes (in white). The connections to these nodes are represented by a sequence of boolean numbers indicating the corresponding adjacency matrix row.

As said, each g-trie node needs to specify the connections of is vertex to all ancestor ones (and to itself). This can be done in several ways, but in our current implementation we just store the correspondent part of the adjacency matrix. If the graphs are undirected, we store in each node the adjacency matrix row up to that vertex. If the graphs are directed we also store the adjacency matrix column up to that vertex, because we must specify ingoing and outgoing connections. In any case, given a path from the root to a node, we have a fully specified graph. The g-trie root node is empty since there are two possible direct child nodes: a vertex with or without a connection to itself.

Considering that we want an unique and univocal representation of a set of graphs, we use a canonical adjacency matrix. This guarantees that any subgraph will always lead to the same path traversing the tree. There are many possible choices here, and we opted for the lexicographically bigger adjacency matrix. This favors the occurrence of more common substructures with higher degree nodes appearing in lower tree depth levels.

This capability of identifying common subtopologies is the main strength of a g-trie. We are compressing information and avoiding redundant storage. But more than that, at a later stage, when using the g-trie to search for sugraphs and when matching a specific vertex in the g-trie, we are matching at the same time all possible descendant subgraphs stored in the g-trie.

In order to avoid subgraph symmetries, g-tries also store symmetry breaking conditions of the form $a < b$ indicating that the vertex in position a should have a graph index smaller than vertex in position b. Similar to what was done in [7], these conditions establish an order for the vertices of the same symmetry group and guarantee that each subgraph can be found only once. More details on this can be seen in our previous work [17].

For the sake of clarity, from now on we will use the term node to refer to a g-trie tree node, and vertex to refer to a vertex of the stored graphs. Given a g-trie node T, we will use $T.vertex$ to refer the new vertex of that node (represented in black in Figure 1), and $T.in[i]$ and $T.out[i]$ to refer to the boolean value of the new vertex having respectively an ingoing or outgoing connection to the vertex with index i, i.e., the new node represented in the ancestor of depth i. Note that if the g-trie stores undirected graphs, then $T.in[i] = T.out[i]$ (and in fact T.out is not even stored in memory). We will also use $T.cond$ to denote the set of conditions that break symmetries for the descendant nodes that correspond to a full graph. $T.root$ denotes the g-trie root node and $T.isGraph$ indicates if the node is the end vertex of a graph (in fig. 1 this corresponds to all leaf nodes).

3.2 Exact Subgraph Frequency

Given a g-trie T and a graph G, the g-trie matching algorithm will find the occurrences of all graphs of T as subgraphs of G, as shown in [17]. The basic idea is to find a set of vertices of G that match completely with a path in T, and we heavily constraint our search by using the information stored about connections and symmetry breaking conditions. For the sake of clarity, we show the matching algorithm, in Algorithm 1, with a subtle modification. It encapsulates some of the work in the `matchingVertices()` function, thus allowing for a logical separation of the recursion calls and the isomorphic matching.

At any stage, V_{used} represents the currently partial match of graph vertices to a g-trie path. We start with the g-trie root children nodes and call the recursive procedure `match()` with an initial empty matched set (line 2). The later procedure starts by creating a set of vertices that completely match the current g-trie node (line 4). We then traverse that set (line 5) and recursively try to expand it through all possible tree paths (lines 7 and 8). If the node corresponds to a full subgraph, then we have found an occurrence of that subgraph (line 6). Note that at this time no isomorphic test is needed, since this was implicitly done as we were matching the vertices.

Generating the set of matching vertices is done in the `matchingVertices()` procedure. The efficiency of the algorithm heavily depends on the above mentioned constraints as they help in reducing the search space. To generate the matching set, we start by creating a set of candidates (V_{cand}). If we are at a root child, then all graph vertices are viable candidates (line 10). If not, we select from the already matched vertices that are connected to the new vertex (line 12), the one with the smallest neighborhood (line 13), reducing the possible candidates (line 14). Then, we traverse the set of candidates (line 16) and if one respects all connections to ancestors (lines 17 to 19), and respects at least one set of symmetry breaking conditions for a possible descendant subgraph (line 19), we add it to the set of matching vertices (line 20).

Algorithm 1. Finding subgraphs of g-trie T in graph G.

```
procedure MATCHALL(T, G)
    for all children c of T.root do MATCH(c, ∅)
procedure MATCH(T, V_used)
    V = MATCHINGVERTICES(T, V_used)
    for all vertex v of V do
        if T.isGraph then FOUNDMATCH()
        for all children c of T do
            MATCH(c, V_used ∪ {v})
function MATCHINGVERTICES(T, V_used)
    if V_used = ∅ then V_cand := V(G)
    else
        V_conn = {v : v = V_used[i], T.in[i] ∨ T.out[i], i ∈ [1..|V_used|]}
        m := m ∈ V_conn : ∀v∈ V_conn, |N(m)| ≤ |N(v)|
        V_cand := {v ∈ N(m) : v ∉ V_used}
    Vertices = ∅
    for all v ∈ V_cand do
        if ∀i∈[1..|V_used|]:
            T.in[i] = G_Adj[V_used[i]][v]   ∧ T.out[i] = G_Adj[v][V_used[i]] then
            if ∃C ∈ T.cond : V_used + v respects C then
                Vertices = Vertices ∪ {v}
    return Vertices
```

3.3 Uniform Sampling

Algorithm 1 creates an exhaustive and complete enumeration of all subgraph occurrences. Our contribution to the existing g-tries methods is to sample only a fraction of all the occurrences. Similarly to what was done in [23], we will be trading accuracy for execution speed. The main idea is that each search branch is only chosen with a certain probability as depicted in Algorithm 2. Note that it is exactly the same as the previous algorithm with the exception of the indicated lines 3 and 9.

Algorithm 2. Sample subgraphs of g-trie T in graph G. Probability of each occurence is P, with $P = \prod P_d$, where P_d is probability of depth d.

```
1: procedure SAMPLEALL(T, G)
2:     for all children c of T.root do
3:         With probability P_0 do SAMPLE(c, ∅)                 ▷ changed line
4: procedure SAMPLE(T, V_used)
5:     V = MATCHINGVERTICES(T, V_used)
6:     for all node v of V do
7:         if T.isGraph then FOUNDMATCH()
8:         for all children c of T do
9:             With probability P_T.depth do SAMPLE(c, V_used ∪ {v})   ▷ changed line
```

In order to follow a probabilistic approach, the algorithm uses a set of probabilities associated to each g-trie depth:, $\{P_0, P_1, \ldots, P_{gtrie\_max\_depth}\}$ where $0 \leq P_i \leq 1$. Any given node of depth d will therefore only be reached with probability $P_0 \times \ldots \times P_{d-1}$. With this, we can produce an unbiased estimator of the frequency count of a single subgraph. Let P_i be the probability associated with depth i and $F_{sample}(G_k)$ be the number of occurrences of the k-subgraph G_k found in G by the `sampleAll()` procedure of Algorithm 2. Then, an unbiased estimator $\hat{F}(G_k, G)$ of the total number of occurrences of G_k in G is given by the following equation:

$$\hat{F}(G_k, G) = \frac{F_{sample}(G_k, G)}{P_0 \times P_1 \times \ldots \times P_{k-1}} \tag{1}$$

We say that the estimator is unbiased because any occurrence of G_k can be found with equal probability, and as we increase the probabilities, the estimator gets closer to the real value. In fact, if we choose $P_i = 1$ for all i, then the result is the same as the original complete algorithm.

As seen, the parameters P_i control the search. Regarding the accuracy, we should avoid small values of probability for lower depths, closer to the root. Its effect is to increase the variance of the result because any disregarded branch in lower depths may correspond to entire parts of the graph, and therefore may correspond to a higher number of subgraph occurrences not found. As to the execution time, the opposite happens. Very high probabilities in the lower depths will increase the execution time, since more parts of the search tree will have to be computed. For example, in the extreme case of having all probabilities equal to one except the last one, in the higher possible depth d, means that in practice we will explore all possible subgraphs of depth $d - 1$.

Picking the parameters is therefore a delicate choice that will influence both the accuracy and speed of our method. Section 4 gives more details on actual useful real parameters. Note that if only k-subgraphs are being sought, than all complete subgraphs of the tree will correspond to leaf nodes and therefore the probability at depth k should always be 1 since when we are at that point, all computation needed to identify the occurrence is already made (no isomorphism test is needed after that), and choosing any value smaller than 1 would only decrease the number of samples without any gain in execution time.

The main benefit of our sampling algorithm regarding previous ones, is that it is able to sample only the desired set of subgraphs (mfinder and ESU can only sample the entire set of possible k-subgraphs and MODA can only sample the occurrences of a particular single subgraph). To our best knowledge, this is the first algorithm doing that.

The quality of the estimation depends on many factors. A fully fledged analytical determination of tight bounds on error margins is very complicated since we do not know beforehand the distribution of the subgraphs that we are looking for. For example, if the subgraph is very well spread in the entire subgraph, we will have less variance than if all occurrences are clustered in a small number of nodes, where a search branch not followed can imply a significant number of occurrences not found.

3.4 Network Motif Discovery

With the algorithms previously defined we can discover all network k-motifs in the following way: first we find all k-subgraphs that occur in the original graph using another algorithm (for example `ESU`). Then we build a g-trie with those k-subgraphs and only search that particular set in the similar ensemble of randomized networks. Eventually, if we have other conditions, like a minimum frequency in the original graph, we can already discard some subgraphs and take advantage of the fact that we can search only for the ones that interest us.

4 Results

In order to evaluate the performance of our proposed algorithm (which from now on we will call `RAND-GTRIE`) we implemented it using C++. Isomorphisms and canonical labellings were computed using the `nauty` tool [14]. All tests were made on a computer with an Intel Core 2 6600 (2.4GHz) with 2GB of memory. We used four different biological networks from different domains, with varied topological features that are summarized in Table 1.

Table 1. Networks used for experimental testing of RAND-GTRIE

Network	Nodes	Edges	Directed	Description	Source
`Social`	62	159	no	Social network of a community of dolphins	[13]
`Neural`	297	2345	yes	Neural network of *C. elegans*	[22]
`Metabolic`	453	2025	yes	Metabolic network of *C. elegans*	[6]
`Protein`	2361	6646	no	Protein-protein inter. of *S. cerevisiae*	[2]

In all tests the construction of the g-trie in itself was a very small fraction of the execution time, and we could even store and reuse canonical labellings on other program runs. On the network discovery problem, the g-trie can be computed once, at the beginning, and then reused for the ensemble random networks. Given this, we chose to leave the g-trie creation out of the picture when stating execution time. For the purposes of this section, we also limited the choice of probability parameters to three levels of quality. In order to sample a fraction f of all k-subgraph occurrences, we can opt for one of the following levels:

- `high`: $\{P_0 = 1, \ldots, P_{k-3} = 1, P_{k-2} = f, P_{k-1} = 1\}$
- `medium`: $\{P_0 = 1, \ldots, P_{k-4} = 1, P_{k-3} = \sqrt{f}, P_{k-2} = \sqrt{f}, P_{k-1} = 1\}$
- `low`: $\{P_0 = \sqrt[k-1]{f}, \ldots, P_{k-2} = \sqrt[k-1]{f}, P_{k-1} = 1\}$

Our first test was to analyze the speed at which `RAND-GTRIE` is able to generate samples. For that we counted how many k-subgraphs per second it was able to generate, both with a complete enumeration (all $P_i = 1$) and with only 10% of the k-subgraphs obtained by sampling with `high` quality level. We also compared the performance with RAND-ESU, the present most efficient network centric method that also allows sampling in a way similar to ours. For that, the publicly available FanMod tool was used, with the same probabilities at the same depths. FanMod is also implemented in C++ and uses nauty for isomorphism. All sizes between 3 (the minimum acceptable for a subgraph to be taken in account) and 6 (the maximum that guarantees computation in a matter of a few hours) were used. We first used ESU to discover all the k-subgraphs in the original graph, constructed a g-trie with those and then used it to estimate the frequency (as we would do with the randomized networks if we were discovering motifs). Figure 2 details the results obtained.

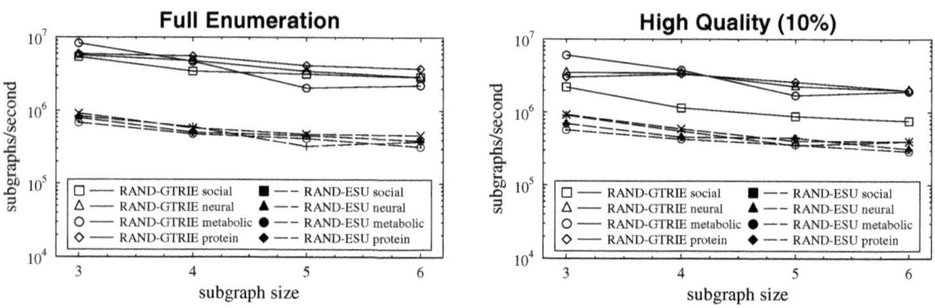

Fig. 2. Sampling speed of RAND-GTRIE and RAND-ESU

The main aspect to note is that RAND-GTRIE is always faster, being an order of magnitude faster. This was also the case for all other networks tested, with the more extreme speedup bigger than 100×, for a power grid network [22]. RAND-GTRIE also appears to scale well with an increasing subgraph size, as is the case with RAND-ESU, since the sampling rate is sustained. Mfinder, the other major alternative for sampling, was shown to be much slower than RAND-ESU and it does not scale well [23].

In order to test the accuracy of our algorithm, we applied all levels of sampling quality, while increasing the fraction of subgraphs being sampled, taking note of the percentage of subgraphs correctly identified. We considered an estimate to be accurate when it was within a 20% error margin of the correct perfect value. We took 100 samples for each fraction and level and only considered the estimate correct when at least 80 of those samples were accurate. The results for two of the networks are shown on fig. 3. As expected, higher probabilities in lower depths correspond to better sampling quality (less variance).

If we measure the execution time for the exact same tests, we can see that the opposite happens, with better quality sampling taking more execution time as detailed in fig. 4. All quality levels have an execution growth proportional to the percentage of samples, but higher quality levels have a minimum time bigger than lower quality minimum time. For example, on the protein network, sampling just 0.1% of the subgraphs in high level of quality takes more then 6% of the time it takes to do a full enumeration. This is because we are traversing the entire tree up to depth $k-2$. Judging by our empirical tests, 10% on high level exhibits a good balance between execution time and sampling quality, but depending on the situation, any other values can be used.

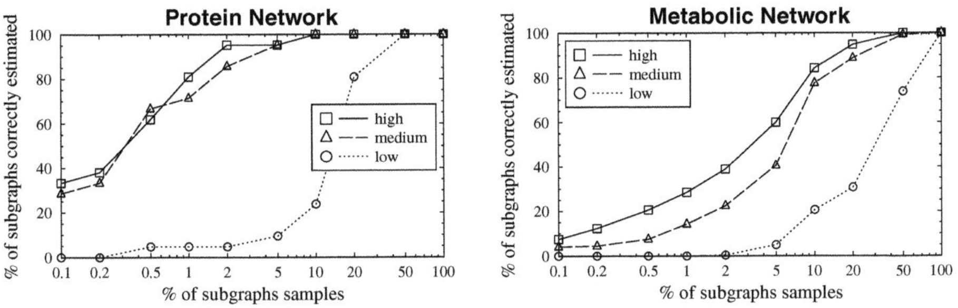

Fig. 3. Accuracy of RAND-GTRIE for 5-subgraphs, measured in percentage of correctly estimated subgraphs as the percentage of samples grows

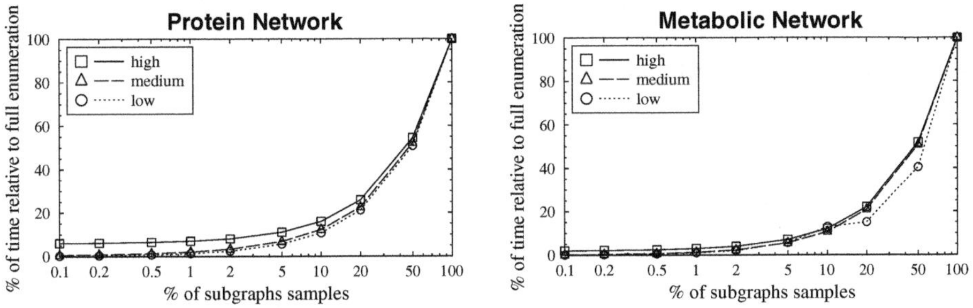

Fig. 4. Execution time of RAND-GTRIE for 5-subgraphs, relative to the time a full enumeration with g-tries takes (i.e., with $P_i = 1$ for every i)

If we take a closer look to what RAND-GTRIE is computing, we can see that the more rare subgraphs are the ones with less estimating quality. This is because a smaller number of occurrences will obviously imply more variance in the estimated values (a "miss" has more weight). For example, with `high` level setting and 10% of samples on the `metabolic` network we have 84.27% subgraph classes estimated correctly. Almost all of the ones not identified appear less than 100 times in the sample, and therefore are estimated to appear less than 1000 times in the original network. On the other hand, with the same `high` level setting and only 0.1% of samples, the 7.49% that were estimated correctly correspond to subgraph classes that were sampled at least 1000 times, which means that they are estimated to occur more than one million times in the original graph.

Finally, regarding motifs, we experimented to discover all motifs of sizes 3 to 6 in the four networks, using 10% sampling with `high` quality level, and we were able to find more than 90% of the motifs that a full enumeration would find. More than that, we spend on average less than 20% of the time it would take using g-tries full enumeration. If we take into account that g-tries are themselves a more efficient data structure than previous methods, we can magnify even more the speedup and potentially reach previously unfeasible network and subgraph sizes. Note that since we can choose the subgraphs that we are looking for, we can even experiment with different probability parameters for different subgraphs, thus paving the way for a more adaptive algorithm.

5 Conclusion

In this paper we presented a novel sampling algorithm for discovering network motifs. It employs as a basis the g-trie data structure, an efficient specialized tree that uses common topologies in subgraphs in order to heavily constraint the search. By associating a probability with each tree depth, it is able to uniformly traverse a fraction of the whole search space. With this it provides an unbiased estimator for the real frequency of the associated subgraphs, and a way of efficiently discovering motifs.

Our algorithm offers many parametrization choices and it is also capable of sampling subgraphs solely from a predefined set, in opposition to having to sample among all of the subgraphs of a determined size, or only sampling one individual subgraph. This has a direct beneficial impact on the execution time and we are able to produce accurate results spending less execution time than previously existent methods.

In the future we intend to exploit even more this property and create an adaptive version of our sampling algorithm that is able to make an initial estimation and then keep refining it for the subgraphs that do not have enough estimation quality. For example, one could remove all frequent subgraphs from the g-trie and only repeat the search for the more rare ones, with an higher fraction of samples. We also intend to study the impact of the original graph labeling on the sampling quality, since our symmetry breaking conditions rely on this order.

Finally, we will also apply our methodology in real-life problems, analyzing networks at scales that were not possible before, attempting to unveil new larger network motifs.

Acknowledgments

Pedro Ribeiro is funded by an FCT Research Grant (SFRH/BD/19753/2004).

References

1. Albert, I., Albert, R.: Conserved network motifs allow protein-protein interaction prediction. Bioinformatics 20(18), 3346–3352 (2004)
2. Bu, D., Zhao, Y., Cai, L., Xue, H., Zhu, X., Lu, H., Zhang, J., Sun, S., Ling, L., Zhang, N., Li, G., Chen, R.: Topological structure analysis of the protein-protein interaction network in budding yeast. Nucl. Acids Res. 31(9), 2443–2450 (2003)
3. Ciriello, G., Guerra, C.: A review on models and algorithms for motif discovery in protein-protein interaction networks. Brief Funct. Genomic Proteomic 7(2), 147–156 (2008)
4. da Costa Luciano, F., Oliveira Jr., O.N., Travieso, G., Rodrigues, F.A., Villas Boas, P.R., Antiqueira, L., Viana, M.P., da Rocha, L.E.C.: Analyzing and modeling real-world phenomena with complex networks: A survey of applications. ArXiv e-prints 0711(3199) (2007)
5. Dobrin, R., Beg, Q.K., Barabasi, A., Oltvai, Z.: Aggregation of topological motifs in the escherichia coli transcriptional regulatory network. BMC Bioinformatics 5, 10 (2004)

6. Duch, J., Arenas, A.: Community identification using extremal optimization. Phys. Rev. E. (Stat. Nonlin. Soft Matter Phys.) 72, 027104 (2005)
7. Grochow, J., Kellis, M.: Network motif discovery using subgraph enumeration and symmetry-breaking. Research in Computational Molecular Biology, 92–106 (2007)
8. Itzkovitz, S., Levitt, R., Kashtan, N., Milo, R., Itzkovitz, M., Alon, U.: Coarse-graining and self-dissimilarity of complex networks. Phys. Rev. E (Stat. Nonlin. Soft Matter Phys.) 71(1 Pt. 2) (January 2005)
9. Juszczyszyn, K., Kazienko, P., Musial, K.: Local topology of social network based on motif analysis. In: Lovrek, I., Howlett, R.J., Jain, L.C. (eds.) KES 2008, Part II. LNCS (LNAI), vol. 5178, pp. 97–105. Springer, Heidelberg (2008)
10. Kashani, Z., Ahrabian, H., Elahi, E., Nowzari-Dalini, A., Ansari, E., Asadi, S., Mohammadi, S., Schreiber, F., Masoudi-Nejad, A.: Kavosh: a new algorithm for finding network motifs. BMC Bioinformatics 10(1), 318 (2009)
11. Kashtan, N., Itzkovitz, S., Milo, R., Alon, U.: Efficient sampling algorithm for estimating subgraph concentrations and detecting network motifs. Bioinformatics 20(11), 1746–1758 (2004)
12. Kuramochi, M., Karypis, G.: Frequent subgraph discovery. In: IEEE International Conference on Data Mining, p. 313 (2001)
13. Lusseau, D., Schneider, K., Boisseau, O.J., Haase, P., Slooten, E., Dawson, S.M.: The bottlenose dolphin community of doubtful sound features a large proportion of long-lasting associations. can geographic isolation explain this unique trait? Behavioral Ecology and Sociobiology 54(4), 396–405 (2003)
14. McKay, B.: Practical graph isomorphism. Cong. Numerantium 30, 45–87 (1981)
15. Milo, R., Shen-Orr, S., Itzkovitz, S., Kashtan, N., Chklovskii, D., Alon, U.: Network motifs: simple building blocks of complex networks. Science 298(5594), 824–827 (2002)
16. Omidi, S., Schreiber, F., Masoudi-Nejad, A.: Moda: An efficient algorithm for network motif discovery in biological networks. Genes & genetic systems 84(5), 385–395 (2009)
17. Ribeiro, P., Silva, F.: G-tries: an efficient data structure for discovering network motifs. In: ACM Symposium on Applied Computing (2010)
18. Ribeiro, P., Silva, F., Kaiser, M.: Strategies for network motifs discovery. In: 5th IEEE International Conference on e-Science. IEEE CS Press, Oxford (2009)
19. Schreiber, F., Schwobbermeyer, H.: Towards motif detection in networks: Frequency concepts and flexible search. In: Proc. of the Int. Workshop on Network Tools and Applications in Biology (NETTAB 2004), pp. 91–102 (2004)
20. Sporns, O., Kotter, R.: Motifs in brain networks. PLoS Biology 2 (2004)
21. Valverde, S., Solé, R.V.: Network motifs in computational graphs: A case study in software architecture. Phys. Rev. E (Stat. Nonlin. Soft Matter Phys.) 72(2) (2005)
22. Watts, D.J., Strogatz, S.H.: Collective dynamics of small-world networks. Nature 393(6684), 440–442 (1998)
23. Wernicke, S.: Efficient detection of network motifs. IEEE/ACM Trans. Comput. Biol. Bioinformatics 3(4), 347–359 (2006)

Accuracy Guarantees for Phylogeny Reconstruction Algorithms Based on Balanced Minimum Evolution

Magnus Bordewich[1,*] and Radu Mihaescu[2]

[1] School of Engineering and Computing Sciences, Durham University, U.K.
[2] Dept. of Computer Science, U.C. Berkeley, U.S.A.

Abstract. Distance based phylogenetic methods attempt to reconstruct an accurate phylogenetic tree relating a given set of taxa from an estimated matrix of pair-wise distances between taxa. This paper examines two distance based algorithms (GreedyBME and FastME) which are based on the principle of trying to minimise the balanced minimum evolution (BME) score of the output tree in relation to the given estimated distance matrix. We show firstly that these algorithms will both reconstruct the correct tree if the input data is *quartet consistent*, and secondly that if the maximum error in any individual distance estimate is ϵ, then both algorithms will output trees containing all edges of the true tree that have length at least 3ϵ. That is to say the algorithms have *edge safety radius* 1/3. In contrast, quartet consistency of the data is not sufficient to guarantee Neighbor Joining (NJ) reconstructs the correct tree, and moreover NJ has edge safety radius of 1/4, which is strictly worse.

1 Introduction

We analyse two algorithms for inferring phylogenetic trees from distance matrices, based on the *balanced minimum evolution (BME)* principle [4]. The optimality criterion (BME score) used is to minimize Pauplin's tree-length estimate [13] relative to the given distance matrix. The algorithms we consider are GreedyBME and FastME. GreedyBME is the greedy algorithm for reconstructing a tree taxon-by-taxon. We start with three taxa arranged in a star tree topology and iteratively add each remaining taxon, inserting them at the location that minimises the BME score of the tree restricted to the taxa inserted so far. So GreedyBME builds a binary phylogenetic tree taxon-by-taxon, starting from the star on three taxa, greedily minimising the BME score at each step. It is interesting to note that Gascuel and Steel, in an excellent review [8], have shown that NJ [14] also is a greedy algorithm which minimises the BME score at each step. The difference is that NJ starts at a star topology on *all* taxa and iteratively chooses two leaves and agglomerates them (effectively regrafts them

* The first author was supported in part by EPSRC grant number EP/D063574/1. The work began while both authors were visiting the University of Canterbury, New Zealand.

V. Moulton and M. Singh (Eds.): WABI 2010, LNBI 6293, pp. 250–261, 2010.

as a cherry attached to the root), where the leaves are chosen to minimise the BME score of the resulting tree on all taxa which still has a central vertex of high degree.

FastME is a local topology search hill-climbing algorithm that takes the output of GreedyBME and iteratively searches through local topologies (those trees differing from the current tree by one topological rearrangement operation) and moves to the topology that minimises the BME score. This approach is implemented in a software called FastME [4]. The two topological rearrangement operations available in the latest release of FastME are: the *balanced subtree prune and regraft (BSPR) algorithm* [9] and the *balanced nearest neighbor interchange (BNNI) algorithm* [4]. FastME has been shown experimentally [4,5] to be a fast and accurate method for tree inference, compared to other popular distance-based methods such as NJ, BIONJ [7], FITCH [6] or WEIGHBOR [3]. The results in this paper provide further theoretical support for using this approach.

Atteson [1] studied the NJ algorithm and gave a condition, the *safety radius* of the algorithm, for accurate reconstruction of the true tree. Atteson showed that NJ has safety radius of 1/2, *i.e.* if the maximum error in the estimated distance matrix is at most 1/2 the minimum edge length in the true tree, then NJ will correctly reconstruct the entire tree. Moreover, no distance based method can have safety radius greater than 1/2. More recently Bordewich *et al.* [2] analysed FastME and showed that it has safety radius at least 1/3 (when using BSPR) and Shigezumi [16] has shown that GreedyBME has safety radius 1/2. Note that FastME and GreedyBME are heuristics for finding a tree that minimises BME score. Pardi *et al.* [12] have shown that the BME principle itself, or equivalently the algorithm that returns the tree that globally minimises BME score, has safety radius 1/2.

The results described above relate the minimum edge length in the entire tree to the maximum allowed error, so a single short edge can massively affect the permitted error across all estimated distances for the guarantee of correct reconstruction to hold. In contrast, in this paper we consider the *edge safety radius*, which guarantees that sufficiently long edges will be correctly reconstructed even if other edges of the true tree are very short, relative to the error. An algorithm that is guaranteed to output a tree topology containing all those edges of the true tree that have length at least l, whenever the the maximum error in the estimated distance matrix is at most rl, is said to have *edge safety radius* r. Atteson conjectured that NJ has an edge safety radius of 1/4, which has recently been proved by Mihaescu *et al.* [11]. The main result of this paper is to show that GreedyBME and FastME each have edge safety radius 1/3. We also show that under a weaker condition than safety radius 1/2, namely quartet consistency (which we will define below), GreedyBME and FastME will correctly reconstruct the true tree. Note that having maximum error at most 1/2 the minimum edge length guarantees quartet consistency, but a distance matrix may be quartet consistent with the true tree while not satisfying the safety radius condition. In related work, Mihaescu [10] has shown that the

BME principle also reconstructs the true tree on quartet consistent inputs, but is strictly weaker than its two heuristic versions (GreedyBME and NJ) in edge safety radius, having an asymptotic edge safety radius of $1/(2n)$ on n taxa.

Our results show strict superiority of GreedyBME over NJ in two ways. It has been shown that quartet consistency is not a sufficient condition for NJ to correctly reconstruct the true tree [11]. Thus GreedyBME will correctly reconstruct the *whole* true tree under a weaker condition than NJ. Also, GreedyBME will correctly reconstruct all *edges* of the true tree having length at least 3 times the maximum error in the input matrix, whereas NJ sometimes fails to reconstruct edges up to 4 times the maximum error [1].

2 Basics, Definitions and Notation

The notation and terminology largely follows [15]. A *phylogenetic tree* is a tree whose leaves are bijectively labelled by the elements of some finite set X. A *binary phylogenetic tree* is a phylogenetic tree in which every internal vertex has degree exactly three. The set X usually denotes a set of species or taxa, and the tree T represents the evolutionary relationships between them. Unless stated otherwise, from now on X will denote a finite set and all trees considered will be phylogenetic trees. Throughout we consider phylogenetic trees as unweighted, i. e. they do not have intrinsic edge lengths, with the exception of the true tree T^* which does have edge lengths (or weights). Furthermore, capital letters will be used in all figures to represent subtrees.

A matrix of pair-wise distances $\delta^* = [\delta^*_{ij}]$ is a *tree-metric* if there is a unique phylogenetic tree T^* with positive edge lengths $l(e)$ so that, for each $x, y \in X$, the distance δ^*_{xy} is the sum of the lengths of edges on the path between x and y in T^*. The input to our algorithms is an estimated pair-wise distance matrix $\delta = [\delta_{ij}]$, and the error ϵ of δ with respect to δ^* is $\max_{x,y \in X}(|\delta_{xy} - \delta^*_{xy}|)$.

We next define the BME score of a phylogenetic tree T. This is equivalent to the tree length formula of Pauplin [13]. Given two nodes $i, j \in V(T)$, we denote by $\mathcal{P}^T(i,j)$ the set of all the internal nodes of T which lie on the (closed) path between i, j in T. In particular, if i or j are internal, then they also belong to $\mathcal{P}^T(i,j)$. Similarly for two nodes i, j of T, internal or not, we let

$$p^T_{ij} = \prod_{v \in P^T(i,j)} (deg(v) - 1)^{-1}.$$

The *Balanced Minimum Evolution score* of T relative to δ is the quantity

$$BME(\delta, T) = \sum_{i,j \in X} p^T_{ij} \delta_{ij}.$$

A *split* $S = \{A, B\}$ on a taxa set X is a bipartition of X into two non-empty disjoint subsets $A, B \subseteq X$ whose union is X. For ease of notation, we will write $A|B$ or, equivalently $B|A$ for the split $\{A, B\}$. In general, a collection of splits of X is called a *split system* of X.

Suppose that T is a phylogenetic tree on X. Each edge e of T corresponds to a split of $A|B$ of X, which may be obtained by deleting e and letting A be the leaf-label set of one of the resulting connected components and B be the leaf-label set of the other. The set of splits corresponding to edges of T are said to the be the splits of T. A *clade* of T is any subset $C \subset X$ such that $C|X-C$ is a split of T. The clade is said to be rooted at the vertex c in T which is the end of the edge e which induces split $C|X-C$ closest to the leaves in C.

A quartet of T is a partial split $ab|cd$, where $a,b,c,d \in X$ and there is a split $A|D$ of T such that $a,b \in A$ and $c,d \in D$. We say that a dissimilarity map δ is *consistent with a quartet* $(ab|cd)$ if $\delta_{ab}+\delta_{cd} < \delta_{ac}+\delta_{bd}, \delta_{ad}+\delta_{bc}$. We say that δ is *consistent with an edge* $e = A|D$ of T, if δ is consistent with all quartets $ab|cd$ of T such that $a,b \in A$ and $b,c \in D$. We say that δ is *quartet consistent* with a phylogenetic tree T if δ is consistent with all the quartets of T.

Given a dissimilarity map δ and two disjoint clades A, B of tree T, rooted at $a,b \in V(T)$ respectively, we define the *average clade distance*, or *clade distance* for short, as

$$\delta_{AB} = \sum_{i\in A, j\in B} \delta_{ij} p^T_{ij} / \sum_{i\in A, j\in B} p^T_{ij} = \sum_{i\in A, j\in B} \delta_{ij} p^T_{ia} p^T_{jb}.$$

Note that the clade distance thus only depends on the rooted topologies $T|A$ and $T|B$ and not on the entire topology T.

3 Results

We now formally state the main results of this paper. The first concerns the algorithm GreedyBME, and gives sufficient conditions for accurate reconstruction of edges of the true tree.

Theorem 1. *Let T^* be a binary phylogenetic tree with induced distance matrix δ^*. Let input matrix δ have error ϵ with respect to δ^*. Then the algorithm* GreedyBME *will return a binary phylogenetic tree T such that*

1. *T contains an edge with split $A|B$ for all edges $e = A|B$ in T^* with $l(e) > 3\epsilon$, i.e.* GreedyBME *has edge-safety radius* 1/3*. Furthermore, this bound is asymptotically tight.*
2. *if δ is quartet consistent with T^* then $T = T^*$.*

The second result concerns the local topology search phase of FastME. We show that if a local search is conducted from a tree T that already contains certain edges from the true tree T^*, then the end result is guaranteed to also contain these edges of T^*. The two forms of local topology search considered are those topologies within one NNI or one SPR operation of the current tree; for details of the definitions of NNI and SPR operations as used in FastME see for example [2].

Theorem 2. *Let T^* be a binary phylogenetic tree with induced distance matrix δ^*. Let input matrix δ have error ϵ with respect to δ^*. Let T be a binary phylogenetic tree and let $e = A|B$ be an edge common to T and T^*. Then*

1. *if $l(e) > 2\epsilon$ then for any T' that may be obtained in one NNI operation from T such that $BME(\delta, T') < BME(\delta, T)$, e must be an edge of T';*
2. *if $l(e) > 3\epsilon$ and T' is the tree at most one SPR operation from T which minimises $BME(\delta, T')$ then e must be an edge of T'; and*
3. *if δ is consistent with e then for any T' that may be obtained in one NNI operation from T such that $BME(\delta, T') < BME(\delta, T)$, e must be an edge of T'.*

Combining the above two theorems we obtain the following immediate corollary.

Corollary 1. FASTME *using an initial tree generated by* GREEDYBME *and a local search based on NNI or SPR operations has edge safety radius 1/3.*

Note that in the local topology using NNI operations, it would not matter if the algorithm jumped immediately to the first tree found with shorter length, or completed the search of all neighbouring trees and moved to the best one. However for an SPR based local topology search we have the restriction that all neighbouring trees are checked, and the best of those selected for the next iteration.

4 Proofs

4.1 GreedyBME

We shall make use of the following lemmas. The first is Lemma 5.1 of [2], which gives a formula for the difference in BME score between two trees of certain structure. The second lemma gives a five point condition on the distance matrix δ, which is extended to clades.

Lemma 1 (Lemma 5.1 of [2]). *Let T^A and T^B be the trees given in Fig. 2. Then $BME(\delta, T^A) - BME(\delta, T^B) =$*

$$\left(\frac{1}{2} - \frac{1}{2^t}\right)(\delta_{A'x_k} - \delta_{x_k B_t}) + \sum_{i=1}^{t-1}\left[\frac{1}{2^{t-i+1}}(\delta_{B_i B_t} - \delta_{B_i x_k}) - \frac{1}{2^{i+1}}(\delta_{A' B_i} - \delta_{B_i x_k})\right].$$

Lemma 2. *Let T^* be a phylogenetic tree on X and let T^* have a split $e = A|B$ of length $l(e)$. Let δ be a matrix of pairwise distances that has error at most $\epsilon < l(e)/3$. Then for A_0, A_1 disjoint subsets of A, and B_i, B_j, B_k disjoint subsets of B, and for any tree T with clades A_0, A_1, B_i, B_j, B_k we have:*

$$(\delta_{A_0 A_1} + \delta_{A_0 B_i} - \delta_{A_1 B_i}) - (\delta_{A_0 B_j} + \delta_{A_0 B_k} - \delta_{B_j B_k}) < 0.$$

Proof. Let a_0, a_1, b_i, b_j, b_k be leaves in clades A_0, A_1, B_i, B_j, B_k respectively. The true topology T^* restricted to these five leaves must be one of the three topologies shown in Fig. 1, where the edge e' depicted represents a path in T^* that contains edge e.

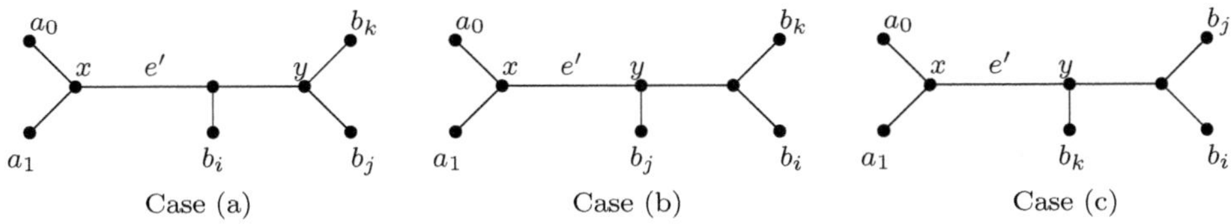

Fig. 1. Possible true topologies relating the leaves a_0, a_1, b_i, b_j, b_k

In each case the true distances satisfy

$$\begin{aligned}(\delta^*_{a_0a_1} + \delta^*_{a_0b_i} - \delta^*_{a_1b_i}) - (\delta^*_{a_0b_j} + \delta^*_{a_0b_k} - \delta^*_{b_jb_k}) &= 2\delta^*_{xy} \\ &< 2l(e),\end{aligned}$$

where x and y are the corresponding internal vertices of T^*. Since each entry in the estimated distance matrix has an error of at most ϵ, we have

$$\begin{aligned}(\delta_{a_0a_1} + \delta_{a_0b_i} - \delta_{a_1b_i}) - (\delta_{a_0b_j} + \delta_{a_0b_k} - \delta_{b_jb_k}) &< \; -2l(e) + 6\epsilon \\ &< \; 0.\end{aligned}$$

Note that for any clade C with root r_c we have $\sum_{c \in C} p^T_{cr_c} = 1$. So we may sum all terms over the leaves in all clades. Thus,

$$\begin{aligned}&(\delta_{A_0A_1} + \delta_{A_0B_i} - \delta_{A_1B_i}) - (\delta_{A_0B_j} + \delta_{A_0B_k} - \delta_{B_jB_k}) \\ &\quad = \sum p^T_{a_0r_0} p^T_{a_1r_1} p^T_{b_ir_i} p^T_{b_jr_j} p^T_{b_kr_k} [(\delta_{a_0a_1} + \delta_{a_0b_i} - \delta_{a_1b_i}) - (\delta_{a_0b_j} + \delta_{a_0b_k} - \delta_{b_jb_k})] \\ &\quad < 0,\end{aligned}$$

where the summation is over all leaves a_0, a_1, b_i, b_j, b_k in A_0, A_1, B_i, B_j and B_k respectively, and r_0, r_1, r_i, r_j, r_k are the roots of the clades A_0, A_1, B_i, B_j and B_k in T respectively. □

We can now present the proof of Theorem 1.

Proof (Theorem 1). First we prove that the edge safety radius of the algorithm GreedyBME is at least 1/3. Let T^* be the true tree, and let $e = A|B$ be a split in T^* of length $l(e) > 3\epsilon$. Let δ be an estimated pairwise distance matrix with maximum error at most ϵ. The proof is by induction on the size of X. If $|X| = 3$, then trivially GreedyBME will return the true tree, since there is only one tree topology on three taxa.

Suppose now that the theorem holds for all trees on taxa sets of size at most $k-1$, and let $|X| = k$. Let x_k be the last taxa added by GreedyBME, and consider the tree T'^* obtained from T^* by removing x_k and its pendent edge. Note that δ' obtained from δ by removing the row and column corresponding to x_k gives a pairwise distance matrix for T'^* with maximum error at most ϵ. Without loss of generality we assume $x_k \in A$. If $A = \{x_k\}$, then trivially GreedyBME will construct a tree containing the split $A|B$. Now we assume $A' = A - x_k \neq \emptyset$. Then T'^* has split $e' = A'|B$ and the length of e' is at least

$l(e)$. By the inductive hypothesis GREEDYBME applied to δ' will construct a tree containing the split $e' = A'|B$. Thus GREEDYBME applied to δ will also construct a tree T' on $X - x_k$ containing the split $e' = A'|B$ after $k-1$ steps.

We must now show that after the addition of x_k the split $A|B$ is present in the resulting tree T. GREEDYBME will position x_k at the point which minimises $BME(\delta, T)$. Suppose for contradiction that there is some position in clade B of T' which minimises this, as depicted in Fig. 2(b), where $B = B_1 \cup B_2 \cup \ldots \cup B_t$, resulting in tree T^B. We will show that tree T^A obtained by attaching x_k between the clades A' and B, as depicted in Fig. 2(a), must obtain a smaller BME score, giving the required contradiction.

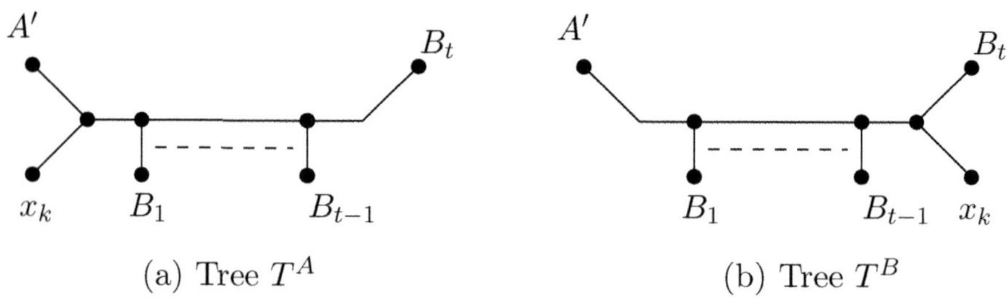

Fig. 2. Possible positions for attaching the final leaf x_k

By Lemma 1, we can express the difference in BME score between T^A and T^B as follows: $BME(\delta, T^A) - BME(\delta, T^B) =$

$$\left(\frac{1}{2} - \frac{1}{2^t}\right)(\delta_{A'x_k} - \delta_{x_k B_t}) + \sum_{i=1}^{t-1}\left[\frac{1}{2^{t-i+1}}(\delta_{B_i B_t} - \delta_{B_i x_k}) - \frac{1}{2^{i+1}}(\delta_{A'B_i} - \delta_{B_i x_k})\right]$$

$$= \sum_{i=1}^{t-1}\frac{1}{2^{i+1}}\left[(\delta_{A'x_k} + \delta_{B_i x_k} - \delta_{A'B_i}) - (\delta_{x_k B_t} + \delta_{B_{t-i} x_k} - \delta_{B_{t-i} B_t})\right]. \quad (1)$$

We may now apply Lemma 2 to each term in the summation (setting $A_0 = x_k, A_1 = A', B_i = B_i, B_j = B_{t-i}$ and $B_k = B_t$) to see that each term is less than zero, and hence $BME(\delta, T^A) < BME(\delta, T^B)$. This contradicts the assumption that x_k will be inserted in clade B, and completes the inductive step.

We now show that the edge safety radius of GREEDYBME is no more than 1/3. For contradiction assume the edge safety radius is $r > 1/3$. Consider a caterpillar tree T^* in which a', x_k is a cherry separated from the rest of the tree by an edge of length l, as depicted in Fig. 2(a), taking the clades $A', B_i, i = 1, \ldots, t$ to be singleton leaves $a', b_1, \ldots, b_t$, and t to be odd. Let $\epsilon = rl$. We will assume the sum of all other edge lengths in the tree is at most ν for some very small $\nu > 0$. Define $\delta = [\delta_{xy}]$ as follows: $\delta_{a'x_k} = \delta^*_{a'x_k} + \epsilon$; for $i = 1$ to $t-1$ we define $\delta_{a'b_i} = \delta^*_{a'b_i} - \epsilon$ and $\delta_{b_i b_t} = \delta^*_{b_i b_t} + \epsilon$; for $i = 1$ to $(t-1)/2$ we define $\delta_{x_k b_i} = \delta^*_{x_k b_i} + \epsilon$; and for $i = (t-1)/2 + 1$ to $t-1$ we define $\delta_{x_k b_i} = \delta^*_{x_k b_i} - \epsilon$.

For brevity we omit the details, but it can be shown that, saving the insertion of x_k until last, GREEDYBME will build the correct tree on $X - \{x_k\}$. For any $r > 1/3$ we can choose t large enough (and ν small enough) that x_k can then be inserted in an incorrect position at lower BME score than T^*. Hence GREEDYBME will fail to reconstruct the correct tree. This gives part 1 of the theorem.

We now prove part 2. Let T be a binary phylogenetic tree and let δ be an estimated pairwise distance matrix which is quartet consistent with T. The proof is similar to part 1, and is again by induction on the size of X. If $|X| = 4$, then it is easy to check that GREEDYBME will return the true tree.

Suppose now that the theorem holds for all trees on taxa sets of size at most $k-1$, and let $|X| = k$. Let x_k be the last taxa added by GREEDYBME, and consider the tree T' obtained from T by removing x_k and its pendent edge. Note that δ' obtained from δ by removing the row and column corresponding to x_k, gives a pairwise distance matrix which is quartet consistent with the topology T'. Without loss of generality we assume the correct position for x_k to be inserted is as a pendent leaf regrafted to some edge $e = A'|B$ in T'. By the inductive hypothesis GREEDYBME applied to δ' will construct T' correctly. Thus GREEDYBME applied to δ will also construct T' on $X - x_k$ after $k-1$ steps.

GREEDYBME will position x_k at the point which minimises the $BME(\delta, T)$. Suppose for contradiction that there is some position other than as a pendent leaf grafted to e which minimises BME score. Without loss of generality we may assume that the position is in clade B of T', as depicted in Fig. 2(b), where $B = B_1 \cup B_2 \cup \ldots \cup B_t$, resulting in tree T^B. We will show that tree T, as depicted in Fig. 2(a), must obtain a smaller BME score, giving the required contradiction.

We will first assume that t is odd: a small adjustment will be needed if t is even. Using Equation (1), we may express the difference in BME score between T^A and T^B as $BME(D, T^A) - BME(D, T^B) =$

$$\sum_{i=1}^{(t-1)/2} \frac{1}{2^{i+1}} \left[(\delta_{A'x_k} + \delta_{B_i x_k} - \delta_{A'B_i}) - (\delta_{x_k B_t} + \delta_{B_{t-i} x_k} - \delta_{B_{t-i} B_t}) \right]$$
$$+ \frac{1}{2^{t-i+1}} \left[(\delta_{A'x_k} + \delta_{B_{t-i} x_k} - \delta_{A'B_{t-i}}) - (\delta_{x_k B_t} + \delta_{B_i x_k} - \delta_{B_i B_t}) \right] . \quad (2)$$

Note that in contrast to the proof of Theorem 1 part 1, in this case we know that the internal topology of the clade B is the same in T' as in T. For each set of leaves $a \in A', b_i \in B_i, b_{t-i} \in B_{t-i}$ and $b_t \in B_t$, we define

$$f(a, b_i, b_{t-i}, b_t) = \frac{1}{2^{i+1}} \left[(\delta_{ax_k} + \delta_{b_i x_k} - \delta_{ab_i}) - (\delta_{x_k b_t} + \delta_{b_{t-i} x_k} - \delta_{b_{t-i} b_t}) \right]$$
$$+ \frac{1}{2^{t-i+1}} \left[(\delta_{ax_k} + \delta_{b_{t-i} x_k} - \delta_{ab_{t-i}}) - (\delta_{x_k b_t} + \delta_{b_i x_k} - \delta_{b_i b_t}) \right] .$$

By applying quartet consistency conditions one can show that the first square bracketed term is less than zero. If the second term is also negative, then $f(a, b_i, b_{t-i}, b_t) < 0$. Otherwise

$$f(a, b_i, b_{t-i}, b_t) < \frac{1}{2^{i+1}} \left[(2\delta_{ax_k} + \delta_{b_i b_t} + \delta_{b_{t-i} b_t} - 2\delta_{x_k b_t} - \delta_{ab_i} - \delta_{ab_{t-i}}) \right] .$$

In this case by applying quartet consistency conditions one can again show that $f(a, b_i, b_{t-i}, b_t) < 0$. In either case $f(a, b_i, b_{t-i}, b_t) < 0$, hence (by summing over the leaves in each clade) we see that each term of the main summation (2) above is less than zero. If t is even, then we need to take into account an extra term in Equation (2), corresponding to $i = t - i = t/2$. This term may similarly be shown to be less than zero.

Thus $BME(\delta, T^A) < BME(\delta, T^B)$ which contradicts the placement of x_k in clade B. This completes the proof of Theorem 1. □

4.2 Local Topology Search

We first show that an NNI based search will never destroy an edge which has length at least twice the maximum error.

Lemma 3. *Let T, T^* be binary phylogenetic trees with some common edge $e = A|B$. Let δ have maximum error $\epsilon < l(e)/2$ relative to δ^*. Then any tree T' one NNI operation from T which does not contain e must have larger BME score than T.*

Proof. Consider an NNI move that could destroy the split $A|B$. Let A_1, A_2, B_1 and B_2 be the subclades of A and B obtained by dividing A and B at the point of attachment of e, as in Fig. 3. The only way e can be destroyed by an NNI is, without loss of generality, by swapping clades A_2 and B_2, as shown by T' in Fig. 3. By Lemma 1

$$BME(\delta, T) - BME(\delta, T') = \frac{1}{4}[(\delta_{A_1A_2} + \delta_{B_1B_2}) - (\delta_{A_1B_2} + \delta_{B_1A_2})].$$

For any leaves a_1, a_2, b_1, b_2 in A_1, A_2, B_1, B_2 we have

$$\begin{aligned}(\delta_{a_1a_2} + \delta_{b_1b_2}) - (\delta_{a_1b_2} + \delta_{b_1a_2}) &\leq (\delta^*_{a_1a_2} + \delta^*_{b_1b_2}) - (\delta^*_{a_1b_2} + \delta^*_{b_1a_2}) + 4\epsilon \\ &\leq (\delta^*_{a_1b_2} + \delta^*_{b_1a_2} - 2l(e)) - (\delta^*_{a_1b_2} + \delta^*_{b_1a_2}) + 4\epsilon \\ &= 4\epsilon - 2l(e) \\ &< 0.\end{aligned}$$

Summing over all leaves in A_1, A_2, B_1, B_2 we see that $BME(\delta, T) - BME(\delta, T') < 0$. Thus any NNI which removes e leads to an increase in BME score and will not be accepted. □

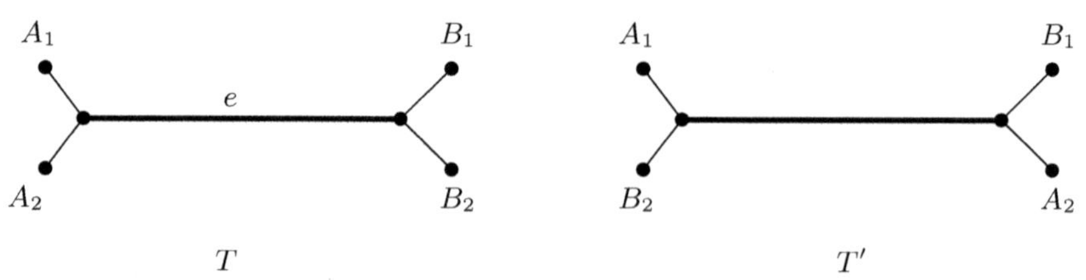

Fig. 3. An NNI operation that breaks split $e = A_1 \cup A_2 | B_1 \cup B_2$

Next we show that as long as we check all trees within one SPR operation of the current tree, and choose the best of these, we will never destroy an edge e which has length at least 3 times the maximum error.

Lemma 4. *Let T, T^* be binary phylogenetic trees with some common edge $e = A|B$. Let δ have maximum error $\epsilon < l(e)/3$ relative to δ^*. Then any tree T' one SPR operation from T which does not contain e must have larger BME score than a tree T^A which preserves e and is also within one SPR operation of T.*

Proof. This proof follows the proof that GreedyBME has safety radius 1/3, and is omitted for reasons of brevity. □

We finally show that if δ is consistent with some edge e in T, then an NNI based local topology search will not remove e.

Lemma 5. *Let T, T^* be binary phylogenetic trees with some common edge $e = A|B$. Let δ be a distance matrix consistent with e. Then NNI starting from T will never break the edge $A|B$.*

Proof. As in the proof of Lemma 3, consider an NNI move that could destroy the split $A|B$. Let A_1, A_2, B_1 and B_2 be the subclades of A and B obtained by dividing A and B at the point of attachment of e, as in Fig. 3. The only way e can be destroyed by an NNI is, without loss of generality, swapping clades A_2 and B_2, as shown by T' in Fig. 3. By Lemma 1

$$BME(\delta, T) - BME(\delta, T') = \frac{1}{4}[(\delta_{A_1A_2} + \delta_{B_1B_2}) - (\delta_{A_1B_2} + \delta_{B_1A_2})].$$

For any leaves a_1, a_2, b_1, b_2 in A_1, A_2, B_1, B_2 we have

$$(\delta_{a_1a_2} + \delta_{b_1b_2}) - (\delta_{a_1b_2} + \delta_{b_1a_2}) < 0.$$

by the consistency of δ with any quartet spanning $A|B$. Summing over all leaves in A_1, A_2, B_1, B_2 we see that $BME(\delta, T) - BME(\delta, T') < 0$. Thus any NNI which removes e leads to an increase in BME score and will not be accepted. □

Theorem 2 follows from Lemmas 3, 4 and 5.

5 Conclusion

In this work we have shown that GreedyBME is a more robust method of inferring a phylogeny than Neighbor Joining in two rigorous senses. Firstly GreedyBME has an edge safety radius of 1/3 and secondly GreedyBME will correctly reconstruct the true tree given a distance matrix that is quartet consistent with the true tree. Both conditions are strict improvements over Neighbor Joining.

The significance of proving bounds on the *edge* safety radius is that *any sufficiently long edge* is correctly reconstructed from the distance matrix (edges longer than three times the maximum error), even in the presence of very short edges elsewhere in the tree. In contrast, with results on safety radius we can only guarantee that the whole tree is correctly reconstructed if *all* edges are sufficiently long, otherwise we cannot guarantee anything.

Minimum evolution, and in particular balanced minimum evolution, has been proposed by several authors as a basic principle for inferring phylogenies (for references and discussion see [4]). Moreover the underlying reason for the accuracy of certain phylogenetic algorithms, including Neighbor Joining, has been attributed to their relationship to the balanced minimum evolution principle [8,11]. It is therefore counterintuitive that a heuristic for minimising BME score, GreedyBME, has an edge safety radius of 1/3, when the underlying principle (i.e. an algorithm that selects the tree of globally minimum BME score) has a weaker edge safety radius, which even approaches zero for large trees [10]. Further work on understanding this issue, as well as extending the robustness guarantees to more reasonable models of error in distance matrices, will help improve distance based phylogenetic inference in the future.

References

1. Atteson, K.: The performance of the neighbor-joining methods of phylogenetic reconstruction. Algorithmica 25, 251–278 (1999)
2. Bordewich, M., Gascuel, O., Huber, K., Moulton, V.: Consistency of Topological Moves Based on the Balanced Minimum Evolution Principle of Phylogenetic Inference. IEEE/ACM Transactions on Computational Biology and Bioinformatics 6(1), 110–117 (2009)
3. Bruno, W.J., Socci, N.D., Halpern, A.L.: Weighted Neighbor Joining: A likelihood based approach to distance-based phylogeny reconstruction. Mol. Biol. Evol. 17, 189–197 (2000)
4. Desper, R., Gascuel, O.: Fast and accurate phylogeny reconstruction algorithms based on the minimum evolution principle. J. Comp. Biol. 9, 587–598 (2002), Latest software available at, `http://atgc.lirmm.fr/fastme/`
5. Desper, R., Gascuel, O.: Theoretical foundation of the balanced minimum evolution method of phylogenetic inference and its relationship to weighted least-squares tree fitting. Mol. Biol. Evol. 21, 587–598 (2004)
6. Felsenstein, J.: An alternating least-squares approach to inferring phylogenies from pairwise distances. Syst. Biol. 46, 101–111 (1997)
7. Gascuel, O.: BIONJ: an improved version of the NJ algorithm based on a simple model of sequence data. Mol. Biol. Evol. 14, 685–695 (1997)
8. Gascuel, O., Steel, M.: Neighbor-joining revealed. Mol. Biol. Evol. 23(11), 1997–2000 (2006)
9. Hordijk, W., Gascuel, O.: Improving the efficiency of SPR moves in phylogenetic tree search methods based on maximum likelihood. Bioinformatics 21(24), 4338–4347 (2005)
10. Mihaescu, R.: Reliability results for the general Balanced Minimum Evolution principle (in preparation)
11. Mihaescu, R., Levy, D., Pachter, L.: Why neighbor-joining works. Algorithmica 54(1), 1–24 (2009)
12. Pardi, F., Guillemot, S., Gascuel, O.: Robustness of phylogenetic inference based on minimum evolution. Bulletin of Mathematical Biology (2010) (in press)

13. Pauplin, Y.: Direct calculation of tree length using a distance matrix. J. Mol. Evol. 51, 66–85 (2000)
14. Saitou, N., Nei, M.: The neighbor-joining method: A new method for reconstructing phylogenetic trees. Mol. Biol. Evol. 4, 406–424 (1987)
15. Semple, C., Steel, M.: Phylogenetics. Oxford University Press, Oxford (2003)
16. Shigezumi, T.: Robustness of greedy type minimum evolution algorithms. In: Computational Science –ICCS 2006, pp. 815–821 (2006)

The Complexity of Inferring a Minimally Resolved Phylogenetic Supertree

Jesper Jansson[1], Richard S. Lemence[1], and Andrzej Lingas[2]

[1] Ochanomizu University, 2-1-1 Otsuka, Bunkyo-ku, Tokyo 112-8610, Japan
Jesper.Jansson@ocha.ac.jp, rslemence@gmail.com
Funded by the Special Coordination Funds for Promoting Science and Technology
[2] Department of Computer Science, Lund University, 22100 Lund, Sweden
Andrzej.Lingas@cs.lth.se

Abstract. A recursive algorithm by Aho, Sagiv, Szymanski, and Ullman [1] forms the basis for many modern rooted supertree methods employed in Phylogenetics. However, as observed by Bryant [4], the tree output by the algorithm of Aho *et al.* is not always minimal; there may exist other trees which contain fewer nodes yet are still consistent with the input. In this paper, we prove strong polynomial-time inapproximability results for the problem of inferring a minimally resolved supertree from a given consistent set of rooted triplets (MINRS). We also present an exponential-time algorithm for solving MINRS exactly which is based on tree separators. It runs in $2^{O(n \log k)}$ time when every node is required to have at most k children which are internal nodes and where n is the cardinality of the leaf label set of the input trees.

Keywords: Phylogenetic tree; rooted triplet; minimally resolved supertree; NP-hardness; tree separator.

1 Introduction

Phylogenetic trees are leaf-labeled trees which are used to represent tree-like evolutionary history. To infer a reliable phylogenetic tree containing a large number of leaves is often a difficult task because of the computational complexity of the underlying optimization problems. One approach which has become increasingly popular in recent years is the divide-and-conquer-based *supertree approach* [2,6,9,12], which first uses a computationally intense method to reconstruct highly accurate trees for small, partially overlapping subsets of the leaf label set, and then applies a combinatorial algorithm to merge the small trees into one large tree called a *supertree.*

One of the common criticisms of supertrees is that they make statements about evolutionary relationships among leaves that are not always directly supported by any one of the input trees, which can create false groupings in the form of "spurious novel clades" [2]. Therefore, a natural idea is to try to avoid making more such statements than necessary to obtain a supertree, and thus to introduce as little unsupported branching information as possible. For this

V. Moulton and M. Singh (Eds.): WABI 2010, LNBI 6293, pp. 262–273, 2010.

reason, minimally resolved supertrees, i.e., trees that contain as few internal nodes as possible while still being consistent with the input, are important in Bioinformatics.

Several alternative supertree methods whose formal definitions differ have been developed; see, e.g., the survey paper by Bininda-Emonds [2] for an extensive list of references. A classic algorithm by Aho, Sagiv, Szymanski, and Ullman [1] named BUILD (see Section 2.2 below for a short description) can be used to merge a given set of *non-conflicting* rooted phylogenetic trees. Due to its simplicity and efficiency, it is used as a starting point of many rooted supertree methods such as the ones presented in [8,14,15,16] for combining a set of *conflicting* trees. These methods imitate the behavior of BUILD until a conflict occurs, at which point the so-called auxiliary graph consists of a single connected component which is then split into smaller components by removing some edges (different methods use different rules to do this), and then the method is executed recursively on each newly created component. Typically, such methods will all yield the same output as BUILD in the ideal case where the input set of trees contains no conflicts; hence, to understand these methods, it is important to fully understand the properties of the trees which are output by BUILD.

A surprising fact about BUILD is that it does not always produce a tree with the minimum possible number of internal nodes. This was first observed by Bryant in [4]. We generalize Bryant's example in Section 2.3 below to show that BUILD may in fact output a tree containing $\Omega(n)$ times more internal nodes than necessary, where n is the cardinality of the leaf label set of the input trees.

A binary phylogenetic tree with exactly three leaves is called a *rooted triplet.* Rooted triplets are a special case of phylogenetic trees, so hardness results concerning the computational complexity of inferring supertrees from rooted triplets will directly carry over to the corresponding problems for general inputs. Moreover, as explained in [9], the branching information contained in any rooted, binary phylogenetic tree with m leaves can be represented by a set of $O(m)$ rooted triplets (one rooted triplet per edge in the tree). From here on, we therefore focus on inputs which consist of rooted triplets.

1.1 Our Results and Organization of the Paper

We study the computational complexity of the problem of inferring a minimally resolved supertree from a given consistent set of rooted triplets over a leaf label set of cardinality n, called MinRS for short.

The paper is organized as follows. Section 2 defines the MinRS problem formally and surveys previous work. Section 3 proves two strong negative results: (1) the decision version of MinRS is NP-hard for any fixed number of internal nodes larger than or equal to 4; and (2) MinRS cannot be approximated within $n^{1-\epsilon}$ for any constant $0 < \epsilon < 1$ in polynomial time, unless P = NP. Then, Section 4.1 describes a simple algorithm for the decision version of MinRS which runs in $O^*(f(q) \cdot q^n)$ time, where q is the allowed number of internal nodes and $f(q)$ is the number of rooted, unlabeled trees with q nodes. Section 4.2 presents a more sophisticated exponential-time exact algorithm based on tree separators that runs in $2^{O(n \log k)}$ time, where every node is required to have at most k children which are internal nodes. Section 5 contains some final remarks.

2 Preliminaries

2.1 Basic Definitions

We will use the following definitions and notation.

To simplify the presentation, any node in a tree is considered to be an ancestor as well as a descendant of itself. For any nodes u, v in a tree, in case u is a descendant of v and $u \neq v$ then we write $u \prec v$ and call u a *proper* descendant of v. The *lowest common ancestor of* u *and* v, denoted by $lca(u, v)$, is the node w such that both u and v are descendants of w and $w \prec x$ holds for every other node x which is an ancestor of both u and v. The set of leaves in a tree T is denoted by $\Lambda(T)$.

A *phylogenetic tree* is a rooted, unordered tree whose leaves are distinctly labeled. Since the leaves in a phylogenetic tree are uniquely labeled, we will refer to them by referring to their labels. A *rooted triplet* is a phylogenetic tree with exactly three leaves in which every internal node has exactly two children, and we let $xy|z$ denote the rooted triplet having leaf label set $\{x, y, z\}$ that satisfies $lca(x, y) \prec lca(x, z) = lca(y, z)$.

Let T be a phylogenetic tree. For any $\{x, y, z\} \subseteq \Lambda(T)$, if the relation $lca(x, y) \prec lca(x, z) = lca(y, z)$ holds in T then the rooted triplet $xy|z$ is said to be *consistent with* T. A given set $\mathcal{R}$ of rooted triplets and a given phylogenetic tree T are *consistent* if every $t \in \mathcal{R}$ is consistent with T. Lastly, any given set $\mathcal{R}$ of rooted triplets is called *consistent* if there exists a tree which is consistent with $\mathcal{R}$ (otherwise, $\mathcal{R}$ is called *inconsistent*).

When $\mathcal{R}$ is given, we denote the set of all leaf labels which occur in $\mathcal{R}$ by L, i.e., we define $L = \bigcup_{t \in \mathcal{R}} \Lambda(t)$. Throughout the paper, we use the notation $n = |L|$ and $k = |\mathcal{R}|$. Given an input set $\mathcal{R}$ of rooted triplets, it is possible to efficiently check whether $\mathcal{R}$ is consistent and if so, to construct a phylogenetic tree consistent with $\mathcal{R}$, by a classic algorithm of Aho, Sagiv, Szymanski, and Ullman [1] named BUILD. The algorithm is described in Section 2.2 below.

Finally, for any consistent set $\mathcal{R}$ of rooted triplets, we say that a phylogenetic tree which is consistent with $\mathcal{R}$ and contains as few internal nodes as possible is a *minimally resolved* supertree for $\mathcal{R}$.

2.2 The Algorithm of Aho *et al.* [1] (BUILD)

In this subsection, we briefly review the algorithm of Aho, Sagiv, Szymanski, and Ullman [1]. The algorithm, referred to as BUILD, constructs a phylogenetic tree consistent with an input set $\mathcal{R}$ of rooted triplets over a leaf label set L, if such a tree exists. In case such a tree does not exist, the algorithm outputs *null*.

BUILD is a top-down, recursive algorithm. The main idea of the algorithm is to first partition the leaf set L into *blocks* according to the rooted triplets in $\mathcal{R}$. Then, the algorithm outputs a tree consisting of a root node whose children are the roots of the trees obtained by recursing on each block. The base case of the recursion is when the leaf set consists of a single leaf.

To perform the partitioning into blocks for any subset $L' \subseteq L$ with $|L'| > 1$, BUILD uses an *auxiliary graph* $\mathcal{G}(L')$. The auxiliary graph for any $L' \subseteq L$ is defined as $\mathcal{G}(L') = (L', E)$, where E contains the edge $\{x, y\}$ if and only if there is some rooted triplet of the form $xy|z$ in $\mathcal{R}$ with $x, y, z \in L'$. After constructing $\mathcal{G}(L')$, the algorithm computes the connected components in $\mathcal{G}(L')$ and lets each such connected component define one block of L'. If, at any point of its execution, $|L'| > 1$ yet L' contains just one block, then BUILD terminates and outputs *null*. This approach is motivated by Proposition 1 below together with the key observation that for any rooted triplet $xy|z$ consistent with a phylogenetic tree T, the leaves labeled by x and y cannot descend from two different children of the root of T, i.e., x and y must belong to the same block. (For a formal proof of correctness, see [1].)

Proposition 1 (Aho, Sagiv, Szymanski, and Ullman [1]). *If $\mathcal{G}(L)$ has only one connected component and $|L| > 1$ then $\mathcal{R}$ is not consistent with any phylogenetic tree.*

The running time of the original implementation of BUILD [1] was $O(nk)$, where $n = |L|$ and $k = |\mathcal{R}|$. Henzinger *et al.* [9] later presented a faster implementation of this algorithm, and replacing the dynamic graph connectivity data structure used by [9] by a more recent one [10] further reduces the complexity of the algorithm to $\min\{O(n + k \log^2 n), O(k + n^2 \log n)\}$ time [11].

2.3 Definition of MinRS

Bryant [4] noted that the BUILD algorithm of Aho *et al.* [1] does not always produce a minimally resolved supertree consistent with a given set of rooted triplets. In the example provided in Section 2.5.2 of [4], Bryant considered the set $\mathcal{R} = \{bc|a,\ bd|a,\ ef|a,\ eg|a\}$. As demonstrated in Figure 13 in [4], BUILD will construct a tree consistent with $\mathcal{R}$ which contains *three* internal nodes (a root node along with two internal nodes which are the parents of the leaves b, c, d and e, f, g, respectively), whereas the optimal solution is a tree containing *two* internal nodes (a root node and an internal node to which the leaves b, c, d, e, f, g are directly attached).

We can simplify Bryant's example to $\{bc|a,\ ef|a\}$. Then, if we extend the example as follows:

$$\mathcal{R} = \{x_1 x_2|x_0,\ x_3 x_4|x_0,\ \ldots,\ x_{2i-1} x_{2i}|x_0)\},$$

we obtain a consistent set of rooted triplets for which BUILD will construct a tree having $i+1$ internal nodes. However, the tree consisting of a root node with two children c_1 and c_2, where c_1 is a leaf labeled by x_0 and c_2 is an internal node with $2i$ children which are leaves labeled by $x_1, x_2, \ldots, x_{2i}$, contains exactly two internal nodes and is also consistent with $\mathcal{R}$. This shows that asymptotically, the BUILD algorithm of Aho *et al.* may produce a tree with $\Omega(n)$ times more internal nodes than the minimally resolved supertree, where n is the cardinality of the leaf label set.

From this observation, a natural question arises: When a consistent set of rooted triplets $\mathcal{R}$ is given, how efficiently can one construct a *minimally resolved* supertree consistent with $\mathcal{R}$? Formally, we define:

THE MINIMALLY RESOLVED SUPERTREE CONSISTENT WITH ROOTED TRIPLETS PROBLEM (MINRS)
Instance: A set $\mathcal{R}$ of rooted triplets with leaf set L.
Output: A rooted, unordered tree whose leaves are distinctly labeled by L which has as few internal nodes as possible and which is consistent with every rooted triplet in $\mathcal{R}$, if such a tree exists; otherwise, *null*.

2.4 Related Work

Besides Bryant [4], other authors such as Henzinger, King, and Warnow [9] have also previously considered the problem of inferring a minimally resolved supertree from a set of rooted triplets. Unfortunately, Henzinger *et al.* [9] incorrectly assumed that the BUILD algorithm of Aho *et al.* [1] always constructs a minimally resolved supertree. According to the proof of Theorem 4 in [9], the tree constructed by Algorithm A' of Henzinger *et al.* [9] is identical to the tree constructed by the BUILD algorithm; therefore, our example from Section 2.3 above also implies that Algorithm A' may output a tree with $\Omega(n)$ times more internal nodes than a minimally resolved supertree. This means that the minimality claim in Theorem 4 in [9] is not correct.

A *fan triplet* is a a phylogenetic tree consisting of a root node to which three leaves are directly attached. For any phylogenetic tree T and any $\{x, y, z\} \subseteq \Lambda(T)$, if $lca(x, y) = lca(x, z) = lca(y, z)$ holds in T then the fan triplet with leaves x, y, z is *consistent with* T. In Section 2.6.3 of [4], Bryant studied a kind of "dual" problem to MINRS called MOST RESOLVED COMPATIBLE TREE, in which the input is a consistent set $\mathcal{R}$ of (rooted and fan) triplets on a leaf set L, and the objective is to construct a phylogenetic tree leaf-labeled by L with *the largest possible* number of internal edges which is consistent with $\mathcal{R}$. (Here, trees containing an internal node with a single child are not allowed.) Note that if $\mathcal{R}$ contains rooted triplets only then the problem is trivial since any binary tree which is consistent with $\mathcal{R}$ will give an optimal solution. However, in the general case, MOST RESOLVED COMPATIBLE TREE is NP-hard [4].

For a recent survey of other optimization problems related to rooted triplets consistency (for example, computing a maximum cardinality subset $\mathcal{R}'$ of an inconsistent set $\mathcal{R}$ of rooted triplets such that $\mathcal{R}'$ is consistent), see Section 2 in [5].

3 Polynomial-Time Inapproximability of MinRS

In this section, we establish a strong polynomial-time inapproximability result for MINRS, namely that MINRS cannot be approximated within $n^{1-\epsilon}$ for any constant $0 < \epsilon < 1$ in polynomial time, unless P = NP. We will obtain this result by reducing the CHROMATIC NUMBER problem to MINRS.

First, recall that for any undirected graph $G = (V, E)$ and any positive integer K, a *K-coloring of G* is a partition of V into (possibly empty) disjoint subsets $V_1, V_2, \ldots, V_K$ called *color classes* such that for any $\{v, w\} \in E$, it holds that v and w belong to different color classes. A graph G is called *K-colorable* if there exists a K-coloring of G. The CHROMATIC NUMBER problem is defined as:

> CHROMATIC NUMBER
>
> **Instance:** An undirected graph $G = (V, E)$.
> **Output:** The smallest integer K such that G is K-colorable.

Zuckerman [17] proved that CHROMATIC NUMBER is NP-hard to approximate within $|V|^{1-\epsilon}$ for every $0 < \epsilon < 1$. Moreover, the decision version of the problem, i.e., to determine if an undirected graph G is K-colorable for a particular value of K, is easily solvable in polynomial time when $K = 2$ but known to be NP-hard for any fixed positive integer $K \geq 3$; see, e.g., [7].

We now describe the reduction. Let $G = (V, E)$ be any given instance of CHROMATIC NUMBER. Without loss of generality, we assume that V contains at least two vertices and that G is connected. Construct an instance of MINRS as follows. Let $L = \{v_1, v_2 : v \in V\}$ be a set of $2|V|$ new leaf labels and define $\mathcal{R} = \{v_1v_2|w_1,\ v_1v_2|w_2,\ w_1w_2|v_1,\ w_1w_2|v_2 : \{v, w\} \in E\}$. Clearly, the reduction can be carried out in polynomial time.

Then we have:

Lemma 1. *If G is K-colorable then there exists a tree which is consistent with $\mathcal{R}$ and contains $K + 1$ internal nodes.*

Proof. Since $G = (V, E)$ is K-colorable, we can partition the vertex set V of G into K disjoint color classes $V_1, V_2, \ldots, V_K$. Order the color classes so that $V_1, V_2, \ldots, V_j$ are non-empty and $V_{j+1} = \cdots = V_K = \emptyset$, where $j \leq K$. Define a tree T having exactly $K + 1$ internal nodes as follows. (See Fig. 1 for an illustration.)

- Let the root of T be one end of a path of length $K - j$ and let a_0 be the other end of the path. Let a_0 have j children $a_1, a_2, \ldots, a_j$.
- For each $i \in \{1, 2, \ldots, j\}$ and each $v \in V_i$, attach two leaves labeled by v_1 and v_2 to the node a_i.

Consider any rooted triplet in $\mathcal{R}$. It is of the form $v_1v_2|w_1$, where $\{v, w\} \in E$; furthermore, since $\{v, w\} \in E$, both of the vertices v and w cannot belong to the same color class V_i. Thus, the parent of the leaves v_1 and v_2 in T is different from the parent of the leaves w_1 and w_2, and so $v_1v_2|w_1$ is consistent with T. Therefore, $\mathcal{R}$ is consistent with T. □

Lemma 2. *If there exists a tree which is consistent with $\mathcal{R}$ and contains $K + 1$ internal nodes then G is K-colorable.*

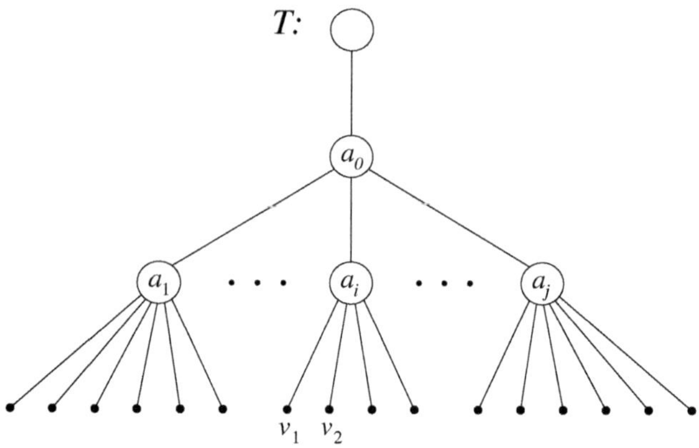

Fig. 1. Illustrating the proof of Lemma 1. In this example, there is one empty color class (i.e., $K - j = 1$), so the path from the root of T to the internal node a_0 has length 1. For each vertex $v \in V$, where v belongs to the color class V_i, the leaves v_1, v_2 in T are directly attached to the internal node a_i.

Proof. Let T be a tree with $K+1$ internal nodes which is leaf-labeled by L and consistent with $\mathcal{R}$. Let c_0 be the root of T and denote the other internal nodes of T by $c_1, c_2, \ldots, c_K$ arbitrarily. For every $i \in \{0, 1, \ldots, K\}$, associate a (possibly empty) subset $C_i \subseteq V$ with the internal node c_i, defined as follows: for each $v \in V$, if $lca(v_1, v_2) = c_i$ in T then let $v \in C_i$. It follows directly that for any $i, j \in \{0, 1, \ldots, K\}$ with $i \neq j$, the subsets C_i and C_j are disjoint.

Observe that every $v \in V$ belongs to at least one edge in E of the form $\{v, w\}$ (otherwise, the graph G would not be connected), and thus, by the construction of $\mathcal{R}$, the rooted triplets $v_1v_2|w_1$, $v_1v_2|w_2$, $w_1w_2|v_1$, and $w_1w_2|v_2$ belong to $\mathcal{R}$. Since $v_1v_2|w_1$ is consistent with T, it holds that $lca(v_1, v_2)$ is a proper descendant of $lca(v_1, w_1)$, i.e., $lca(v_1, v_2)$ cannot be the root of T. We have just shown that $C_0 = \emptyset$.

Next, we claim that for any two vertices $v, w \in V$, if $\{v, w\} \in E$ then v and w cannot belong to the same subset C_i. For the purpose of obtaining a contradiction, suppose that $v, w \in C_i$. Then $lca(v_1, v_2)$ and $lca(w_1, w_2)$ are the same node in T according to the definition of C_i. By transitivity, at least one of $lca(v_1, w_1)$, $lca(v_1, w_2)$, $lca(v_2, w_1)$, and $lca(v_2, w_2)$ is also equal to this node. However, since T is consistent with the rooted triplets $v_1v_2|w_1$, $v_1v_2|w_2$, $w_1w_2|v_1$, and $w_1w_2|v_2$, it follows from the definition of "consistent with" that the node $lca(v_1, v_2)$ is a proper descendant of (and hence different from) $lca(v_1, w_1)$ as well as of $lca(v_1, w_2)$, and in the same way that $lca(w_1, w_2)$ is a proper descendant of $lca(v_2, w_1)$ and of $lca(v_2, w_2)$. This yields a contradiction, so the claim must hold.

Thus, the partition of V into disjoint subsets $C_1, C_2, \ldots, C_K$ gives a K-coloring of G, and so G is K-colorable. □

Theorem 1. MinRS *cannot be approximated within* $n^{1-\epsilon}$ *for any constant* $0 < \epsilon < 1$ *in polynomial time, unless* $P = NP$.

Proof. Follows from Lemmas 1 and 2 together with the fact that CHROMATIC NUMBER is NP-hard to approximate within $|V|^{1-\epsilon}$ for every $0 < \epsilon < 1$ [17]. □

Since the decision version of CHROMATIC NUMBER is NP-hard for any fixed positive integer $K \geq 3$ (see, e.g., [7]), using the above reduction and applying Lemmas 1 and 2 also yields:

Corollary 1. *The decision version of* MINRS *is NP-hard for any fixed positive integer $q \geq 4$, where q is the allowed number of internal nodes.*

4 Exact Algorithms for MINRS

4.1 A Brute-Force Algorithm

We can solve the decision version of MINRS with a simple brute-force algorithm as follows.

- Let q be the allowed number of internal nodes.
- Generate all possible trees having q nodes and for each one try all q^n ways of attaching the n leaves in L to the q different nodes. For each obtained tree, check if it is consistent with $\mathcal{R}$ in polynomial time. If at least one such tree exists then output "yes"; otherwise, output "no".

This yields:

Theorem 2. *For any given positive integer q, the decision version of* MINRS *can be solved in $O^*(f(q) \cdot q^n)$ time, where $f(q)$ is the number of rooted, unlabeled trees with q nodes.*

It is known that $f(q) \sim c \cdot d^q \cdot q^{-3/2}$, where $c = 0.439924\ldots$ and $d = 2.955765\ldots$ [13]. Thus, the algorithm runs in exponential time for $q = O(1)$.

4.2 An Exponential-Time Algorithm for a Restricted Case of MINRS

The derivation of our main result in this section relies on the following variant of the tree separator theorem. A *non-leaf child* of a node v in a tree is a child of v which is an internal node, and for any rooted tree T and node v in T, the notation T_v means the subtree of T rooted at v.

Lemma 3. *Let T be a rooted tree with n leaves. There is a node v such that the subtree T_v contains strictly greater than $\frac{n}{2}$ leaves but for each non-leaf child w of v, the subtree T_w has at most $\frac{n}{2}$ leaves.*

Proof. We start from the root of T and perform the following procedure. If the root satisfies the condition then we set v to it and stop. Otherwise, we set v to the child of the root with the largest number of leaves and iterate the procedure for T_v.

Note that whenever a new iteration is applied to T_v then T_v has to have more than $\frac{n}{2}$ leaves. It follows from the finiteness of T that eventually a node v satisfying the condition will be found. □

We now use Lemma 3 to design a $2^{O(n \log k)}$-time procedure for the variant of MINRS where every internal node is allowed to have at most k non-leaf children.

The procedure is recursive. We enumerate all partitions of the leaf set L corresponding to the condition in Lemma 3. Then, we recursively apply the procedure on the resulting leaf subsets, possibly augmented by a dummy leaf, modifying the rooted triplets accordingly.

A partition Q corresponding to the condition in Lemma 3 has two levels. See Fig. 2. Firstly, it splits the set L of leaves into a set L' corresponding to T_v of size strictly greater than $\frac{n}{2}$ and its complement $L \setminus L'$ corresponding to $T \setminus T_v$. Secondly, Q splits L' into $k' \leq k$ sets $L'_1, \ldots, L'_{k'}$ corresponding to the non-leaf children of v in the condition, each of the sets of size at most $\frac{n}{2}$, and a number of singletons corresponding to the leaves pending on v.

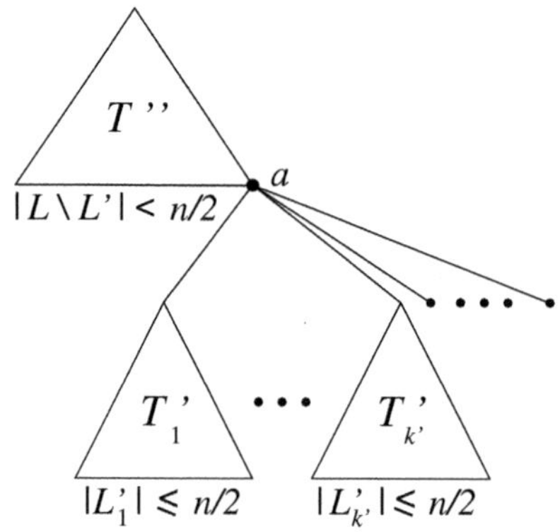

Fig. 2. According to Lemma 3, T has a node whose removal would divide the leaf set L into a subset L' containing strictly more than $\frac{n}{2}$ leaves and its complement $L \setminus L'$, and L' would be further partitioned into subsets $L'_1, \ldots, L'_{k'}$ of at most $\frac{n}{2}$ leaves each as well as a number of singletons.

If there is a rooted triplet $xy|z$ where $x, z \in L'$ and $y \in L \setminus L'$ then we can disregard Q. On the other hand, if $x, y \in L'$ and $z \in L \setminus L'$ then $xy|z$ is satisfied by Q and the triplet can be disregarded. As for the sets $L'_1, \ldots, L'_{k'}$ and the remaining leaf singletons, for each rooted triplet of the form $xy|z$ where $x, y, z \in L'$, if x, y are not in the same set L'_l then we can also disregard Q.

Otherwise, we augment $L \setminus L'$ by a dummy leaf a and for each rooted triplet of the form $xy|z$ where $x,\ z \in L \setminus L'$ and $y \in L'$, we form the rooted triplet $xa|z$. Analogously, for each rooted triplet of the form $xy|z$ where $x,\ y \in L \setminus L'$ and $z \in L'$, we form the rooted triplet $xy|a$.

For each such remaining partition Q, we run our procedure recursively on $L \setminus L' \cup \{a\}$ with the original set $\mathcal{R}$ of rooted triplets restricted to $L \setminus L'$ and the set of additional rooted triplets containing the dummy leaf a. Let T'' be the tree returned by the procedure.

We run also our procedure on each of the sets $L'_1, \ldots, L'_{k'}$ obtaining trees $T'_1, T'_2, \ldots, T'_{k'}$. Then, we make the trees as well as the singleton leaves children of the leaf a in the tree T'', and put the resulting tree on a candidate list.

Finally, we return the tree on the candidate list which has the smallest number of internal nodes.

The correctness of our procedure follows from Lemma 3 and the fact that we can join T'' with $T'_1, T'_2, \ldots, T'_{k'}$ in the described way at the dummy leaf a. The additional rooted triplets with a representing any leaf in L' satisfied by T'' make the join possible.

Let us estimate the time complexity $T(n)$ of our procedure in terms of the number of leaves n. Note that the number of rooted triplets is $O(n^3)$. The number of partitions considered and the time needed to generate them are trivially $O((k+2)^n) = 2^{O(n \log k)}$. Each of the at most $k+1$ recursive calls is applied to a set of leaves of size at most $\frac{n}{2}$. Thus, the total time complexity of processing the partitions is $2^{O(n \log k)}((k+1)T(\frac{n}{2})+n^{O(1)})$. By the inequality $2^{\alpha n \log k} 2^{c(\frac{n}{2}) \log k} \leq 2^{cn \log k}$ for $c \geq 2\alpha$, $k \geq 2$, and $n \geq 3$, we obtain $T(n) = 2^{O(n \log k)}$.

Theorem 3. *The problem of constructing a minimally resolved tree consistent with a set $\mathcal{R}$ of rooted triplets on a leaf set L under the restriction that each node has at most k non-leaf children, where $k \geq 2$, is solvable in $2^{O(n \log k)}$ time.*

Any tree with n leaves which is consistent with $\mathcal{R}$ can easily be converted into a tree consistent with $\mathcal{R}$ where each node has at most k non-leaf children by increasing the number of internal nodes by an $O(\log_k n)$ multiplicative factor. (Simply connect each internal node v to its non-leaf children C_v via a k-ary tree of depth $O(\log_k n)$ having C_v as leaves.) Hence, we obtain the following corollary.

Corollary 2. MinRS *can be approximated within a $O(\log_k n)$ factor in time $2^{O(n \log k)}$.*

5 Concluding Remarks

Recall that at each recursion level, the BUILD algorithm of Aho *et al.* [1] partitions the leaf set into blocks by computing the connected components in the auxiliary graph, and then represents each block by one node in the tree. A simple idea to reduce the number of internal nodes in the tree produced by BUILD is to merge blocks while ensuring that no rooted triplets are violated as follows:

> Proceed as in BUILD, but after computing the blocks (i.e., the connected components in $\mathcal{G}(L')$), construct an undirected graph H whose vertices are the blocks and where $\{A, B\}$ is an edge in H if and only if $\mathcal{R}$ contains some rooted triplet of the form $xy|z$ where either $x, y \in A$ and $z \in B$, or $x, y \in B$ and $z \in A$. Compute a minimum coloring of H and merge all blocks whose vertices in H received the same color. Then, continue the execution of BUILD.

The above step can be implemented in $O^*(2^j)$ time by applying an exact algorithm for CHROMATIC NUMBER [3], where j is the number of vertices in H. Summation over all recursive calls yields a total running time of $O^*(2^n)$. An interesting question is if this method minimizes the number of internal nodes, i.e., whether or not it always gives an optimal solution for MINRS.

Acknowledgments

The authors would like to thank Sylvain Guillemot and Wing-Kin Sung for some helpful comments.

References

1. Aho, A.V., Sagiv, Y., Szymanski, T.G., Ullman, J.D.: Inferring a tree from lowest common ancestors with an application to the optimization of relational expressions. SIAM Journal on Computing 10(3), 405–421 (1981)
2. Bininda-Emonds, O.R.P.: The evolution of supertrees. TRENDS in Ecology and Evolution 19(6), 315–322 (2004)
3. Björklund, A., Husfeldt, T.: Exact graph coloring using inclusion-exclusion. In: Kao, M.-Y. (ed.) Encyclopedia of Algorithms, p. 289. Springer Science+Business Media, LLC, Heidelberg (2008)
4. Bryant, D.: Building Trees, Hunting for Trees, and Comparing Trees: Theory and Methods in Phylogenetic Analysis. PhD thesis, University of Canterbury, Christchurch, New Zealand (1997)
5. Byrka, J., Guillemot, S., Jansson, J.: New results on optimizing rooted triplets consistency. Discrete Applied Mathematics 158(11), 1136–1147 (2010)
6. Chor, B., Hendy, M., Penny, D.: Analytic solutions for three taxon ML trees with variable rates across sites. Discrete Applied Mathematics 155(6-7), 750–758 (2007)
7. Garey, M., Johnson, D.: Computers and Intractability – A Guide to the Theory of NP-Completeness. W. H. Freeman and Company, New York (1979)
8. Gąsieniec, L., Jansson, J., Lingas, A., Östlin, A.: On the complexity of constructing evolutionary trees. Journal of Combinatorial Optimization 3(2-3), 183–197 (1999)
9. Henzinger, M.R., King, V., Warnow, T.: Constructing a tree from homeomorphic subtrees, with applications to computational evolutionary biology. Algorithmica 24(1), 1–13 (1999)
10. Holm, J., de Lichtenberg, K., Thorup, M.: Poly-logarithmic deterministic fully-dynamic algorithms for connectivity, minimum spanning tree, 2-edge, and biconnectivity. Journal of the ACM 48(4), 723–760 (2001)
11. Jansson, J., Ng, J.H.-K., Sadakane, K., Sung, W.-K.: Rooted maximum agreement supertrees. Algorithmica 43(4), 293–307 (2005)
12. Kearney, P.: Phylogenetics and the quartet method. In: Jiang, T., Xu, Y., Zhang, M.Q. (eds.) Current Topics in Computational Molecular Biology, pp. 111–133. The MIT Press, Massachusetts (2002)
13. Otter, R.: The number of trees. Annals of Mathematics 49(3), 583–599 (1948)

14. Page, R.D.M.: Modified mincut supertrees. In: Guigó, R., Gusfield, D. (eds.) WABI 2002. LNCS, vol. 2452, pp. 537–552. Springer, Heidelberg (2002)
15. Semple, C., Steel, M.: A supertree method for rooted trees. Discrete Applied Mathematics 105(1-3), 147–158 (2000)
16. Snir, S., Rao, S.: Using Max Cut to enhance rooted trees consistency. IEEE/ACM Transactions on Computational Biology and Bioinformatics 3(4), 323–333 (2006)
17. Zuckerman, D.: Linear degree extractors and the inapproximability of Max Clique and Chromatic Number. Theory of Computing 3(1), 103–128 (2007)

Reducing Multi-state to Binary Perfect Phylogeny with Applications to Missing, Removable, Inserted, and Deleted Data

Kristian Stevens and Dan Gusfield

Department of Computer Science
University of California, Davis

Abstract. Multi-State Perfect Phylogeny is an extension of Binary Perfect Phylogeny where characters are allowed more than two states. In this paper we consider four problems that extend its utility: In the Missing Data (MD) Problem some entries in the input are missing and the question is whether (bounded) values can be imputed so that the resulting data has a multi-state Perfect Phylogeny; In the Character-Removal (CR) Problem we want to minimize the number of characters to remove from the data so that the resulting data has a multi-state Perfect Phylogeny; In the Missing-Data Character-Removal (MDCR) Problem we want to impute values for the missing data to minimize the solution to the resulting Character-Removal Problem; In the Insertion and Deletion (ID) Problem insertion and deletion mutational events spanning multiple characters are also allowed.

In this paper, we introduce a new general conceptual solution to these four problems. The method reduces k-state problems to *binary* problems with missing data. This gives a new conceptual solution to the multi-state Perfect Phylogeny problem, and conceptual solutions to the MD, CR, MDCR and ID problems for *any* k significantly improving previous work. Empirical evaluations of our implementations show that they are faster and effective for larger input than previously established methods for general k.

1 Introduction and Background

A current and central problem in the study of evolution concerns the reconstruction of a sample's evolutionary history in the form of a *phylogenetic tree*. Extant organisms (*taxa*) correspond to the tree leaves, while internal nodes correspond to hypothetical ancestral taxa. Each node is labeled with a series of features or traits in the form of characters. Each character takes on one of several possible states. Each edge in the tree describes a mutational event.

Under a commonly used model of evolution, the *infinite-sites* model of population genetics, characters are binary and only mutate once in the history of the sample. Phylogenetic reconstruction under this model is referred to as the *Perfect Phylogeny* problem (PP) and can be solved in linear time [5].

V. Moulton and M. Singh (Eds.): WABI 2010, LNBI 6293, pp. 274–287, 2010.

Fig. 1. A Perfect Phylogeny for a 3-state matrix M with 3 characters from [4]. Extant taxa (square) are given in M and ancestral taxa (rounded) are inferred.

This paper concerns an important generalization of the binary Perfect Phylogeny problem that relaxes the constraint on the number of states and the number of mutational events per character. Under the *infinite-alleles* model of population genetics a character may take on up to k-states but the mutation that originates a state different from the root occurs only once in the sample's evolutionary history. This mutational constraint is referred to as *convexity*, which we will more formally define. In the biological literature, violating convexity is referred to as *homoplasy*. We refer to the phylogeny under this model as a *k-state Perfect Phylogeny*. Historically this problem has been motivated by qualitative data mostly of morphological origin. However, today informative multi-state data of molecular and genomic origin is widely available.

We are given a matrix M of input data for m characters on n taxa. We will use $\mathbf{c}$ to denote a particular character, $\mathcal{A}_\mathbf{c}$ to denote the set of observed states for $\mathbf{c}$, and α to denote a particular state in $\mathcal{A}_\mathbf{c}$. We use $\mathbf{t}$ to denote the length m vector of states corresponding to a particular taxon, and $\mathbf{c}(\mathbf{t})$ denote the state of $\mathbf{c}$ for taxon $\mathbf{t}$. Similarly, we use $\mathbf{c}(v)$ to denote the state of $\mathbf{c}$ for node v in a tree. We use the symbol ? to represent that the state is unknown. A character is said to be *incomplete* if it $\mathbf{c}(\mathbf{t}) = ?$ for some extant taxon, otherwise it is *complete*. To denote the set of taxa labeled with state α of character $\mathbf{c}$ we use $\mathbf{c}(\alpha)$.

Definition 1 (k-State Perfect Phylogeny). *Assume $|\mathcal{A}_\mathbf{c}| \leq k$ for every character $\mathbf{c}$ in M, a k-state Perfect Phylogeny for M is a tree T with n leaves, where each leaf is labeled by a distinct taxon $\mathbf{t}$ in M. Each internal node of T is labeled by a distinct taxon $\mathbf{t}$, which might not appear in M, where $\mathbf{c}(\mathbf{t}) \in \mathcal{A}_\mathbf{c}$ for each character $\mathbf{c}$. Furthermore, all characters must satisfy the following convexity requirement on T. A character $\mathbf{c}$ is* convex *with respect to T, if for every state α in $\mathcal{A}_\mathbf{c}$, the subgraph induced by the nodes labeled with α is a connected subtree of T which we denote as $T_{\mathbf{c}(\alpha)}$.*

Figure 1 shows an example 3-state Perfect Phylogeny. The *k-State Perfect Phylogeny problem* is to find and construct a k-State Perfect Phylogeny for M or determine there is none. If neither k nor n nor m is fixed, so k may grow with n, then the k-state Perfect Phylogeny problem is NP-complete [18]. In contrast, if k is any fixed integer, independent of n and m, then the k-state Perfect Phylogeny Problem can be solved in time that is polynomial in n and m [1,11,12].

In this paper we consider the following four variants that extend the biological utility of the basic k-state Perfect Phylogeny model.

Missing Data Problem (MD). In this realistic variant, some of the characters are incomplete. For an $m \times n$ input matrix M with incomplete characters, a *fill* is a setting of the missing values that replaces each ? of every character $\mathbf{c}$ with a state in $\mathcal{A}_{\mathbf{c}}$. We are interested in determining if any fill of the missing values of M gives data with a k-state Perfect Phylogeny. We will refer to such a fill as a *convex fill*.

The MD problem is motivated by the reality of missing entries in biological datasets. In many phylogenetic applications, missing data in the 30% range is not uncommon. While genotyping platforms usually produce missing data at a rate less than 10%, meta-genomic and population-genomic applications using whole genome shotgun for obtaining genotypes can have very high missing data rates. We will show the method presented here can handle missing entries at substantially higher rates than those considered in [6]. Even for binary data this problem is NP-complete, although the directed version of it can be solved in polynomial time [14]. The problem we address here has the biologically meaningful and computationally challenging constraint that missing values be selected from the set of observed states. The MD problem has a simple formulation if O(n) additional states can be assigned [16]. Solutions to the binary MD problem were shown in [7] using integer linear programming (ILP) and in [15] using a more specialized algorithm. A general solution for the k-state MD problem using chordal graph theory was shown in [6] but is practical for much smaller problems.

Character Removal Problem (CR). If M has no missing entries and does not have a k-state Perfect Phylogeny, what is the minimum number of characters to remove so that the resulting matrix does have a k-state Perfect Phylogeny? Even for binary data this problem is NP-complete.

Missing Data Character Removal Problem (MDCR). If M contains missing entries and does not have a k-state Perfect Phylogeny, how should the missing values be set in order to minimize the solution to the resulting CR problem? Another way of looking at the objective function is to minimize the number of characters removed such that the resulting MD problem has a Perfect Phylogeny.

The aforementioned CR and MDCR problems are motivated by the common practice in phylogenetics of removing characters when the existing data does not fit the Perfect Phylogeny model. This is most often done when the data is binary, but the problem and practice also arise for non-binary data [4,6]. An ILP formulation to the CR problem for the specific cases of k=3, 4, 5 was presented in [6]. An ILP formulation to the binary MDCR problem was shown in [7] and for k=3 in [6]. Recently it was shown that the chordal-graph approach of [6] can be extended to the CR and MDCR problems for general k [8]. Here we present a new conceptually simple solution to the CR and MDCR problems for general k and a corresponding implementation that is effective for large matrices.

Insertions and Deletions (ID). Frequently in nature, molecular sequences will undergo insertion and deletion events, the effect of which is to either insert a novel substring between two existing characters in a biological sequence or to delete an existing interval of characters from the sequence. We collectively refer to these intervals of spaces as *indels*, consistent with the fact that we do not know if one is caused by an insertion or a deletion, unless the known tree is rooted and the ancestral taxon is known. Indels can be used as phylogenetic characters. The importance of these characters to phylogenetic reconstruction arises from the fact that they occur frequently enough to be informative, and yet there is a wide acceptance that indels have a lower potential for homoplasy than do point mutations [13]. Both large and small indels are becoming a readily available source of mutational information from resequencing data. Indels are an example of the utility of our reduction for imposing restrictions on state transitions. Additional utility for restricted state transitions on multi-state characters is shown in [2] for the gain and loss of characters in the evolution of gene structure.

2 Reducing k-States to Binary

In this section we formally present a central result of the paper on which the other results are based, that every instance of k-state Perfect Phylogeny can be reduced to an instance of binary Perfect Phylogeny with missing data.

We now give an alternative definition of convexity for a character $\mathbf{c}$ with respect to a more general class of trees with labels only on the leaves. Given a leaf-labeled tree T, a character $\mathbf{c}$, and a state α of $\mathbf{c}$, and let $I_{\mathbf{c}(\alpha)}$ denote the minimal subtree of T that connects the set of leaves labeled by the taxa in set $\mathbf{c}(\alpha)$. Then $\mathbf{c}$ is convex with respect to T if for any two states α_i and α_j in $\mathcal{A}_{\mathbf{c}}$, subtrees $I_{\mathbf{c}(\alpha_i)}$ and $I_{\mathbf{c}(\alpha_j)}$ are node disjoint. Any leaf labeled tree that satisfies this definition also has an internal labeling that satisfies our previous definition.

The following lemmas directly follow from our two definitions of convexity and apply to the more general class of leaf labeled trees:

Lemma 1 (Convexity). *A character* $\mathbf{c}$ *is convex with respect to a leaf-labeled tree* T *if for every unique pair of nodes in* T *labeled with the same state* α *of* $\mathbf{c}$, *all labeled nodes on the unique path between them are labeled with state* α *or not labeled by a state. More formally, if* $v_i, \ldots v_k \ldots v_j$ *is a path on tree* T, $\mathbf{c}(v_i) \neq ?$, $\mathbf{c}(v_j) \neq ?$, *and* $\mathbf{c}(v_k) \neq ?$ *then* $\mathbf{c}(v_i) = \mathbf{c}(v_j) \implies \mathbf{c}(v_i) = \mathbf{c}(v_j) = \mathbf{c}(v_k)$.

Lemma 2 (Non Convexity). *If* $v_i, \ldots v_k \ldots v_j$ *is a path on the leaf-labeled tree* T, $\mathbf{c}(v_i) \neq ?$, $\mathbf{c}(v_j) \neq ?$, $\mathbf{c}(v_k) \neq ?$, *and* $\mathbf{c}(v_i) = \mathbf{c}(v_j)$ *but* $\mathbf{c}(v_k) \neq \mathbf{c}(v_j)$ *then the character* $\mathbf{c}$ *is* not convex *with respect to* T *and* T *can not be Perfect Phylogeny for* M.

A *state tree* can be constructed for $\mathbf{c}$ from a completely labeled tree T by merging all nodes that have state α for each $\alpha \in \mathcal{A}_{\mathbf{c}}$. A state tree can be constructed for $\mathbf{c}$ from a leaf labeled tree T by merging all nodes in the subtree with leaves labeled with state α for each $\alpha \in \mathcal{A}_{\mathbf{c}}$.

Lemma 3 (State Tree). *If a character* $\mathbf{c}$ *is convex with respect to a tree* T*, the* state tree *has exactly* $|\mathcal{A}_c|$ *nodes, and each node corresponds to* $T_{c(\alpha)}$ *for any* α *in* $\mathcal{A}_c$.

2.1 Reducing k-State Perfect Phylogeny to Binary Perfect Phylogeny with Missing Data

Given an instance of the k-state Perfect Phylogeny problem in matrix M we will derive a matrix M' of binary characters such that a tree T can be labeled to be a Perfect Phylogeny for M' if and only if it can also be labeled to be a Perfect Phylogeny for M. In Figure 2, we present our algorithm for reducing k-state problems which forms the basis of Theorem 1 and its subsequent proof.

Theorem 1. *Every instance of the k-state Perfect Phylogeny problem corresponds to an instance of the binary Perfect Phylogeny problem with missing data.*

$M' = \emptyset$
FOREACH character $\mathbf{c} \in M$
  FOREACH pair of states $(\alpha_i, \alpha_j) \in \mathcal{A}_\mathbf{c}$ such that $\alpha_i < \alpha_j$.
    FOREACH taxon $\mathbf{t} \in M$

$$\mathbf{c}'(\mathbf{t}) = \begin{cases} 0 \text{ if } \mathbf{c}(\mathbf{t}) = \alpha_i \\ 1 \text{ if } \mathbf{c}(\mathbf{t}) = \alpha_j \\ ? \text{ otherwise} \end{cases}$$

      Append $\mathbf{c}'$ to M'.
    ENDFOR
  ENDFOR
ENDFOR

Fig. 2. Algorithm: Convert k-state M to binary M'

Proof. For the $m \times n$ k-state input matrix M and the binary matrix M', we denote the finite and countable space of all candidate trees with n leaves as $\mathcal{T}^n$.

First we show that if a tree $T \in \mathcal{T}^n$ can be a Perfect Phylogeny for M it can also be a Perfect Phylogeny for M'. If T can be a Perfect Phylogeny for M then each character $\mathbf{c}$ in M is convex with respect to T and a *state tree* exists for $\mathbf{c}$ with respect to T. We now describe how to fill each $\mathbf{c}'$ in the expanded binary matrix M' so that they will also satisfy the convexity condition with respect to T. For the state pair α_i and α_j corresponding to the binary character $\mathbf{c}'$, choose an arbitrary edge on the path between α_i and α_j in the state tree for $\mathbf{c} \in M$ and remove it. We then label all taxa $\mathbf{t} \in \mathbf{c}'$ where $\mathbf{c}'(\mathbf{t}) = ?$ as follows: If $\mathbf{t}$ appears in the component containing α_i it is labeled with 0 otherwise if $\mathbf{t}$ appears in the component containing α_j it is labeled with 1. Now we prove that the filled character $\mathbf{c}'$ is convex by contradiction. If $\mathbf{c}'$ was not convex then by Lemma 2 there must exist a path $v_i \ldots v_k \ldots v_j$, where v_i, v_j, and v_k are labeled with states, $\mathbf{c}'(v_i) = \mathbf{c}'(v_j) = \alpha$, and $\mathbf{c}'(v_k) \neq \alpha$. This would require that $\mathbf{c}(v_k)$

be in the other component of the state tree, a violation of the convexity of $\mathbf{c}$. This proves we can fill in each $\mathbf{c}'$ in such a way that is convex with respect to T.

Next we show that if a tree $T \in \mathcal{T}^n$ can not a Perfect Phylogeny for M it can also not be a Perfect Phylogeny for any fill of M'. If T can not be a Perfect Phylogeny for M then by Definition 1 there must exist a character $\mathbf{c}$ that is *not convex*. By Lemma 2 there must exist a path $v_i, \ldots v_k \ldots v_j$, where v_i, v_j, and v_k are labeled with states, such that $\mathbf{c}(v_i) = \mathbf{c}(v_j)$ but $\mathbf{c}(v_k) \neq \mathbf{c}(v_j)$. Without loss of generality, assume $\alpha_i = \mathbf{c}(v_k)$ and $\alpha_j = \mathbf{c}(v_j)$ and $\alpha_i < \alpha_j$. By the reduction algorithm there is a character c' in M' corresponding to the relabeling of α_i as 0 and α_j as 1. Because these states are fixed, regardless of the fill of $\mathbf{c}'$, the following will also hold $\mathbf{c}'(v_i) = \mathbf{c}'(v_j)$ but $\mathbf{c}'(v_k) \neq \mathbf{c}'(v_j)$ proving that $\mathbf{c}'$ cannot be convex and T cannot be a Perfect Phylogeny for M'.

It remains to prove that if there is a Perfect Phylogeny for a fill of M' there always exists a Perfect Phylogeny for M. This also introduces the labeling algorithm for the missing taxa in M. Suppose there is a Perfect Phylogeny T' for M'. We will create a Perfect Phylogeny T for M where the unlabeled, undirected trees T and T' are identical, and where the mapping from leaves of T' to taxa of M' is identical to the mapping from leaves of T to taxa of M. We start with the tree T, with each leaf only labeled by the taxon mapped to it. Next, for any character $\mathbf{c}$ of M and every state α_i of $\mathbf{c}$, label every leaf f of T with state α_i of character $\mathbf{c}$ if and only if taxon f has state α_i for character $\mathbf{c}$, in M. Let $I_{\mathbf{c}(\alpha_i)}$ be the induced subtree of T that spans the leaves with state α_i for character $\mathbf{c}$ in M. We claim that for any character $\mathbf{c}$ and two states α_i and α_j of $\mathbf{c}$, $I_{\mathbf{c}(\alpha_i)}$ and $I_{\mathbf{c}(\alpha_j)}$ do not intersect. If they do intersect at a node v in T, then consider character $\mathbf{c}'$ in M' corresponding to the state pair (α_i, α_j) of $\mathbf{c}$, and suppose $\alpha_i < \alpha_j$. Then all of the leaves with state α_i for $\mathbf{c}$ in M will have state 1 for character $\mathbf{c}'$ of M', and all of the leaves with state α_j for $\mathbf{c}$ in M will have 0 for character $\mathbf{c}'$ of M'. But then the subtree of T' that spans all of the leaves with state 1 for $\mathbf{c}'$ will contain node v, as will the subtree of T' that spans all of the leaves with state 0 for c', and so character $\mathbf{c}'$ will not be convex in T'. Hence $I_{\mathbf{c}(\alpha_i)}$ and $I_{\mathbf{c}(\alpha_j)}$ do not intersect in T. So, for each state α of each character $\mathbf{c}$ of M, label each of the nodes in $I_{\mathbf{c}(\alpha)}$ with state α of $\mathbf{c}$. To complete the labeling of T to create a convex fill of M, let u be a node of T which is not labeled with a state for a character $\mathbf{c}$. Do a breadth first search in T from u until some node v is reached which is labeled by a state α_k of character $\mathbf{c}$, and label all of the nodes on the shortest path from u to v with state α_k. Repeat until all of the nodes are labeled with a state of each character. The result is a Perfect Phylogeny T for M. □

The following corollary of Theorem 1 arises as a generalization for arbitrary k of the often practiced removal of redundant characters for binary data.

Theorem 2 (k-state Character Redundancy). *A character $\mathbf{c}$ contains redundant information and can be removed from M for the purposes of solving the Perfect Phylogeny problem if for every pair of states α_i and α_j in the set of allowed states $\mathcal{A}_c$, the partitions $\mathbf{c}(\alpha_i)$ and $\mathbf{c}(\alpha_j)$ appear together in characters remaining in M.*

Proof. When the reduction algorithm is applied to M, characters in M' created by character $\mathbf{c}$ are created from remaining characters in M. □

2.2 Extension to k-State Perfect Phylogeny with Missing Data

Theorem 3. *Theorem 1 can be extended to the case of incomplete characters.*

Proof. For an $m \times n$ input matrix M° with incomplete characters, a fill is a setting of the missing values, that independently replaces ?'s with states from $\mathcal{A}_\mathbf{c}$ for every incomplete character $\mathbf{c}$ in M°. The reduction algorithm from M to M' applies without modification to M°. In Theorem 1 we established that if there is no Perfect Phylogeny for M there is no Perfect Phylogeny for M' and this can be applied to all fills of M°. In Theorem 1 we also established if there is a Perfect Phylogeny for M there is a Perfect Phylogeny for M' and this can be applied toall fills of M°.

It remains to show that we cannot have a convex fill for M' and not M°. By contradiction, suppose there is a convex fill for M' that results in a Perfect Phylogeny T'. Because T' is not a Perfect Phylogeny for M° there must exist some character $\mathbf{c} \in M^\circ$ that is non-convex with respect to T'. We showed previously in Theorem 1 the existence of a non-convex character in the fill of M° with respect to a tree T' implies the existence of one or more non convex characters in M' with respect to T' violating Perfect Phylogeny. Hence if T' is a Perfect Phylogeny for M' it is also a Perfect Phylogeny for M° and using the tree T' and the labeling algorithm of Theorem 1, we can fill the missing entries of M° so that all characters are convex with respect to T'. □

3 Solving the MD Problem for Arbitrary k

In this section we will describe how to effectively solve k-state MD and present our empirical results. To restate, our objective is to fill the binary matrix M' containing missing values in such a way that that it has a Perfect Phylogeny or note that it is not possible. This problem is NP-hard, but it has previously been shown to be solvable in an effective manner via ILP [7].

Before proceeding we first note a well-known necessary and sufficient condition for a Perfect Phylogeny on complete binary characters commonly referred to as the *four-gamete condition.* A *gamete* is an ordered state pair $\mathbf{c}_i(\mathbf{t}_k), \mathbf{c}_j(\mathbf{t}_k)$ for characters $\mathbf{c}_i$ and $\mathbf{c}_j$ over taxon $\mathbf{t}_k$.

Theorem 4 (Splits-Equivalence Theorem [3,16]). *There is an tree for a collection of binary characters if and only if no pair of characters contain all four of the gametes* $\{0,0;0,1;1,0;1,1\}$.

A pair of binary characters containing all four gametes is referred to as *incompatible.* Our objective is to fill the matrix M' in such a way that Theorem 4 is satisfied and no pair of characters is incompatible. We briefly describe the program from the top level down.

ILP Formulation. For each missing value in M', we create a variable $Y(i,j)$ that indicates the imputed state in cell $M'(i,j)$. For each pair of characters (p,q) where, because of missing data, there is a potential for four gametes we have a variable $C(p,q)$. For each of the missing gametes g in the character pair (p,q) we add a variable $B(p,q,g)$ and an inequality that forces it to 1 for settings of the Y variables that create the missing gamete. We add an inequality for each C variable that forces it to 1 when all missing gametes are present as indicated by its corresponding B variables. The overall optimization objective is find a setting of the Y variables that minimizes the number of incompatible character pairs given by $\sum_{p,q \in M'} C(p,q)$.

Early Termination and Problem Size Reduction

Once the problem has been reduced to a binary matrix a number of effective heuristics based on the four gamete condition can be applied to decrease the problem size or terminate if it is observed early that no Perfect Phylogeny exists. The heuristics below are repeatedly applied in succession until the binary matrix M' can no longer be modified.

Forbidden Gametes. We examine each pair of characters $\mathbf{c}_i$ and $\mathbf{c}_j$ in M' where $i < j$. If a pair of columns contains the four gametes $\{0,0;0,1;1,0;1,1\}$ we may terminate early knowing no Perfect Phylogeny exists. Let $\mathbf{c}_i(\mathbf{t}_k), \mathbf{c}_j(\mathbf{t}_k)$ be the gamete for taxon $\mathbf{t}_k$. If a pair of columns contains exactly three gametes then we can uniquely identify the remaining forbidden gamete f_i, f_j that must not appear if there is a Perfect Phylogeny for M'. We apply the following two rules to impute missing values for each taxon $\mathbf{t}_k$:

$$\mathbf{c}_i(\mathbf{t}_k) \text{ equal to } f_i \text{ implies } \mathbf{c}_j(\mathbf{t}_k) \text{ must be } 1 - f_j$$
$$\mathbf{c}_j(\mathbf{t}_k) \text{ equal to } f_j \text{ implies } \mathbf{c}_i(\mathbf{t}_k) \text{ must be } 1 - f_i$$

Logic similar to this has been previously used to impute missing values [9,15].

Character Nesting. For every unordered pair of characters $\mathbf{c}_i$ and $\mathbf{c}_j$ in M'. If $\mathbf{c}_i(\alpha_i) \subseteq \mathbf{c}_j(\alpha_j)$ and $\mathbf{c}_i(1-\alpha_i) \subseteq \mathbf{c}_j(1-\alpha_j)$ for some $\alpha_i \in 0,1$ and $\alpha_j \in 0,1$ we can remove $\mathbf{c}_i$ from M'. To see this, consider a third character $\mathbf{c}_k$. After imputation, any setting of the missing values for $\mathbf{c}_j$ and $\mathbf{c}_k$ that results in fewer that four gametes can be used to compute a valid setting for $\mathbf{c}_i$ with fewer that four gametes when paired with $\mathbf{c}_k$. In detail, the missing entries are set so that either $\mathbf{c}_i$ is identical to $\mathbf{c}_j$ or for every 1 in $\mathbf{c}_i$ we have a 0 in $\mathbf{c}_j$ and for every 0 in $\mathbf{c}_i$ we have a 1 in $\mathbf{c}_j$.

Taxon Nesting. For every unordered pair of taxa $\mathbf{t}_k$ and $\mathbf{t}_l$ in M'. If $\mathbf{c}_i(\mathbf{t}_k) = \mathbf{c}_i(\mathbf{t}_l)$ or $\mathbf{c}_i(\mathbf{t}_l) = ?$ for all $1 \leq i \leq m$ then $\mathbf{t}_l$ can be removed from from M'. To see this, for any pair of characters $\mathbf{c}_i$ and $\mathbf{c}_j$ the gamete for $\mathbf{t}_l$ is either identical to the gamete for $\mathbf{t}_k$ or it can be set to be identical to $\mathbf{t}_k$ without increasing the total number of gametes in the character pair.

Table 1. Illustrative execution times for our implementation of k-state Perfect Phylogeny problems. We compare to the implementation alternative for general k using chordal graphs described in [6]. We report the median of 10 executions on a 2.8 GHz Intel Core 2 Duo with 8 Gb of memory. A dash "–" indicates an instance where the program did not terminate with a result.

Problem Size	Missing = 0%			Missing = 25%				Missing = 50%			
chars×taxa states	MD [6]	MD $k \to 2$	CR $k \to 2$	MD [6]	MD $k \to 2$	*imput. error*	MDCR $k \to 2$	MD [6]	MD $k \to 2$	*imput. error*	MDCR $k \to 2$
100 × 100											
$k = 3$	6.7s	0.0s	0.1s	6.9s	0.0s	1.6%	0.1s	13s	0.0s	2.7%	0.1s
$k = 5$	10s	0.1s	0.3s	9.8s	0.1s	1.9%	0.3s	75s	0.1s	4.0%	0.2s
$k = 10$	25s	0.2s	2.1s	1m3s	0.6s	5.0%	1.9s	–	0.8s	7.8%	1.3s
100 × 200											
$k = 3$	1m14s	0.1s	0.3s	1m21s	0.1s	1.3%	0.3s	2m09s	0.1s	2.1%	0.2s
$k = 5$	1m34s	0.5s	1.2s	2m12s	0.4s	2.2%	1.0s	–	0.3s	3.4%	0.7s
200 × 400											
$k = 3$	–	0.6s	1.4s	–	0.5s	0.5%	1.3s	–	0.5s	0.9%	1.1s
$k = 5$	–	4.4s	9.1s	–	3.7s	1.0%	9.6s	–	2.6s	1.3%	5.6s
1000 × 1000											
$k = 3$	–	20s	45s	–	18s	0.1%	42s	–	17s	0.2%	34s
$k = 5$	–	2m22s	5m48s	–	2m16s	0.2%	5m7s	–	1m36s	0.3%	4m0s

Trivial Characters. Characters that do not contain more than one 0 or more than one 1 can be set in such a way that there is only one 0 or one 1. Such a character can never contain all four gametes when paired with any other character. These *trivial* characters are removed from M' for later re-insertion if a Perfect Phylogeny is found.

Empirical Results. Multi-state data was simulated with R. Hudson's `ms` program [10] using an application developed in [6] for converting instances of $k = 2$ to realistic instances where $k > 2$ both with and without Perfect Phylogenies. Table 1 presents our results and compares them to the alternative method for general k introduced in [6]. A few trends are noteworthy. In comparison to the method for solving k-state MD described in [6] based on chordal graphs, our implementation is faster and effective for much larger problems. We increasingly outperform the alternative method as the size of the matrix and the amount of missing data increases. The space requirements of the alternative method rapidly increase with the number of states and the amount of missing data eventually exhausting available resources. Both methods slow down for a matrix of fixed size as the number of states k increases. We also include the error rate for imputed states as an indication of how this method might perform when used to impute missing values when the data originates from a tree. The results indicate that the ability to recover the correct state is diminished as the number of states increases, the number of characters decreases, and the amount of missing data increases.

4 Solving the CR and MDCR Problems for Arbitrary k

Our reduction allows us to adapt, to the case of general k, the ILP formulation for the binary CR and MDCR problems introduced in [7].

ILP Formulation. $D(i)$ is a binary variable used to indicate whether or not character $\mathbf{c}_i$ in M' will be removed. For each pair of characters (p, q) that is already incompatible we add the inequality $D(p) + D(q) \geq 1$ which requires at least one of them be removed.

For each pair of characters (p, q) where, because of missing data, there is a potential for incompatibility we add the inequality $D(p) + D(q) - C(p, q) > 0$. This forces either $D(p)$ or $D(q)$ to 1 if missing values are imputed such that the pair is incompatible as indicated by a value of 1 for $C(p, q)$.

Next we formalize the concept of a character group $\mathcal{G}$. In the reduction from M to M' each k-state character $\mathbf{c}_i$ in M creates a group $\mathcal{G}_i$ of up to $\binom{k}{2}$ binary characters in M'. Instead of minimizing the number of characters from M' being removed, we are interested in minimizing the number of character groups associated with binary characters removed from M'. For each character group $\mathcal{G}_i$ we create a variable $G(i)$ that will be forced to 1 if any of its associated binary characters are removed from M'. To do this we use an inequality that is essentially a logical OR: $|\mathcal{G}_i|G(i) \geq \sum_{j \in \mathcal{G}_i} D(j)$. The overall objective function for the ILP is to minimize $\sum_{i \in M'} G(i)$. The setting of the $G(i)$ variables tell us which characters to remove from M.

The *taxon nesting* and *trivial character* heuristics for reducing the problem size can be applied as described in Section 3. The remaining two heuristics described in Section 3 are not applied because it is not known which characters will ultimately be present in the subsequent MD problem. Because not all the reduction tools are available, the resulting size of an MDCR ILP is much larger in practice than a comparable MD ILP. Most of the size of an MDCR ILP comes from B and C inequalities associated with potential pairwise incompatibilities. For instance, in our empirical results for 5 state 100x100 MDCR ILPs these inequalities always account for over 95% of its total size. This allows us to use an effective approach for determining the optimal solution to large MDCR problems utilizing our ability to quickly solve problem MD. Recall that the MDCR problem is to find the minimum number of characters to remove from M such that the resulting matrix has a Perfect Phylogeny as determined by problem MD. We also note that a solution to a relaxed MDCR minimization problem provides a lower bound on the number of characters that must be removed as well as a candidate list of characters to remove. Rather than solve the full MDCR ILP, we successively solve more constrained relaxations of the full MDCR ILP. To generate a relaxed MDCR ILP we impose a minimum cutoff q on number of gametes observed in a pair of characters before the associated B and C inequalities for that pair are generated. We can test if a candidate list of removal characters is an optimal solution using our approach to problem MD. We decrease q from 4 to 0, applying successively greater constraint, until we verify a solution is optimal.

Empirical Results. Illustrative total execution times are given in Table 1. Strikingly the times are of the same order of magnitude as the corresponding MD problems. The rate of homoplasy was set so that roughly 10% of the characters were removed in the optimal solution. Asserting the validity of our approach, the program only went below the cutoff $q = 3$ in the case of 10 states with 25% or more missing data.

5 Insertions and Deletions as Phylogenetic Characters

In the k-state Perfect Phylogeny problem with Insertions and Deletions (ID) we are given data with indels and want to construct a Perfect Phylogeny or determine that one does not exist, considering the indels as phylogenetic characters. The locations of insertions or deletions are ultimately revealed in a multiple alignment of the sequences. In this setting insertions and deletions are identically coded as a contiguous interval of spaces '–' in the taxa that do not contain the inserted or deleted characters. Figure 3 shows a data matrix containing indels encoded by spaces derived from a hypothetical multiple alignment, where identical characters have been removed. Such a matrix is the input to problem ID.

If each space in an indel is treated as an independent character, rather than considering the contiguous interval of spaces as a one single character, then common ancestry will be incorrectly inferred when two indels overlap. For an example of this see character 2 in Figure 3. To address this problem, Simmons and Ochoterena describe an encoding of indels using multi-state characters (MCIC) that is maximally informative for phylogenetic construction by parsimony [17]. However, a multi-state character alone is not enough to represent all the phylogenetic information, and MCIC provides additional information to the phylogenetic reconstruction algorithm in the form of a transition matrix. Six rules that implement the MCIC encoding for all situations with up to 3 overlapping indels are described in [17]. We present here a more rigorous multi-state encoding of indels for the ID problem that captures all the phylogenetic information in the MCIC model. Our multi-state encoding is simple to describe and implement. Moreover through reduction we will ultimately derive a binary encoding and demonstrate an equivalency between binary and multi-state encodings under the Perfect Phylogeny model. We show that one additional binary character per indel is a sufficiently informative encoding for the Perfect Phylogeny problem and that Perfect Phylogeny with indels is inherently a missing data problem. Finally, we provide a method for obtaining the optimal solution that explicitly handles the missing data.

We are given a matrix M of non-redundant characters where some entries contain spaces '–'. We can assume, because of their low homoplasy, that the interval of spaces is unique to each indel event. To encode indels as multi-state characters, we first assign each unique contiguous interval of spaces to a state $\underline{\alpha}$ that is not yet present in M, and denote its interval as the set span($\underline{\alpha}$). We then assign state $\underline{\alpha}$ to the associated spaces in all taxa containing that indel. This corrects the problem that overlapping spaces incorrectly suggest shared ancestry. We note that even though the state $\underline{\alpha}$ associated with a unique indel event may span multiple characters, we do not need to explicitly require that the state arise at the same location in the tree for all characters in span($\underline{\alpha}$). If a

Perfect Phylogeny exists, this constraint arises naturally from the edge-induced partitions of the taxa. The tree will have $n-1$ edges each corresponding to a unique subset of the taxa. Convexity ensures one edge must correspond to the set of taxa $\mathbf{c}(\underline{\alpha})$ which is the same for all characters $\mathbf{c}$ in span($\underline{\alpha}$).

At this point we have a multi-state Perfect Phylogeny problem, but we have not yet restricted the possible state trees for characters with indel states. For the most part, an indel should be a leaf in the state tree, since under the model a character does not exist before it is inserted or after it is deleted. The exception to this is when one indel contains another, an example of which is when one sequence of characters is inserted within another inserted sequence of characters. To be specific, one indel is said *contain* another indel, denoted by $\underline{\alpha}_i \supset \underline{\alpha}_j$, if span($\alpha_i$) $\supset$ span(α_j). An indel state must be a leaf in the state tree or adjacent to an indel state in which it can be contained. This imposes a partial ordering restriction on the state tree. We can use the reduction presented in Figure 2 and apply post-processing to the binary matrix M' to accomplish the desired restrictions.

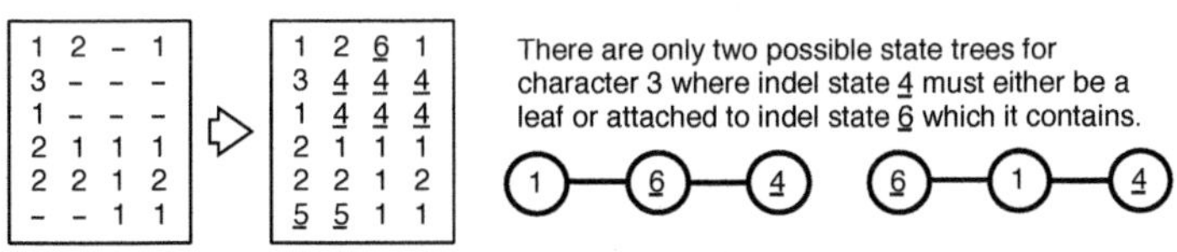

Fig. 3. For problem ID, an example of the original data, a multi-state encoded matrix, and the state tree restrictions on character 3

We now describe our algorithm for reducing the search space using M'. Recall that for any character $\mathbf{c}'$ in M' we have a corresponding character $\mathbf{c}$ in M and a pair of states α_i and α_j in $\mathcal{A}_{\mathbf{c}}$ corresponding to filled taxa. We visit each character $\mathbf{c}'$ in M' where either α_i or α_j corresponds to an indel state. Without loss of generality, assume that α_i is an indel state and that α_i in $\mathbf{c}$ corresponds to 0 in $\mathbf{c}'$ for all taxa $\mathbf{c}(\alpha_i)$. We first assume the simple case where α_i is a leaf on the state tree and that by convexity all taxa not in $\mathbf{c}(\alpha_i)$ can be set to 1 in $\mathbf{c}'$. We then examine each taxon $\mathbf{t}$ of character $\mathbf{c}'$ to find any containing indel states. If $\mathbf{c}(\mathbf{t})$ is an indel state which can contain α_i, we set $\mathbf{c}'(\mathbf{t}) = ?$.

Theorem 5. *k-state ID reduces to k-state MD.*

Proof. Note that the aforementioned algorithm can be implemented by modifying the matrix M as follows: For each unique indel state $\underline{\alpha}$ we add a binary character $c_{\underline{\alpha}}$ indicating it's presence or absence. Then we set $\mathbf{c}_{\underline{\alpha}}(\mathbf{t}) = ?$ for all taxa $\mathbf{t}$ with an indel that contains $\underline{\alpha}$. Finally, all spaces '–' in M are set to ?. Both the original and modified matrices will reduce to the same matrix M'. □

Discussion of the ID Results. We see that Problem ID must be posed as a missing data problem. The missing data in this case corresponds to the actual sequence of states that were deleted or inserted which, by the convexity requirement, can be filled into the missing entries. Perhaps not surprisingly the additional binary characters introduced by indels into the reduced matrix correspond to a simple indel encoding (SIC) proposed by Simmons and Ochoterena as a less phylogenetically informative alternative to MCIC. By showing an equivalency between the binary and multi-state encoding schemes, we clarify that this assertion does not apply under the Perfect Phylogeny model. Using this conceptually simple extension of our implementation of problem MD we provide a rigorous method of determining the optimal solution to problem ID for general k. We leave open the complete characterization of the associated MD, CR, and MDCR problems in the presence of indels.

Acknowledgements

This research is supported in part by NIH grant R01-HG002942 (KS) and NSF grants SEI-BIO 0513910 (DG) and IIS-0803564 (DG).

References

1. Agarwala, R., Fernandez-Baca, D.: A polynomial-time algorithm for the perfect phylogeny problem when the number of character states is fixed. SIAM Journal on Computing 23(6), 1216–1224 (1994)
2. Alekseyenko, A.V., Lee, C.J., Suchard, M.A.: Wagner and dollo: a stochastic duet by composing two parsimonious solos. Syst. Biol. 57(5), 772–784 (2008)
3. Buneman, P.: The recovery of trees from measures of dissimilarity. Mathematics in the archaeological and historical sciences, 387–395 (1971)
4. Fernández-Baca, D.: The perfect phylogeny problem. In: Du, D.Z., Cheng, X. (eds.) Steiner Trees in Industries. Kluwer Academic Publishers, Dordrecht (2001)
5. Gusfield, D.: Efficient algorithms for inferring evolutionary trees. Networks 21(1), 19–28 (1991)
6. Gusfield, D.: The multi-state perfect phylogeny problem with missing and removable data: Solutions via integer-programming and chordal graph theory. In: Batzoglou, S. (ed.) RECOMB 2009. LNCS, vol. 5541, pp. 236–252. Springer, Heidelberg (2009)
7. Gusfield, D., Frid, Y., Brown, D.: Integer Programming Formulations and Computations Solving Phylogenetic and Population Genetic Problems with Missing or Genotypic Data. In: Lin, G. (ed.) COCOON 2007. LNCS, vol. 4598, p. 51. Springer, Heidelberg (2007)
8. Gysel, R., Gusfield, D.: Extensions and Improvements to the Chordal Graph Approach to the Multi-state Perfect Phylogeny Problem. In: Borodovsky, M., Gogarten, J.P., Przytycka, T.M., Rajasekaran, S. (eds.) Bioinformatics Research and Applications. LNCS, vol. 6053, pp. 52–60. Springer, Heidelberg (2010)
9. Halperin, E., Karp, R.: Perfect phylogeny and haplotype assignment. In: Proceedings of the eighth annual international conference on Resaerch in computational molecular biology, pp. 10–19. ACM, New York (2004)
10. Hudson, R.: Generating samples under a Wright-Fisher neutral model of genetic variation. Bioinformatics 18(2), 337–338 (2002)

11. Kannan, S., Warnow, T.: Inferring evolutionary history from DNA sequences. In: Proceedings of 31st Annual Symposium on Foundations of Computer Science, pp. 362–371 (1990)
12. Kannan, S., Warnow, T.: A fast algorithm for the computation and enumeration of perfect phylogenies when the number of character states is fixed. In: Proceedings of the sixth annual ACM-SIAM symposium on Discrete algorithms, pp. 595–603. Society for Industrial and Applied Mathematics, Philadelphia (1995)
13. Lloyd, D.: Multi-residue gaps, a class of molecular characters with exceptional reliability for phylogenetic analyses. Journal of Evolutionary Biology 4(1), 9–21 (2002)
14. Pe'er, I., Pupko, T., Shamir, R., Sharan, R.: Incomplete directed perfect phylogeny. SIAM Journal on Computing 33(3), 590–607 (2004)
15. Satya, R., Mukherjee, A.: The undirected incomplete perfect phylogeny problem. IEEE/ACM Transactions on Computational Biology and Bioinformatics 5(4), 618–629 (2008)
16. Semple, C., Steel, M.: Phylogenetics. Oxford University Press, USA (2003)
17. Simmons, M., Ochoterena, H.: Gaps as characters in sequence-based phylogenetic analyses. Systematic Biology 49(2), 369–381 (2000)
18. Steel, M.: The complexity of reconstructing trees from qualitative characters and subtrees. Journal of Classification 9(1), 91–116 (1992)

An Experimental Study of Quartets MaxCut and Other Supertree Methods

M. Shel Swenson[1,2], Rahul Suri[1], C. Randal Linder[3], and Tandy Warnow[1]

[1] Department of Computer Science, The University of Texas at Austin
[2] Department of Mathematics, The University of Texas at Austin
[3] Section of Integrative Biology, The University of Texas at Austin

Abstract. Although many supertree methods have been developed in the last few decades, none has been shown to produce more accurate trees than the popular Matrix Representation with Parsimony (MRP) method. In this paper, we evaluate the performance of several supertree methods based upon the Quartets MaxCut method of Snir and Rao. We show that two of these methods usually outperform MRP and all other supertree methods we studied under many realistic model conditions. In addition, we show that the popular criterion of minimizing the total topological distance to the source trees is only weakly correlated with topological accuracy, and therefore that evaluating supertree methods on biological datasets is problematic.

1 Introduction

Supertree methods comprise one approach to reconstructing large molecular phylogenies given a set (called a *profile*) of estimated trees (called *source trees*) for overlapping subsets of the entire set of taxa. Source trees are combined into a single supertree on the full set of taxa using various algorithmic techniques. Because of the computational difficulties in estimating large phylogenies, many computational biologists think that the only feasible strategy to estimating the Tree of Life will involve a divide-and-conquer approach where trees are estimated on subsets of taxa and a supertree method is used to assemble a tree on the entire taxon set from the source trees. While there are many supertree methods, only MRP is used regularly in supertree constructions on biological datasets (4); furthermore, no other supertree method has been shown to produce trees that are comparable in accuracy to MRP under the standard bipartition metric (5).

One version of the supertree estimation problem uses quartet amalgamation methods. Each estimated source tree is encoded by an *appropriately chosen subset* of its induced quartet trees, and the set of quartets (the union of the chosen subsets for each source tree) is used to estimate a supertree. Quartet amalgamation methods can thus be used to assemble supertrees from source trees of arbitrary size.

The Maximum Quartet Consistency (MQC) problem is a natural optimization problem, in which the input is a set of quartet trees and a supertree is sought that *displays* the maximum number of quartet trees. MQC is NP-hard, and generally hard to approximate except in special cases (3; 15; 6; 11). Theoretical results and heuristics for the special case where the input set contains a tree on every quartet appear in (24; 20; 14; 16; 22). In a recent paper (21), Snir and Rao presented *Quartets MaxCut* (QMC), a heuristic for

V. Moulton and M. Singh (Eds.): WABI 2010, LNBI 6293, pp. 288–299, 2010.

MQC that can be applied to arbitrary sets of quartet trees (i.e., ones that may not contain a tree on every quartet). Snir and Rao showed that by encoding the source trees as quartet trees, QMC could be used as a generic supertree method. They then constructed supertrees using this QMC-based supertree method for a number of biological supertree profiles. Since the true supertree was not known, they could not evaluate the topological accuracy of the supertrees they constructed; instead, they compared the QMC supertree to the source trees to produce two different average similarity measures for each supertree. A comparison between QMC-based supertrees and MRP supertrees showed that QMC had higher average similarity to the source trees under one criterion, and lower average similarity with respect to another. QMC's failure to outperform MRP as a supertree method with respect to the average similarity to the source trees should not be considered a serious limitation for two reasons. First, average similarity to the source trees is not the same as accuracy with respect to the true tree (a phenomenon we investigate directly in this paper). Second, QMC depends critically upon the specific technique used to encode each source tree as a set of quartet trees. In other words, QMC might be producing highly accurate trees even though the average similarity is lower than MRP, and it might produce more accurate trees if other encodings of the source trees were used.

In this paper, we report results from a study in which we employ several encodings of the source trees by quartet trees and apply QMC to the resultant sets of quartet trees. We compare the accuracy of QMC using different encodings to MRP and five other supertree methods: Robinson-Foulds Supertrees (1), Q-Imputation (13), MinFlip (8; 7; 9), SFIT (10), and PhySIC (19). We find that the topological accuracy of QMC supertrees computed on different encodings varies substantially. Two QMC-based supertree methods, QMC(All) and QMC(Exp+TSQ) (differing only in how the source trees are encoded), perform similarly and *outperform* all the other supertree methods under many realistic model conditions, and have comparable accuracy under most others. However, MRP outperforms all QMC methods on the largest (1000-taxon) datasets. Finally, we find that using topological similarity to source trees as a proxy for topological accuracy with respect to the true tree is of limited use, and can be misleading. Thus, evaluating supertree methods on biological datasets is problematic, and supertree methods that seek to minimize topological distance to source trees may not have the best accuracy.

2 Basics

Supertree Datasets. Because of the taxon sampling strategies used by biologists, source trees tend to be focused either on intensively sampled, smaller subgroups, like big cats, or on larger, sparsely sampled groups, like all vertebrates. The first type is called a *clade* source tree, and the second type is called a *scaffold*. Supertree profiles include scaffolds to ensure sufficient overlap among the clade trees.

The input to the supertree problem is a set of source trees, $\{t_1, t_2, \ldots, t_k\}$, on subsets of a set S of taxa. Source trees are often estimated using biomolecular sequence datasets. Each source tree is estimated on its aligned sequence dataset using computationally intensive methods–e.g., maximum parsimony or maximum likelihood heuristics like RAxML (23). A supertree method combines the source trees into a tree on the full dataset.

Matrix Representation with Parsimony. Matrix representation with parsimony (MRP) (2; 18) is currently the most widely used supertree method. It encodes source trees as a matrix of *partial binary characters*: all entries in the matrix are 0, 1, or ?, with each column in the matrix defined by a single edge in a source tree. The matrix is then analyzed using a heuristic for the NP-hard maximum parsimony problem (12).

Quartets MaxCut (QMC). QMC is a quartet amalgamation method, operating in polynomial time and providing no guarantees with respect to its optimization problem, MQC. The source trees are encoded by sets of quartet trees, and QMC is applied to the union of these sets.

Quartet Encodings of Source Trees. Here, we explore several techniques for representing source trees by sets of quartet trees. Two of these techniques use random sampling strategies (21), which are based upon computation of the topological distance between leaves in the source tree. The *topological diameter* of a quartet tree q with respect to a source tree t is the maximum of its leaf-to-leaf topological distances within the source tree and is denoted $diam_t(q)$. The quartet encoding strategies used in (21) also include calculation of the *Topologically-Short Quartet* (TSQ) trees, defined as follows: For each edge in a source tree, pick the topologically nearest leaves in each of the subtrees around the edge. If two or more leaves within a subtree have the same topological distance to the edge, pick all such leaves. The set of quartet trees formed by picking one such leaf from each subtree forms the TSQs around that edge. The union of all these is the set of TSQ trees.

We tested five strategies for encoding a source tree t by a set of quartet trees:

All quartets: include all induced four-taxon trees.

k-short: a generalization of the TSQs: for each edge in a source tree, pick the k topologically nearest leaves in each of the subtrees around that edge. The (approximately) k^4 quartets of leaves are the k-short quartet trees around that edge, and the set of all such k-short quartet trees (unioning over all the internal edges) forms the set k-short. In this study, we let $k = 5$ and $k = 25$.

Geo+TSQ: include a quartet q with probability d^{-3} where $d = diam_t(q)$, and add the TSQ trees (this was studied in (21)).

Exp+TSQ: compute the topological distance between every pair of leaves, include a quartet with probability 1.5^{-d} where $d = diam_t(q)$, and add the TSQ trees (this was also studied in (21)).

3 Performance Study

We performed a study using simulated datasets to evaluate QMC-based supertree methods in comparison to MRP and other supertree methods. Simulations are used to evaluate phylogeny estimation methods, because the true tree is known exactly. For our simulations, we used the SMIDGen (25) methodology, and used datasets with 100, 500 and 1000 taxa. We used SMIDGen to produce supertree datasets of *mixed* source trees, consisting of one scaffold dataset (produced by a random selection of taxa from the entire dataset) and many clade-based datasets (focused, dense taxon sampling within a rooted subtree).

Simulation Study Design: For this study, we used simulated datasets generated for another study (25), and, therefore, describe the methodology only in brief. The simulated datasets are produced by simulating evolution under a GTR+Gamma+I process, down pure-birth model trees, deviated from a clock, and containing up to 1000 leaves. We generated 30 replicates for each 100- and 500-taxon model condition, and 10 replicates for each 1000-taxon model condition. Each model condition is indicated by the density of the scaffold dataset, which is the percentage of the entire taxon set in the scaffold dataset, with scaffold densities ranging from 20% to 100%. We used RAxML (23) to estimate phylogenetic trees. We performed the MP search in the MRP analyses, using a very effective heuristic search technique called *the Ratchet* (17), and computed a greedy consensus (gMRP) tree for the set of most parsimonious trees found during this search. We also computed supertrees based upon five ways of encoding the source trees as sets of quartet trees and then applying QMC, as described above. Finally, we computed supertrees using several other methods, including Q-imputation (Q-Imp), Robinson-Foulds Supertrees (RFS), MinFlip, SFIT, and PhySIC, all in their default settings. For RFS, MinFlip, and PhySIC, methods that require rooted trees, we used mid-point rooting to root the source trees, a method commonly used to root unrooted trees and particularly appropriate because our source trees were not strongly deviated from ultrametricity. We computed three types of topological error rates for each estimated supertree when compared with the model tree: false positive rates, false negative rates, and Robinson-Foulds rates. We also computed the total topological distance of each supertree to the estimated source trees, using FN (false negative), FP (false positive) and RF (bipartition distance) errors modified so that we could handle trees on different taxon datasets. We restricted the supertree to the subset of taxa for the source tree, and then compute the topological distances between the two trees. We note that the bipartition distance, also known as the "Robinson-Foulds" (RF) distance, is the standard metric used in most studies. In our study, we show both FN and FP as well, thus providing a more nuanced description of error. Because QMC failed to return trees on some inputs, we restricted our results to datasets for which all the reported methods returned trees. This reduced the number of replicates for some model conditions. We also recorded the running time of each method on each dataset. Because the analyses were run under Condor (a distributed software environment (26)), running times (for the larger datasets, especially) are inexact and are larger than if they had been run on a dedicated processor. Running times are, therefore, an approximation of the time needed to perform these analyses.

4 Results

4.1 Exploring QMC under Different Source Tree Encodings

We compared the performance of the QMC variants and gMRP (Fig. 1). For a given model condition, we include only those methods that successfully completed on at least one third of the replicates, and display results for only those replicates on which every selected method successfully completed. We report performance with respect to FN rates, but the performance with respect to FP and RF rates is almost identical.

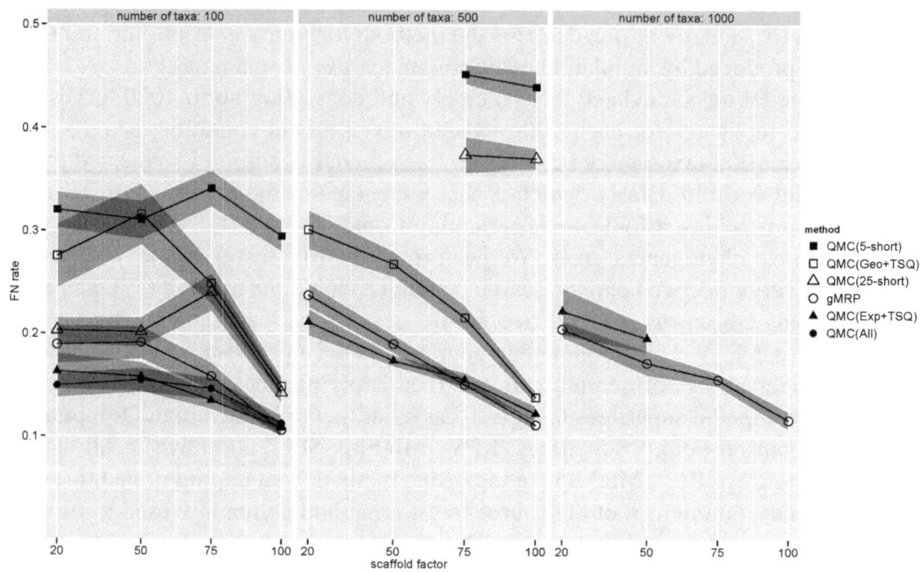

Fig. 1. Average topological error (False Negative (FN) rates) with standard error regions on *mixed* source-tree datasets. We use shaded regions in place of standard error bars as it better demonstrates overlap; however, the shading between data points for a method is not intended as an interpolation of error for scaffold factors not tested. Results are reported for the QMC variants and gMRP, as a function of the scaffold factor and by number of taxa. Points are graphed for a method if it had at least six datasets that completed in common with all other methods.

On the mixed 100-taxon datasets, QMC(All) and QMC(Exp+TSQ) were essentially tied as the best methods, followed by gMRP. Furthermore, QMC(All) and QMC(Exp+TSQ) had the greatest advantage over gMRP for the sparse scaffold cases. The other QMC variants had worse accuracy. On a large number of the 500- and 1000-taxon datasets, many of the QMC variants failed to complete, indicating that computational requirements can limit QMC's utility. On the 500-taxon datasets for which QMC(Exp+TSQ) could be run, it produced topologically more accurate trees than gMRP, giving the biggest advantage on the sparse scaffold datasets. For the 1000-taxon datasets, gMRP outperformed all the QMC variants that completed. However, most QMC variants failed to return trees on most inputs.

4.2 Comparing QMC(Exp+TSQ) to Other Supertree Methods

We compared QMC(Exp+TSQ) to six other supertree methods: gMRP, Q-Imp, SFIT, MinFlip, PhySIC, and Robinson-Foulds Supertrees (RFS).

All of these methods could be run on the 100-taxon datasets, but some failed to run on the larger datasets. For this reason, we obtained results for all seven methods on the 100-taxon datasets, but only five methods on the 500-taxon datasets (where SFIT and Q-Imp failed to run, due to computational limitations), and only four methods on the

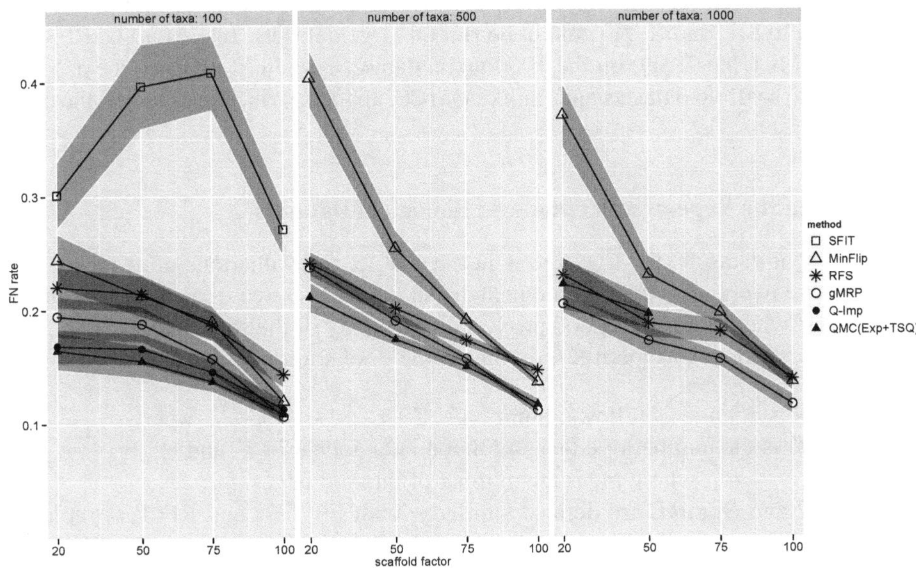

Fig. 2. We report False Negative (FN) rates (means with standard error regions) for QMC (Exp+TSQ), gMRP, SFIT, MinFlip, RFS, and Q-Imp, as a function of the scaffold factor, for 100-, 500- and 1000-taxon model conditions

1000-taxon datasets (where we did not try to run PhySIC, since it was computationally intensive for the 500-taxon datasets). In addition, QMC(Exp+TSQ) failed to run on some datasets; we therefore only report results for those datasets on which all reported methods were able to run. PhySIC gives by far the worst results, producing completely unresolved trees except when the scaffold density is 100%, at which point it produces results that are still worse than the other methods. Because of it is not competitive with other methods, we omit PhySIC from our graphs.

The experiments show that three methods–QMC(Exp+TSQ), Q-Imp, and gMRP–generally outperform the remaining methods with respect to topological accuracy (Fig. 2). As with Fig. 1, in Fig. 2 we only include results for replicates for which all displayed methods were able to complete. Since Fig. 2 includes a different collection of methods, the results for a different collection of replicates are used. On the 100-taxon datasets, QMC(Exp+TSQ) and Q-Imp both gave higher accuracy than gMRP and all other methods (except on the 100% scaffold datasets, where they were equal to gMRP). On the 500-taxon datasets with sparse scaffolds, QMC(Exp+TSQ) performed better than all methods, with only a slight advantage over gMRP. On the 500-taxon datasets with dense (75% and 100%) scaffolds, QMC(Exp+TSQ) and gMRP were the most accurate, and had essentially the same accuracy. On the 1000-taxon datasets, gMRP had an advantage over QMC(Exp+TSQ) and other methods, and QMC(Exp+TSQ) failed to run on the dense scaffold datasets (QMC fails to run on profiles with large source trees, due to computational reasons). The remaining methods–PhySIC, SFIT, MinFlip, and RFS–are generally less accurate than QMC(Exp+TSQ), Q-Imp, and gMRP, and some (i.e., PhySIC and SFIT) cannot be run on large datasets. Interestingly, RFS outperforms QMC(Exp+TSQ) on the 1000 taxon datasets, where it matches the accuracy on the sparse scaffold datasets and (unlike QMC(Exp+TSQ)) is able to run on the dense scaffold datasets.

4.3 Evaluating Supertree Methods on Biological Datasets

For biological datasets, the true tree is not available, so evaluations of accuracy have tended to use average or total topological distance to the source trees (for example, (1; 21)). To test whether this is a good proxy for the quality of the supertree, we computed three distances for each supertree T to the profile $\mathcal{T}$ of source trees:

- **SumFN** is defined as follows: SumFN$(T, \mathcal{T}) = \frac{\sum_{t \in \mathcal{T}} (\mathrm{FN}(T,t))}{M}$, where FN$(T,t)$ is the number of edges in t that do not appear in T, and $M = \sum_{t \in \mathcal{T}} m_t$, where m_t is the number of internal edges in t.
- **SumFP** and **SumRF** are defined similarly, with FP(T,t) and RF(T,t) replacing FN(T,t), respectively. Here, FP denotes the false positive distance and RF denotes the Robinson-Foulds ("bipartition") distance. Each distance is normalized to produce a value between 0 and 1. The false positive distance between a supertree T and a source tree t in the profile $\mathcal{T}$ is the number of edges in T that do not appear in t, and the Robinson-Foulds distance is the total number of missing and false positive edges.

Note that if the supertree and all source trees are binary, then for each source tree t, RF(T,t) = 2FN(T,t) = 2FP(T,t), and after normalization all three distances are equal.

We examined how closely measurements of this sort are correlated to actual topological accuracy, that is, how closely SumFN, SumFP, or SumRF are correlated to the FN, FP or RF distance to the true tree. We found the correlations to be largely independent of the choice of topological distance to source trees (SumFN, SumFP, or SumRF) or topological error (FN, FP or RF). The reason for this was that the true supertree was fully resolved or nearly so, and all the computed supertrees were either fully resolved or nearly so. We therefore present results focusing on the correlation between SumFN (topological distance to the source trees) and FN (topological distance to the true tree).

To assess whether SumFN, SumFP or SumRF is a good optimality criterion, we calculated Spearman rank-correlations for each of the 100-taxon simulated datasets for the six supertree methods that consistently perform reasonably well (MinFlip, gMRP, Q-Imp, QMC(All), QMC(Exp+TSQ), and RFS). Correlations were calculated for each of these measures of distance to source trees and each of FN, FP and RF (calculated by comparing the supertree estimated by each of the methods with the true tree). The statistics were calculated this way to test whether the rank-order of the topological distances to source trees correlated strongly with the true rank-order of the supertrees, in terms of topological accuracy with respect to the true tree.

Table 1. Results of Spearman rank-order correlations of SumFN, SumFP, and SumRF with the true FN, FP, and RF measures of supertrees estimated using six supertree methods

		FN		FP		RF	
scaffold factor	optimality criterion	mean	range	mean	range	mean	range
25	SumFN	0.401	-0.890, 0.939	0.376	-0.890, 0.926	0.391	-0.890, 0.926
	SumFP	0.421	-0.890, 0.939	0.421	-0.890, 0.926	0.426	-0.890, 0.926
	SumFN	0.406	-0.890, 0.939	0.395	-0.890, 0.926	0.406	-0.890, 0.926
50	SumFN	0.544	-0.203, 1.000	0.536	-0.348, 0.971	0.541	-0.203, 0.971
	SumFP	0.546	-0.143, 1.000	0.539	-0.257, 0.971	0.543	-0.143, 0.971
	SumRF	0.546	-0.143, 1.000	0.539	-0.257, 0.971	0.543	-0.143, 0.971
75	SumFN	0.593	-1.000, 0.986	0.589	-1.000, 0.986	0.591	-1.000, 0.986
	SumFP	0.593	-1.000, 0.986	0.589	-1.000, 0.986	0.591	-1.000, 0.986
	SumRF	0.593	-1.000, 0.986	0.589	-1.000, 0.986	0.591	-1.000, 0.986
100	SumFN	0.447	-0.789, 1.000	0.447	-0.789, 1.000	0.447	-0.789, 1.000
	SumFP	0.447	-0.789, 1.000	0.447	-0.789, 1.000	0.447	-0.789, 1.000
	SumRF	0.447	-0.789, 1.000	0.447	-0.789, 1.000	0.447	-0.789, 1.000

The results (Table 1) show clearly that attempting to optimize the total distance to the source trees is of limited use in producing accurate supertrees. None of the optimality criteria averaged better than 60% correlation with measures of true accuracy for a given scaffold factor, and for some datasets, the criteria were negatively correlated with the true quality of the supertrees that were estimated.

Thus, the correlation between topological distance to source trees and topological error (i.e., distance to the true tree) tends to be only weakly positive, so that while, in general, supertrees with smaller topological distance to the source trees are more accurate, there can be more accurate supertrees with higher topological distance to the source trees. These results suggest that the highest accuracy supertrees may not optimize SumFN (or any other topological distance to source trees).

This observation has two consequences for supertree analyses. First, directly trying to optimize the topological distance is not likely to produce the most accurate trees, since better trees are being produced through other means. Secondly, because the true tree is not known for biological supertree datasets, it is difficult to evaluate supertree methods using biological datasets.

These conclusions are clearly based upon the conditions of this experiment, in which the source trees were reasonably, but not extremely, accurate. However, when source trees have no error at all, the true tree is guaranteed to minimize the distance to the source trees. Under this condition, MRP will also be guaranteed to return the true tree as one of the solutions. Thus, for very highly accurate source trees, both MRP and minimizing the total topological distance may be very good optimality criteria; the issue is how well supertree methods perform under more realistic conditions, where source trees have error.

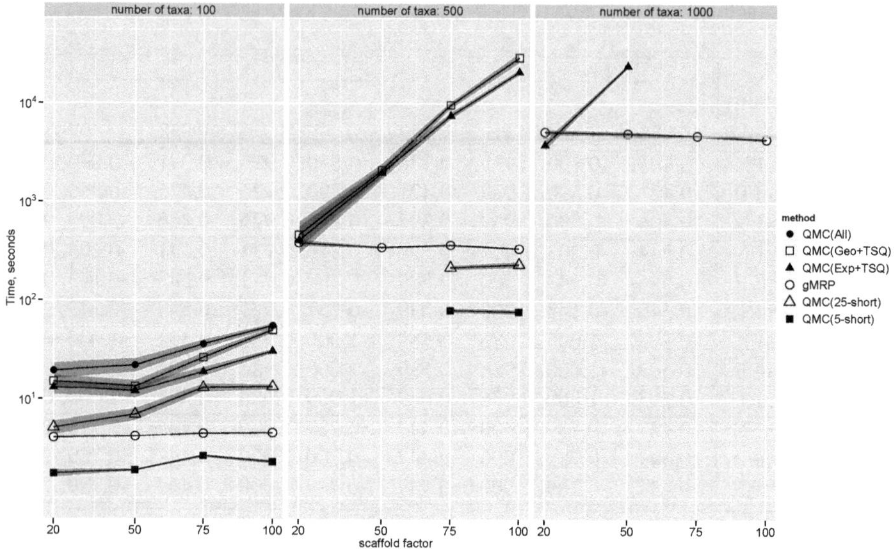

Fig. 3. Running times in seconds (means with standard error regions) of QMC supertree methods on mixed datasets; the y-axis is given with a logarithmic scale

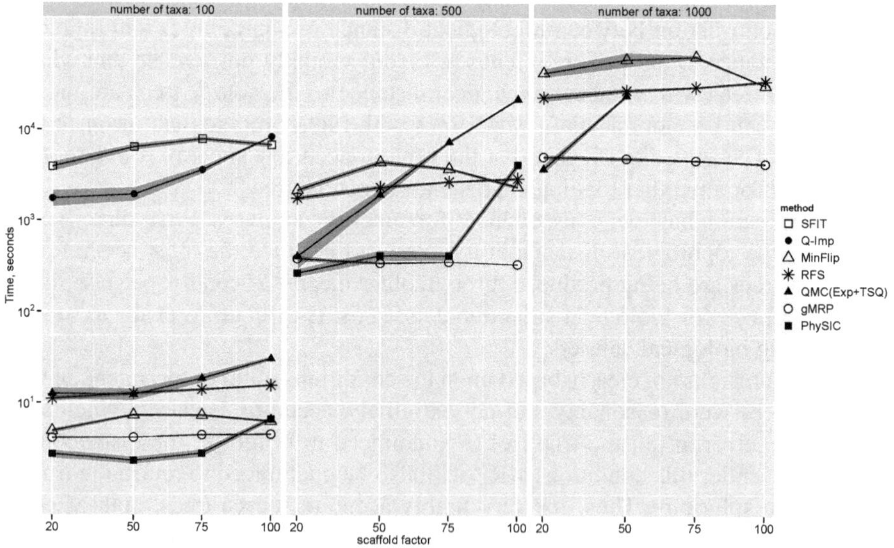

Fig. 4. Running times in seconds (means with standard error regions) of supertree methods on mixed datasets; the y-axis is given with a logarithmic scale

4.4 Scalability

We now discuss running time issues on simulated data. Fig. 3 gives the results for the QMC variants and gMRP, and Fig. 4 gives results for QMC(Exp+TSQ), gMRP, and the other (not QMC-based) supertree methods.

Supertree methods on the simulated datasets showed some differences in running times. First, gMRP was faster than the accurate QMC variants for most of the model conditions, and the degree of improvement ranged from very small (a few seconds) to several hours. In general, we saw that profiles with large source trees were particularly difficult for QMC(Exp+TSQ) and QMC(All), and that for such datasets, gMRP had a running time advantage.

We note that the running times of QMC(Geo+TSQ), QMC(Exp+TSQ), and QMC(All) are directly impacted by the size of the source trees, since each four-tuple of taxa must be examined to produce the quartet trees. Thus, for large source trees, we expect these three methods to suffer computational limitations.

5 Conclusions

This study makes several important contributions. First, we show that while MRP is still the most accurate supertree method for the largest datasets, both QMC(Exp+TSQ) and Q-Imp produce more accurate supertrees than MRP and other supertree methods for the smaller (100- and 500-taxon) datasets. Therefore, an effort should be made to produce scalable and robust implementations of the quartet methods, QMC(Exp+TSQ) and Q-Imp. Each of these methods produces, at some point, a quartet encoding of the source trees. Scalable implementations of these methods will require *not* using all the quartets in these encodings, as such approaches simply will fail on large datasets.

The second important contribution of this study is that the total topological distance to the source trees only provides limited information about topological accuracy, and that reliable comparisons can only be made between supertrees that have very different total topological distances. Consequently, previous studies that have explored performance of supertree methods using total topological distance to the source trees need to be revisited.

Acknowledgments

We thank Sagi Snir for assistance with using the QMC code and for providing the software for generating the quartet encodings Exp+TSQ, Geo+TSQ, and AllQuartets. This work was supported by the US National Science Foundation ITR-0331453, for the CIPRES project.

References

[1] Bansal, M., Burleigh, J.G., Eulenstein, O., Fernández-Baca, D.: Robinson-foulds supertrees (2009)

[2] Baum, B.R.: Combining trees as a way of combining data sets for phylogenetic inference, and the desirability of combining gene trees. Taxon 41, 3–10 (1992)

[3] Ben-dor, A., Chor, B., Graur, D., Ophir, R., Pelleg, D.: Constructing phylogenies from quartets: Elucidation of eutherian superordinal relationships. Journal of Computational Biology 5(3), 377–390 (1998), Earlier version appeared in RECOMB 1998 (1998)
[4] Bininda-Emonds, O.R.P.: The evolution of supertrees. Trends in Ecology and Evolution 19, 315–322 (2004)
[5] Bininda-Emonds, O.R.P.: Phylogenetic Supertrees: Combining Information To Reveal The Tree Of Life. Computational Biology. Kluwer Academic, Dordrecht (2004)
[6] Bolaender, H., Fellows, M., Warnow, T.: Two strikes against perfect phylogeny. In: Kuich, W. (ed.) ICALP 1992. LNCS, vol. 623, pp. 273–283. Springer, Heidelberg (1992)
[7] Burleigh, J.G., Eulenstein, O., Fernández-Baca, D., Sanderson, M.J.: MRF supertrees. In: Bininda-Emonds, O.R.P. (ed.) Phylogenetic Supertrees: Combining Information To Reveal The Tree Of Life, pp. 65–86. Kluwer Academic, Dordrecht (2004)
[8] Chen, D., Diao, L., Eulenstein, O., Fernández-Baca, D., Sanderson, M.J.: Flipping: a supertree construction method. In: Bioconsensus. DIMACS: Series in Discrete Mathematics and Theoretical Computer Science, vol. 61, pp. 135–160. American Mathematical Society-DIMACS, Providence (2003)
[9] Chen, D., Eulenstein, O., Fernández-Baca, D., Burleigh, J.G.: Improved heuristics for minimum-flip supertree construction. Evol. Bioinform. 2, 401–410 (2006)
[10] Creevey, C.J., McInerney, J.O.: Clann: investigating phylogenetic information through supertree analyses. Bioinformatics 21(3), 390–392 (2005)
[11] Dress, A., Steel, M.: Convex tree realizations of partitions. Applied Mathematics Letters 5(3), 3–6 (1992)
[12] Foulds, L.R., Graham, R.L.: The steiner problem in phylogeny is NP-complete. Adv. in Appl. Math. 3(43-49), 299 (1982)
[13] Holland, B., Conner, G., Huber, K., Moulton, V.: Imputing supertrees and supernetworks from quartets. Syst. Biol. 57(1), 299–308 (2007)
[14] Jiang, T., Kearney, P., Li, M.: Orchestrating quartets: approximation and data correction. In: Motwani, R. (ed.) Proceedings of the 39th IEEE Annual Symposium on Foundations of Computer Science, Los Alamitos, CA., pp. 416–425 (1998)
[15] Jiang, T., Kearney, P., Li, M.: A polynomial-time approximation scheme for inferring evolutionary trees from quartet topologies and its applications. SIAM J. Comput. 30(6), 1924–1961 (2001)
[16] John, K.S., Warnow, T., Moret, B.M.E., Vawter, L.: Performance study of phylogenetic methods: (unweighted) quartet methods and neighbor-joining. Journal of Algorithms 48, 173–193 (2003)
[17] Nixon, K.C.: The parsimony ratchet, a new method for rapid parsimony analysis. Cladistics 15, 407–414 (1999)
[18] Ragan, M.A.: Phylogenetic inference based on matrix representation of trees. Mol. Phylo. Evol. 1, 53–58 (1992)
[19] Ranwez, V., Berry, V., Criscuolo, A., Fabre, P.H., Guillemot, S., Scornavacca, C., Douzery, E.J.P.: PhySIC: a veto supertree method with desirable properties. Syst. Biol. 56(5), 798–817 (2007)
[20] Ranwez, V., Gascuel, O.: Quartet-Based phylogenetic inference: Improvements and limits. Mol. Biol. Evol. 18(6), 1103–1116 (2001)
[21] Snir, S., Rao, S.: Quartets MaxCut: a divide and conquer quartets algorithm. IEEE/ACM Trans. Comput. Biol. Bioinform. (2008)
[22] Snir, S., Warnow, T., Rao, S.: Short quartet puzzling: A new Quartet-Based phylogeny reconstruction algorithm. J. Comput. Biol. 15(1), 91–103 (2008)
[23] Stamatakis, A.: RAxML-NI-HPC: Maximum likelihood-based phylogenetic analyses with thousands of taxa and mixed models. Bioinformatics 22, 2688–2690 (2006)

[24] Strimmer, K., von Haeseler, A.: Quartet puzzling: A quartet maximim-likelihood method for reconstructing tree topologies. Molecular Biology and Evolution 13(7), 964–969 (1996)
[25] Swenson, M.S., Barbançon, F., Warnow, T., Linder, C.R.: A simulation study comparing supertree and combined analysis methods using SMIDGen. Algorithms for Molecular Biology 5, 8 (2010)
[26] Thain, D., Tannenbaum, T., Livny, M.: Distributed computing in practice: the Condor experience. Concurrency and Computation: Practice and Experience 17, 323–356 (2005)

An Efficient Method for DNA-Based Species Assignment via Gene Tree and Species Tree Reconciliation

Louxin Zhang* and Yun Cui

Department of Mathematics, National University of Singapore
10 Lower Kent Ridge Road, Singapore 119076
{matzlx,matcy}@nus.edu.sg

Abstract. DNA-based species assignment and delimitation are two important problems in systematic biology. In a recent work of O'Meara, species delimitation is investigated through coupling it with species tree inference in the framework of gene tree and species tree reconciliation. We present a polynomial time algorithm for splitting individuals into species to minimize the deep coalescence cost of the gene tree and species tree reconciliation, a species assignment problem arises from species delimitation via gene tree and species tree reconciliation. How to incorporate this proposed algorithm into the heuristic search strategy of O'Meara for species delimitation is also discussed. The proposed algorithm is implemented in C++.

Keywords: DNA barcoding, species delimitation, gene tree and species tree reconciliation, deep coalescence, dene duplication and loss.

1 Introduction

DNA sequencing of living organisms holds great promise in species identification and delimitation, two important but difficult tasks in taxonomy. Based on the premise that genetic difference between species exceeds that within species, DNA information is currently being used for species assignment ("DNA barcoding", [12,13]). To identify which species an individual belongs in, one retrieves a short DNA sequence - the barcode - from some gene region from the individual and compares it with reference barcoding sequences for known species under the current Linnean system. The species membership of the individual is determined based on the degree of sequence similarity using pairwise similarity [29] or a statistical method [12,21,22].

DNA information also provides a way for discovering new species (equivalent to species delimitation) from poorly studied individuals [5,27,16,30]. However, establishing the association of DNA sequence information with a species (which is a group of individuals) is extremely challenging due to intraspecies DNA

---

* To whom correspondence may be addressed. The work was financially supported by AcRF R146-000-109-112.

V. Moulton and M. Singh (Eds.): WABI 2010, LNBI 6293, pp. 300–311, 2010.

variation [20,31], the possible discord of gene and species histories [8,9,10,25], and inconsistency of species concepts [1,3,6]. For DNA species delimitation, one needs to adopt a species concept and choose a criterion to apply this concept to DNA sequence information.

A set of methods have been developed for deciding whether new individuals belong in existing species [12,21,22,29]. These methods typically use DNA information at a single locus or multiple loci where the assignment of individuals to species is only partially known, but the species history is of little interest. In this paper, inspired by a work [23] of O'Meara, we suggest a new combinatorial method for DNA barcoding via gene tree and species tree reconciliation. One main cause for the gene tree and species tree discord is lack of coalescence of intraspecific sequences between speciation event. Hence, the total number of "extra" gene lineages that fails to coalesce on a species tree is proposed to measure gene tree and species tree difference [19]. If gene tree is considered as a neutral coalescent tree, the most probable gene tree matches the species tree except for extreme cases involving short internal branches [7,14]. Hence, we propose to assign species to individuals by minimizing deep coalescence events in the reconciliation of the gene tree, estimated from sampled sequences, and the species tree. Such a method takes advantage of the species history and is also efficient.

Gene trees are often used to discover new species from poorly studied organisms [3,14,23,26,27]. However, unlike other works, O'Meara proposed to infer species boundaries by coupling species delimitation with species tree inference [23]. In the approach, gene trees are estimated from sampled sequences and then the considered individuals are split into putative species by minimizing the structure cost of the gene and species tree reconciliation. As joint species delimitation and species tree inference is more general than species tree inference, the complexity study in [18] and [33] suggests that this problem is unlikely solvable in polynomial time. In this work, we shall also discuss how to improve the heuristic strategy presented in [23] using our species assignment algorithm.

2 Species Assignment Problems

Assume that there are a set of species whose phylogenetic tree is known and a set of individuals whose taxonomic classification is unknown. To identify which species each individual belongs in, one obtains a gene sequence from these individuals as well as from some other individuals that belong to the species under consideration and builds a gene tree over these sampled sequences. The gene tree is partially leaf labeled in the sense that a leaf representing a sequence sampled from an individual whose species is known is labeled by the species whereas a leaf representing a sequence from an individual whose species needs to be identified is unlabeled. Applying the parsimony principle, we split the individuals into species to minimize the cost of the reconciliation of the resulting fully leaf labeled gene tree and the species tree. Formally, the species identification problem is modeled as the following algorithmic problem.

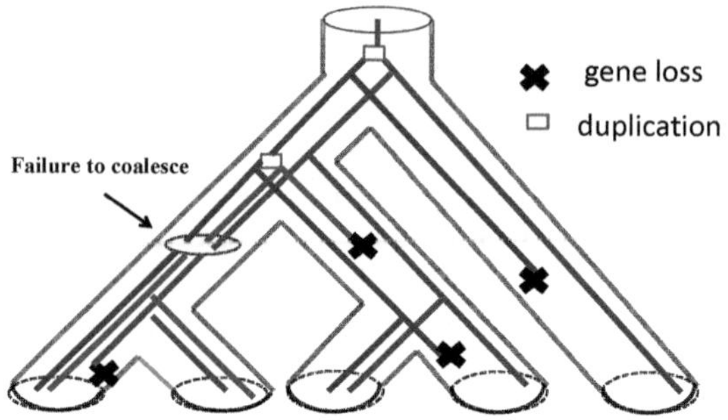

Fig. 1. Illustration of gene tree and species tree reconciliation

Species Assignment Problem
Input: A species tree S for a set of species and a partially leaf-labeled gene tree G, and a reconciliation cost function $c(,)$.
Solution: A labeling $\mathcal{L}$ of unknown leaves of G that minimizes the reconciliation cost $c(G_{\mathcal{L}}, S)$.

The discord of gene tree and the containing species tree is caused by lineage sorting, gene duplication and loss, or horizontal gene transfer shown in Figure 1. The importance of these causes depends on the considered genes and species. The gene duplication/loss [10,24] and deep coalescence costs [19] (to be defined later) are proposed to study the gene tree and species tree relationship. In species identification content, the deep coalescence cost is probably more suitable as multiple genes' persistence without coalescence before the divergence of the containing species is a key mutational process to be considered. However, the duplication-plus-loss cost is closely related to the deep coalescence cost [33]. Therefore, we shall study the species assignment problem for each of them.

When multilocus sequence data are used for species assignment, the **multilocus species assignment** problem arises, in which there are a set of partially leaf-labeled gene trees and a species tree S and the goal is to find a labeling that minimizes the total reconciliation cost.

In the study of environmental samples of fungi or other group of unknown individuals for which there are no taxonomic hypotheses [27], the species tree assumed in the Species Assignment problem simply does not exist and so species delimitation is formulated as the joint species delimitation and species tree inference problem in [23], which we call:

Species Assignment With Unknown Species
Input: A set of partially leaf-labeled gene tree $G^{(1)}, G^{(2)}, \cdots, G^{(t)}$, and a reconciliation cost function $c(,)$.
Solution: A species tree S and a labeling $\mathcal{L}$ minimizing $\sum_i c(G^{(i)}_{\mathcal{L}}, S)$.

3 Basic Concepts and Notation

Both gene tree and species tree are rooted fully binary trees. In a species tree, the leaves represent extant species and are labeled with the corresponding species.

In a gene tree, each leaf may or may not be labeled; if a leaf is labeled, its label represents the species of the individual from which the corresponding gene sequence is sampled; if a leaf is unlabeled, the corresponding gene sequence is sampled from an individual whose species is to be identified. Let T be a gene or species tree,

- Leaf(T) denotes the set of leaves of T;
- $L(T)$ denotes the set of leaf labels. If no leaf of T is labeled, we simply write $L(T) = \phi$;
- $|T|$ denotes the number of the nodes of T;
- $\| T \|$ denotes the number of the leaves of T;
- the notation $t \in T$ denotes that t is a node of T;
- the notation $A \subseteq L(T)$ denotes that A is a subset of the label set $L(T)$; and
- finally, t_a and t_b denote the two children of an internal node $t \in T$.

For a node $t \in T$, any node in the unique path from the root of T to t is called an ancestor of t; and any node below t is called a descendant of t. For any two nodes $t', t'' \in T$, their *least common ancestor*, denoted by lca(t', t''), is the ancestor of t' and t'' whose children are not an ancestor of either t' or t''.

For a node $x \in T$, we use T_x to denote the subtree consisting of x and its descendants.

For a subset $A \subseteq$ Leaf(T), the restriction of T on A is the smallest subtree of T containing A as its leaf set, denoted by $T|_A$. In general, $T|_A$ may not be a full binary tree as it may contain non-root degree-2 nodes and its root is the least common ancestor of the nodes of A. Let $L \subseteq L(T)$ be a leaf label subset. For simplicity, we use $T|_L$ to denote the restriction of T on the leaves having a label in L.

4 Measuring Gene Tree and Species Tree Reconciliation

Let G be a fully leaf labeled gene tree and S a species tree such that $L(G) \subseteq L(S)$. To reconcile G and S, each node $g \in G$ is mapped to a unique node $M(g) \in S$ as

$$M(g) = \begin{cases} \text{the corresponding leaf with the same label}, & \text{if } g \in \text{Leaf}(G), \\ \text{lca}\left(M(g_a), M(g_b)\right), & \text{if } g \notin \text{Leaf}(G). \end{cases}$$

The node $M(g)$ with which g is associated is called the image of g (with respect to M). The mapping M was first considered in [10] and then formulated in [24]. We call M the *reconciliation* (*map*) of G within S.

4.1 Gene Duplication and Loss

Let G be a gene tree and $g \in G$. For any descendant g' of g, $M(g')$ is equal to either $M(g)$ or a descendant of $M(g)$ in the reconciliation M of G within a species tree S such that $L(G) \subseteq L(S)$. If $M(g_a) = M(g)$ or $M(g_b) = M(g)$, then we say that a duplication occurs at $M(g)$ (or more exactly in the lineage entering $M(g)$) in S and also say that it is associated with g. The total number of duplications is defined as the *duplication cost* of the reconciliation of G within S, which is denoted by $c_{dup}(G, S)$. Note that the duplication cost is an asymmetric measure for tree comparison.

For any two nodes s' and s'' such that s'' is an ancestor of s', we use $P(s', s'')$ to denote the unique path from s' to s'' and define

$$d(s', s'') = (\text{No. of nodes } s \ (\neq s', s'') \text{ on the path } P(s', s'')) . \qquad (1)$$

Then, the number of losses l_g associated to g is defined as

$$l_g = \begin{cases} 0, & \text{if } M(g) = M(g_a) = M(g_b), \\ d\left(M(g_a), M(g)\right) + 1, & \text{if } M(g_a) \subset M(g) = M(g_b), \\ \sum_{h=g_a,g_b} d\left(M(h), M(g)\right), & \text{if } M(g_a), M(g_b) \subset M(g). \end{cases}$$

The *gene loss cost* of the reconciliation of G within S, denoted by $c_{loss}(G, S)$, is defined as the total number of losses $\sum_{g \in G} l_g$ (see [11,24]).

The *gene duplication plus loss* cost is equal to the sum of the gene duplication and gene loss costs, is denoted by $c_{dl}(G, S)$. The gene duplication plus loss cost is also called the mutation cost in [11].

4.2 Deep Coalescence

Let G be a gene tree and S a species tree such that $L(G) \subseteq L(S)$. In the reconciliation M of G within S, if a branch e of S is in the $k(\geq 2)$ paths from $M(g_i)$ to $M(g'_i)$, where $g_i \in G$ $(1 \leq i \leq k)$ and g'_i is a child of g_i, then we say that there are $k-1$ 'extra' lineages on e failing to coalesce on e. The *deep coalescence cost* is defined as the total number of the 'extra' lineages on all the branches of S in the reconciliation M of G within S (see [19]), which is denoted as $c_{dc}(G, S)$.

The deep coalescence cost is closely related to the gene duplication and loss costs. It is not hard to see that, in the reconciliation of G within S such that $L(G) \subset L(S)$, G is mapped onto $S|_{L(G)}$. We have the following equation.

Theorem 1. ([33]) *Let G be a fully leaf-labeled gene tree and S a species tree such that $L(G) \subseteq L(S)$. Then, for the reconciliation of G within S,*

$$c_{dc}(G, S) = c_{loss}(G, S) - 2c_{dup}(G, S) + \left(|G| - |\, S|_{L(G)}\,|\right), \qquad (2)$$

where $|T|$ is the number of the nodes of T for $T = G,\ S|_{L(G)}$.

5 Solving the Species Assignment Problem in Poly. Time

In this section, we shall present a polynomial-time algorithm for the species assignment problem for both the deep coalescence and duplication plus loss costs.

Here, we use the latter only as a reconciliation cost function as inferred duplication or loss events have no biological meaning when gene tree and species tree are reconciled for the purpose of species assignment and delimitation. Additionally, the algorithm for the duplication plus loss cost could find applications in other fields where gene tree and species tree are compared.

5.1 Algorithm for the Duplication Plus Loss Cost

Let S be a species tree and G a partially leaf-labeled gene tree such that $L(G) \subseteq L(S)$. We use $\mathrm{UL}(G)$ to denote the set of unlabeled leaves of G. Hence, $\mathrm{Leaf}(G) - \mathrm{UL}(G)$ is the set of labeled leaves in G, denoted as $LF(G)$.

Let $x \in G$ and $y \in S$. Obviously, each labeling $\mathcal{L}$: $\mathrm{UL}(G_x) \rightarrow L(S)$ induces a reconciliation $M_{\mathcal{L}}$ of G_x within S_y if and only if x is mapped to y under $M_{\mathcal{L}}$. For such a labeling $\mathcal{L}$: $\mathrm{UL}(G_x) \rightarrow L(S)$, we use $c_{dl,\mathcal{L}}(G_x, S_y)$ to denote the duplication-plus-loss cost of the resulting reconciliation of G_x within S_y, and further define

$$C_{dl}(x, y) = \min_{\mathcal{L}: M_{\mathcal{L}}(x)=y} c_{dl,\mathcal{L}}(G_x, S_y). \tag{3}$$

If it is impossible to reconcile G_x within S_y with the condition that x is mapped to y, we simply write $C_{dl}(x, y) = \infty$.

If each labeled leaf in G_x has the same label as y, then labeling all unlabeled leaves in G_x by the label $l(y)$ results in a reconciliation with the gene duplication cost $\parallel G_x \parallel -1$ but no gene loss. Hence, for any leaf $y \in S$,

$$C_{dl}(x, y) = \begin{cases} \parallel G_x \parallel -1, & \text{if any } f \in LF(G_x) \text{ has the label } l(y), \\ \infty, & \text{otherwise} \end{cases} \tag{4}$$

where $LF(G_x)$ is the set of the labeled leaves of G_x. In particular, if x is an unlabeled leaf of G and y a leaf of S, we have that $C_{dl}(x, y) = 0$.

Let y be an arbitrary internal node of S and $x \in G$. Recall that t_a and t_b denote the children of t for $t = x, y$. In a reconciliation $M_{\mathcal{L}}$ of G_x within S_y such that x is mapped to y, one of the following cases holds:

C1. Both x_a and x_b are mapped to y;
C2. The node x_a is mapped to y, but x_b is mapped to a descendant of y;
C3. The node x_b is mapped to y, but x_a is mapped to a descendant of y;
C4. The node x_a is mapped to a descendant of y in S_{y_a}, and x_b is mapped to a descendant of y in S_{y_b};
C5. The node x_a is mapped to a descendant of y in S_{y_b}, and x_b is mapped to a descendant of y in S_{y_a}.

Assume that $M_{\mathcal{L}}$ has the minimum duplication plus loss cost over all the reconciliations of G_x within S_y. If the case C1 holds, one duplication and 0 loss are associated with x in $M_{\mathcal{L}}$. Define $d_g = 1$ if a duplication is associated with $g \in G$ and 0 otherwise. Then,

$$\begin{aligned} C_{dl}(x,y) &= c_{loss,\mathcal{L}}(G_x, S_y) + c_{dup,\mathcal{L}}(G_x, S_y) \\ &= \sum_{g \in G_{x_a}} (l_g + d_g) + \sum_{g \in G_{x_b}} (l_g + d_g) + (l_x + d_x) \\ &= C_{dl}(x_a, y) + C_{dl}(x_b, y) + 1. \end{aligned}$$

If the case C2 holds, we may assume that x_b is mapped to z, a descendant of y. Then, a duplication is associated with x and the gene loss l_x associated with x is $1 + d(z, y)$, where $d(,)$ is defined by Eqn. (1) in Section 4.1, and so

$$\begin{aligned} C_{dl}(x,y) &= \sum_{g \in G_{x_a}} (l_g + d_g) + \sum_{g \in G_{x_b}} (l_g + d_g) + (l_x + d_x) \\ &= C_{dl}(x_a, y) + C_{dl}(x_b, z) + l(z, y) + 1, \end{aligned}$$

in this case.

If the C3 holds, we similarly have

$$C_{dl}(x,y) = C_{dl}(x_a, z) + C_{dl}(x_b, y) + d(z, y) + 2,$$

where z is the image of x_a.

If the case C4 or C5 hold, no duplication occurs at y and we have

$$C_{dl}(x,y) = C_{dl}(x_a, z_1) + C_{dl}(x_b, z_2) + d(z_1, y) + d(z_2, y),$$

where z_1 and z_2 are the images of x_a and x_b, respectively.

In summary, we obtain that

$$C_{dl}(x,y) = \min \begin{cases} C_{dl}(x_a, y) + C_{dl}(x_b, y) + 1, \\ C_{dl}(x_a, y) + \min_{y \neq z \in S_y} (C_{dl}(x_b, z) + d(z, y)) + 2, \\ \min_{y \neq z \in S_y} (C_{dl}(x_a, z) + d(z, y) + 2) + C_{dl}(x_b, y), \\ \min_{z_1 \in S_{y_a}} f(x_a, z_1) + \min_{z_2 \in S_{y_b}} f(x_b, z_2), \\ \min_{z_1 \in S_{y_b}} f(x_a, z_1) + \min_{z_2 \in S_{y_a}} f(x_b, z_2), \end{cases} \quad (5)$$

where $f(x', z) = C_{dl}(x', z) + d(z, y)$ for an internal node z below y and a child x' of x.

Eqn. (4)-(5) lead to a dynamic programming algorithm of time complexity $O(\| G \| \cdot \| S \|^2)$.

It is not hard to see that the above idea can be used to obtain an efficient algorithm for the species assignment problem with the duplication cost. Actually, a quadratic-time algorithm for this problem is known for the duplication cost [2] due to the fact that duplication events can be counted effectively.

5.2 Algorithm for the Deep Coalescence Cost

The dynamic programming algorithm presented in Section 5.1 is based on the following two facts:

(a) For a node $g \in G$, labeling the leaves below a child of g is independent of labeling the leaves below the other child of g.
(b) For the duplication-plus-loss cost, finding the best labeling for the leaves below g is equivalent to finding the best labeling for the leaves below each child of g given that the images of the children are fixed.

Unfortunately, the fact (b) does not hold for the deep coalescence cost since different labelings for the leaves below the two children of g may induce different number of extra lineages on each branch in the subtree below the image of g. However, Eqn. (2) can be used to develop a dynamic programming algorithm which takes polynomial time for the gene tree G and species tree S such that $L(G) = L(S)$, that is, if a reference sequence is present in the gene tree for each species under consideration.

Let S be a species tree and G a partially leaf-labeled gene tree such that $L(G) = L(S)$. Assume that $x \in G$ and $y \in S$. Similar to the duplication plus loss cost case, we define

$$C_{dc}(x, y) = \min_{\mathcal{L}:M_{\mathcal{L}}(x)=y} c_{dc,\mathcal{L}}(G_x, S_y). \tag{6}$$

Again, we write $C_{dc}(x, y) = \infty$ if it is impossible to reconcile G_x within S_y with the condition that x is mapped to y.

Let r_G and r_S be the root of G and S respectively. By assumption that $L(G) = L(S)$, we have that

$$S|_{L(G)} = S, \;\; |G| - | S|_{L(G)} | = 2(\| G \| - \| S \|).$$

Thus, Eqn. (2) becomes

$$C_{dc}(r_G, r_S) = \min_{\mathcal{L}:M_{\mathcal{L}}(r_G)=r_S} \left[c_{loss}(G, S) - 2c_{dup}(G, S) + 2(\| G \| - \| S \|)\right]. \tag{7}$$

Hence, we only need to compute $\min_{\mathcal{L}:M_{\mathcal{L}}(r_G)=r_S} c_{loss}(G, S) - 2c_{dup}(G, S)$ since $\| G \| - \| S \|$ is fixed and independent of labelings.

For each $x \in G$ and $y \in S$, we define

$$F(x, y) = \min_{\mathcal{L}:M_{\mathcal{L}}(x)=y} \left[c_{loss}(G_x, S_y) - 2c_{dup}(G_x, S_y)\right]. \tag{8}$$

Recall that we use $l(y)$ to denote the label of $y \in \text{Leaf}(S)$. If y is a leaf of S and each labeled leaf in G_x has the same label as y, then labeling all unlabeled leaves in G_x by $l(y)$ results in a reconciliation having 0 gene loss and $\| G_x \| -1$ gene duplications. Hence, for any $x \in G$ and $y \in \text{Leaf}(S)$

$$F(x, y) = \begin{cases} 2(1- \| G_x \|), & \text{if any } f \in LF(G_x) \text{ has the label } l(y), \\ \infty, & \text{otherwise.} \end{cases} \tag{9}$$

Furthermore, we can obtain the following recurrence formula for computing $F(x, y)$ in the same way as for the duplication plus loss cost:

$$F(x,y) = \min \begin{cases} F(x_a, y) + F(x_b, y) - 2, \\ F(x_a, y) + \min_{y \neq z \in S_y} \left(F(x_b, z) + d(z, y) - 1\right), \\ \min_{y \neq z \in S_y} \left(F(x_a, z) + d(z, y) - 1\right) + F(x_b, y), \\ \min_{z_1 \in S_{y_a}} V(x_a, z_1) + \min_{z_2 \in S_{y_b}} V(x_b, z_2), \\ \min_{z_1 \in S_{y_b}} V(x_a, z_1) + \min_{z_2 \in S_{y_a}} V(x_b, z_2), \end{cases} \tag{10}$$

where $V(x', z) = F(x', z) + d(z, y)$ for an internal node z below y and a child x' of x.

Eqn. (7)-(10) give us a desired dynamic programming algorithm that takes $O(\| G \| \cdot \| S \|^2)$ basic operations for a gene tree G and a species trees S such that $L(G) = L(S)$.

Finally, we remark that it is not clear whether there is a polynomial time algorithm for the species assignment problem for an arbitrary partially leaf labeled gene tree G and a species tree S such that $L(G) \subseteq L(G)$ or not.

5.3 Implementation Issues of the Dynamic Algorithms

We employ a two-dimensional matrix to implement each of the dynamic programming algorithms presented in this section. The detail description of this part can be found in the journal version of this paper.

5.4 Generalization to the Species Phylogenetic Networks

Genomes evolved vertically and horizontally [4,32]. Acyclic directed network is sought as a model of the evolution of genomes or species, called a *phylogenetic network*, to capture not only speciation events but also recombination and horizontal gene transfer events.

In a phylogenetic network, non-leaf tree nodes correspond to speciation events whereas reticulation nodes correspond to recombination or horizontal gene transfer events. In this study, we will not distinguish these two events. Note that a tree is a network not containing reticulation nodes. By analogy to trees, we say that a node u is an ancestor of a node v or equivalently v is a descendant of u if u is on a path from the root to v in the network; we also say that u is a common ancestor of a set of nodes if it is an ancestor of each node in the set.

In a tree, there is the least common ancestor for any set of nodes. However, there may be two or more common ancestors that do not contain one another in a network. We say a common ancestor u of a set of nodes is minimal if any descendant of u is not a common ancestor of the nodes.

Given a gene tree G and a phylogenetic network $\mathcal{N}$ such that $L(G) \subseteq L(\mathcal{N})$, where $L(\mathcal{N})$ denotes the set of leave labels in $\mathcal{N}$. We can reconcile G within $\mathcal{N}$ by mapping a leaf of G to the corresponding leave of N and an internal node v to a minimal common ancestor of the images of the leaves in G_v. It is possible that more than one reconciliation exist for G and $\mathcal{N}$. For a reconciliation M of G within $\mathcal{N}$, its duplication, duplication-plus-loss and deep coalescence costs can be defined in the same way as for gene tree and species tree except for not counting reticulation nodes when the gene loss associated with an internal node is computed [33], which are denoted by $c_{dup}(M)$, $c_{dl}(M)$ and $c_{dc}(M)$ respectively. Let $R(G,\mathcal{N})$ denote all of the possible reconciliations of G within $\mathcal{N}$. For each cost function $c(,)$, we define the cost of the reconciliation of G within $\mathcal{N}$ as

$$c(G,\mathcal{N}) = \min_{M \in R(G,\mathcal{N})} c(M). \tag{11}$$

Since recurrence equations similar to (4)–(5) can be established for the duplication and duplication-plus-loss costs in the case of gene tree vs species phylogenetic network reconciliation, the dynamic programming technique leads to a polynomial time algorithm for

(a) Computing $c_{dup}(G,\mathcal{N})$ or $c_{dl}(G,\mathcal{N})$ for a gene tree G and a species network N, and
(b) Solving the species assignment problem for a partially leaf-labeled gene tree and a species network with the duplication or duplication-plus-loss cost.

These algorithms are described in detail in the journal version of this work. However, it is not clear whether the same result hold for the deep coalescence cost or not.

6 Species Assignment with Unknown Species Is Hard

The species tree inference problem is, given a set of fully labeled gene trees G_i $(1 \leq i \leq n)$, to find a species tree S minimizing the total cost $\sum_{1 \leq i \leq n} c(G_i, S)$ for a reconciliation cost function $c(,)$. In [18] and [33], the species tree inference problem is proved to be NP-hard for each of the duplication, duplication-plus-loss, and deep coalescence costs. As a matter of fact, the species tree inference problem is a special case of the species assignment with unknown species problem in which an instance contains just fully leaf-labeled gene trees. By proof by restriction, the problem of species delimitation with unknown species is NP-hard for each of the reconciliation costs mentioned above.

7 Conclusion

We have presented an efficient systematic approach for species identification. In this approach, identifying species membership for a set of individuals is modeled as the species assignment problem in the framework of gene tree and species tree reconciliation, in the spirit of the work [23]. The dynamic programming technique has been applied to different problems in tree comparison [2,17,28]. Using this powerful technique, we develop a cubic-time algorithm for solving the species assignment problem for the duplication-plus-loss and deep coalescence costs. These algorithms have been implemented in C++ and the program package is available upon request. Hence, our proposed method takes advantage of species history and is efficient. As our on-going work, we shall evaluate the performance of this bioinformatic approach on some benchmark DNA barcoding data.

We remark that the species assignment with unknown species is NP-hard. This means that there is unlikely a polynomial time algorithm for it. We have discussed how to incorporate our solution to the species assignment problem into the heuristic search method proposed in [23] for the problem. Without doubt, this problem deserves more investigation.

The species assignment problem also arises from tree comparison in ligand and receptor pairing [2]. Exploring its application in it and other research fields is definitely worthy further study.

Finally, several algorithmic problems arise from this work. First, we have presented a polynomial time algorithm for the species assignment problem when the deep coalescence cost is used if the input gene tree contains at least one reference sequence for each considered species. This is probably good enough for the species identification purpose. However, it is of theoretical interest whether the species assignment problem is polynomial time solvable or not for arbitrary gene trees and species trees for the deep coalescence cost. Secondly, we have generalized the polynomial time algorithm for the species assignment problem from species tree to species network. The study of various cost functions for gene network and species network reconciliation and algorithms for the species assignment problem in the network case are open for investigation.

References

1. Agapow, P.M.: The impact of species concept on biodiversity studies. Q. Rev. Biol. 79, 161–179 (2004)
2. Bafna, V., Hannenhalli, S., Rice, K., Vawter, L.: Ligand-Receptor pairing via tree comparison. J. Comput. Biol. 7, 59–70 (2000)
3. Baum, D.A., Shaw, K.L.: Genealogical perspectives on the species problem. In: Hoch, P.C., Stephenson, A.C. (eds.) Experimental and molecular approaches to plant biosystematics, pp. 289–303. Missouri Botanical Garden, Saint Louis (1995)
4. Brockelman, W.Y., Gittins, S.P.: Natural hybridization in the Hylobates lar species group: implications for speciation in gibbons. In: Preushcoft, H., Chivers, D.J., Brockelman, W.Y., et al. (eds.) The Lesser Apes. Evolutionary and Behavioral Biology. Edinburgh University Press, Edinburgh (1984)
5. Burns, J.M., Janzen, D.H., Hajibabaei, M., Hallwachs, W., Hebert, P.D.: DNA barcodes and cryptic species of skipper butterflies in the genus Perichares in Area de Conservacion Guanacaste, Costa Rica. Proc. Nat'l. Acad. Sci. USA 105, 6350–6355 (2008)
6. De Queiroz, K.: Species concepts and species delimitation. Syst. Biol. 56, 879–886 (2007)
7. Degnan, J.H., Rosenberg, N.A.: Discordance of species trees with their most likely gene trees. PLoS Genet. 2, e68 (2006)
8. Doyle, J.J.: Gene trees and species trees: molecular systematics as one-character taxonomy. Syst. Bot. 17, 144–163 (1992)
9. Fitch, W.: Distinguishing homologous from analogous proteins. Syst. Zool. 19, 99–113 (1970)

10. Goodman, M., et al.: Fitting the gene lineage into its species lineage, a parsimony strategy illustrated by cladograms constructed from globin sequences. Syst. Zool. 28, 132–163 (1979)
11. Guigó, R., Muchnik, I., Smith, T.: Reconstruction of ancient molecular phylogeny. Mol. Phylogenet Evol. 6, 189–213 (1996)
12. Hebert, P.D.N., Cywinska, A., Ball, S.L., DeWaard, J.R.: Biological identification through DNA barcodes. Proc. R. Soc. B 270, 313–321 (2003)
13. Hebert, P.D.N., Gregory, T.R.: The promise of DNA barcoding for taxonomy. Syst. Biol. 54, 852–859 (2005)
14. Knowles, L.L., Carstens, B.C.: Delimiting species without monophyletic gene trees. Syst. Biol. 56, 887–895 (2007)
15. Leaché, A.D., et al.: Quantifying ecological, morphological, and genetic variation to delimit speceis in the coast horned lizard speceis complex (*Phrynosoma*). Proc. Nat'l Acad. Sci. USA 106, 12418–12423 (2009)
16. Leliaert, F., Verbruggen, H., Wysor, B., De Clerck, O.: DNA taxonomy in morphologically plastic taxa: algorithmic species delimitation in the Boodlea complex (Chlorophyta: Cladophorales). Mol. Phylogenet. Evol. 53, 122–133 (2009)
17. Libeskind-Hadas, R., Charleston, M.A.: On the computational complexity of the reticulate cophylogeny reconstruction problem. J. Comput. Biol. 16, 105–117 (2009)
18. Ma, B., Li, M., Zhang, L.X.: From gene trees to species trees. SIAM J. Comput. 30, 729–752 (2000)
19. Maddison, W.P.: Gene trees in species trees. Syst. Biol. 46, 523–536 (1997)
20. Mallet, J., Willmott, K.: Taxonomy: Renaissance or tower of babel? Trends Ecol. Evol. 18, 57–59 (2003)
21. Matz, M.V., Nielsen, R.: A likelihood ratio test for species membership based on DNA sequence data. Phil. Trans. R. Soc. B 360, 1969–1974 (2005)
22. Nielsen, R., Matz, M.: Statistical approaches for DNA barcoding. Syst. Biol. 55, 162–169 (2006)
23. O'Meara, B.C.: New heuristic methods for joint species delimitation and species tree inference. Syst. Biol. 59, 59–73 (2010)
24. Page, R.: Maps between trees and cladistic analysis of historical associations among genes, organisms, and areas. Syst. Biol. 43, 58–77 (1994)
25. Pamilo, P., Nei, M.: Relationship between gene trees and species trees. Mol. Biol. Evol. 5, 568–583 (1988)
26. Petit, J.R., Excoffier, L.: Gene flow and species delimitation. Trends Ecol. Evol. 24, 386–393 (2009)
27. Pons, J., et al.: Sequence-based species delimitation for the DNA taxonomy of undescribed insects. Syst. Biol. 55, 595–609 (2006)
28. Ronquist, F.: Three-dimensional cost matrix optimisation and maximum cospeciation. Cladistics 14, 167–172 (1998)
29. Steinke, D., Vences, N., Salzburger, W., Meyer, A.: TaxI: a software tool for DNA barcoding using distance methods. Phil. Trans. R. Soc. B 360, 1975–1980 (2005)
30. Tautz, D., et al.: DNA points the way ahead in taxonomy. Nature 418, 479 (2002)
31. Will, K.W., Rubinoff, D.: Myth of the molecule: DNA barcodes for species cannot replace morphology for identification and classigication. Cladistics. 20, 47–55 (2004)
32. Xu, S.: Phylogenetic analysis under reticulate evolution. Mol. Biol. Evol. 17, 897–907 (2000)
33. Zhang, L.X.: From gene trees to species trees II: Species tree inference in the deep coalescence model. Manuscript (2010)

Effective Algorithms for Fusion Gene Detection

Dan He and Eleazar Eskin

Dept. of Comp. Sci., Univ. of California Los Angeles, Los Angeles, CA 90095, USA
{danhe,eeskin}@cs.ucla.edu

Abstract. Chromosomal rearrangements which shape the genomes of cancer cells often result in fusion genes. Several recent studies have proposed using oligo microarrays targeting fusion junctions to detect fusion genes present in a sample. These approaches design a microarray targeted to discover known fusion genes by using a probe for each possible fusion junction. The hybridization of a sample to one of these probes suggests the presence of a fusion gene. Application of this approach is impractical to detect de-novo gene fusions due to the tremendous number of possible fusion junctions. In this paper we develop a novel approach related to string barcoding which reduces the number of probes necessary for de-novo gene fusion detection by a factor of 3000. The key idea behind our approach is that we utilize probes which match multiple fusion genes where each fusion gene is represented by a unique combination of probes.

Keywords: Fusion Gene, Suffix Tree, Minimum Set Cover, Integer Linear Programming.

1 Introduction

Chromosomal rearrangements which shape the genomes of cancer cells often result in fusion genes. These genes arise when a rearrangement occurs within genomic regions of two distinct genes creating a novel gene which contains a mixture of the exons of the two genes. The cancer cell transcribes the "fused" genomic regions of the two genes and then the splicing mechanism of the cell removes the introns resulting in a mRNA transcript consisting of several exons from one gene followed by exons from the other gene. Since as long as the rearrangement happens within the same introns which are often considerably sized genomic regions, the same fusion gene will be created. Characteristic fusion genes are found in a variety of cancers including hematological cancers, sarcomas and prostate cancer [10], [17].

Discovery of fusion genes in cancer cells is a difficult process traditionally involving karyotyping analysis to identify regions where a fusion gene may occur and then following up these regions using FISH and RT-PCR[16]. This process is both time consuming and difficult and can often fail for many cancer samples. Recently, several groups have proposed using high throughput sequencing technologies to identify fusion genes[9]. Unfortunately, despite tremendous advances in sequencing technologies, the cost of sequencing a tumor sample is still expensive and time consuming.

V. Moulton and M. Singh (Eds.): WABI 2010, LNBI 6293, pp. 312–324, 2010.

Several recent studies have proposed using oligo microarrays targeting fusion junctions to detect fusion genes present in a sample[16,15,4,8,11,12]. The basic idea behind these studies is that they attempt to identify a presence of a junction between two exons which are originally in different genes. Extracted mRNA from a cancer sample is hybridized to an array which contains probes spanning the boundaries of each possible pair of fused exons. Most of these studies focus on a small set of candidate fusions[15,4,8,11,12]. In their study, Skotheim et al[16], generate a microarray which can identify all currently known fusion genes (371 at the time of their publication) using 59,381 probes spanning the junctions. Each potential fusion gene corresponds to one of these probes. Unfortunately, direct application of this technique to discover new fusion genes is impractical because it requires too many probes. A direct application of this approach to cover all possible fusion genes requires approximately 25 billion probes, each covering a specific pair of fused exons.

We want to design a smaller set of probes to identify all these possible junctions of fusion exons. The fusion junction detection problem is very challenging in the following aspects:

- The probes need to detect all possible fusion junctions, which is a very large number. Given n exons, the number of possible fusion junctions is $O(n^2)$. This number is very large if the number of exons $n > 100,000$. Even just iterating through the 25 billion possible fusion junctions is not feasible.
- The probes can only detect the potential fusion junctions but due to possible contamination of the sample, must not match the normal genes, including normal junctions between exons of the same gene. This is because if the probe also detects the normal gene, the prediction of the probe will always hybridize in the experiments and thus won't provide any useful information.
- The probes must detect each fusion junctions uniquely.

The fusion junction detection problem is similar to the well-known String Barcoding problem, introduced by Rash and Gusfield [13]. In their problem settings, they have a large amount of genomic sequences and they want to identify certain unidentified sequences using a minimum set of probes. Given a set of probes and an unidentified genomic sequence, a probe reports the presence of the substring if it matches exactly the substring of the sequence and reports absence if not. If we consider "presence" and "absence" as bits $1, 0$, respectively, the unidentified sequence can be detected by a binary string, or barcode, as outputs from the set of probes. The barcode for each unidentified sequence needs to be unique such that each sequence can be detected uniquely, namely for each pair of genomic sequences s_i, s_j, there exists at least one probe which is a substring of either s_i or s_j but not of both.

Many algorithms have been proposed for this problem. Rash and Gusfield [13] proposed an integer programming algorithm to try to minimize the number of probes. They also use suffix tree to conduct efficient probe construction. Borneman et al. [1] used Lagrangian relaxation and simulated annealing to achieve similar results. DasGupta et al. [6] described a greedy algorithm which has high scalability to the whole-genome sequence. Their algorithm selects the probe in each iteration which distinguishes the largest number of not yet distinguished pairs of genomic sequences. Lancia and Rizzi [7] showed the string barcoding problem is NP-complete for even binary alphabets and the problem is as hard to approximate as the Set Cover problem.

There are three key differences between the string barcoding problem and our fusion gene detection problem: (1) The transcriptome (the set of possible mRNA sequences) is relatively small and with the requirement of probes of a minimum length due to technological constraints, most probes will match the transcriptome uniquely. (2) All probes are potentially valid in the string barcoding problem, while in the fusion gene detection problem, probes matching normal junctions or sequences within exons can not be used. (3) Probes with gap or mismatch positions are allowed. This leads to much more flexibility for our problem.

In this paper we present a novel approach to this problem which requires significantly fewer probes to cover all possible fusion genes. Our approach reduces the number of required probes to discover all putative fusion genes in the genome to only 9 million probes, a reduction by a factor close to 3,000. The key idea behind our approach is that we make use of "unbalanced" probes which unevenly span the junction of two exons, uniquely matching the exon on one side of the junction and extending only a few bases onto the exon on the other side.

2 Methods

2.1 Problem Statement

Assume gene i contains 5 exons $(e^i_1, e^i_2, e^i_3, e^i_4, e^i_5)$. For illustration purpose, here we assume each gene contains only 5 exons, each typically is of length less than a few hundred bases. There are 4 junctions between adjacent exons $(e^i_1|e^i_2, e^i_2|e^i_3, e^i_3|e^i_4, e^i_4|e^i_5)$. We call these junctions *normal junctions*. Here we exclude the probability of alternative splicing which can be easily addressed by adding additional normal junctions. Given any pair of genes i and j with exons $(e^i_1, e^i_2, e^i_3, e^i_4, e^i_5)$, $(e^j_1, e^j_2, e^j_3, e^j_4, e^j_5)$, where e^i_k denotes the kth exon of gene i, one exon of a gene may be fused, or concatenated with an exon of the other gene, for example, e^i_2 is concatenated with e^j_3, we call the resulting gene $(e^i_1, e^i_2, e^j_3, e^j_4, e^j_5)$ *fusion gene*, call the junction "$e^i_2|e^j_3$" *fusion junction*, call "e^i_2" the *left exon* of the fusion junction (left exon for simplicity), call "e^j_3" the *right exon* of the fusion junction (right exon). The same two exons can be fused in two different ways to form two different fusion junctions, namely "$e^i_k|e^j_l$" and "$e^j_l|e^i_k$". Any exon of a gene can be fused to any exon of another gene, with the exception that the first exon of a gene can not be fused as the right exon and the last exon of a gene can not be fused as the left exon. A length h probe is a string of length h. We say a probe *identifies* a junction if the probe spans the junction and the left half of the probe (we call it *left probe*) matches exactly the suffix of the left exon, the right half of the probe (we call it *right probe*) matches exactly the prefix of the right exon. A probe identifies a normal string if the probe matches the string exactly. The left probe and the right probe don't need to be of the same length, but the sum of their length needs to be h. Given all this notation, the fusion gene probe selection problem can be formally stated as following:

Given f genes, each having various number of exons, a solution to the fusion gene probe selection problem is a set of probes which may contain gaps such that each fusion junction can be identified by the set of probes uniquely, under the constraint that all the non-junction regions as well as normal junctions of the genes (we call these regions *normal regions*) will not match any probe sequences. We define an optimal solution as the solution using the smallest number of probes.

We note that in practice the constraint of a minimum length h has a large effect on the problem. For microarrays probes to perform effectively, often h has to be in the range of $30-40$. Such a long probe will most likely be unique in the transcriptome making it difficult to construct probes which match in multiple locations.

2.2 Naive Algorithm

Assuming there are totally n exons, the number of fusion junctions is $O(n^2)$ since any exon of a gene can be fused with any exon of another gene. A naive way of designing probes is to design one probe to uniquely identify one fusion junction for each fusion junction. The probe simply spans the fusion junction and matches both the left and right exons. This algorithm requires $O(n^2)$ probes. We shown an example in Figure 1 where we have two genes $(ACTCA|GTCAG|AGTAG|AAACT)$ and $(GCTCG|CGTGA|TAGCA|CCTAT)$. For illustration purpose, each gene contains only 4 exons, each of length 5. There are 8 possible fusion junctions as shown in Figure 1 (remember that the first exon of a gene can not be the right exon and the last exon of a gene can not be the left exon). Assuming the probe length is 6, according to the naive algorithm, we design one probe for each fusion junction and these probes are shown in Figure 1 as the "Naive Probe Selection" column. As we can see, in this example, these probes identify one fusion junction uniquely and they don't match the normal regions of the genes.

Fusion Junction Set	Naive Probe Selection	Unbalanced Probe Selection (Greedy)
$ACTCA\|CGTGA$	$TCA\|CGT$	$TCA\|C--;--A\|CGT$
$ACTCA\|TAGCA$	$TCA\|TAG$	$TCA\|T--;-CA\|TAG$
$ACTCA\|CCTAT$	$TCA\|CCT$	$TCA\|C--;TCA\|CCT$
$GTCAG\|CGTGA$	$CAG\|CGT$	$CAG\|C--;-AG\|CGT$
$GTCAG\|TAGCA$	$CAG\|TAG$	$CAG\|T--;--G\|TAG$
$GTCAG\|CCTAT$	$CAG\|CCT$	$CAG\|C--;--G\|CCT$
$AGTAG\|CGTGA$	$TAG\|CGT$	$TAG\|C--;-AG\|CGT$
$AGTAG\|TAGCA$	$TAG\|TAG$	$TAG\|T--;--G\|TAG$
$AGTAG\|CCTAT$	$TAG\|CCT$	$TAG\|C--;--G\|CCT$
$GCTCG\|GTCAG$	$TCG\|GTC$	$TCG\|G--;--G\|GTC$
$GCTCG\|AGTAG$	$TCG\|AGT$	$TCG\|A--;-CG\|AGT$
$GCTCG\|AAACT$	$TCG\|AAA$	$TCG\|A--;-CG\|AAA$
$CGTGA\|GTCAG$	$TGA\|GTC$	$TGA\|G--;-GA\|GTC$
$CGTGA\|AGTAG$	$TGA\|AGT$	$TGA\|A--;--A\|AGT$
$CGTGA\|AAACT$	$TGA\|AAA$	$TGA\|A--;--A\|AAA$
$TAGCA\|GTCAG$	$GCA\|GTC$	$GCA\|G--;GCA\|GTC$
$TAGCA\|AGTAG$	$GCA\|AGT$	$GCA\|A--;--A\|AGT$
$TAGCA\|AAACT$	$GCA\|AAA$	$GCA\|A--;--A\|AAA$

Fig. 1. An example of probe selection for two genes $(ACTCA|GTCAG|AGTAG|AAACT)$ and $(GCTCG|CGTGA|TAGCA|CCTAT)$, where "$-$" denotes a gap

2.3 Unbalanced Probe Selection Algorithm

The naive algorithm is very simple, however, the number of probes by the naive algorithm is $O(n^2)$ which is obviously very large for large n greater than 100,000. Therefore the naive algorithm is too expensive for practical use.

However, since each probe must be length h, usually larger than $30 - 40$ for current microarray technologies, even most sequences of length $h/2$ will occur only at most once within the set of exons. This greatly constrains our ability to select probes that match multiple fusion junctions. Our key idea is to take advantage of "unbalanced" junction probes which span a junction in a way that the majority of the probe is on one exon and a small portion of the probe is on the second exon. The probe will uniquely match one of the exons, but will match many possible fusion genes that include the first exon and any exons that match the small portion of the probe. We note that if we use balanced probes with $h/2$ length sequences on each side of the junction, it is likely that both the right and left probes match the sequence uniquely and these probes are equivalent to the probes used in the naive solution.

For simplicity, we first consider the case where the probes are contiguous, meaning that we disallow gaps or mismatch positions within the probes. Thus the left probe must exactly match the suffix of the left exon and the right probe must exactly match the prefix of the right exon.

In our unbalanced solution, each true fusion will match two probes. The first probe will match the suffix of the left exon uniquely and matches a set of possible right exons. The second probe will match the prefix of the right exon uniquely and a set of possible left exons. Observing the combination of these two probes uniquely defines a fusion junction.

For example, consider the probe $TCAC$ for the example above. The prefix of the probe, TCA, matches the exon $ACTCA$ uniquely. The suffix of the probe C matches both the exons $CGTGA$ and $CCTAT$. The probe $TCAG$ would not be valid because it matches the normal junction between exons $ACTCA$ and $GTCAG$. Similarly, consider the probe $ACGT$, where the suffix CGT matches the prefix of exon $CGTGA$ and the prefix A matches both exons $ACTCA$ and $CGTGA$. A pair of unbalanced probes uniquely identifies a single fusion junction. If we observe hybridization at both probes $TCAC$ and $ACGT$, then that uniquely identifies the presence of the fusion between exons $ACTCA$ and $CGTGA$. We refer to the different type of probes as $forward$ and $backward$ unbalanced probes.

Now consider the fusion between exons $ACTCA$ and $CCTAT$. The same probe from above $TCAC$ will match the suffix of $ACTCA$ and the prefix of $CCTAT$. However, care must be taken to construct the probe that matches the suffix of $CCTAT$ and the prefix of $ACTCA$. The probe $ACCT$ unfortunately is not valid since it matches the normal junction between exons $TAGCA$ and $CCTAT$. Similarly $CACCT$ is also invalid for the same reason. Instead, we use the probe $TCACCT$ to represent this fusion junction. Any shorter probe would match the normal exon. Specifically, the length of the shorter portion of the fusion junction probe is the shortest length that does not match the normal junction.

Greedy Algorithm for Contiguous Probes. Instead of directly examining each of the 25 billion fusion junctions and create probes to identify these junctions, we leverage these insights to directly efficiently construct the optimal probeset for contiguous unbalanced probes of the type described above using a greedy algorithm.

It is obvious that for contiguous probes, we must construct a set of forward probes for each left exon that can be part of a fusion and backward probes for each right exon that can be part of a fusion. This set of probes must contain a portion that uniquely matches the exon and contains prefixes (or suffixes) that match each possible fusion exon but none of the normal exons. The algorithm below constructs such a minimal set.

For each exon, without loss of generality, assume it is the left exon in the normal junction. We construct the unbalanced probes where the long portion matches the exon uniquely. In the meanwhile, the short portions of the probes match all the exons which can be potentially fused to the current exon and do not match the corresponding right exon in the same normal junction. To construct the short portions of the unbalanced probes, we build a suffix tree for the prefix of all potential fusion exons as well as the right exon in the same normal junction. In the suffix tree, the leaf nodes are the exons and a path from the root to the leaf corresponds to the sequence of the prefixes of the corresponding exon sequences. The depth of the suffix tree is bounded by the length of the right probe. We then search along the suffix tree for the normal junction. Whenever we see a new branch differ from the current search path, we create a right probe $(n_1, n_2, \ldots, n_k, f_{k+1}, -, -, \ldots)$, where $n_1, n_2, \ldots, n_k$ are the symbols on the current search path and f_{k+1} is the first symbol in the new branch. This probe covers all the fusion exons in the subtree of the branch since all these vertices have the same prefix $(n_1, n_2, \ldots, n_k, f_{k+1})$. We keep on searching along the suffix tree until the *normal destination vertex* is reached, which corresponds to the last symbol of the normal junction exon in the suffix tree. All fusion junction exons will thus be covered. Therefore, for each normal junction, we need to construct a set of right probes to match all potential fusion exons but not the right exon in the normal junction. We show an example in the left graph of Figure 2. The right exon of the normal junction has prefix as *ATTGAA*. The symbol besides the triangle denotes the first symbol of a branch. Each triangle denotes a subtree of the corresponding branch. According to our algorithm, we design 5 right probes $(G{-}{-}{-}{-}{-})$, $(ATC{-}{-}{-})$, $(ATTGC{-}{-})$, $(ATTGAT{-})$, $(ATTGAG{-})$. As we can see, these right probes cover all the fusion junction exons in the subtrees of their corresponding branches but not the normal destination vertices. We need to repeat the same process for each exon to generate both forward and backward unbalanced probes for the case the exon is left exon and the case the exon is right exon, respectively.

For each internal node in the suffix tree along the path from root to the normal destination vertex, there can be at most 3 new branches, which leads to at most 3 new right probes. Therefore, the number of right probes is bounded by $3 \times$ length(right probe) in our algorithm. As we need to repeat the process for every

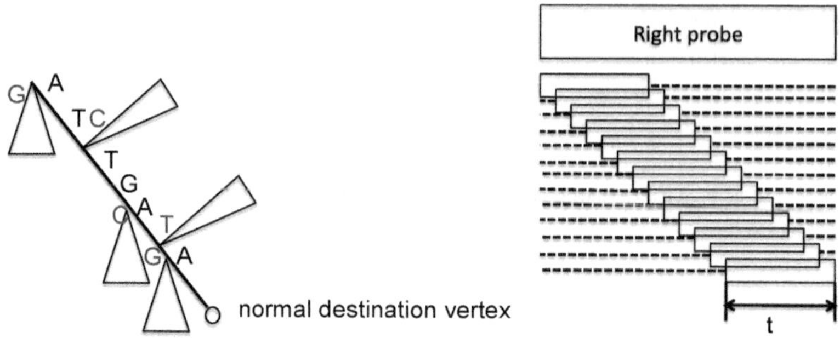

Fig. 2. (left) Example of probe selection with a suffix tree. (right) The initial set of probes starting at position 0, 1, ..., $s - t + 1$ in the short portion of the probe, which is the right probe in this example. The minimum non-gap symbols threshold is t. The boxes are the t contiguous non-gap symbols. The dot lines are gaps.

exon as a forward and backward unbalanced probe, the total number of probes is bounded by $O(3 \times 2 \times \frac{h}{2} \times n)$, where n is the number of exons, h is the length of the probe and we assume the left probe and the right probe are of equal length. Since only two suffix trees are needed for the whole selection process, one for all left exons and one for all right exons, the probe selection process for each exon is linear with the size of the probe. Therefore, our algorithm takes $O(n \times h)$ time complexity, where n is the number of exons, h is the length of the probe.

When a forward and a backward unbalanced probe are used to identify uniquely a fusion junction, the greedy algorithm can produce the optimal solution for contiguous sequences. This is because any longer probes will require adding additional probes without covering any additional fusion junctions since they would already be covered by these shorter probes. Shorter probes other than the ones we already picked will be invalid since they match normal junctions.

We summarize the probes required by our unbalanced probe algorithm for each fusion junction in the above example in Figure 1. As we can see, the number of probes required is 25, which actually exceeds the number of probes required by the naive algorithm as 18. This is because as we showed before, the number of probes of the naive algorithm is $O(n^2)$, while the the number of probes of the greedy algorithm is $O(3 \times 2 \times \frac{h}{2} \times n)$, where n is the number of exons and h is the length of the probes. In the example of Figure 1, the length of probes h is 6, while number of exons n is 8. Therefore, the number of probes of the greedy algorithm is greater than the corresponding number of the naive algorithm. Thus when the total number of exons is too small, the probe sets from the "unbalanced" probe design may not be optimal. However, for much longer exons, as shown later in our experiments, the unbalanced probe algorithm requires many fewer probes than the naive algorithm does.

One problem of this design is cross-hybridization, namely if the short portion of a probe is too short, it may bind to other exon junctions. To address this problem, we may need to set a minimum length threshold for the short portion.

The greedy algorithm can be easily adapted to integrate this threshold t such that the left or right probe need to contain no fewer than t symbols.

Minimum Set Cover Conversion. While the greedy algorithm can produce the optimal solution for contiguous sequences, although not necessarily when we allow for gaps in the probe sequences. To allow for gaps in the sequence, we reformulate the problem. We construct a graph where each exon is a vertex, each probe is then a set covering certain vertices, corresponding to the exons the probe matches, the problem of seeking minimum number of probes matching all exons can be converted to the problem of minimum set cover on the graph. Minimum Set Cover problem is known to be NP-complete. We thus format the problem into the following ILP (integer linear programming) problem and solve it optimally with an ILP solver CPLEX [5].

$$Minimize : \sum_{i}^{N} S_i$$

$$Subject \ \ to : \sum_{j \in Cov_k} S_j \geq 1 \ \ for \ \ each \ \ V_k$$

$$S_j \in \{0, 1\}$$

Where S_i is the ith set, corresponding to the ith probe, V_k is the kth vertex, corresponding to the kth exon, Cov_k is the set of sets covering vertex V_k, and all S_j's are binary. Therefore we require each vertex to be covered by at least one set.

The Minimum Set Cover problem, however, requires knowing the initial set of probes which must first be determined. Since we allow gaps, assuming the short portion of the probe is of length s and the minimum number of non-gap symbols in the short portion is l_2. There are a total of s choose l_2 ways of choosing non-gap symbols for the probes, where the non-gap symbols are intervened with gaps such that the total length of the short portion is s. However, to avoid cross-hybridization, we require the non-gap symbols in the short portion to be contiguous. Gaps are only allowed between the non-gap symbols of the long portion and the non-gap symbols of the short portion. Therefore, given the short portion of the probe with length s, the minimum number of non-gap symbols in the short portion as l_2, the possible ways of choosing non-gap symbols is $s - l_2 + 1$ instead of s choose l_2. Again, we can set a minimum non-gap symbols threshold t. Then the initial set of probes are the probes having t contiguous non-gap symbols which start at position 0, 1, ..., $s - l_2 + 1$ in the short portion of the probe, respectively. We show an example in the right graph of Figure 2, where the non-gap symbols need to be contiguous.

We do not use probes with more than t non-gap symbols because a probe always covers more vertices than another probe which contains all the non-gap symbols contained by the first probe as well as some extra non-gap symbols. For example, the probe $ACTT$ always matches more exons than the probe $ACTTC$ or $TACTT$. The t contiguous non-gap symbols of a probe must have at least one different symbol from the length t substring of the normal junction exon starting at the same position. This is to guarantee that the probes matches to all fusion junction exons but not the normal junction exon. For example, for normal junction exon $ACTACCCTT$, assuming $t=2$, then at position 0, we can have all possible length two probes $AA, AT, AG, \ldots$, etc., but not AC. Similarly, at position 1, we can have all possible length two probes but not CT. Therefore, for each normal junction, we need to initialize a set of probes whose size is $(s-t+1) \times (4^t - 1)$, where $4^t - 1$ means the probe can be any combination of the four characters but not the one from the normal junction exon. Then CPLEX can be applied to find the optimal set of probes.

3 Experimental Results

3.1 Synthetic Gene Sequences

We first evaluate the performance of our algorithm on synthetic gene sequences. We randomly generate a set of gene sequences, each with 5 exons. The exon length is 100. We vary the number of genes as 500, 1000, 2000, 4000, 8000, 16000. and for each set of parameters we conduct our experiments 10 times and show the average performance. We set the probe length as 30 and we set left probe and right probe both as length 15 such that the probes are long enough to uniquely identify any exon. We set the minimum non-gap symbols threshold as 1. We compare the number of probes generated by our greedy algorithm and by the naive algorithm. We also show the run time of our algorithm. The results are shown in Figure 3.

As we can see in Figure 3, as gene number increases, the exon number increases and the probe number required by both algorithms increases. However,

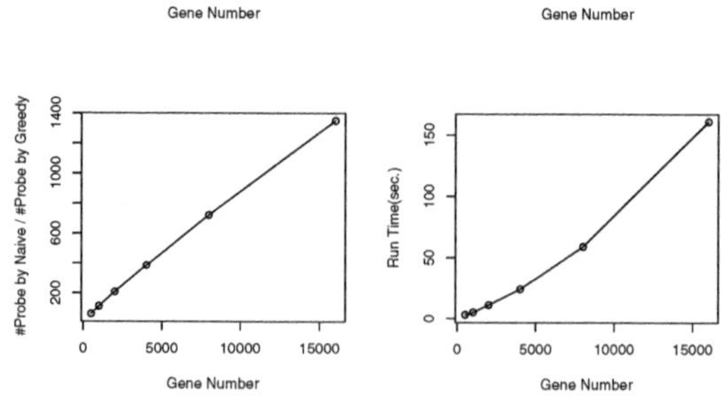

Fig. 3. Comparison between the naive algorithm and the unbalanced probe algorithm for randomly generated gene sequences. The gene numbers vary as 500, 1000, 2000, 4000, 8000, 16000. Each gene contains 5 exons of length 100 each.

the probe number by the naive algorithm increases quadratically, while the probe number by our unbalanced probe algorithm increases linearly. The probe number by naive algorithm far exceeds that by our unbalanced probe algorithm. The ratio of the probe number by naive algorithm versus the probe number by unbalanced probe algorithm increases linearly as gene number increase, indicating unbalanced probe algorithm is more efficient when the number of exons is large. For example, for 16000 genes, namely 80000 exons, the number of probes using the naive algorithm is 1400 times larger than the number of probes using the unbalanced probe algorithm. The run time increases almost linearly as the number of genes increases, indicating the unbalanced probe algorithm is time efficient and is scalable to large data set.

3.2 Complete Human Gene Sequences

We next download the complete human gene sequences from the UCSC genome browser [2]. The gene number is 23,754 and the total exon number is 183,249. Each gene roughly contains 7.7 exons. Again, we set probe length to 30 and left and right probe length to 15. We set the minimum non-gap symbols threshold as 1. Our greedy algorithm runs on these gene sequences for 30 minutes and generates approximately 9 million probes. The naive algorithm, however, generates approximately 26 billion probes, which is approximately 3000 times larger than the number of probes generated by our algorithm. This again indicates our method generates a lot fewer probes and is time efficient.

To address cross-hybridization, in the following experiment, we vary the minimum non-gap symbols threshold for the short portion of the probe from 1 to 5 and the number of probes required for each threshold are 8,944,202, 11,836,694, 26,299,154, 87,041,486, 330,814,516, respectively. As we can see, the numbers of required probes for non-gap symbols thresholds 1,2, and 3 are close to each other, all very low, but the number increases dramatically for thresholds 4 and 5, indicating the feasible non-gap symbols threshold needs to be under 5. This again confirms our previous claim that we need to design unbalanced probe where the short portion needs to be short enough, otherwise the number of required probes is too large. Given a non-gap symbols threshold of 5, our method still requires only 1/60 of the probes required by the naive method.

We can convert the problem into ILP problem and solve it with the solver CPLEX. However, in our experiments, the number of exons is too large. CPLEX can not find the optimal solution in a reasonable amount of time, especially when the minimum non-gap symbols threshold is greater than one. Therefore, we set a time limit for CPLEX as 20 minutes for each ILP problem. With a longer time limit, the results may be better. Again, we vary the minimum non-gap symbols threshold for the short portion of the probe from 1 to 5. Since for each normal junction, we need to design a set of probes, to illustrate the effects of error correction on the number of required probes, we randomly select 15 normal junctions. For each junction, we compare the number of required probes for the greedy algorithm and for CPLEX. Again, due to cross-hybridization, we require

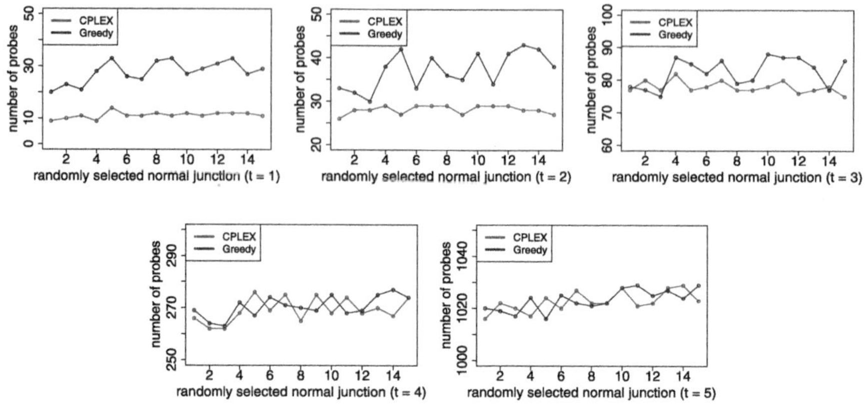

Fig. 4. Comparison of the number of required probes for t (minimum non-gap symbols threshold) for the short portion of the probe as 1, 2, 3, 4, 5, on randomly selected normal junctions in human gene sequences for the greedy algorithm and CPLEX

all contiguous portions of the probe to be of length no less than the minimum non-gap symbols threshold t.

We show the results in Figure 4. As we can see, CPLEX successfully reduced the number of required probes to around half of what would be needed compared to the greedy algorithm. However, as t increases, the differences between the number of required probes by CPLEX and by the greedy algorithm decrease, especially when $t = 5$, CPLEX performs almost the same as the greedy algorithm does. This behavior is reasonable for the following two reasons: (1) We set time limit for CPLEX as only 20 minutes. As t gets bigger, almost all the solutions from CPLEX are not optimal. Using a longer time limit should lead to better performance of CPLEX. (2) As t increases, the number of vertices, or exons, covered by each probe, decreases sharply. Therefore, there is not much flexibility with introducing gaps. The gaped probes perform almost the same well as the none-gaped probes.

4 Discussion

In this paper, we study the fusion gene probe selection problem where each possible fusion junctions needs to be uniquely identified by a set of probes and none of the normal regions including normal junctions of the genes can be identified by these probes. We first propose an efficient greedy algorithm which is linear time in the number of exons. We show our algorithm requires many fewer probes than the naive algorithm does. The number of probes generated by our algorithm is small enough for practical use. We next convert the problem of seeking minimum number of probes into the Minimum Set Cover problem and we show the problem can be solved optimally with an ILP (Integer Linear Programming) solver such as CPLEX.

There are several technological limitations related to the use of microarrays for detection of fusion genes. First, the technology is limited to identify fusion genes which are expressed in the cancer sample. Fusion genes which have very low expression will not be detected with any probe in the array. Second, due to differing hybridization properties of the probes, different probes will have different baseline levels of intensity. One possible approach to mitigate this problem is to hybridize both a tumor sample as well as a normal sample to the array and compare the differential expression. Finally, single nucleotide polymorphisms will affect our approach if they are present within the probe sequences. However, databases such as dbSNP now contain the majority of these polymorphisms and it is straightforward to exclude these positions from our probe design.

We also tried introducing error-correction mechanism in the probe selection process to address the problem of possible failures of the probes due to cross-hybridyzation. Given at most k failures are allowed, the ILP formula can be easily revised to incorporate error-corrections, where each vertex (probe) needs to be covered at least k times, and each set can be selected multiple times. Due to space limit, we do not show the experimental results for error-correction in the paper.

An alternate strategy for detecting fusion genes is to take advantage of high-throughput sequencing technologies[14] to directly sequencing the transcriptome of a cancer sample and then look for fusion junctions among the sequence reads[3]. Although these technologies have great promise, they are still being developed and it is not clear how soon they will be ready for wide scale clinical use due to several logistical issues. The sequencing machines are extremely expensive and require large computational and bioinformatics infrastructure for analysis. In addition, to reduce the cost, many samples must be combined in the same sequencing run at the same time which creates logistical challenges.

Acknowledgements

D.H. and E.E. are supported by National Science Foundation grants 0513612, 0731455, 0729049 and 0916676, and NIH grants K25-HL080079 and U01-DA024417. This research was supported in part by the University of California, Los Angeles subcontract of contract N01-ES-45530 from the National Toxicology Program and National Institute of Environmental Health Sciences to Perlegen Sciences.

References

1. Borneman, J., Chrobak, M., Della Vedova, G., Figueroa, A., Jiang, T.: Probe selection algorithms with applications in the analysis of microbial communities. Bioinformatics 17(suppl. 1), S39–S48 (2001)
2. UCSC Genome Browser, `http://genome.ucsc.edu/`
3. Chen, W., et al.: Mapping translocation breakpoints by next-generation sequencing. Genome Res. 18(7), 1143–1149 (2008)
4. Chun, S.-M.M., et al.: Identification of leukemia-specific fusion gene transcripts with a novel oligonucleotide array. Mol. Diagn. Ther. 11(1), 21–28 (2007)

5. CPLEX, `http://www.ilog.com/products/cplex/`
6. Dasgupta, B., Konwar, K.M., Mandoiu, I.I., Shvartsman, A.A.: Highly scalable algorithms for robust string barcoding. Int. J. Bioinform. Res. Appl. 1(2), 145–161 (2005)
7. Lancia, G., Rizzi, R.: The approximability of the string barcoding problem. Algorithms Mol. Biol. 1, 12 (2006)
8. Lu, Q., et al.: A sensitive array-based assay for identifying multiple tmprss2:erg fusion gene variants. Nucleic Acids Res. 36(20), e130 (2008)
9. Maher, C.A., et al.: Transcriptome sequencing to detect gene fusions in cancer. Nature 458(7234), 97–101 (2009)
10. Mitelman, F., Johansson, B., Mertens, F.: The impact of translocations and gene fusions on cancer causation. Nat. Rev. Cancer 7(4), 233–245 (2007)
11. Nasedkina, T.V., et al.: Identification of chromosomal translocations in leukemias by hybridization with oligonucleotide microarrays. Haematologica 87(4), 363–372 (2002)
12. Nasedkina, T.V., et al.: Clinical screening of gene rearrangements in childhood leukemia by using a multiplex polymerase chain reaction-microarray approach. Clin. Cancer Res. 9(15), 5620–5629 (2003)
13. Rash, S., Gusfield, D.: String barcoding: uncovering optimal virus signatures. In: RECOMB 2002: Proceedings of the sixth annual international conference on Computational biology, pp. 254–261. ACM, New York (2002)
14. Reis-Filho, J.S.: Next-generation sequencing. Breast Cancer Res 11(suppl. 3), S12 (2009)
15. Shi, R.Z., Morrissey, J.M., Rowley, J.D.: Screening and quantification of multiple chromosome translocations in human leukemia. Clin. Chem. 49(7), 1066–1073 (2003)
16. Skotheim, R.I., et al.: A universal assay for detection of oncogenic fusion transcripts by oligo microarray analysis. Mol. Cancer 8, 5 (2009)
17. Teixeira, M.R.: Recurrent fusion oncogenes in carcinomas. Crit. Rev. Oncog. 12(3-4), 257–271 (2006)

Swiftly Computing Center Strings

Franziska Hufsky[1,3], Léon Kuchenbecker[2], Katharina Jahn[2], Jens Stoye[2], and Sebastian Böcker[1]

[1] Lehrstuhl für Bioinformatik, Friedrich-Schiller-Universität Jena, Ernst-Abbe-Platz 2, Jena, Germany
{franziska.hufsky,sebastian.boecker}@uni-jena.de
[2] AG Genominformatik, Technische Fakultät, Universität Bielefeld, Germany
kjahn@cebitec.uni-bielefeld.de, {lkuchenb,stoye}@techfak.uni-bielefeld.de
[3] International Max Planck Research School, Jena, Germany

Abstract. The center string (or closest string) problem is a classical computer science problem with important applications in computational biology. Given k input strings and a distance threshold d, we search for a string within Hamming distance d to each input string. This problem is NP-complete. In this paper, we focus on exact methods for the problem that are also fast in application. First, we introduce data reduction techniques that allow us to infer that certain instances have no solution, or that a center string must satisfy certain conditions. Then, we describe a novel search tree strategy that is very efficient in practice. Finally, we present results of an evaluation study for instances from a biological application. We find that data reduction is mandatory for the notoriously difficult case $d = d_{\text{opt}} - 1$.

1 Introduction

The Center String problem (also called Closest String problem) is defined as follows: Given k strings of length L over an alphabet Σ and a distance threshold d, find a string of length L that has Hamming distances at most d to each of the given strings.

The Center String problem has been studied extensively in theoretical computer science and particularly in computational biology [5,9], and has various applications such as degenerate PCR primer design [10] or motif finding [2,5]. We are particularly interested in its application as part of finding approximate gene clusters: The increasing speed of genome sequencing and the resulting number of available data offers the possibility of comparing gene order of whole genomes. During the course of evolution, speciation results in the divergence of genomes that initially have the same gene order and content. Conserved gene order is evidence for some biological signal [11]. Approximate gene cluster models account for reordering inside the gene cluster, as well as additional and missing genes in the compared genomes [8, 1]. The *center gene cluster model* limits the distance between the gene cluster and each of the approximate occurrences. For given approximate occurrences, finding the center gene cluster is equivalent to finding a center string for binary input strings.

V. Moulton and M. Singh (Eds.): WABI 2010, LNBI 6293, pp. 325–336, 2010.

Previous Work. The CENTER STRING problem is NP-complete even for three strings [3, 5], hence no polynomial time algorithm can exist unless P = NP. Different approaches have been studied for the problem: Ma and Sun [7] presented a polynomial time approximation scheme with time complexity $O(n^{O(\epsilon^{-2})})$ for an approximation ratio of $1+\epsilon$ for any $\epsilon > 0$. Also, heuristics and parallel implementations with good practical running times have been developed [6]. The drawback of these approaches is that they cannot guarantee to find an exact solution.

Parameterized algorithms use a parameter to describe the complexity of a problem instance and restrict the running time using this parameter, while at the same time guarantee to find optimal solutions. Parameters that have been studied in the literature for the CENTER STRING problem are the distance threshold d and the number of input strings k. For the latter parameter, Gramm *et al.* [4] showed that the problem is fixed-parameter tractable using an Integer Linear Program. Evaluations indicate that this approach is of theoretical interest only and impractical for $k \geq 5$. Regarding the distance threshold d, in the same paper an algorithm was given with running time $O(kL+kd^{d+1})$. Later, Ma and Sun [7] presented an algorithm with running time $O\big(kL + kd \cdot 16^d\,(|\Sigma| - 1)^d\big)$. Recently, Wang and Zhu [9] improved the running time to $O\big(kL+kd\cdot 9.53^d\,(|\Sigma|-1)^d\big)$. All of these algorithms are based on the search tree paradigm. Note that for binary strings the term $(|\Sigma| - 1)^d$ vanishes.

Our Contribution. In this paper, we focus on exact methods that are also swift in application. We have developed an advanced preprocessing to quickly filter out unsolvable instances. Additionally, we compute rules that can be used within search tree algorithms to bound the search space, excluding unsolvable instances. We show how to integrate this information into the algorithms from [4, 7]. We then present a new search tree strategy called *MismatchCount* that, despite its bad worst case running time, works extremely well in practice. We implemented all three algorithms to evaluate their performance in combination with our preprocessing. We then present results of our experimental evaluation, showing that preprocessing and the novel algorithm improve running times by several orders of magnitude. We find that particularly the case $d = d_{\text{opt}} - 1$ is notoriously difficult for all approaches, where d_{opt} is the smallest distance value for which a solution exists.

2 Preliminaries

Given a string s over a finite alphabet Σ, let $s[i]$ indicate the ith character of s and $s[i,j]$ the substring of s starting at position i and ending at position j. The length of s is denoted by $|s|$.

The *Hamming distance* $d_{\text{H}}(s,t)$ of two strings s and t of the same length L is the number of positions p with $s[p] \neq t[p]$. Let $R = \{p_1, \ldots, p_m\} \subseteq \{1, \ldots, L\}$ be a set of positions such that $p_i < p_{i+1}$ for all $1 \leq i < m$. Then $s|_R := s[p_1] \ldots s[p_m]$ denotes the subsequence of s restricted to the positions in R. We define the Hamming distance of two strings s and t restricted to R as $d_{\text{H}}^R(s,t) :=$

$d_{\mathrm{H}}(s|_R, t|_R)$. For two strings s and t, let $D_{s,t} := \{p : s[p] \neq t[p]\} \subseteq \{1, \ldots, L\}$ denote the set of positions where s and t differ, and let $E_{s,t} := \{p : s[p] = t[p]\} = \{1, \ldots, L\} \setminus D_{s,t}$ be the set of positions where s and t are identical. Note that $d_{\mathrm{H}}^{D_{s,t}}(s,t) = d_{\mathrm{H}}(s,t)$. For k input strings $s_1, \ldots, s_k$, we write $D_{i,j} := D_{s_i,s_j}$ and $E_{i,j} := E_{s_i,s_j}$. For strings over the binary alphabet $\Sigma = \{0,1\}$, which is our default, we define $\overline{s[p]} = 1 - s[p]$.

The CENTER STRING problem is defined as follows: Given strings $s_1, \ldots, s_k$ of length L over an alphabet Σ, and a distance threshold d, find a string $\hat{s}$ of length L, called *center string*, that has Hamming distances at most d to each of the given strings.

For k strings $s_1, \ldots, s_k$ and distance threshold d, we can construct a *naïve kernel* as follows [4]: A position p is called *clean* if all sequences coincide at this position, i.e. $s_i[p] = s_j[p]$ for all $1 \leq i < j \leq k$, otherwise it is called *dirty.* One can easily see that there can be at most kd dirty positions if an instance allows for a center string of distance d. If a position is not dirty, then all strings share the same character at this position, and the center string will also share this character. So, we can remove all positions but the dirty ones, and get an instance of length $L \leq kd$.

In our algorithms, we assume a distance threshold d to be given. In applications, we might not know the distance threshold d in advance, but instead search for a center string minimizing d. We can do so by calling our algorithms repeatedly, increasing $d = 0, 1, 2, \ldots$ until a solution is found for $d = d_{\mathrm{opt}}$. Both in theory and in our experimental evaluation, we find that the running time of this iteration is governed by the last subroutine calls with $d = d_{\mathrm{opt}} - 1$ and $d = d_{\mathrm{opt}}$. That is why in our evaluations we will put special focus on these two cases.

In the following, we present a data reduction that will often allow us to conclude that no solution can exist for a particular distance threshold d. But in case we cannot rule out the existence of a center string by data reduction (what is obviously the case when $d = d_{\mathrm{opt}}$) we still have to decide whether a valid center strings exists. All algorithms for doing so, such as [4,7,9] and the *MismatchCount* algorithm presented below, are based on the search tree paradigm: In principle, we scan through all 2^L possible binary strings and test whether any such string is a center string of the input. The algorithms differ in the order in which they process the 2^L strings and, in particular, how they constrain the search space to speed up computations.

3 Data Reduction

Our data reduction is based on the pairwise comparison of the input strings. Given an instance $s_1, \ldots, s_k$ and d of the CENTER STRING problem, we can divide all pairs of strings $\{s_i, s_j\}$ into three groups: pairs with distance less than $2d - 1$, greater than $2d$, or equal to $2d$ or $2d - 1$. If there exist two strings s_i, s_j with Hamming distance $d_{\mathrm{H}}(s_i, s_j) > 2d$, then the instance has no solution. This follows from the fact that a center string $\hat{s}$ can have at most distance d

to each of s_i and s_j and, hence, $d_{\mathrm{H}}(s_i, s_j) \leq d_{\mathrm{H}}(s_i, \hat{s}) + d_{\mathrm{H}}(\hat{s}, s_j) \leq 2d$. So, $d \geq \frac{1}{2}\max_{i,j} d_{\mathrm{H}}(s_i, s_j)$ must hold for the instance to have a solution.

Solving Trivial Positions. Some positions of the solution string can be trivially solved. This is based on the following observation:

Lemma 1. *Given strings $s_1, \ldots, s_k$ and a center string $\hat{s}$ with distance d. For two strings s_i, s_j such that $d_{\mathrm{H}}(s_i, s_j) = 2d$ or $d_{\mathrm{H}}(s_i, s_j) = 2d - 1$, we have*

$$\hat{s}[p] = s_i[p] = s_j[p] \text{ for all } p \in E_{i,j} \ .$$

Proof. A center string with distance at most d to all strings is located central between the two strings s_i and s_j with distance $2d$ and hence has distance d to both of them. Thus, all positions fixed between s_i and s_j must also be fixed in $\hat{s}$. Our reasoning can be extended to string pairs with distance $2d - 1$: We need to change, in at least one of the strings, d positions and $E_{i,j}$ is the set of equal positions between *both* strings, hence we are still not allowed to change any position $p \in E_{i,j}$. □

As a reduction rule, if we find two strings s_i, s_j with $d_{\mathrm{H}}(s_i, s_j) \geq 2d - 1$, then we can set $\hat{s}[p] := s_i[p]$ for all $p \in E_{i,j}$ and mark these positions as "permanent". Let $\mathcal{P}$ denote this set of permanent positions. By doing this for all s_i, s_j with $d_{\mathrm{H}}(s_i, s_j) = 2d$ or $d_{\mathrm{H}}(s_i, s_j) = 2d - 1$, we may run into conflicting situations where we have to permanently set a certain position to '0' and '1' simultaneously. We call such a situation a *conflict* and infer that the instance has no solution for the current choice of d. If we do not have a conflict, then applying this data reduction results in a partially solved solution string $\hat{s}$ with $\hat{s}[p] = c \in \Sigma$ fixed for all $p \in \mathcal{P}$, whereas all positions not in $\mathcal{P}$ still have to be decided.

Computation of Position Subsets. We next focus on pairs of strings s_i, s_j with $d_{\mathrm{H}}(s_i, s_j) = \delta < 2d - 1$. For a given center string $\hat{s}$ we define

$$X_{i,j}(\hat{s}) := \{p \in E_{i,j} \ : \ s_i[p] = s_j[p] \neq \hat{s}[p]\}$$

as the set of positions where s_i and s_j agree, but disagree with the center string $\hat{s}$. We extend the reasoning behind Lemma 1 as follows:

Lemma 2. *Given strings $s_1, \ldots, s_k$ and a center string $\hat{s}$ with distance d. For two strings s_i, s_j such that $d_{\mathrm{H}}(s_i, s_j) < 2d - 1$, we have*

$$|X_{i,j}(\hat{s})| \leq d - \tfrac{1}{2} d_{\mathrm{H}}(s_i, s_j) \ .$$

Proof. Let $D := D_{i,j}$. Regarding the distances between $\hat{s}|_D$ and $s_i|_D$ as well as $s_j|_D$, we can state that $\hat{s}|_D$ has to at least one of the strings $s_i|_D$ or $s_j|_D$ a distance at least $\frac{1}{2} d_{\mathrm{H}}(s_i, s_j)$:

$$\max\{d_{\mathrm{H}}(s_i|_D, \hat{s}|_D), d_{\mathrm{H}}(s_j|_D, \hat{s}|_D)\} \geq \tfrac{1}{2} d_{\mathrm{H}}(s_i, s_j) \ .$$

This is true since d_{H} is a metric and the triangle inequality holds, $d_{\mathrm{H}}(s_i|_D) \leq d_{\mathrm{H}}(s_i|_D, \hat{s}|_D) + d_{\mathrm{H}}(s_j|_D, \hat{s}|_D)$. Since we need a distance of at least $\frac{1}{2} d_{\mathrm{H}}(s_i, s_j)$ to solve the positions from D, a distance of at most $d - \frac{1}{2} d_{\mathrm{H}}(s_i, s_j)$ remains to solve the positions from E. □

Lemma 2 implies that the maximum number of positions $p \in E_{i,j}$ we are allowed to choose in the center string with $\hat{s}[p] \neq s_i[p]$ is bounded by $d - \frac{1}{2}d_{\mathrm{H}}(s_i, s_j)$. We can transform this observation into a reduction rule as follows: When, during search tree traversal or by other reduction rules, we have a partially solved solution string $\hat{s}$ such that

$$|X_{i,j}(\hat{s})| > d - \tfrac{1}{2}d_{\mathrm{H}}(s_i, s_j)$$

for any pair s_i, s_j, then we can infer that $\hat{s}$ cannot be extended to a solution for the current choice of d. For each pair s_i, s_j, we therefore set $x_{i,j} := d - \frac{1}{2}d_{\mathrm{H}}(s_i, s_j)$ and store all tuples $(E_{i,j}, x_{i,j})$ in an array $\mathcal{T}$.

Removing redundant information from $\mathcal{T}$ may lead to further trivially solved positions. This is done by removing, for all $1 \leq i < j \leq k$, all positions $p \in \mathcal{P} \cap E_{i,j}$ from $E_{i,j}$. Moreover, if $\hat{s}[p] \neq s_i[p]$ then we decrease $x_{i,j}$ by one.

For $x_{i,j} = 0$ we set all positions p from $E_{i,j}$ to "permanent" and include them in $\mathcal{P}$. Since $\mathcal{P}$ has changed, we continue our data reduction again until there is no tuple $(E_{i,j}, x_{i,j})$ with $x_{i,j} = 0$ in $\mathcal{T}$. For $x_{i,j} < 0$ we can easily infer that there must exist a conflict and, hence, the instance has no valid solution for this distance threshold d.

Cascading. To further enlarge the number of solved positions we consider all pairs of strings s_i, s_j with $x_{i,j} = 1$ and use *cascading*. A valid center string $\hat{s}$ has to agree with s_i in at least $|E_{i,j}| - 1$ positions from $E_{i,j}$, hence for binary strings at most one position $p \in E_{i,j}$ can be set to $\hat{s}[p] = \overline{s_i[p]}$.

To this end, we test for all positions $p \in E_{i,j}$ what we can infer from setting $\hat{s}[p] = \overline{s_i[p]}$. This implies $x_{i,j} = 0$, hence the remaining positions $q \in E_{i,j}$, $q \neq p$, are added to $\mathcal{P}$ and the tuple set $\mathcal{T}$ is reduced. If we run into a conflict during this reduction, we know that setting $\hat{s}[p] = \overline{s_i[p]}$ cannot result in a valid solution. In this case, we infer $\hat{s}[p] = s_i[p]$ and permanently set position p.

Unfortunately, if not running into a conflict, setting $\hat{s}[p] = s_i[p]$ is not mandatory. However, we get a partially solved solution string $\hat{s}_{p,v}$ and a set of "potentially permanent" positions $\mathcal{P}_{p,v}$ depending on the position p and the value $v = \overline{s_i[p]}$. We store this information in a set of rules $\mathcal{R}$.

We can use the set of rules $\mathcal{R}$ when solving the remaining instance by, say, a search tree algorithm. If, during the search tree traversal, we decide to set $\hat{s}[p] = v$ for the solution string $\hat{s}$, then we can immediately start the above data reduction: For all positions $q \in \mathcal{P}_{p,v} \setminus \mathcal{P}$, we set the solution string $\hat{s}[q] = \hat{s}_{p,v}[q]$. For the remaining positions $q \in \mathcal{P}_{p,v} \cap \mathcal{P}$, the condition $\hat{s}[q] = \hat{s}_{p,v}[q]$ must be met, otherwise we run into a conflict and, thus, this branch of the search tree does not lead to a valid solution.

4 Integration into Search Tree Algorithms

We can use the information derived during preprocessing, stored in the sets $\mathcal{P}, \mathcal{T}, \mathcal{R}$, to speed up the algorithms of Ma and Sun [7] and Gramm *et al.* [4].

Integrating the set of solved positions $\mathcal{P}$ into the algorithm of Ma and Sun is straightforward, as this algorithm tackles the more general Neighbor String problem. In all other cases, it is necessary to interweave the use of $\mathcal{P}, \mathcal{T}, \mathcal{R}$ with the actual search tree algorithms. Here, we use the information from $\mathcal{P}, \mathcal{T}, \mathcal{R}$ to shrink the search tree, by excluding search tree branches which cannot lead to a valid solution. To do so, we simply test if the (partial) string candidate of the search tree is already conflicting with this information. The integration of $\mathcal{P}, \mathcal{T}, \mathcal{R}$ is somewhat different for the algorithms of Ma and Sun and Gramm *et al.*, we defer the technical details to the full version of this paper. Unfortunately, the use of $\mathcal{P}, \mathcal{T}, \mathcal{R}$ does not change the worst-case running times of both algorithms. But our preprocessing, as an algorithm engineering technique, allows us to speed up the algorithms in practice, as demonstrated in Sect. 6.

5 Algorithm MismatchCount

After we have applied our data reduction rules, we have to solve the remaining instance using a search tree algorithm, like those from [7, 4]. In this section we present another such procedure, *MismatchCount*, that is very efficient in practice, as we will show below. Given binary strings $s_1, \ldots, s_k$ of length L and a distance threshold d, the MismatchCount algorithm solves the Closest String problem as follows: We iterate through all strings s with distance at most d to a chosen string s_i — without loss of generality, we may choose that string to be s_1. This leaves us with a search space of size $\sum_{d'=0}^{d} \binom{L}{d'}$. We present an enumeration scheme for those s that allows efficient testing for the center condition on each candidate, and that makes it possible to skip large areas of the search space based on information gained while checking those candidates.

The mismatch positions for d mismatches in s_1 (and therefore the center string candidates s) are enumerated, equivalently to generating all binary numbers of length m with d bits set to 1, in reverse order.

An example for the placement of at most three mismatches is shown in Fig. 1. For every s, its Hamming distance to the remaining strings $s_2, s_3, \ldots, s_k$ has to be checked. Rather than recomputing these distances entirely new for each candidate, the Hamming distances from the previous candidate s' are updated by increasing (resp. decreasing) the distances according to the changed positions. The running time for verifying a center candidate s is therefore bounded by $O(g \cdot k)$, where g is the number of positions changed from s' to s.

The overall number of changes performed during the enumeration of all center candidates can be determined like this: using the presented enumeration scheme, each position p in s is changed once to '1' and once to '0' for every configuration of $s[1, p-1]$ with at most d mismatches to $s_1[1, p-1]$. There are $\binom{p-1}{d'}$ such configurations for each $d' = 1, 2, \ldots, d$. Summing over all possible combinations of p and d', the number of bit changes performed can be bounded by $O(2^L)$. Since for each character change in s, k Hamming distances need to be updated, the overall worst-case running time of the algorithm is bounded by $O(k \cdot 2^L)$.

$d_H(s, s_1) = 0$	$d_H(s, s_1) = 1$	$d_H(s, s_1) = 2$	$d_H(s, s_1) = 3$
0 0 0 0 0	1 0 0 0 0	1 1 0 0 0	1 1 1 0 0
	0 1 0 0 0	1 0 1 0 0	1 1 0 1 0
	0 0 1 0 0	1 0 0 1 0	1 1 0 0 1
	0 0 0 1 0	1 0 0 0 1	1 0 1 1 0
	0 0 0 0 1	0 1 1 0 0	1 0 1 0 1
		0 1 0 1 0	1 0 0 1 1
		0 1 0 0 1	0 1 1 1 0
		0 0 1 1 0	0 1 1 0 1
		0 0 1 0 1	0 1 0 1 1
		0 0 0 1 1	0 0 1 1 1

Fig. 1. Enumeration scheme for all strings s with Hamming distance at most 3 to a bit string s_1 of length 5. The '0's denote matches between s and s_1 at the respective positions, while '1's denote mismatches.

However, this worst-case analysis refers to the exploration of all legal mismatch configurations of s. As already mentioned above, the enumeration scheme enables us to skip large areas of that search space. Using the maximum Hamming distance $d_{\max} = \max_{i=2,\ldots,k}(d_H(s, s_i))$ computed in each iteration, we can derive a lower bound for the number of positions that have to be changed in s in order to fulfill the center condition. Therefore, for each candidate s taken into consideration, we compute $c_{\min} = \lceil \frac{d_{\max}-d}{2} \rceil$, where $2 \cdot c_{\min}$ is the minimum number of positions in s that have to be changed when its successor is generated. This bound can be used in two ways: We cannot change $2 \cdot c_{\min}$ positions in s by changing the positions of less than $c_{\min}$ mismatches. Therefore, if currently all candidates s with $d_H(s_1, s) = d$ are enumerated and we encounter a candidate that reveals a $c_{\min} > d$, we can proceed to the generation of candidates with $d_H(s_1, s) = c_{\min}$, omitting the enumeration for all s with $d_H(s_1, s) \in \{d, d+1, \ldots, c-1\}$.

Furthermore, even if $c_{\min}$ does not exceed d for a currently observed candidate, we can use that bound to skip the enumeration of certain candidates. Since we know that we have to change at least $2 \cdot c_{\min}$ positions in s, we can omit all enumeration steps that involve less than $c_{\min}$ mismatch positions.

Applying the data reduction to this algorithm is straightforward. Let $\mathcal{Q} := \{1, \ldots, L\} \setminus \mathcal{P}$ be the set of positions that are not permanent. Then, the reduced instance is $s_1|_{\mathcal{Q}}, \ldots, s_k|_{\mathcal{Q}}$. When estimating for every candidate s its Hamming distance to each remaining string s_i, the additional amount $d_H(\hat{s}|_{\mathcal{P}}, s_i|_{\mathcal{P}})$ has to be added to the distances of the reduced strings. This is done only once at the beginning, since the Hamming distances are updated during the algorithm.

6 Computational Results

We performed our tests on a data set obtained by applying the approximate gene cluster algorithm described in [1] to five γ-proteobacteria genomes from

Table 1. Five γ-proteobacteria from the NCBI Genome database, used for detection of approximate gene clusters to generate biological instances of the center string problem. 'Refseq' denotes reference sequence from NCBI Genome database, 'PC' number of protein-coding genes.

Species name	Refseq	Genes	PC
Buchnera aphidicola str. APS	NC_002528	607	564
Escherichia coli str. K-12 substr. MG1655	NC_000913	4493	4149
Haemophilus influenzae Rd KW20	NC_000907	1789	1657
Pasteurella multocida subsp. multocida str. Pm70	NC_002663	2092	2015
Xylella fastidiosa 9a5c	NC_002488	2838	2766

the NCBI Genome database[1], see Tab. 1. The gene classification is based on COG[2] functional categories.

The generation of center string instances from gene cluster predictions works as follows: Each gene cluster consists of five approximate occurrences, one on each genome, that are transformed into binary strings based on their gene composition. Since the instances generated from a single cluster are too short to evaluate the performance of our algorithms, larger instances are created by concatenation until the length L is reached. Additional strings are constructed in the same fashion, incorporating further cluster occurrences.

We created 50 instances for each combination of k and L with $k = 20, 30, 40, 50$ and $L = 250, 300, \ldots, 500$. The origin of our data, based on finding approximate gene clusters, results in many clean columns that are trivially solved. We keep only the dirty columns, representing the "hard part" of the instances. In our dataset, there were between 36.2% and 57.7% dirty columns. We stress that results in the following sections are reported for this naïve kernel. In the further evaluation we examine only the 567 instances with $d_{\text{opt}} \leq 40$ and we concentrate on the computation of center strings for $d = d_{\text{opt}}$ and $d = d_{\text{opt}} - 1$, since these are the computationally hard instances, see Fig. 3 below. For the search tree algorithms evaluated below, the search tree size grows (super-) exponentially with increasing d, hence the algorithms' running times are usually dominated by these cases.

Excluding Unsolvable Instances by Preprocessing. Our preprocessing allows us to exclude unsolvable instances more efficiently than the naïve kernel, when d is too small for a center string to exist. This is of particular interest as here search tree algorithms have to scan the complete search tree to ensure that no solution exists. Recall that the naïve kernel tests if there exist more than kd dirty columns, in which case the instance cannot have a solution for this choice of d. Table 2 shows the number of excluded instances via preprocessing, for $d = d_{\text{opt}} - 1$. Our improved preprocessing always filters out more instances than the naïve kernel does. For different k, we can exclude between 15.9 % and 44.4 % of instances that have not been filtered by the naïve kernel. We note that for

[1] `http://www.ncbi.nlm.nih.gov/sites/entrez?db=genome`
[2] `http://www.ncbi.nlm.nih.gov/COG/`

Table 2. Percentage of instances excluded by preprocessing, for $d = d_{\text{opt}} - 1$

number of sequences k	20	30	40	50
naïve kernel (%)	56.5	59.1	67.4	68.4
our preprocessing, from remaining (%)	24.3	15.9	36.4	44.4
total excluded instances (%)	67.1	65.6	79.3	82.4

$d = d_{\text{opt}} - 2$, more than 99 % of the instances are rejected by the naïve kernel or since $d < \frac{1}{2}\max_{i,j} d_{\text{H}}(s_i, s_j)$. Clearly, no instances are rejected for $d = d_{\text{opt}}$.

Solving Trivial Positions by Preprocessing. The second advantage of our method is the computation of positions that can be trivially solved during preprocessing, see Fig. 2. The percentage of fixed positions is especially high for the important case $d = d_{\text{opt}}$. In fact, an average of 41.0 % of the positions was fixed for these instances during preprocessing. We also observe that there is no "twilight zone" of fixed positions: In 57.8 % of the instances, more than 40 % of positions were fixed; in 38.5 % the data reduction did not fix any positions; and in less than 3.7 % of the instances we observed a fixation of 0–40 % of positions.

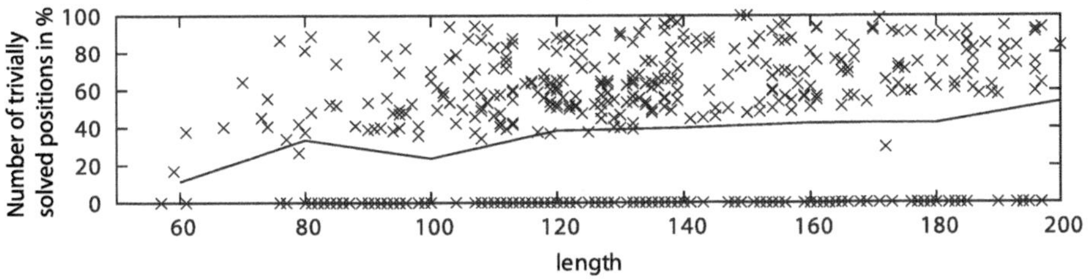

Fig. 2. Percentage of trivially solved positions in $\mathcal{P}$ for $d = d_{\text{opt}}$, plotted against the length L of the instance. Crosses represent individual instances, solid line is average percentage for intervals of width 20.

Running Times. We have implemented the algorithms of Gramm *et al.* [4], Ma and Sun [7], and the *MismatchCount* algorithm from Sect. 5, referred to as "*Gramm*", "*MaSun*" and "*MismatchCount*", respectively. These algorithms do not include any preprocessing beyond the naïve kernel. Name suffix "*RT*" indicates that preprocessing, algorithm engineering, and the use of $\mathcal{R}$ and $\mathcal{T}$ are enabled. For the *MismatchCount* algorithm, only the information from $\mathcal{P}$ is used, denoted as name suffix "*P*".

All algorithms have been implemented in Java and compiled with the Sun Java Standard Edition compiler version 1.6. All computations were done on a quad-core 2.2 GHz AMD Opteron processor with 5 GB of main memory under the Solaris 10 operating system. The presented running times are the core running

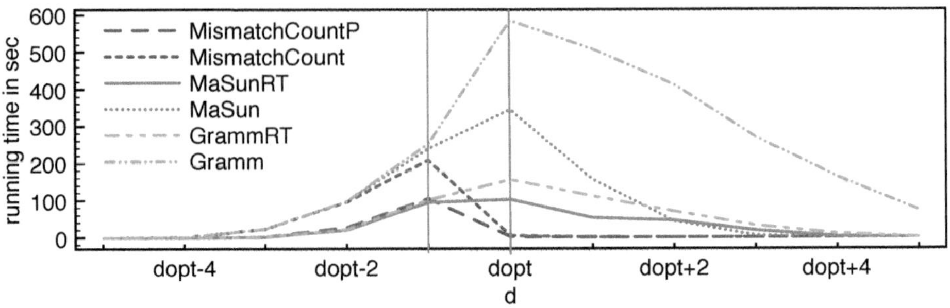

Fig. 3. Average running times for the 395 instances with $d_{\text{opt}} \leq 35$. Running times are depicted in dependency on varying d around d_{opt}.

times of the algorithms and do not include I/O or removal of clean columns. We set a time limit of ten minutes per instance.

We first show that running times of all algorithms are truly dominated by the cases $d = d_{\text{opt}} - 1$ and $d = d_{\text{opt}}$. To this end, we consider the 395 instances with $d_{\text{opt}} \leq 35$ of length $57 \leq L \leq 243$ after removing clean columns. Results are shown in Fig. 3. It is clearly visible that it is sufficient to concentrate on the two cases $d = d_{\text{opt}} - 1$ and $d = d_{\text{opt}}$. Algorithms *MaSun* and *Gramm* show large running times for both of these cases, whereas *MismatchCount* reaches its maximum running times for $d = d_{\text{opt}} - 1$ while it is faster for $d = d_{\text{opt}}$. Note that we cannot circumvent calling the algorithm with $d = d_{\text{opt}} - 1$ to ensure that d_{opt} is truly optimal.

We now show the dependency of running times on the parameter d_{opt}. Therefore, we pooled the instances with respect to the optimum center distance d_{opt}. For $d = d_{\text{opt}} - 1$ we excluded all instances where $d < L/k$ after removing clean columns, or $d < \frac{1}{2} \max_{i,j} d_{\text{H}}(s_i, s_j)$, as these obviously have no solution, leaving us with 241 instances. In Tab. 3, we show the percentage of instances that were rejected in less than 600 s, all other instances remain undecided by the algorithm. Running times for both $d = d_{\text{opt}} - 1$ and $d = d_{\text{opt}}$ are depicted in Fig. 4. Note that the unmodified algorithms of Gramm *et al.* and Ma and Sun usually run into the time limit at 600 s, true running times are expected to be much higher. We also see that the *MismatchCount* algorithm is much faster for the case $d = d_{\text{opt}}$ than for $d = d_{\text{opt}} - 1$.

Table 3. Percentage of instances rejected within different time limits, for $d = d_{\text{opt}} - 1$. '*MC*' denotes *MismatchCount* algorithm.

	MCP	*MC*	*MaSunRT*	*MaSun*	*GrammRT*	*Gramm*
time limit 600 s (%)	49.4	14.9	51.9	5.4	48.1	0
time limit 1 s (%)	45.6	12.4	44.0	4.1	43.6	0

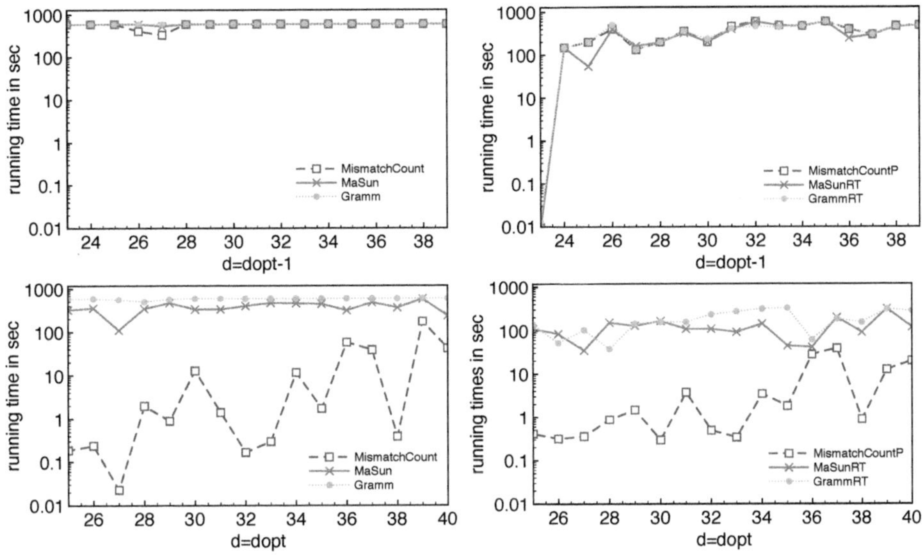

Fig. 4. Average running times for varying d_{opt}. Running times for the original (left) and improved implementations (right) for $d = d_{\text{opt}} - 1$ (top) and $d = d_{\text{opt}}$ (bottom). Logarithmic scale for running times.

7 Conclusion

We have presented an improved preprocessing for the CENTER STRING problem. This is based on the observation that for strings with an optimal center at distance d, there usually exist many pairs of strings with distance close or equal to $2d$. Our data reduction allows us to reject more instances that do not have a valid center string, and to draw conclusions about certain positions of a center string. We show how this information can be used in the search tree algorithms of Gramm *et al.* and Ma and Sun. We have also presented the *MismatchCount* algorithm for binary alphabets. In our experimental evaluation, we could show that our data reduction is very efficient and that the *MismatchCount* algorithm outperforms the other two in practice. Our data reduction is particularly helpful for tackling the case $d = d_{\text{opt}} - 1$, where the *MismatchCount* algorithm has maximum running times, as we can exclude more instances.

Currently, the *MismatchCount* algorithm does not use information encoded in $\mathcal{R}$ and $\mathcal{T}$. We are working on a modified version of the algorithm that will allow us to approach even larger instances in reasonable running time, as it will speed up computations for the "neuralgic" case $d = d_{\text{opt}} - 1$.

Acknowledgments

This research was partially funded by DFG grant STO 431/5.

References

1. Böcker, S., Jahn, K., Mixtacki, J., Stoye, J.: Computation of median gene clusters. J. Comput. Biol. 16(8), 1085–1099 (2009)
2. Davila, J., Balla, S., Rajasekaran, S.: Fast and practical algorithms for planted (l, d) motif search. IEEE/ACM Trans. Comput. Biol. Bioinformatics 4(4), 544–552 (2007)
3. Frances, M., Litman, A.: On covering problems of codes. Theory Comput. Systems 30(2), 113–119 (1997)
4. Gramm, J., Niedermeier, R., Rossmanith, P.: Fixed-parameter algorithms for closest string and related problems. Algorithmica 37(1), 25–42 (2003)
5. Lanctot, J.K., Li, M., Ma, B., Wang, S., Zhang, L.: Distinguishing string selection problems. Information and Computation 185(1), 41–55 (2003)
6. Liu, X., He, H., Sykora, O.: Parallel genetic algorithm and parallel simulated annealing algorithm for the closest string problem. In: Li, X., Wang, S., Dong, Z.Y. (eds.) ADMA 2005. LNCS (LNAI), vol. 3584, pp. 591–597. Springer, Heidelberg (2005)
7. Ma, B., Sun, X.: More efficient algorithms for closest string and substring problems. SIAM J. Comput. 39(4), 1432–1443 (2009)
8. Rahmann, S., Klau, G.W.: Integer linear programming techniques for discovering approximate gene clusters. In: Mandoiu, I., Zelikovsky, A. (eds.) Bioinformatics Algorithms: Techniques and Applications. Wiley Series on Bioinformatics: Computational Techniques and Engineering, ch. 9, pp. 203–222. Wiley, Chichester (2008)
9. Wang, L., Zhu, B.: Efficient algorithms for the closest string and distinguishing string selection problems. In: Deng, X., Hopcroft, J.E., Xue, J. (eds.) FAW 2009. LNCS, vol. 5598, pp. 261–270. Springer, Heidelberg (2009)
10. Wang, Y., Chen, W., Li, X., Cheng, B.: Degenerated primer design to amplify the heavy chain variable region from immunoglobulin cdna. BMC Bioinformatics 7(suppl. 4), S9 (2006)
11. Yanai, I., DeLisi, C.: The society of genes: networks of functional links between genes from comparative genomics. Genome Biol. 3(11), research0064 (2002)

Speeding Up Exact Motif Discovery by Bounding the Expected Clump Size

Tobias Marschall and Sven Rahmann

Bioinformatics for High-Throughput Technologies
at the Chair of Algorithm Engineering,
Computer Science Department, TU Dortmund,
44221 Dortmund, Germany
{tobias.marschall,sven.rahmann}@tu-dortmund.de

Abstract. The overlapping structure of complex patterns, such as IUPAC motifs, significantly affects their statistical properties and should be taken into account in motif discovery algorithms. The contribution of this paper is twofold. On the one hand, we give surprisingly simple formulas for the expected size and weight of motif clumps (maximal overlapping sets of motif matches in a text). In contrast to previous results, we show that these expected values can be computed without matrix inversions. On the other hand, we show how these results can be algorithmically exploited to improve an exact motif discovery algorithm. First, the algorithm can be efficiently generalized to arbitrary finite-memory text models, whereas it was previously limited to i.i.d. texts. Second, we achieve a speed-up of up to a factor of 135. Our open-source (GPL) implementation is available at `http://www.rahmannlab.de/software`.

1 Introduction

The *motif discovery problem* consists of finding extraordinary patterns (motifs) in a given set of strings (texts). A common application in computational biology is the discovery of transcription factor binding sites. Much attention has been given to this problem, resulting in hundreds of published (mostly heuristic) algorithms. For an overview, refer to the review articles [1,2,3,4]. Most algorithms can be classified as being either *alignment driven* or *pattern driven*. The former algorithms align a (fixed or variable) number of motif occurrences from which a position weight matrix (PWM) is constructed [5,6,7]; that means, the search space is the space of all alignments of motif occurrences. In contrast, pattern-driven algorithms search the space of all patterns [8,9,10,11]. To allow approximate matches, either a generalized alphabet (including wildcards) like the IUPAC alphabet is used or a Hamming neighborhood is added to the patterns. Pattern-driven algorithms theoretically allow computing a globally optimal motif by enumerating all patterns. The benefit of knowing the global optimum, however, varies with the used objective function. In a recent article (ISMB'09, [11]), we verified that the compound Poisson approximation to the pattern's p-value is accurate, and we developed a pattern-driven motif discovery algorithm that

V. Moulton and M. Singh (Eds.): WABI 2010, LNBI 6293, pp. 337–349, 2010.

returns the optimal motif with respect to this objective. We demonstrated the algorithm's performance on a benchmark set [12] and on the non-coding regions of *Mycobacterium tuberculosis'* genome. Here, we substantially speed up this algorithm and generalize it to higher-order text models.

Notation. For concreteness, we define $\Sigma := \{\mathtt{A}, \mathtt{C}, \mathtt{G}, \mathtt{T}\}$ as the nucleotide alphabet. All indexing starts at 0, that means $w = w_0 \dots w_{\ell-1}$ for $w \in \Sigma^\ell$. Define $\Lambda := 2^\Sigma \setminus \{\emptyset\}$, where 2^Σ is the power set of Σ, and note that each $g \in \Lambda$ uniquely maps to a IUPAC one-letter code; e.g. $\{\mathtt{A}, \mathtt{G}\}$ corresponds to the IUPAC code R. Each $m \in \Lambda^*$ is called *generalized string.* We define a *motif* of length ℓ to be an element of Λ^ℓ. We say $w \in \Sigma^\ell$ matches $m \in \Lambda^\ell$, written $w \lhd m$, if $w_i \in m_i$ for $0 \leq i < \ell$. The distribution of a random variable X is denoted by $\mathcal{L}(X)$. Row and column vectors are written $\langle x|$ and $|y\rangle$, respectively. All elements of the column vector $|1\rangle$ are 1. The identity matrix is denoted by $\mathbf{1}$. For a matrix A, the row-sum norm is denoted by $\|A\|_\infty := \max_{i=1,\dots,n} \sum_{j=1}^n |A_{ij}|$.

Random Texts. Let a word $w \in \Sigma^*$ be given. Assume that w occurs k times in a given text (or genome). Then, its p-value is the probability that it occurs k or more times in a random text of the same length. To compute this p-value, a text model has to be given. The simplest possibility is an i.i.d. model, that is, a distribution over the alphabet Σ. For genomic sequences, more elaborate models are necessary. In Markovian text models, the character distribution depends on a finite history. Character-emitting hidden Markov models (HMMs) are even more sophisticated. These three types of text models are all special cases of finite-memory text models. Formally, a random text is a stochastic process $(S_t)_{t\in\mathbb{N}_0}$, where S_t is a random variable giving the t-th character.

Definition 1 (Finite-memory text model). *A finite-memory text model is a tuple $(\mathcal{C}, c_0, \Sigma, \varphi)$, where $\mathcal{C}$ is a finite state space (called* context space*), $c_0 \in \mathcal{C}$ a start context, Σ an alphabet, and $\varphi : \mathcal{C}\times\Sigma\times\mathcal{C} \to [0,1]$ with $\sum_{\sigma\in\Sigma,\, c'\in\mathcal{C}} \varphi(c,\sigma,c') = 1$ for all $c \in \mathcal{C}$. The random variable giving the context after t steps is denoted C_t with $C_0 :\equiv c_0$. A probability measure is now induced by stipulating*

$$\mathbb{P}(S_0 \dots S_{n-1} = s,\, C_1 = c_1, \dots, C_n = c_n) := \prod_{i=0}^{n-1} \varphi(c_i, s_i, c_{i+1}) \tag{1}$$

for all $n \in \mathbb{N}_0$, $s \in \Sigma^n$, and $(c_1, \dots, c_n) \in \mathcal{C}^n$. We require the context process $(C_t)_{t\in\mathbb{N}_0}$ to converge to an equilibrium distribution.

The intuition behind this definition is that when in context c, we make a random choice to emit the next text character σ and transit to the next context c', the corresponding probability being $\varphi(c,\sigma,c')$. It follows that the probability for a text of length n is given by $\mathbb{P}(S_0 \dots S_{n-1} = s) := \sum_{c_1,\dots,c_n} \prod_{i=0}^{n-1} \varphi(c_i, s_i, c_{i+1})$, where c_0 is the fixed start context. This sum can be efficiently computed by dynamic programming. Similar models are used in [13] where they are called probability transducers. In this article, all analyses and algorithms are valid for arbitrary finite-memory text models.

Compound Poisson Distribution. The difficulty in computing the distribution of the number of word occurrences lies in possible self-overlaps. Although exact algorithms, for example using finite automata [14,15,16], are known, it remains infeasible to compute the exact p-values for a large set of motifs. This problem can be overcome by using a compound Poisson approximation. The idea is to decompose the set of motif occurrences into clumps of overlapping occurrences. Formally, a clump is a maximal set of overlapping motif occurrences. For example, the string `AAACACACTGACAT` contains two clumps of the word `ACA` (underlined). The random variable Z_i gives the size, i.e. the number of contained occurrences, of the i-th clump in the random text $(S_t)_{t\in\mathbb{N}_0}$. If no i-th clump exists, we set $Z_i := 0$; this happens with zero probability for infinite texts. The number of occurrences is now given by $\sum_{i=0}^{N_n-1} Z_i$ (ignoring border effects), where N_n is the number of clumps in a text of length n. The *compound Poisson approximation* assumes that N_n is Poisson distributed. This approximation is accurate for rare words [17].

Example 1 (Influence of clumping). Consider a random text of length 10,000 over $\Sigma = \{\mathtt{A}, \mathtt{C}, \mathtt{G}, \mathtt{T}\}$ where each character is distributed independently and uniformly. The expected number of occurrences for the two patterns `AAAAAAAAAA` and `AAAAAAAAAC` equals 0.0095 in both cases, but the probabilities of observing 10 or more occurrences differ greatly and amount to $2.982 \cdot 10^{-8}$ and $1.546 \cdot 10^{-27}$, respectively. The different behavior is reflected in the expected clump sizes that equal 4/3 and 1, respectively. When computing the compound Poisson approximations of these p-values, we obtain $2.986 \cdot 10^{-8}$ and $1.685 \cdot 10^{-27}$, respectively.

Definition 2 (Expected Clump Size). *The* expected clump size ψ *is*

$$\psi := \lim_{i\to\infty} \mathbb{E}(Z_i)\,. \tag{2}$$

Note that ψ is well-defined as the existence of limit (2) follows from the assumed convergence of the text model to an equilibrium distribution.

Results. Clumps and compound Poisson approximations to the occurrence count distribution of words have been studied extensively [18,19,20,21]. For the first time, however, we give a simple and general formula for the expected clump size that holds for sets of patterns and arbitrary finite-memory text models that are equivalent to character-emitting HMMs (Theorem 1). The formula is short and involves no laborious operations like matrix inversions. Furthermore, it can readily be generalized to the case where each pattern is given a weight (Theorem 2), which is useful to handle reverse complements in DNA motifs.

In [11], we introduced a motif discovery algorithm that finds (within a given motif space) the optimal motif with respect to its p-value (in compound Poisson approximation). Based on the new formula for the expected clump size, we derive bounds for the expected clump size for partially known patterns. These bounds allow us to improve our motif discovery algorithm in two respects. First, finding the globally optimal motif with respect to arbitrary finite-memory text models is now possible; before, the search was restricted to i.i.d. models. Second, the algorithm is substantially faster than before.

Before we compute and bound the expected clump size, Section 2 explains how motif discovery can benefit from such a procedure. In Section 3, formulas for expected clump size and expected clump weight are given and proved. Subsequently, in Section 4, bounds for partially known motifs are derived. Experiments showing the practical benefit are presented in Section 5.

2 Efficient Motif Discovery with Clump Size Bounds

The idea of performing a pattern-driven search by walking a suffix tree has long been known [8]. We took up this idea in [11] and furthered it to optimize p-values of degenerate motifs rather than their number of occurrences. The basic idea is to enumerate all candidate motifs from a given motif space in lexicographic order and skip parts of the search space that cannot contain motifs of interest. This is done by examining the suffix tree nodes that correspond to the prefixes of the current motif. If a prefix does not occur frequently enough to be interesting, all motifs sharing this prefix can be skipped. When optimizing p-values, we need to answer the following question: Given a motif space $\mathcal{M}$ to be searched, a motif prefix and the number of occurrences of this prefix, compute a lower bound for the p-value of all $m \in \mathcal{M}$ with this prefix. If this lower bound is too large, we can discard the prefix along with all its continuations.

Obviously, the number of occurrences is greater than or equal to the number of clumps. By compound Poisson approximation, the number of clumps has a Poisson distribution with motif-specific mean λ_m. Write $\mathcal{P}_\lambda(i) := \mathrm{e}^{-\lambda}\lambda^i/i!$. Assume that the motif m is observed n times, then

$$\text{p-value}(m) \geq \sum_{i=n}^{\infty} \mathcal{P}_{\lambda_m}(i) \geq \sum_{i=n}^{\infty} \mathcal{P}_{\lambda'}(i) \text{ for } \lambda_m \geq \lambda' ,$$

where the right inequality follows from the monotonicity of the cumulative Poisson distribution function in its parameter. Therefore, the problem is reduced to computing a lower bound $\lambda' \leq \lambda_m$ for the expected number of clumps

$$\lambda_m = \frac{\text{expected number of occurrences of } m}{\text{expected clump size of } m} .$$

To find a $\lambda' \leq \lambda_m$, we require a lower bound for the expected number of occurrences and an upper bound for the expected clump size.

In [11], the first problem was approached by partitioning the motif space into subsets containing motifs with equal expectation and computing the expectation exactly, which is easy for i.i.d. text models, but not practical for more complex text models. For the expected clump size, a global upper bound of 3 was used, which lead to long runtimes. Here, we present a more efficient strategy based on motif prefixes instead of whole motifs. Better individual bounds on the clump size will allow to prune larger parts of the search space.

A lower bound for the expected number of occurrences for all continuations of a partially known motif from $\mathcal{M}$ can be obtained by computing the equilibrium probability of the prefix and using the lowest possible continuation probability for each unknown character. This probability can be efficiently precomputed by examining, for all combinations of characters σ and text model contexts c, the probability that, starting from context c, character σ is generated.

We now derive a formula for the exact expected clump size and use it in Section 4 to derive the desired bound.

3 Computing the Expected Clump Size

In the following, we assume a length $\ell \geq 2$ and a pattern set $\mathcal{W} \subset \Sigma^\ell$ to be given. Each IUPAC motif can be represented as such a set $\mathcal{W}$. A length-ℓ substring of the random text $(S_t)_{t\in\mathbb{N}_0}$ is written $S_t^\ell := S_t \dots S_{t+\ell-1}$.

In finite-memory text models, it is not sufficient to look at strings S_t^ℓ only, but we simultaneously need to keep track of the context. To shorten notation, we define $X_t := (S_t^\ell, C_t)$; so $X_t = (w, c)$ means that word $w \in \Sigma^\ell$ starts at position t in S, and before generating its first letter, we are in context c, i.e., $S_t^\ell = w$ and $C_t = c$. We say that "word w occurs in context c at position t". Furthermore, we define $\mathcal{X} := \{(w,c) \in \mathcal{W} \times \mathcal{C} : \lim_{t\to\infty} \mathbb{P}(X_t = (w,c)) > 0\}$. Restricting attention to $\mathcal{X}$ is sufficient as pairs (w, c) with zero occurrence probability do not contribute to the expected clump size ψ (cf. Definition 2). To derive a formula for ψ, we need several definitions.

Definition 3 (Overlap probability function). *The* overlap probability function $\kappa : \mathcal{X} \times \mathcal{X} \to [0,1]$ *is defined by*

$$\kappa\big((w_1,c_1),(w_2,c_2)\big) :=$$
$$\sum_{i=1}^{\ell-1} \mathbb{P}\big(X_{t+i} = (w_2,c_2), S_{t+i-1}^\ell, \dots, S_{t+1}^\ell \notin \mathcal{W} \,|\, X_t = (w_1,c_1)\big).$$

By definition of finite-memory text models, the involved conditional probabilities do not depend on t.

Intuitively, $\kappa\big((w_1,c_1),(w_2,c_2)\big)$ is the probability that, having seen w_1 in context c_1, there follows another word from $\mathcal{W}$ in the same clump and that the next such word is w_2 in context c_2.

Example 2 (Overlap probability function). Let $\Sigma = \{\mathtt{A}, \mathtt{B}\}$ and consider an i.i.d. text model, that means a finite-memory text model with only one context c_0. Let the character probabilities be given by $p_\mathtt{A} = 0.1$, $p_\mathtt{B} = 0.9$. Let further $\mathcal{W} := \{\mathtt{AAA}, \mathtt{AAB}, \mathtt{ABA}\}$. We obtain $\kappa\big((\mathtt{AAA}, c_0), (\mathtt{AAA}, c_0)\big) = 0.1$, $\kappa\big((\mathtt{AAA}, c_0), (\mathtt{AAB}, c_0)\big) = 0.9$, and $\kappa\big((\mathtt{AAA}, c_0), (\mathtt{ABA}, c_0)\big) = 0$. The last probability is zero as ABA cannot right-overlap AAA without first creating an occurrence of AAB.

It is useful to view κ as a matrix. Thus, we define a bijective mapping $\iota : \mathcal{X} \to \{0, \dots, |\mathcal{X}| - 1\}$ and denote the resulting matrix as *overlap matrix* $K = \big(k_{\iota(w,c),\iota(w',c')}\big) \in \mathbb{R}^{|\mathcal{X}|\times|\mathcal{X}|}$, where $k_{\iota(w,c),\iota(w',c')} := \kappa\big((w,c),(w',c')\big)$. As all results hold independent of the choice of ι, we use pairs $(w,c) \in \mathcal{X}$ as indices.

Definition 4 (Word and clump start distributions). *The* word distribution vector, *denoted* $|p\rangle = \big(p_{(w,c)}\big) \in \mathbb{R}^{|\mathcal{X}|}$, *is given by*

$$p_{(w,c)} := \lim_{t\to\infty} \mathbb{P}\left(X_t = (w,c) \,|\, S_t^\ell \in \mathcal{W}\right).$$

It is the equilibrium probability to see word w in context c, given that a word from $\mathcal{W}$ is seen. The clump start distribution vector $|p^{start}\rangle = \left(p^{start}_{(w,c)}\right) \in \mathbb{R}^{|\mathcal{X}|}$ *is given by*

$$p^{start}_{(w,c)} := \lim_{t\to\infty} \mathbb{P}\left(X_t = (w,c) \,|\, S_t^\ell \in \mathcal{W}, S_{t-1}^\ell, \dots, S_{t-\ell+1}^\ell \notin \mathcal{W}\right). \tag{3}$$

It is the equilibrium probability to see word w in context c, given that a word from $\mathcal{W}$ is seen and starts a clump of such words.

Theorem 1 (Expected clump size). *Let a pattern set $\mathcal{W} \subset \Sigma^\ell$ be given such that $\|K\|_\infty < 1$. Then, its expected clump size is finite and given by*

$$\psi \;=\; \langle p^{start} | (\mathbf{1} - K)^{-1} | 1 \rangle \;=\; \frac{1}{1 - \langle p | K | 1 \rangle}.$$

The first equality is not surprising, and similar formulas have been known. The second equality, to our knowledge, is new and our first main result. To prove the theorem, we need additional definitions and an auxiliary lemma.

Definition 5 (Clump end vector). *The* clump end vector, *denoted* $|f\rangle = \big(f_{(w,c)}\big) \in \mathbb{R}^m$, *is given by*

$$f_{(w,c)} := \mathbb{P}\big(S_{t+1}^\ell, \dots, S_{t+\ell-1}^\ell \notin \mathcal{W} \,|\, X_t = (w,c)\big),$$

which is the conditional probability that, when seeing word w in context c, no further word from $\mathcal{W}$ follows in the same clump. Here $f_{(w,c)}$ does not depend on t due to conditioning on $C_t = c$.

We express f in terms of the overlap probability matrix:

$$f_{(w,c)} = 1 - \sum_{(w',c')\in\mathcal{X}} k_{(w,c),(w',c')} \quad \text{or} \quad |f\rangle = (\mathbf{1} - K)|1\rangle. \tag{4}$$

Definition 6 (Backward overlap function and matrix). *The* backward overlap function $\overleftarrow{\kappa} : \mathcal{X} \times \mathcal{X} \to [0,1]$ *is defined by*

$$\overleftarrow{\kappa}\big((w_1,c_1),(w_2,c_2)\big) :=$$
$$\lim_{t\to\infty} \sum_{i=1}^{\ell-1} \mathbb{P}\big(X_{t-i} = (w_2,c_2), S_{t-i+1}^\ell, \dots, S_{t-1}^\ell \notin \mathcal{W} \,|\, X_t = (w_1,c_1)\big).$$

Intuitively, $\overleftarrow{\kappa}\big((w_1,c_1),(w_2,c_2)\big)$ is the equilibrium probability that, observing w_1 in context c_1, there exists a preceding word from $\mathcal{W}$ in the same clump and that the immediately preceding such word is w_2 in context c_2. Note the symmetry to Definition 3; however, the conditional probability may depend on t, so we take the equilibrium limit. The *backward overlap matrix* $\overleftarrow{K}$ is defined accordingly.

Lemma 1. *For all* $(w_1, c_1), (w_2, c_2) \in \mathcal{X}$,

$$p_{(w_1,c_1)} k_{(w_1,c_1),(w_2,c_2)} = p_{(w_2,c_2)} \overleftarrow{k}_{(w_2,c_2),(w_1,c_1)}.$$

This is a "detailed balance" between p, K and $\overleftarrow{K}$: The equilibrium probability of seeing a word-context pair (w_1, c_1) followed by (w_2, c_2) in the same clump equals the equilibrium probability of (w_2, c_2) preceded by (w_1, c_1). It can be verified directly from the definitions.

Proof (Theorem 1, first part). Every clump can be uniquely decomposed into a sequence of overlapping occurrences of words from $\mathcal{W}$: A clump of size z starts with a word-context pair $x_1 = (w_1, c_1) \in \mathcal{X}$, and makes $z-1$ transitions to following word-context pairs $x_j = (w_j, c_j)$. The transition probabilities are given by the corresponding entries of K. The clump ends with a word-context pair $x_z = (w_z, c_z)$. In equilibrium,

$$\lim_{i\to\infty} \mathbb{P}(Z_i = z) = \sum_{x_1} \cdots \sum_{x_z} p^{start}_{x_1} k_{x_1,x_2} \cdots k_{x_{z-1},x_z} f_{x_z} = \langle p^{start} | K^{z-1} | f \rangle;$$

$$\psi = \lim_{i\to\infty} E(Z_i) = \sum_{z=1}^{\infty} z \cdot \langle p^{start} | K^{z-1} | f \rangle = \langle p^{start} | \Big(\sum_{z=1}^{\infty} z K^{z-1} \Big) | f \rangle \tag{5}$$

$$= \langle p^{start} | \Big(\sum_{z=0}^{\infty} K^z \Big)^2 | f \rangle = \langle p^{start} | \left(\mathbf{1} - K\right)^{-2} | f \rangle \tag{6}$$

$$= \langle p^{start} | \left(\mathbf{1} - K\right)^{-1} | 1 \rangle, \tag{7}$$

where the rearrangement from (5) to (6) is allowed as both series converge absolutely because $\|K\|_\infty < 1$. Equation (6) uses the value of a geometric series of matrices (see [22], Proposition 9.4.13) and (6)=(7) follows from (4). □

Proof (Theorem 1, second part). We now rewrite $|p^{start}\rangle$:

$$\begin{aligned}
p^{start}_{(w,c)} &= \lim_{t\to\infty} \mathbb{P}\left(X_t = (w,c) \,\middle|\, S^\ell_t \in \mathcal{W}, S^\ell_{t-1}, \ldots, S^\ell_{t-\ell+1} \notin \mathcal{W}\right) \\
&= \lim_{t\to\infty} \frac{\mathbb{P}\left(X_t = (w,c),\, S^\ell_{t-1}, \ldots, S^\ell_{t-\ell+1} \notin \mathcal{W}\right)}{\mathbb{P}\left(S^\ell_t \in \mathcal{W},\, S^\ell_{t-1}, \ldots, S^\ell_{t-\ell+1} \notin \mathcal{W}\right)} \\
&= \lim_{t\to\infty} \frac{\mathbb{P}\left(S^\ell_{t-1}, \ldots, S^\ell_{t-\ell+1} \notin \mathcal{W} \,\middle|\, X_t = (w,c)\right) \mathbb{P}\big(X_t = (w,c)\big)}{\sum_{(w_1,c_1)} \mathbb{P}\left(S^\ell_{t-1}, \ldots, S^\ell_{t-\ell+1} \notin \mathcal{W} \,\middle|\, X_t = (w_1,c_1)\right) \mathbb{P}\big(X_t = (w_1,c_1)\big)} \\
&= \frac{\left(1 - \sum_{(w',c')} \overleftarrow{k}_{(w,c),(w',c')}\right) \cdot p_{(w,c)}}{\sum_{(w_1,c_1)} \left(1 - \sum_{(w_2,c_2)} \overleftarrow{k}_{(w_1,c_1),(w_2,c_2)}\right) \cdot p_{(w_1,c_1)}}
\end{aligned} \tag{8}$$

$$= \frac{p_{(w,c)} - \sum_{(w',c')} p_{(w',c')} k_{(w',c'),(w,c)}}{\sum_{(w_1,c_1)} \left(p_{(w_1,c_1)} - \sum_{(w_2,c_2)} p_{(w_2,c_2)} k_{(w_2,c_2),(w_1,c_1)}\right)}, \tag{9}$$

where (8)=(9) follows from Lemma 1. Thus

$$\langle p_{start}| = \frac{\langle p|(\mathbf{1}-K)}{\langle p|(\mathbf{1}-K)|1\rangle} = \frac{\langle p|(\mathbf{1}-K)}{1-\langle p|K|1\rangle}.$$

The proof is completed by combining the above expression with (7) and noting that $\langle p|1\rangle = 1$, since p is a probability distribution. □

Weighted Patterns. We defined the clump size as the number of occurrences of words from $\mathcal{W}$ in a clump. That means, we assigned a weight of 1 to each occurrence. In this section, we permit individual weights for each $w \in \mathcal{W}$ by defining a weight function $\nu : \mathcal{W} \to \mathbb{R}$. This weighted case is important when patterns over the DNA alphabet are considered and the reverse complement is taken into account.

Example 3. Consider the DNA alphabet $\Sigma = \{\mathtt{A},\mathtt{C},\mathtt{G},\mathtt{T}\}$ and the pattern set $\mathcal{W} = \{\mathtt{AAT},\mathtt{ACT},\mathtt{ATT}\}$. Adding all reverse complements, we obtain the multiset $\mathcal{W}' = \{\{\mathtt{AAT},\mathtt{ACT},\mathtt{ATT},\mathtt{ATT},\mathtt{AGT},\mathtt{AAT}\}\}$ and thus weights $\nu(\mathtt{AAT}) = 2$, $\nu(\mathtt{ATT}) = 2$, $\nu(\mathtt{ACT}) = 1$, and $\nu(\mathtt{AGT}) = 1$.

Definition 7 (Weight vector and expected clump weight). *We define the* weight vector $|v\rangle \in \mathbb{R}^{|\mathcal{X}|}$ *by* $v_{(w,c)} := \nu(w)$ *for* $(w,c) \in \mathcal{X}$. *For a given weight vector* $|v\rangle$, *the random variable* W_i *denotes the weight of the i-th clump, i.e. the sum of the weights of the words forming the clump. The* expected clump weight *is defined as* $\psi_v := \lim_{i\to\infty} \mathbb{E}(W_i)$.

Theorem 2 (Expected clump weight). *Given a pattern set* $\mathcal{W} \subset \Sigma^\ell$ *such that* $\|K\|_\infty < 1$ *and a weight vector* $|v\rangle$, *then the expected clump weight is*

$$\psi_v = \langle p|v\rangle \cdot \psi = \frac{\langle p|v\rangle}{1-\langle p|K|1\rangle} < \infty.$$

Proof (Sketch). The idea is to combine a scaling and a homogeneity argument: The (asymptotic) expected number of occurrences of $\mathcal{W}$ per text character, say $\mu = \langle p|1\rangle \cdot \mu$, changes to $\langle p|v\rangle \cdot \mu$ when weights are assigned, as p is the equilibrium distribution restricted to $\mathcal{W}$. In equilibrium, all clumps have the same stochastic properties, and the distribution of words from $\mathcal{W}$ in clumps is p as well, because by definition words only occur in clumps. Hence the expected clump weight is $\langle p|v\rangle \cdot \psi$, and the result follows from Theorem 1. □

4 Bounding the Expected Clump Size of Motifs

We assume a motif space $\mathcal{M} \subseteq \Lambda^\ell$ to be defined by giving constraints on the number of allowed wildcard characters. For example, we might use the motif space given in [11], defined by $\ell = 10$ and allowing at most six $g \in \Lambda$ with $|g| = 2$ (IUPAC codes R, Y, W, S, K, M), zero characters with $|g| = 3$ (IUPAC codes B, D, H, V), and at most two characters with $|g| = 4$ (IUPAC code N). This space covers many biologically relevant motifs (see [11]).

The following theorem translates the results from the last section into bounds for the expected clump size of IUPAC motifs. We assume that text model and all motifs are such that the overlap probability matrix K for each motif satisfies $\|K\|_\infty < 1$, which means that there is zero probability for infinitely large clumps.

Theorem 3. *Consider a text model that converges to an equilibrium context distribution. Let a motif $m \in \Lambda^\ell$ and a bound $P < 1$ be given such that*

$$B_m := \max_{c \in \mathcal{C}} \sum_{i=1}^{\ell-1} \mathbb{P}\left(S_{t+i}^\ell \lhd m \mid S_t^\ell \lhd m, C_t = c\right) \leq P, \tag{10}$$

where B_m is independent of t due to conditioning on C_t.

Then, the expected clump size ψ_m satisfies $\psi_m \leq 1/(1 - B_m) \leq 1/(1 - P)$.

Here B_m, and thus also P, are upper bounds for the conditional probability that a given occurrence of m is right-overlapped by another occurrence.

Proof. Let $\mathcal{W} := \{w \in \Sigma^\ell : w \lhd m\}$. Applying the definitions of $|p\rangle$ and K,

$$\begin{aligned}
\langle p|K|1\rangle &= \sum_{(w,c)\in\mathcal{X}} \sum_{(w',c')\in\mathcal{X}} p_{(w,c)}\, k_{(w,c),(w',c')} \\
&= \lim_{t\to\infty} \sum_{(w,c)} \sum_{(w',c')} \sum_{i=1}^{\ell-1} \mathbb{P}\big(X_{t+i} = (w',c'), S_{t+i-1}^\ell, \dots, S_{t+1}^\ell \notin \mathcal{W} \,\big|\, X_t = (w,c)\big) \\
&\qquad\qquad \cdot \mathbb{P}\big(X_t = (w,c) \,\big|\, S_t^\ell \lhd m\big) \\
&\leq \lim_{t\to\infty} \sum_{(w,c)} \sum_{(w',c')} \sum_{i=1}^{\ell-1} \mathbb{P}\big(X_{t+i} = (w',c') \,\big|\, X_t = (w,c)\big)\, \mathbb{P}\big(X_t = (w,c) \,\big|\, S_t^\ell \lhd m\big) \\
&= \lim_{t\to\infty} \sum_{(w,c)} \sum_{(w',c')} \sum_{i=1}^{\ell-1} \mathbb{P}\big(X_{t+i} = (w',c'),\, X_t = (w,c) \mid S_t^\ell \lhd m\big) \\
&= \lim_{t\to\infty} \sum_{i=1}^{\ell-1} \mathbb{P}\big(S_{t+i}^\ell \lhd m \mid S_t^\ell \lhd m\big) \\
&= \lim_{t\to\infty} \sum_{c\in\mathcal{C}} \left(\sum_{i=1}^{\ell-1} \mathbb{P}\big(S_{t+i}^\ell \lhd m \mid S_t^\ell \lhd m, C_t = c\big) \right) \cdot \mathbb{P}\big(C_t = c \mid S_t^\ell \lhd m\big) \\
&\leq B_m \cdot \lim_{t\to\infty} \sum_{c\in\mathcal{C}} \mathbb{P}\big(C_t = c \mid S_t^\ell \lhd m\big) \;=\; B_m \cdot 1 \leq P < 1 .
\end{aligned}$$

Applying Theorem 1 yields the claimed result, as $x \mapsto 1/(1-x)$ is increasing. □

Given only a length-ℓ' motif prefix $m_0 \dots m_{\ell'-1}$, we derive a bound P for Theorem 3 that is valid for all possible continuations of this prefix within $\mathcal{M}$. To be useful in motif discovery as sketched in Section 2, fast calculation of the bound must be possible. We approach the problem by computing bounds $P_1, \dots, P_{\ell-1}$ for each possible shift separately such that

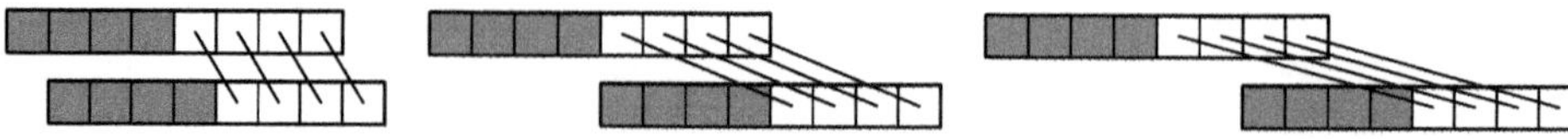

Fig. 1. Different overlap situations of partially known motifs are illustrated. The first part of the motif (gray) is known, while the second part (white) is unknown. Left: Shift of 1, the aligned known IUPAC characters must be compatible. Middle: Shift of 4, no known characters align, but the diagonal lines emphasize that, although unknown, the characters in the top motif are the same as the ones in the bottom motif. Right: Shift of 6, the known characters right of the previous occurrence must be present.

$$\max_{m_{\ell'},\ldots,m_{\ell-1}\in\Lambda:\, m\in\mathcal{M}} \max_{c\in\mathcal{C}} \mathbb{P}\left(S_{t+i}^{\ell} \lhd m \mid S_t^{\ell} \lhd m, C_t = c\right) \leq P_i \,.$$

Then, $P := P_1 + \ldots + P_{\ell-1}$ is a valid bound, i.e., $P \geq B_m$ for all continuations $m_{\ell'} \ldots m_{\ell-1}$ of $m_0 \ldots m_{\ell'-1}$.

Computing the bound P_i for shift i. There are different strategies for obtaining good bounds for different shifts.

For short shifts, where many known IUPAC characters overlap (Figure 1, left), we can bound the conditional probability that text character S_t matches the intersection of the overlapping IUPAC characters. For this, we use a precomputed table $P_{\max}$ of worst-case conditional probabilities (with respect to all possible contexts), and define for $g, g' \in \Lambda$

$$P_{\max}(g \mid g') := \max_{c\in\mathcal{C}} \mathbb{P}(S_t \lhd g \mid S_t \lhd g',\, C_t = c)\,,$$

a quantity that does not depend on t by Definition 1. Note that for incompatible IUPAC characters, i.e. $g \cap g' = \emptyset$, an overlap is impossible and $P_{\max}(g \mid g') = 0$.

The strategy of considering every position separately does not work effectively for longer shifts, as shown in Figure 1 (middle). For each position, the unknown character can be chosen unfavorably, such that a probability bound equals 1. To obtain a meaningful bound, we have to take into account that the top and bottom motifs are the same, and therefore unknown positions cannot be chosen independently. The idea is to partition the columns into groups such that each top character is the bottom character in the next column in that group, as illustrated in Figure 2. The groups defined in that way can then be bounded jointly. To this end, we precompute so-called *telescope bounds* for each (initial) IUPAC character: For all $g \in \Lambda$, define

$$P_{\mathrm{tel}}(g) := \max_{M\in\Lambda^*:\, M_0=g} \left(\prod_{i=0}^{|M|-2} P_{\max}(M_i|M_{i+1}) \right) P_{\max}(M_{|M|-1}|\Sigma)\,.$$

In the example in Figure 2, we bound the telescope product for $M = m_2m_4m_6$ by $P_{\mathrm{tel}}(m_2)$ and the product for $M = m_3m_5m_7$ by $P_{\mathrm{tel}}(m_3)$. To compute these bounds, we need not check all $M \in \Lambda^*$, but can restrict our attention to those M with $M_i \subsetneq M_{i+1}$ for $0 \leq i < |M|-1$. It can be shown that the maximum over all

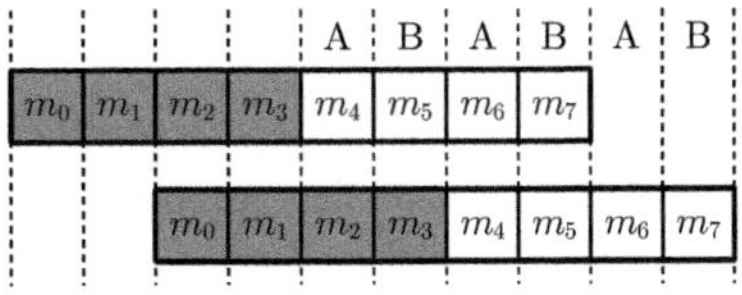

Fig. 2. For each set of columns labeled with the same letter (A or B), a bound can be obtained. For group A, a bound is given by $P_{\max}(m_2 \mid m_4)P_{\max}(m_4 \mid m_6)P_{\max}(m_6 \mid \Sigma)$. For group B, it is given by $P_{\max}(m_3 \mid m_5)P_{\max}(m_5 \mid m_7)P_{\max}(m_7 \mid \Sigma)$. Assuming that the characters $m_4, \ldots, m_7$ are unknown (white background), the maximum over all possible values of $m_4, \ldots, m_7$ must be used.

$M \in \Lambda^*$ is attained for an M with this property (proof omitted). Thus, $P_{\text{tel}}(g)$ can be precomputed for all $g \in \Lambda$. In the i.i.d. model, the telescope bound takes a particularly simple form because of cancellations: $P_{\text{tel}}(g) = \lim_{t\to\infty} \mathbb{P}(S_t \lhd g)$, the equilibrium probability of g.

In the right part of Figure 1, some of the *known* characters do not overlap the previous motif occurrence. For such a character g, the bound $P_{\max}(g|\Sigma)$ can be used. Based on the motif space and its constraints on the multiplicity of wildcard characters, the approach may also apply to columns where *unknown* characters do not overlap the previous occurrence. When, for example, the known characters are `ANNT` and at most two wildcards are allowed, the unknown characters must be from $\Sigma = \{\texttt{A}, \texttt{C}, \texttt{G}, \texttt{T}\}$ and the bound $\max_{\sigma\in\Sigma} P_{\max}(\sigma|\Sigma)$ can be used.

Depending on the motif prefix and the shift i, the above three ideas can be combined in different ways to produce a bound P_i. There are several case distinctions, which we omit in this extended abstract. Instead, we exemplarily consider the situation in Figure 2 (i.e. $\ell = 8$ and shift 2) for the motif prefix `ARRA` and calculate bound P_2. Recall that $\texttt{R} = \{\texttt{A}, \texttt{G}\}$. In two columns, known characters overlap and we use bounds from $P_{\max}$ for them. For the groups A and B, as shown in Figure 2, we employ the telescope bounds. Assuming an i.i.d. text model with uniform character distribution, we get

$$P_2 = P_{\max}(\texttt{A}|\texttt{R}) \cdot P_{\max}(\texttt{R}|\texttt{A}) \cdot P_{\text{tel}}(\texttt{R}) \cdot P_{\text{tel}}(\texttt{A}) = \frac{1}{2} \cdot 1 \cdot \frac{1}{2} \cdot \frac{1}{4} = \frac{1}{16}.$$

Calculations for the other shifts yield $P_1 = 1/8$, $P_3 = 1/16$, and $P_4 = P_5 = P_6 = P_7 = 1/64$. We obtain $P = P_1 + \ldots + P_7 = 5/16$ and a bound for the expected clump size of $1/(1-P) = 16/11 \approx 1.45$. For all possible continuations, the largest exact expected clump size is 1.3144 (for the motif `ARRANNNN`).

So far, we discussed only bounds for single IUPAC motifs. By virtue of Theorem 2, jointly handling a motif and its reverse complement poses no principal difficulties but involves several more case distinctions. Again, we omit the details in this extended abstract.

5 Experiments and Conclusions

Using the new bounds, we improve the IUPAC motif discovery algorithm presented in [11], where we assessed the quality of results and compared it to other algorithms. We only discuss runtimes here. Because of new compute cluster hardware (quad-core CPUs at 2.66 GHz), we re-ran the old experiment (twice as fast) to report comparable results in Table 1.

Table 1. Running times of *M. tuberculosis* motif discovery in non-coding regions using different algorithms, text models and restrictions. Times are single-core hours, i.e. running the algorithm on a single core would have taken this time. The p-value bound was given as 10^{-50}. In some cases, the search was restricted to motifs with 100 expected occurrences to avoid highly degenerate motifs and to reduce runtime. The motif space was chosen as in [11], defined by a motif length of 10, allowing at most six $g \in \Lambda$ with $|g| = 2$, zero with $|g| = 3$, and at most two with $|g| = 4$. (Runtime with * is estimated based on searching 2% of the motif space.)

Text model	Expectation restriction	Runtime of original alg.	Runtime of improved alg.	Speed-up factor
i.i.d.	≤ 100	127.2 h	21.2 h	6x
i.i.d.	none	*56,000.0 h	412.5 h	135x
Markov order 1	≤ 100	not possible	118.3 h	—
Markov order 2	≤ 100	not possible	800.3 h	—
Markov order 3	≤ 100	not possible	6,364.4 h	—

Formerly, one key to feasibility was bounding the expected number of motif occurrences by 100 in addition to setting a strict p-value bound of 10^{-50}. This restriction is no longer necessary. Even with the restriction, a speed-up factor of 6 is observed. Without the restriction, time is reduced by a factor of 135 with the new bounds. This shows that the new bounds are especially effective for more degenerate motifs. Additionally, direct optimization under a Markov text model was not feasible before.

As far as we are aware, our method is the only practically efficient algorithm that finds provably optimal IUPAC motifs in general finite-memory text models. Runtimes of several hundred single-core hours pose no practical problem on a medium sized cluster. Topics for future work are further engineering and fine-tuning the implementation, e.g. on massively-parallel hardware like GPUs.

References

1. Tompa, M., Li, N., Bailey, T.L., et al.: Assessing computational tools for the discovery of transcription factor binding sites. Nature Biotechnology 23(1), 137–144 (2005)
2. Sandve, G.K., Drabløs, F.: A survey of motif discovery methods in an integrated framework. Biology Direct 1(1), 11 (2006)
3. Das, M., Dai, H.K.: A survey of DNA motif finding algorithms. BMC Bioinformatics 8(suppl. 7), S21 (2007)

4. Narlikar, L., Ovcharenko, I.: Identifying regulatory elements in eukaryotic genomes. Briefings in Functional Genomics and Proteomics 8(4), 215–230 (2009)
5. Bailey, T.L., Williams, N., Misleh, C., Li, W.W.: MEME: discovering and analyzing DNA and protein sequence motifs. Nucleic Acids Research 34(suppl.2), W369–W373 (2006)
6. Hertz, G.Z., Stormo, G.D.: Identifying DNA and protein patterns with statistically significant alignments of multiple sequences. Bioinformatics 15(7-8), 563–577 (1999)
7. Rahmann, S., Marschall, T., Behler, F., Kramer, O.: Modeling evolutionary fitness for DNA motif discovery. In: Rothlauf, F. (ed.) Genetic and Evolutionary Computation Conference (GECCO), Montreal, Québec, Canada, pp. 225–232. ACM, New York (2009)
8. Sagot, M.F.: Spelling approximate repeated or common motifs using a suffix tree. In: Lucchesi, C.L., Moura, A.V. (eds.) LATIN 1998. LNCS, vol. 1380, pp. 374–390. Springer, Heidelberg (1998)
9. Pavesi, G., Mauri, G., Pesole, G.: An algorithm for finding signals of unknown length in DNA sequences. Bioinformatics 17(suppl. 1), S207–S214 (2001)
10. Sinha, S., Tompa, M.: A statistical method for finding transcription factor binding sites. In: Proceedings of the 8th International Conference on Intelligent Systems for Molecular Biology (ISMB), pp. 344–354 (2000)
11. Marschall, T., Rahmann, S.: Efficient exact motif discovery. Bioinformatics 25(12), i356–i364 (2009)
12. Sandve, G.K., Abul, O., Walseng, V., Drabløs, F.: Improved benchmarks for computational motif discovery. BMC Bioinformatics 8, 193 (2007)
13. Kucherov, G., Noé, L., Roytberg, M.: A unifying framework for seed sensitivity and its application to subset seeds. Journal of Bioinformatics and Computational Biology 4(2), 553–569 (2006)
14. Nicodème, P., Salvy, B., Flajolet, P.: Motif statistics. Theoretical Computer Science 287, 593–617 (2002)
15. Marschall, T., Rahmann, S.: Probabilistic arithmetic automata and their application to pattern matching statistics. In: Ferragina, P., Landau, G.M. (eds.) CPM 2008. LNCS, vol. 5029, pp. 95–106. Springer, Heidelberg (2008)
16. Nuel, G.: Pattern Markov chains: optimal Markov chain embedding through deterministic finite automata. Journal of Applied Probability 45, 226–243 (2008)
17. Stefanov, V., Robin, S., Schbath, S.: Waiting times for clumps of patterns and for structured motifs in random sequences. Discrete Appl. Math. 155(6-7), 868–880 (2007)
18. Schbath, S.: Compound Poisson approximation of word counts in DNA sequences. ESAIM: Probability and Statistics 1, 1–16 (1995)
19. Reinert, G., Schbath, S.: Compound Poisson and Poisson process approximations for occurrences of multiple words in Markov chains. Journal of Computational Biology 5(2), 223–253 (1998)
20. Pape, U.J., Rahmann, S., Sun, F., Vingron, M.: Compound Poisson approximation of the number of occurrences of a position frequency matrix (PFM) on both strands. Journal of Computational Biology 15(6), 547–564 (2008)
21. Bassino, F., Clément, J., Fayolle, J., Nicodème, P.: Constructions for clumps statistics. In: Proceedings of the Fifth Colloquium on Mathematics and Computer Science. Discrete Mathematics and Theoretical Computer Science, pp. 179–194 (2008)
22. Bernstein, D.S.: Matrix mathematics, 2nd edn. Princeton University Press, Princeton (2009)

Pair HMM Based Gap Statistics for Re-evaluation of Indels in Alignments with Affine Gap Penalties

Alexander Schönhuth[1,⋆], Raheleh Salari[2,⋆], and S. Cenk Sahinalp[2]

[1] Department of Mathematics, University of California at Berkeley
alexsch@math.berkeley.edu
[2] School of Computing Science, Simon Fraser University, Burnaby

Abstract. Although computationally aligning sequence is a crucial step in the vast majority of comparative genomics studies our understanding of alignment biases still needs to be improved. To infer true structural or homologous regions computational alignments need further evaluation. It has been shown that the accuracy of aligned positions can drop substantially in particular around gaps. Here we focus on re-evaluation of score-based alignments with affine gap penalty costs. We exploit their relationships with pair hidden Markov models and develop efficient algorithms by which to identify gaps which are significant in terms of length and multiplicity. We evaluate our statistics with respect to the well-established structural alignments from SABmark and find that indel reliability substantially increases with their significance in particular in worst-case twilight zone alignments. This points out that our statistics can reliably complement other methods which mostly focus on the reliability of match positions.

1 Introduction

Having been introduced over three decades ago [20] the sequence-alignment problem has remained one of the most actively studied topics in computational biology. While the vast majority of comparative genomics studies crucially depend on alignment quality inaccuracies abundantly occur. This can have detrimental effects in all kinds of downstream analyses [16]. Still, our understanding of the involved biases remains rather rudimentary [14, 17]. That different methods often yield contradictory statements [8] further establishes the need for further investigations into the essence of alignment biases and their consequences [14].

While the sequence-alignment problem virtually is that of inferring the correct placement of gaps, insertions and deletions (indels) have remained the most unreliable parts of the alignments. For example, Lunter et al. [17], in a whole-genome alignment study, observe 96% alignment accuracy for alignment positions which are far away from gaps while accuracy drops down to 56% when considering positions closely surrounding gaps. They also observe a downward bias in the number of inferred indels which is due to effects termed gap attraction and gap annihilation. Decreased numbers of inferred indels were equally observed in other recent studies [15, 23]. This points out that *numbers and size* of computationally inferred indels can make statements about alignment quality.

⋆ Joint first authorship.

V. Moulton and M. Singh (Eds.): WABI 2010, LNBI 6293, pp. 350–361, 2010.

The *purpose of this paper* is to *systematically address* such questions. We develop a statistical framework by which to efficiently compute probabilities of the type

$$\mathbb{P}(I_{d,\mathcal{A}}(x,y) \geq k \,|\, L_{\mathcal{A}}(x,y) = n, \mathrm{Sim}_{\mathcal{A}}(x,y) \in [\sigma_1, \sigma_2]) \qquad (1)$$

where (x, y) has been randomly sampled from an appropriate pool of protein pairs. In the following pools contain protein pairs which have a (either false or true positive) structural SABMark [27] (see below) alignment. In case of, for example, all pairs of human proteins, (1) would act as null distribution for human. $\mathcal{A}$ is a local or global optimal, score-based alignment procedure with affine gap penalties such as the affine gap cost version of the Needleman-Wunsch (NW) algorithm [20,12] or the Smith-Waterman (SW) algorithm [29,31], $L_{\mathcal{A}}(x,y)$ is the length of the alignment, $\mathrm{Sim}_{\mathcal{A}}(x,y)$ denotes alignment similarity that is the fraction of perfectly matching and "well-behaved" mismatches vs. "bad" mismatches (as measured in terms of biochemical affinity [21]) and gap positions. $I_{d,\mathcal{A}}(x,y)$ finally denotes the length of the d-th longest gap in the alignment. In summary, (1) can be read as the probability that a NW resp. SW alignment of length n and similarity between σ_1 and σ_2 contains at least d gaps of length k and the reasoning is that gaps which make part of significant such gap combinations are more likely to reflect true indels. Significance is determined conditioned on the length $L(x, y)$ of the alignment as well as alignment similarity $\mathrm{Sim}(x, y)$. The reasoning behind this is that longer alignments are more likely to accumulate spurious indels such that only increased gap length and multiplicity are significant signs of true indels. Increased similarity $\mathrm{Sim}(x, y)$, however, indicates that already shorter and less gaps are more likely to reflect true indels simply because an alignment of high similarity is an overall more trustworthy statement. In summary, we provide a *statistically sound, systematic* approach to answering questions such as "Am I to believe that 4 gaps of size at least 6 in an alignment of length 200 and similarity 50 are likely to reflect true indels" as motivated by the recent studies [15,17,23].

We opted to address these questions for score-based alignments with affine gap costs for two reasons:

1. To employ score-based such alignments still is a most popular option among most bioinformatics practitioners.

2. Such alignments can be alternatively viewed as Viterbi paths in pair HMMs. While exact statistics on Viterbi paths are hard to obtain and beyond the scope of this study we obtain reasonable approximations by "Viterbi training" sensibly modified versions of the hidden Markov chains which underlie the pair HMMs.

We evaluate our statistics on the well-established SABmark [27] alignments. SABmark is a database of structurally related proteins which cover the entire known fold space. The "Twilight Zone set" was particularly designed to represent the worst case scenario for sequence alignment. While we obtain good results also in the more benign "Superfamilies set" of alignments it is that worst case scenario of twilight zone alignments where our statistics prove their particular usefulness. Here significance of gap multiplicity is crucial while significance of indel length alone does not necessarily indicate enhanced indel quality.

Related Work: [18] re-evaluate match (but not indel) positions in global score-based alignments by obtaining reliability scores from suboptimal alignments. Similarly, [28] derive reliability scores also for indel positions in global score-based alignments. However, the method presented in [28] reportedly only works in the case of more than 30% sequence identity. Related work where structural profile information is used is [30] whereas [6] re-align rather than re-evaluate.

Posterior decoding algorithms (see e.g. [9, 17, 3] for most recent approaches) are related to re-evaluation of alignments insofar as posterior probabilities can be interpreted as reliability scores. However, how to score indels as a whole by way of posterior decoding does not have a straightforward answer. We are aware of the potential advantages inherent to posterior decoding algorithms—it is work in progress of ours to combine the ideas of pair HMM based posterior decoding aligners with the ideas from this study[1].

To assess statistical significance of alignment phenomena is certainly related to the vastly used Altschul-Dembo-Karlin statistics [13, 7, 1] where score significance serves as an indicator of protein homology.

To devise computational indel models still remains an area of active research (e.g. [24, 5, 4, 17, 19]). However, the community has not yet come to a final conclusion.

Last but not least, the algorithms presented here are related to the algorithms developed in [25] where the special case of $d = 1$ for only global alignments in (2) was treated to explore the relationship of indel length and functional divergence. The advances achieved here are to provide null models also for the more complex case of local alignments and to devise a dynamic programming approach also for the case $d > 1$ which required to develop generalized inclusion-exclusion arguments.

Just like in [25] note that *empirical statistics approaches fail* for the same reasons that have justified the development of the Altschul-Dembo-Karlin statistics: sizes of samples are usually much too small. Here samples (indels in alignments) are subdivided into bins of equal alignment similarity and then further into bins of equal length n and d-th longest indel size k.

1.1 Summary of Contributions

As above-mentioned, our work is centered around computation of probabilities

$$\mathbb{P}(I_d(x,y) \geq k \mid L(x,y) = n, \text{Sim}(x,y) \in [\sigma_1, \sigma_2]). \tag{2}$$

We refer to this problem as *Multiple Indel Length Problem (MILP)* in the following. Our contributions then are as follows:

1. We are the first ones to address this problem and derive appropriate Markov chain based null models from the pair HMMs which underlie the NW resp. SW algorithms to yield *approximations* for the probabilities (2).

2. Despite having a natural formulation, the inherent Markov chain problem had no known efficient solution. We present the first efficient algorithm to solve it.

3. We demonstrate the usefulness of such statistics by showing that significant gaps in both global and local alignments indicate increased reliability in terms of identifying true structural indel positions. This became particularly obvious for worst-case twilight zone alignments of at most 25% sequence identity.

---

[1] Note that although we derive statistical scores for indels as a whole our evaluation in the Results section will refer to counting individual indel positions.

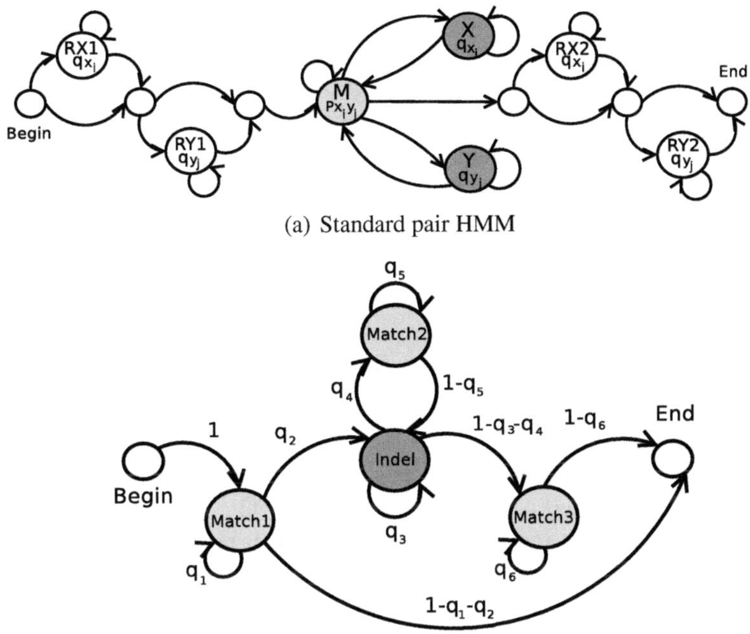

(a) Standard pair HMM

(b) Markov Chain

Fig. 1. Standard pair HMM corresponding to local Smith-Waterman alignments and the Markov Chain whose generative statistics, after Viterbi training, approximate the Viterbi statistics of the pair HMM for local alignments

4. Thereby we deliver statistical evidence of that computational alignments are biased in terms of numbers and sizes of gaps as described in [17, 23]. In particular too little numbers of gaps can reflect alignment artifacts.

5. Re-evaluation of indels in score-based both local and global alignments had not been explicitly addressed before, in particular, reliable solutions for worst-case twilight zone alignments were missing. Our work adds to (rather than competes with) the above-mentioned related work.

In *summary*, we have complemented extant methods for score-based alignment re-evaluation. Note that none of the existing methods explicitly addresses indel reliability but rather focus on the reliability of substitutions.

2 Methods

2.1 Pair HMMs and Viterbi Path Statistics

In the following we only treat the more complex case of local Smith-Waterman alignments. See [25] for the case of global Needleman-Wunsch like alignments where in the following the statistical models derived in [25] have to be, mutatis mutandis, plugged into the computations of the subsequent subsections.

A local Smith-Waterman alignment with affine gap penalties of two sequences $x = x_1 ... x_w, y = y_1 ... y_z$ is associated with the most likely sequence of hidden states (i.e. the

Viterbi path) in the pair HMM of Fig. 1(a) [10]. The path of hidden states translates to an alignment of the two sequences by emitting the necessary symbols along the run. Statistics on Viterbi paths in HMMs pose hard mathematical problems and have not been fully understood. In analogy to [25], we construct a Markov chain whose common, generative statistics mimick the Viterbi statistics of interest here. Hence probabilities derived from this Markov chain serve as approximations of (2). We do this by the following steps:

1. We take the Markov chain of the pair HMM in Fig. 1(a) as a template.
2. We add two match states $M1$, $M3$. The original match state is $M2$.
3. We merge the initial resp. terminal regions into one start resp. end state.
4. We collapse states X and Y into one indel state I.

The Markov chain approach is justified by the fact that consecutive runs in Viterbi paths are approximately governed by the *geometric distribution* which is precisely what a Markov chain reflects. The second point is to take into account the non-stationary character of the original Markov chain. Note that in local alignments, initial and final consecutive stretches of (mis)matches are longer than intermediate (mis)match stretches which translates to $q_1, q_6 > q_5$ in Fig. 1(a). See an extended version of this paper [26] for more detailed discussions and tables. Point 3 merely reflects that we are only interested in statistics on alignment regions. Point 4 finally accounts for that we do not make a difference between insertions and deletions due to the involved symmetry (relative to exchanging sequences).

2.2 Algorithmic Solution of the MILP

We define $C_{n,k,d}$ to be the set of sequences over the alphabet B, M1, I, M2, M3, E (for Begin, Match1, Indel, Match2, Match3 and End) of length n that contain at least d consecutive I stretches of length at least k. Let $A_n := \{X_n = \mathrm{M3}, X_{n+1} = \mathrm{E}\}$ be the set of sequences with an alignment region of length n. We then suggest the following procedure to compute approximations of the probabilities (2) where $T(\sigma_1, \sigma_2)$ is supposed to be a pool of protein pairs (x, y) whose alignments exhibit alignment similarity $\mathrm{Sim}(x, y) \in [\sigma_1, \sigma_2]$.

1: Compute alignments for all sequence pairs in $T(\sigma_1, \sigma_2)$.
2: Infer parameters $q_1, q_2, q_3, q_4, q_5, q_6$ of the Markov chain by Viterbi training it with the alignments.
3: $n \leftarrow$ length of the alignment of x and y
4: Compute $\mathbb{P}(C_{n,k,d} \cap A_n)$ as well as $\mathbb{P}(A_n)$, the probabilities that the Markov chain of Fig. 1(b) generates sequences from $C_{n,k,d} \cap A_n$ and A_n
5: *Output*

$$\mathbb{P}(C_{n,k,d} \mid A_n) = \frac{\mathbb{P}(C_{n,k,d} \cap A_n)}{\mathbb{P}(A_n)} \tag{3}$$

as an *approximation* for (2).

The idea of step 1 and 2 is to specifically train the Markov chain to generate alignments from the pool $T(\sigma_2, \sigma_2)$. In our setting, Viterbi training translates to counting M_1-to-M_1, M_1-to-I, I-to-I, I-to-M_2, M_2-to-M_2 and M_3-to-M_3 transitions in the alignments under consideration to provide maximum likelihood estimates for q_1, q_2, q_3, q_4, q_5 and q_6.

2.3 Efficient Computation of $\mathbb{P}(C_{n,k,d} \cap A_n)$

The problem of computing probabilities of the type (2) has been made the problem of computing the probability that the Markov chain generates sequences from $C_{n,k,d} \cap A_n$ and A_n. While computing $\mathbb{P}(A_n) = \mathbb{P}(X_n = \mathrm{M}_3) \cdot \mathbb{P}(X_{n+1} = \mathrm{E} \mid X_n = \mathrm{M}_3)$ is an elementary computation, the question of efficient computation and/or closed formulas for probabilities of the type $\mathbb{P}(C_{n,k,d} \cap A_n)$ had not been addressed in the mathematical literature and poses a last, involved problem.

The approach taken here is related to the one taken in [25], which treated the special case of single consecutive runs (i.e. $d = 1$) in the context of the two-state Markov chains which reflect null models for global alignments. We generalize this in two aspects. First, we provide a solution for more than two states (our approach applies for arbitrary numbers of states). Second, we show how to deal with multiple runs.

The probability event design trick inherent to our solution was adopted from that of [22]. The solution provided in [22] can be used for the (rather irrelevant) case of global alignments with linear gap penalties, i.e. gap opening and extension are identically scored. See also [11,2] for related mathematical treatments of the i.i.d. case.

In the following, let $i, j \in \{\mathrm{B}, \mathrm{M}_1, \mathrm{I}, \mathrm{M}_2, \mathrm{M}_3, \mathrm{E}\}$ be indices ranging over the alphabet of Markov chain states. Let $e_i \in \mathbb{R}^6$ be the standard basis vector of $\mathbb{R}^6$ having a 1 in the i-th component and zero elsewhere. For example, $e_{\mathrm{I}} = (0, 0, 1, 0, 0, 0), e_{\mathrm{M}_3} = (0, 0, 0, 0, 1, 0)$. We furthermore denote the standard scalar product on $\mathbb{R}^6$ by $\langle ., . \rangle$.

Efficient computation of the probabilities $\mathbb{P}(C_{n,k,d} \cap A_n)$ is obtained by a dynamic programming approach. As usual, we collect the Markov chain parameters (in accordance with Fig. 1(b)) into a state transition probability matrix

$$P = (p_{ij} := \mathbb{P}(X_t = i \mid X_{t-1} = j))_{i,j \in \{\mathrm{B,M1,I,M2,M3,E}\}} \tag{4}$$

such that, for example $p_{I,M3} = 1 - q_3 - q_4$ and an initial probability distribution vector $\pi = e_B = (1, 0, 0, 0, 0, 0)^T$. The initial distribution reflects that we start an alignment from the 'Begin' state. More formally, $\mathbb{P}(X_0 = \mathrm{B}) = 1$. For example, according to the laws that govern a Markov chain, the probability of being in the indel state I at position t in a sequence generated by the Markov chain is

$$\mathbb{P}(X_t = \mathrm{I}) = \langle e_I, P^t \pi \rangle = \langle e_I, P^t e_B \rangle. \tag{5}$$

It can be seen that naive approaches to computing $\mathbb{P}(C_{n,k,d} \cap A_n)$ result in runtimes that are exponential in n, the length of the alignments, which is infeasible. Efficient computation of these probabilities is helped by adopting the event design trick of [22]. In detail, we define

$$D_{t,k} := \{X_t = \mathrm{I}, ..., X_{t+k-1} = \mathrm{I}, X_{t+k} \neq \mathrm{I}\} \tag{6}$$

to be the set of sequences that have a run of state I of length k that stretches from positions t to $t + k - 1$ and ends at position $t + k - 1$, that is, the run is followed by a visit of state different from I at position $t + k$.

We further define

$$\pi_{\mathrm{I}} := \frac{1}{(1-p_{\mathrm{II}})} \cdot (p_{\mathrm{BI}}, p_{\mathrm{M_1I}}, 0, p_{\mathrm{M_2I}}, p_{\mathrm{M_3I}}, p_{\mathrm{EI}})^T \tag{7}$$

which can be interpreted as the state the Markov chain is in if we know that the Markov chain has left state I at the time step before. Consider $\mathbb{P}(X_{t+s} = \mathrm{I} \mid X_{t-1} = \mathrm{I}, X_t \neq \mathrm{I})$ as the probability that the Markov chain is in state I at period $t+s$ after having been in the state π_{I} at period t (note that this probability is independent of t as we deal with a homogeneous Markov chain). Similarly $\mathbb{P}(A_{t+k+s} \mid D_{t,k})$ is the probability that the Markov chain transits from state M_3 to state E at position $t+k+s+1$ while it has a run of state I of length k that stretches from positions t to $t+k-1$ and ends at position $t+k-1$. Lastly, we introduce the variables

$$Q_{l,m} := \sum_{\substack{1 \le s_1,\ldots,s_m \le l \\ s_1+\ldots+s_m = l}} \mathbb{P}(X_{s_1} = \mathrm{I}) \prod_{i=2}^{m} \mathbb{P}(X_{t+s_i} = \mathrm{I} \mid X_{t-1} = \mathrm{I}, X_t \neq \mathrm{I}) \tag{8}$$

$$R_{L,m} := \sum_{l=m}^{L} Q_{l,m} \mathbb{P}(A_{t+k+L-l} \mid D_{t,k}), \quad 1 \le m \le L \le n \tag{9}$$

for $1 \le m \le l \le n$ where the sum reflects summing over partitions of the integer l into m positive, not necessarily different, integers s_i. We then obtain the following lemma a proof of which needs a generalized inclusion-exlusion argument. See an extended version of this paper [26] for the proof.

Lemma 1.

$$\mathbb{P}(C_{n,k,d} \cap A_n) = \sum_{m=1}^{\lfloor \frac{n}{k+1} \rfloor} (-1)^{m+d} \binom{m-1}{d-1} \cdot \left(p_{\mathrm{II}}^{k-1}(1-p_{\mathrm{II}})\right)^m \cdot R_{n-mk,m}. \tag{10}$$

The consequences of the above considerations can be summarized in the following theorem.

Theorem 1. *A full table of values* $\mathbb{P}(C_{n,k,d} \cap A_n), k \le n \le N$ *can be computed in* $O(N^3)$ *runtime.*

Proof. Observing the recursive relationship

$$Q_{l,m} = \sum_{s=1}^{l-m+1} \mathbb{P}(X_{t+s} = \mathrm{I} \mid X_{t-1} = \mathrm{I}, X_t \neq \mathrm{I}) Q_{l-s,m-1}, \quad m > 1 \tag{11}$$

yields a standard dynamic programming procedure by which the ensemble of the $Q_{l,m}$ and the $R_{L,m}$ $(1 \le m \le l, L \le N)$ can be computed in $O(N^3)$ runtime. This also requires that the values $\mathbb{P}(X_s = \mathrm{I}), \mathbb{P}(X_{t+s} = \mathrm{I} \mid X_{t-1} = \mathrm{I}, X_t \neq \mathrm{I})$ have been precomputed which can be done in time linear in N. After computation of the $Q_{l,m}$ and the $R_{L,m}$, computation of the $\mathbb{P}(C_{n,k,d} \cap A_n), 1 \le k \le n \le N$ then equally requires $O(N^3)$ time which follows from lemma 1. ◇

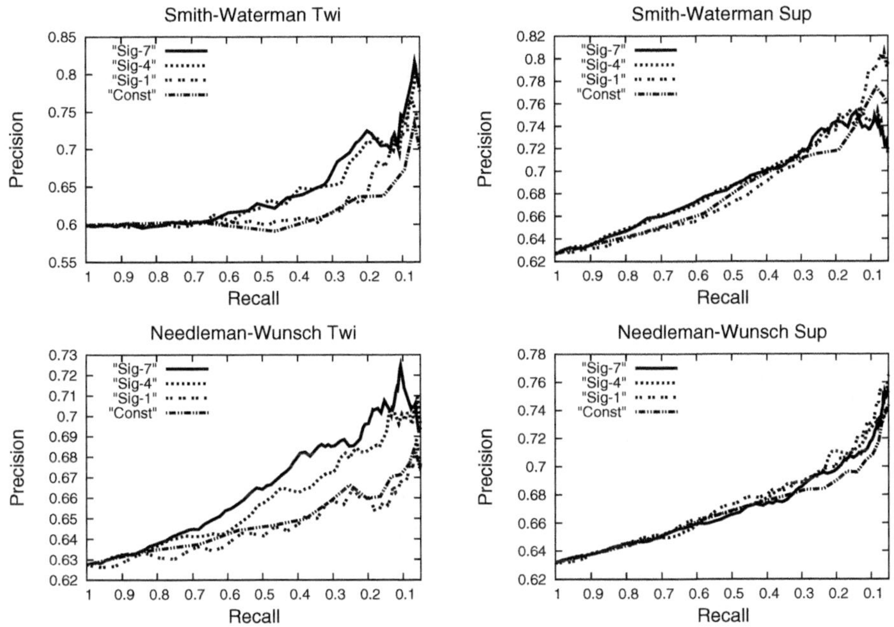

Fig. 2. Precision-Recall curves for the different sets of computational alignments. Recall is lowered through lowering the significance threshold θ for the strategies $\mathrm{Sig_D}(\theta)$ ($\theta = 1.0$ for maximal recall of 1.0) and for raising indel length in the baseline strategy Const (length = 1 for maximal recall of 1.0).

3 Results

Data. We downloaded both the "Superfamilies" (Sup) and "Twilight Zone" (Twi) datasets together with their structural alignment information from SABmark 1.65 [27], including the suggested false positive pairs (that is structurally unrelated, but apparently similar sequences, see [27] for a detailed description). While Sup is a more benign set of structural alignments where protein pairs can be assumed to be homologous and which contains alignments of up to 50% identity, Twi is a worst case scenario of alignments between only 0-25 % sequence identity where the presence of a common evolutionary ancestor remains unclear.

To calculate pairwise global resp. local alignments we used the "GGSEARCH" resp. "LALIGN" tool from the FASTA sequence comparison package [21]. As a substitution matrix, BLOSUM50 (default) was used. GGSEARCH resp. LALIGN implement the classical Needleman-Wunsch (NW) resp. Smith-Waterman (SW) alignment algorithm both with affine gap penalties. We subsequently discarded global resp. local alignments of an e-value larger than 10.0 resp. 1.0, as suggested as a default threshold setting [21], in order to ensure to only treat alignments which can be assumed not to be entirely random.

We then subdivided the resulting 4 groups (NW Twi, NW Sup, SW Twi and SW Sup) of computational alignments into pools of alignments of similarity in $[\sigma, \sigma + 10]$

where σ ranged from 20 to 90. We then trained parameters (using also the false positive SABmark alignments in order to obtain unbiased null models) for the $36 = 4 \times 9$ different Markov chains (2-state as in [25] resp. 6-state as described here for global resp. local) and computed probability tables as described in the Methods section. After computation of probability tables, false positive alignments were discarded. See [26] for Markov chain parameters and plots.

The remaining (non false-positive) NW Twi, NW Sup, SW Twi and SW Sup alignments contained 179018, 407629, 20853 and 86233 gap positions contained in 122701, 276082, 17776 and 68513 gaps. In the global alignments this includes also initial and end gaps.

3.1 Evaluation Strategies

Based on efficient computation of probabilities of the type (2) we devise strategies $\text{Sig}_\text{D}(\theta)$ for predicting indel reliability in NW and SW alignments where $D = 1, 4, 7$. Let K be the length of the L-th longest indel in the NW resp. SW alignment of proteins x, y. In strategy $\text{Sig}_\text{D}(\theta)$, this indel is classified as reliable if

$$\text{Sig}_\text{D}(\theta): \quad \mathbb{P}(I_{\min(D,L)}(x,y) \geq K \mid L(x,y), \text{Sim}(x,y)) \leq \theta. \tag{12}$$

In other words, we look up whether it is significant that an alignment of length $L(x,y)$ and similarity $\text{Sim}(x,y)$ contains at least L resp. D, in case of $D > L$ resp. $D \leq L$, indels of size K. Note that in strategy $\text{Sig}_1(\theta)$ that is for $D = 1$, since $L \geq 1$ hence $\min(D,L) = 1$, an indel of length K is evaluated as reliable if and only if the indel is significantly long without considering its relationship with the other gaps in the alignment. This is different for strategy $\text{Sig}_7(\theta)$ where, for example, the 6-th longest indel is evaluated as reliable if it is significant to have at least 6 indels of that length ($\min(D,L) = 6$) whereas the 8-th longest indel is supposed to be reliable if there are at least 7 indels of that length ($\min(D,L) = 7$). Note that in strategy $\text{Sig}_7(\theta)$ already shorter indels are classified as reliable in case that there are many indels of that length in the alignment which is not the case in strategy $\text{Sig}_1(\theta)$. Clearly, raising D beyond 7 might make sense. For sake of simplicity only, we restricted our attention to $D = 1, 4, 7$.

As a simple baseline method we suggest Const which considers an indel as reliable if its length exceeds a constant threshold. Both raising the constant length threshold in Const and lowering θ in $\text{Sig}_\text{D}(\theta)$ lead to reduced amounts of indels classified as reliable.

Evaluation Measures. We found that for both global and local alignments further evaluation of gaps of length at most 4 and length greater than 30 (global) resp. 20 (local) did not make much sense. See the extended version [26] for some basic statistics on such gaps. However, for gaps of length ranging from 5 to 20 resp. 30 in local resp. global alignments a significance analysis made sense.

We evaluated the indel positions in gaps of length $5 - 20$ resp. $5 - 30$ in local resp. global alignments by defining a true positive (TP) to be a computational gap position which is classified as reliable (meaning that it is found to be significant by $\text{Sig}_\text{D}(\theta), D = 1, 4, 7$ or long enough by Const) and coincides with a true structural

Table 1. Relationship between Recall and θ (displayed as $-\log(\theta)$) for strategies $\mathrm{Sig_D}$ and indel length (= IL) for strategy Const (= Con)

Recall	$-\log(\theta)$			IL	$-log(\theta)$			IL	$-\log(\theta)$			IL	$-log(\theta)$			IL
1.0	0.0	0.0	0.0	5	0.0	0.0	0.0	5	0.0	0.0	0.0	5	0.0	0.0	0.0	5
0.75	2.0	2.0	1.0	5	2.0	2.0	1.0	6	19.0	18.5	6.5	6	21.0	19.5	7.0	6
0.5	2.5	2.5	1.5	6	3.0	3.0	1.5	7	28.0	24.0	10.5	8	30.0	27.0	11.5	8
0.25	3.5	3.5	2.5	8	4.5	4.5	3.0	10	38.5	33.0	16.0	11	41.5	36.0	18.0	11
	$\mathrm{Sig_7}$	$\mathrm{Sig_4}$	$\mathrm{Sig_1}$	Con	$\mathrm{Sig_7}$	$\mathrm{Sig_4}$	$\mathrm{Sig_1}$	Con	$\mathrm{Sig_7}$	$\mathrm{Sig_4}$	$\mathrm{Sig_1}$	Con	$\mathrm{Sig_7}$	$\mathrm{Sig_4}$	$\mathrm{Sig_1}$	Con
	SW Twi				**SW Sup**				**NW Twi**				**NW Sup**			

indel position in the reference structural alignment as provided by SABmark. Correspondingly, a false positive (FP) is a gap position classified as reliable which cannot be found in the reference alignment. A true negative (TN) is a gap position not classified as reliable and not a structural indel position and a false negative (FN) is not classified as reliable but refers to a true structural indel position. Recall, as usual, is calculated as $TP/(TP + FN)$ whereas Precision (also called PPV=Positive Predictive Value) is calculated as $TP/(TP + FP)$.

3.2 Discussion of Results

Results are displayed in Figure 2 where we have plotted Precision vs. Recall while lowering θ for the strategies $\mathrm{Sig_D}(\theta)$ and increasing indel length for the baseline method Const. While Recall $= 1.0$ relates to $\theta = 1.0$ in the strategies $\mathrm{Sig_D}$ maximal recall relates to indel length 5 in the strategy Const. Table 1 displays further supporting statistics on the relationship between choices of θ resp. indel length and Recall.

A first look reveals that indel reliability clearly increases for increasing indel length—longer indels are more likely to contain true indel positions. However, further improvements can be achieved by classifying indels as reliable according to significance. For the Sup alignments improvements over the baseline method are only slight. For both local and global alignments strategy $\mathrm{Sig_1}$ is an option in particular when it comes to achieving utmost precision which can be raised up to 0.8. For the Twi alignments differences are obvious. More importantly, just considering indel length without evaluating multiplicity does not serve to achieve substantially increased Precision. Here, multiplicity is decisive which in particular confirms the findings on twilight zone alignments reported in [23]. In the Twi alignments Precision can be raised up to about 0.7. Note that [28] achieve 0.7 Precision on both match and gap positions for structural alignments (not from SABmark) of $25 - 30\%$ identity while reporting that their evaluation does not work for alignments of less than 25% identity which renders it not applicable for the Twi alignments. The posterior decoding aligner FSA which outperformed all other multiple aligners in terms of Precision on both (mis)match and gaps in the entire SABmark dataset, comprising both Sup and Twi [3] report Precision of 0.52 (all other aligners fall below 0.5) without further re-evaluation of their alignments. This lets us conclude that our statistical re-evaluation makes an interesting complementary contribution to alignment re-evaluation.

4 Conclusion

Most recent studies have again pointed out that computational alignments of all kinds need further re-evaluation in order to avoid detrimental effects in downstream analyses of comparative genomics studies. While exact gap placement is at the core of aligning sequence positive prediction rates are worst within or closely around inferred indels. Here we have systematically addressed that indel size and multiplicity can serve as indicators of alignment artifacts. We have developed a pair HMM based statistical evaluation pipeline which can soundly distinguish between spurious and reliable indels in alignments with affine gap penalties by measuring indel significance in terms of indel size and multiplicity. As a result we are able to reliably identify indels which are more likely to enclose true structural indel positions as provided by SABmark, raising positive prediction rates up to 0.7 even for worst-case twilight zone alignments of maximal 25% sequence identity. Since previous approaches predominantly addressed re-evaluation of match/mismatch positions we think that we have made a valuable, complementary contribution to the issue of alignment re-evaluation. Future work of ours is concerned with re-evaluation of pair HMM based posterior decoding aligners which have proven to be superior over score-based aligners in a variety of aspects.

Acknowledgments. AS is funded by donation from David DesJardins, Google Inc. We would like to thank Lior Pachter for helpful discussions on Viterbi path statistics.

References

1. Altschul, S.F., Gish, W.: Local alignment statistics. Methods in Enzymology 266, 460–480 (1996)
2. Bassino, F., Clement, J., Fayolle, J., Nicodeme, P.: Constructions for Clumps Statistics. In: MathInfo 2008 (2008), www.arxiv.org/abs/0804.3671
3. Bradley, R.K., Roberts, A., Smoot, M., Juvekar, S., Do, J., Dewey, C., Holmes, I., Pachter, L.: Fast statistical alignment. PLoS Computational Biology 5(5), e1000392 (2009)
4. Cartwright, R.A.: Logarithmic gap costs decrease alignment accuracy. BMC Bioinformatics 7, 527 (2006)
5. Chang, M.S.S., Benner, S.A.: Empirical analysis of protein insertions and deletions determining parameters for the correct placement of gaps in protein sequence alignments. Journal of Molecular Biology 341, 617–631 (2004)
6. Cline, M., Hughey, R., Karplus, K.: Predicting reliable regions in protein sequence alignments. Bioinformatics 18 (2), 306–314 (2002)
7. Dembo, A., Karlin, S.: Strong limit theorem of empirical functions for large exceedances of partial sums of i.i.d. variables. Annals of Probability 19, 1737–1755 (1991)
8. Dewey, C.N., Huggins, P.M., Woods, K., Sturmfels, B., Pachter, L.: Parametric alignment of Drosophila genomes. PLoS Computational Biology 2, e73 (2006)
9. Do, C.B., Mahabhashyam, M.S., Brudno, M., Batzoglou, S.: ProbCons: Probabilistic consistency-based multiple sequence alignment. Genome Research 15, 330–340 (2005)
10. Durbin, R., Eddy, S., Krogh, A., Mitchison, G.: Biological sequence analysis. Cambridge University Press, Cambridge (1998)
11. Fu, J.C., Koutras, M.V.: Distribution theory of runs: a Markov chain approach. Journal of the American Statistical Association 89(427), 1050–1058 (1994)

12. Gotoh, O.: An improved algorithm for matching biological sequences. Journal of Molecular Biology 162, 705–708 (1982)
13. Karlin, S., Altschul, S.F.: Methods for assessing the statistic significance of molecular sequence features by using general scoring schemes. Proceedings of the National Academy of Sciences of the USA 87, 2264–2268 (1990)
14. Kumar, S., Filipski, A.: Multiple sequence alignment: In pursuit of homologous DNA positions. Genome Research 17, 127–135 (2007)
15. Loeytynoja, A., Goldman, N.: An algorithm for progressive multiple alignment of sequences with insertions. Proceedings of the National Academy of Sciences of the USA 102 (30), 10557–10562 (2005)
16. Loeytynoja, A., Goldman, N.: Phylogeny-aware gap placement prevents errors in sequence alignment and evolutionary analysis. Science 320, 1632–1635 (2008)
17. Lunter, G., Rocco, A., Mimouni, N., Heger, A., Caldeira, A., Hein, J.: Uncertainty in homology inferences: Assessing and improving genomic sequence alignment. Genome Research 18 (2007), doi:10.1101/gr.6725608
18. Mevissen, H., Vingron, M.: Quantifying the local reliability of a sequence alignment. Stochastic Models of Sequence Evolution including Insertion-Deletion Events. Protein Engineering 9(2), 127–132 (1996)
19. Miklos, I., Novak, A., Satija, R., Lyngso, R., Hein, J.: Stochastic Models of Sequence Evolution including Insertion-Deletion Events. In: Statistical Methods in Medical Research 2009 (2008), doi:10.1177/096228020809950
20. Needleman, S.B., Wunsch, C.D.: A general method applicable to the search for similarities in the amino acid sequence of two proteins. Journal of Molecular Biology 48, 443–453 (1970)
21. Pearson, W.R., Lipman, D.J.: Improved tools for biological sequence comparison. Proc. Natl. Acad. Sci. USA 85, 2444–2448 (1988)
22. Peköz, E.A., Ross, S.M.: A simple derivation of exact reliability formulas for linear and circular consecutive-k-of-n F systems. Journal of Applied Probability 32, 554–557 (1995)
23. Polyanovsky, V.O., Roytberg, M.A., Tumanyan, V.G.: A new approach to assessing the validity of indels in algorithmic pair alignments. Biophysics 53(4), 253–255 (2008)
24. Qian, B., Goldstein, R.A.: Distribution of indel lengths. Proteins: Structure, Function and Bioinformatics 45, 102–104 (2001)
25. Schönhuth, A., Salari, R., Hormozdiari, F., Cherkasov, A., Sahinalp, S.C.: Towards improved assessment of functional similarity in large-scale screens: an indel study. Journal of Computational Biology 17(1), 1–20 (2010)
26. Schönhuth, A., Salari, R., Sahinalp, S.C.: Pair HMM based gap statistics for re-evaluation of indels in alignments with affine gap penalties—Extended Version (2010), `http://arxiv.org/abs/1006.2420`
27. Van Walle, I., Lasters, I., Wyns, L.: SABmark - a benchmark for sequence alignment that covers the entire known fold space. Bioinformatics 21, 1267–1268 (2005)
28. Schlosshauer, M., Ohlsson, M.: A novel approach to local reliability of sequence alignments. Bioinformatics 18 (6), 847–854 (2002)
29. Smith, T.M., Waterman, M.: Identification of common molecular subsequences. Journal of Molecular Biology 147, 195–197 (1981)
30. Tress, M.L., Jones, D., Valencia, A.: Predicting reliable regions in protein alignments from sequence profiles. Journal of Molecular Biology 330 (4), 705–718 (2003)
31. Waterman, M.S., Eggert, M.: A new algorithm for best subsequences alignment with application to tRNA-rRNA comparisons. J. MoL. BioL. 197, 723–728 (1987)

Quantifying the Strength of Natural Selection of a Motif Sequence

Chen-Hsiang Yeang

Institute of Statistical Science, Academia Sinica, Taipei, Taiwan
chyeang@stat.sinica.edu.tw

Abstract. Quantification of selective pressures on regulatory sequences is a central question in studying the evolution of gene regulatory networks. Previous methods focus primarily on single sites rather than motif sequences. We propose a method of evaluating the strength of natural selection of a motif from a family of aligned sequences. The method is based on a Poisson process model of neutral sequence substitutions and derives a birth-death process of the motif occurrence frequencies. The selection coefficient is treated as a penalty for the motif death rate. We demonstrate that the birth-death model closely approximates statistics generated from simulated data and the Poisson process assumption holds in mammalian promoter sequences. Furthermore, we show that a considerably higher portion of known transcription factor binding motifs possess high selection coefficients compared to negative controls with high occurrence frequencies on promoters. Preliminary analysis supports the potential applications of the model to identify regulatory sequences under selection.

Summary

Motivation

Many sequence motifs –such as transcription factor binding sites – are present in multiple locations of the genomes. Due to their functional constraints, selective pressures are often exerted on the evolution of sequence motifs. Quantification of selective pressures on motifs is a powerful tool to study the evolution of biological systems and to identify functionally important motifs from sequence data. However, most existing methods either target specific sites (rather than motif sequences) or apply only to two species. Consequently, a quantitative model and algorithm to evaluate the selective strength of motif sequences from multiple species need to be developed.

Main results

We propose a simple model of neutral (i.e., selection-free) evolution of motif sequences. The model hypothesizes that each position undergoes an independent random sequence substitution. The occurrence of a motif results from stochastic additions (birth) and removals (death) based on sequence substitutions. We derive the neutral and selective models of motif evolution and propose an algorithm to quantify the selective strength of a motif based on the two models.

V. Moulton and M. Singh (Eds.): WABI 2010, LNBI 6293, pp. 362–373, 2010.

To validate its utility we demonstrate a considerably higher portion of known transcription factor binding motifs possess strong selective strengths compared to random controls. In contrast, a conservation score fails to separate functional motifs and the short random motifs that occur frequently on promoters.

Significance
Initial analysis indicates that the birth-death model is adequate for the neutral evolution of motifs. Furthermore, selective coefficients outperform conservation scores in separating functional motifs from random sequences. The results justify the use of our model and algorithm in studying the evolution of functional motifs and identifying de novo functional motifs.

1 Introduction

High sequence similarity of protein-coding genes between distant species has led to the shift of focus in studying the evolution of non protein-coding regions. One central issue in this area is to gauge the selective pressure of a sequence motif. Cis-regulatory elements or regulatory RNAs may possess strong sequence specificity and resist random drifts. It is therefore possible to identify these elements from the sequences of multiple genes and organisms. One can align the promoter sequences of orthologous genes and apply motif-finding algorithms to identify the conserved motifs [1]. Conservation alone, however, may not confer natural selection since it also depends on the rate of neutral evolution, sequence length and complexity, population structure, and other factors. A variety of methods have been proposed to detect/quantify natural selection from sequences, including the ratios of non-synonymous to synonymous substitution rates $\frac{K_a}{K_s}$, [2], likelihood scores from a background sequence substitution model [3], comparison of intra-specific variation versus inter-specific divergence [4], deviation between heterozygosity and number of segregation sites [5], and comparison of SNP frequencies in distinct haplotype groups [6]. Despite the rich literature in detecting natural selection from sequences, the majority of the studies consider the evolution of single sites instead of motifs. Furthermore, most of these models require intra-specific polymorphism data which may not be available, and the $\frac{K_a}{K_s}$ test applies only to protein-coding regions.

To overcome these drawbacks, we propose a method of evaluating the strength of natural selection of a motif from aligned sequences. The method is based on a simple neutral model of sequence substitution: what is the distribution of motif occurrences in a sequence of fixed length if each position undergoes an independent sequence substitution? The rate of sequence substitution, the entire sequence length, evolutionary distances of sampled species and sequence complexity of the motif determine the rates of addition (birth) and deletion (death) of motifs in neutral evolution. In contrast, a motif under purifying selection such as a transcription factor binding site often populates on promoters and resists deletions. We quantify natural selection by a coefficient penalizing the rate of motif deletions, and develop an algorithm to estimate the maximum-likelihood selection coefficient.

Our model resembles the probabilistic model of promoter evolution in [7] as both models define a motif as a collection of fixed-length sequences and employ continuous-time Markov processes on sequence substitutions. However, our model differs from [7] by discarding sequence-specific substitution rates, considering the evolution of the motif occurrence frequencies, and being applied to the aligned sequences of more than two species.

The birth-death model approximates the empirical distributions derived from simulated data. Analysis on the 5kb upstream promoters of 34 mammalian genomes also validates the underlying hypothesis of the model – Poisson process of sequence substitution. We then calculate the selection coefficients of 388 known transcription factor binding motifs and random sequences. The selection coefficient distribution of known motifs is significantly tilted to high values compared to random controls, suggesting the tendency of positive selection of many transcription factor binding motifs. In contrast, the magnitudes of conservation (fraction of species containing the motif) on transcription factor binding motifs are not higher than random controls.

2 Methods

2.1 Overview

Our method is based on a neutral model of independent sequence substitution in each position. The distribution of motif counts depends on (1)the rate of sequence substitution, (2)the time interval of interest, (3)the promoter sequence length, (4)the degeneracy and complexity of the motif in the sequence space.

In the neutral model a Poisson process is employed to the sequence substitution of each position. The instantaneous rates of additions (birth) and deletions (death) of the motif can be derived from the sequence substitution model. In contrast, if purifying selection occurs to the motif then the death rate is penalized by a constant. The evolution of motif occurrence frequency distributions is thus expressed as a system of differential-difference equations parameterized by the penalty constant and the four factors described above. We can calculate the motif count distributions by simulating the differential-difference equations. Furthermore, according to the simulated distributions we apply binary search to find the penalty constant that maximizes the likelihood score of aligned sequences. The penalty constant characterizes the strength of natural selection.

2.2 A Poisson Process Model of Sequence Substitution

A Poisson process is probably the simplest model of sequence substitution [8]. In an infinitesimal time interval dt the nucleotide sequence of a position transitions to another base with probability λdt. Denote $n_P(t)$ the cumulative number of sequence changes at time t. The transitions from t to $t + dt$ follow

$$\begin{aligned} P(n_P(t+dt) = N+1|n_P(t) = N) &= \lambda dt. \\ P(n_P(t+dt) = N|n_P(t) = N) &= 1 - \lambda dt. \end{aligned} \tag{1}$$

and the conditional probability at a finite time interval t is

$$P(n_P(t) = N | n_P(0) = 0) = \frac{(\lambda t)^N}{N!} e^{-\lambda t}. \tag{2}$$

Poisson processes are Markovian as conditional probabilities are invariant with time shifts. Suppose $n_P(t)$ is observed at time points $t_i's$ with interval $T_i's$:

$$t_{i+1} = t_i + T_i. \tag{3}$$

Denote $m_i \equiv n_P(t_{i+1}) - n_P(t_i)$ as the number of sequence changes in time interval $(t_i, t_{i+1}]$. The log likelihood of the data is

$$\begin{aligned} L(\lambda) &= \sum_i \log P(n_P(t_{i+1}) - n_P(t_i) = m_i) \\ &= \sum_i m_i \log(\lambda T_i) - \lambda T_i + C. \end{aligned} \tag{4}$$

By taking the derivative of $L(\lambda)$ with respect to λ the maximum likelihood rate is

$$\hat{\lambda} = \frac{N}{T}. \tag{5}$$

where N is the total number of changes along each time interval and T is the sum of all time intervals.

In this work we assume the Poisson process rate λ is identical in all positions and across all lineages and estimate λ from a family of aligned sequences and their phylogenetic tree. The parsimonious sequences of the internal nodes of the tree are inferred by a dynamic programming algorithm [9]. In brief, for the aligned sequences at each position we construct a cost function of assigning a sequence configuration to the internal nodes of the tree as the total number of sequence substitutions along each branch of the tree. The cost function along a tree can be recursively computed. Denote v as an internal node and $u \in \mathcal{B} \equiv \{A, C, G, T\}$ as a nucleotide. The cost function of $v = u$ becomes

$$C(v = u) = \sum_i \max_{u_i \in \mathcal{B}} [C(v_i = u_i) + d(u, u_i)]. \tag{6}$$

where each v_i is a child of v and $d(u, u_i)$ denotes the distance between bases u and u_i. The reconstructed ancestral sequences maximize equation 6 and can be recursively computed using dynamic programming.

2.3 A Birth-Death Model for the Neutral Evolution of Motif Occurrences

The major contribution of this study is a neutral model of motif evolution. Motif occurrences can be modeled as a birth-death process [10]. The birth and death rates are determined by the sequence substitution rate and the degeneracy of the motif sequences.

A motif $\mathcal{M} \subset \mathcal{B}^l$ is defined as a collection of nucleotide sequences of length l. Degenerate symbols in IUPAC format are allowed in $\mathcal{M}$. For instance, R denotes purines (A or G) and Y denotes pyrimidines (C or T). Given a promoter sequence $\mathcal{S}$ of length l_s and the sequence substitution rate λ at each position, we want to model the distributions of $n(t)$, the frequency of motif occurrences in sequence $\mathcal{S}$ at time t.

We first consider the sequence evolution in a window of length l. There are 4^l possible sequences, and each sequence $s \in \mathcal{B}^l$ can be labeled as either a member of the motif ($s \in \mathcal{M}$) or not ($s \notin \mathcal{M}$). These sequences comprise an undirected graph $G = (V, E)$, where a node $v \in V$ denotes a sequence and an edge $e = (v_1, v_2)$ denotes the two sequences v_1 and v_2 differing at one position. $\mathcal{M}$ constitutes a subset of nodes in G, and the evolution of the sequences in an l-mer window can be viewed as a Markov random walk on G. In an infinitesimal time interval a sequence is allowed only to transition to neighboring nodes in G. The overall rate of transitions is the sequence substitution rate of the entire window λl. With an independent and identically distributed (iid) assumption this rate is equally divided among all the neighboring nodes.

We are interested in the transition rate from a non-motif sequence to a motif sequence or vice versa. In principle this rate depends on the initial and final states of each transition and is quite complicated. To simplify the model we use two numbers to characterize the average fraction of motif $\rightarrow$ non-motif transitions and vice versa.

$$\begin{aligned} r_{01} &= \frac{|\{(v_1,v_2)\in E: v_1 \notin \mathcal{M}, v_2 \in \mathcal{M}\}|}{|\{(v_1,v_2)\in E: v_1 \notin \mathcal{M}\}|}. \\ r_{10} &= \frac{|\{(v_1,v_2)\in E: v_1 \in \mathcal{M}, v_2 \notin \mathcal{M}\}|}{|\{(v_1,v_2)\in E: v_1 \in \mathcal{M}\}|}. \end{aligned} \quad (7)$$

r_{01} is the fraction of all non-motif $\rightarrow$ motif transitions among all transitions from non-motifs. For simplicity we expect the non-motif $\rightarrow$ motif transitions and non-motif $\rightarrow$ non-motif transitions are distributed by a ratio $\frac{r_{01}}{1-r_{01}}$. A reciprocal argument applies to the motif $\rightarrow$ non-motif transitions for r_{10}.

An equal transition rate to each sequence may not be an adequate assumption as the distribution of vertebrate genes has a strong bias in the CpG islands [11]. Consequently, we calibrate the ratios r_{01} and r_{10} by the background frequencies of nucleotides (P_A, P_C, P_G, P_T):

$$\begin{aligned} r_{01} &= \frac{\sum_{\{(v_1,v_2)\in E: v_1 \notin \mathcal{M}\}} w(v_1,v_2)\delta(v_2 \in \mathcal{M})}{\sum_{\{(v_1,v_2)\in E: v_1 \notin \mathcal{M}\}} w(v_1,v_2)}. \\ r_{10} &= \frac{\sum_{\{(v_1,v_2)\in E: v_1 \in \mathcal{M}\}} w(v_1,v_2)\delta(v_2 \notin \mathcal{M})}{\sum_{\{(v_1,v_2)\in E: v_1 \in \mathcal{M}\}} w(v_1,v_2)}. \end{aligned} \quad (8)$$

where $w(v_1, v_2)$ is the nucleotide background probability of v_2 at the position where v_1 and v_2 differ. For instance, $w(AGGC, AGTC) = P_T$. $\delta(.)$ is an indicator function. r_{01} and r_{10} are weighted by the background nucleotide frequencies such that more transitions are allocated to GC-rich sequences.

Summarizing the discussions above the transitions of motif occurrence of an l-mer window conform with the following equations:

$$\begin{aligned} P(n(t+dt) = 1 | n(t) = 0) = \lambda l r_{01} dt. \\ P(n(t+dt) = 0 | n(t) = 1) = \lambda l r_{10} dt. \end{aligned} \quad (9)$$

We then extend the analysis to the entire promoter sequence of length l_s. In an infinitesimal time interval dt, $n(t) = n$ can only increase/decrease by 1 or remain intact. Assuming the motif instances on the promoter do not overlap, there are ln positions occupied by existing motifs and $l_s - ln$ free positions. The ln occupied positions are divided into n independent windows, and the motif $\rightarrow$ non-motif transitions of each window follow equation (9.2). Thus the "death rate" of motif occurrence on the entire sequence is multiplied by n:

$$P(n(t+dt) = n-1|n(t) = n) = \lambda l r_{10} n dt. \quad (10)$$

The "birth rate" of motif occurrence is more difficult to analyze because the number of windows depends on the actual positions of existing motifs and these windows are not independent. For simplicity we approximate the number of independent l-mer windows among free positions by $l_s - ln + l + 1$, the number of l-mer windows in $l_s - ln$ consecutive positions. Thus the "birth rate" of motif occurrence on the entire sequence is multiplied by $l_s - ln + l + 1$:

$$P(n(t+dt) = n+1|n(t) = n) = \lambda l r_{01}(l_s - ln + l + 1)dt. \quad (11)$$

Equations (11) and (10) specify the birth and death rates of motif occurrences in an infinitesimal time interval. The distribution $P_n(t) \equiv P(n(t) = n)$ of motif occurrences over time can be expressed as the differential-difference equations:

$$\begin{array}{ll} \frac{dP_0(t)}{dt} & = \mu(1)P_1(t) - \lambda(0)P_0(t). \\ \frac{dP_n(t)}{dt} & = \lambda(n-1)P_{n-1}(t) + \mu(n+1)P_{n+1}(t) - (\lambda(n) + \mu(n))P_n(t). \\ \lambda(n) & = \lambda l r_{01}(l_s - ln + l + 1). \\ \mu(n) & = \lambda l r_{10} n. \end{array} \quad (12)$$

2.4 A Birth-Death Model of the Selective Evolution of Motif Occurrences

The purpose of constructing a neutral model of motif evolution is to identify the motif sequences that undergo purifying selection. Intuitively, purifying selection penalizes decrements of a functional motif on the promoter. Thus we divide the death rates by a selection coefficient:

$$\mu'(n) = \frac{\mu(n)}{s}. \quad (13)$$

When $s > 1$, the process of motif deletion slows down and more motifs are accumulated. This phenomenon is consistent with purifying selection. s is different from the conventional definition of selection coefficients in population genetics, which denotes the deviation of genotype frequencies from the neutral model.

2.5 Evaluating the Selection Coefficient of the Birth-Death Model

To evaluate the strength of purifying selection we apply both neutral and selective models to aligned sequences. Figure 1 outlines the algorithm of evaluating the selection coefficient.

Inputs: Motif $\mathcal{M}$, phylogenetic tree $\mathcal{T} = (V_T, E_T)$ of k species, n orthologous families of aligned promoter sequences, s_{ij} denotes the aligned sequence of gene i in species j.
Outputs: Selection coefficient of $\mathcal{M}$ on the aligned sequences.

1. Reconstruct the ancestral sequences of the internal nodes using equation (6).
2. Infer the Poisson rate λ using equation (5).
3. Split a promoter sequence into multiple segments of fixed length $l_s = 30l$.
4. Count the motif occurrence in each segment of both terminal and internal species.
5. Count the empirical conditional frequency along each branch.
6. Apply binary search to find the selection coefficient that maximizes the log likelihood score.

Fig. 1. Evaluating the selection coefficient of a motif

The inputs of the algorithm are the phylogenetic tree $\mathcal{T} = (V_T, E_T)$ of k species, n orthologous families of aligned sequences, and a sequence motif $\mathcal{M}$. We first apply dynamic programming to reconstruct sequences of internal nodes of $\mathcal{T}$ and infer the Poisson rate λ using equation (5). For simplicity we assume the sequence substitution rates of all families are identical.

The birth-death models in equations (11) and (10) apply to sequences of any lengths as long as $l_s \gg l$. In practice, longer sequences are computationally challenging for the following reasons. First, due to frequent recombinations, insertions and deletions, more gaps will appear in a long stretch of aligned sequences. Gaps add complexities in evaluating likelihood scores hence are undesirable. Second, longer sequences accommodate more motif instances by random sequence substitution. Hence more terms in equation (12) need to be considered. We divide the promoter sequence into segments of length $l_s = 30l$. A segment with more than 10% gaps in a species is treated as a missing data and discarded.

Motif occurrences in a segment are counted by sliding a window of length l along the segment. Denote s_{ijk} the aligned sequence of the kth segment of gene i in species j, and n_{ijk} the motif count of the corresponding segment. The motif occurrences of internal nodes can be inferred from their reconstructed sequences.

The joint log likelihood of the observed and reconstructed motif counts is

$$\mathcal{L} = \sum_i \sum_k \sum_{(v,w)\in E_T} \log P(n(t_{(v,w)}) = n_{iwk} | n(0) = n_{ivk}) + C. \quad (14)$$

where summation is over indices of gene i, segment k and edge (v,w) in $\mathcal{T}$. $t_{(v,w)}$ denotes the branch length of edge (v,w). Resembling EM, our method fills the missing data of internal nodes with reconstructed motif counts and avoids the cumbersome evaluation of the marginal likelihood.

The log likelihood can also be expressed as

$$\mathcal{L} = \sum_t \sum_{n_0} \sum_{n_1} f(t, n_0, n_1) \log P(n(t) = n_1 | n(0) = n_0) + C. \quad (15)$$

where $f(t, n_0, n_1)$ denotes the frequency of the instances where the motif counts in the parent and child nodes are n_0 and n_1 and the branch length is t. These empirical frequencies can be directly obtained from the observed and reconstructed data. $P(n(t) = n_1|n(0) = n_0)$ is the conditional probability derived from the birth-death model of the neutral or selective evolution (equations (12) and (13)). In this work we solve the transient responses $P_n(t)$ numerically by simulating the differential-difference equations. Given the relatively short segments ($30l$) only the first few equations in equation (12) are needed.

The only free parameter of the log likelihood is the selection coefficient s. We want to find the s that maximizes equation (15). Because s is integrated in equation (15) in a complex form and $P(n(t) = n_1|n(0) = n_0)$ has no analytic solutions, we apply a binary search to find the optimum value of s over the interval $[0, 20]$.

3 Results

3.1 The Birth-Death Model Agrees with Simulation Data

We first verified that the birth-death model approximated the motif count distribution derived from a Poisson sequence substitution process with simulation data. A random 100-base initial sequence and 20 random 4-base motifs were constructed. 1000 instances with the identical initial sequence underwent independent Poisson sequence substitutions with $\lambda = 0.2$ and $T = 4.0$. The empirical data were compared to the conditional probabilities predicted by equations (12) and (13). The birth-death model strongly agreed with the simulated data. Figure 2 shows the time evolution of motif count distributions of 3 motifs with 0, 1 and 2 instances in the initial sequence respectively. The predicted models (dashed lines) closely follow the empirical distributions (solid lines) in each case. The results indicate the birth-death model accurately describes motif count distributions in a Poisson process of sequence substitution.

3.2 Sequence Substitutions on Mammalian Promoters Follow a Poisson Process

Aligned 5kb upstream sequences of 27667 orthologous gene families from 34 mammalian species were extracted from the UCSC Genome Browser [12]. The maximum likelihood rate of the Poisson process was obtained from the procedures described in the Method Section. Figure 3 compares the distributions of sequence substitutions from the data and the Poisson process with the maximum-likelihood rate $\lambda = 0.8937$. For each position, the empirical number of sequence substitutions between two species is the number of sequence changes along the path connecting the two species in the phylogenetic tree. The predicted Poisson distributions closely resembled the empirical distributions at various time intervals, suggesting that promoter sequence substitutions in mammals follow a Poisson process.

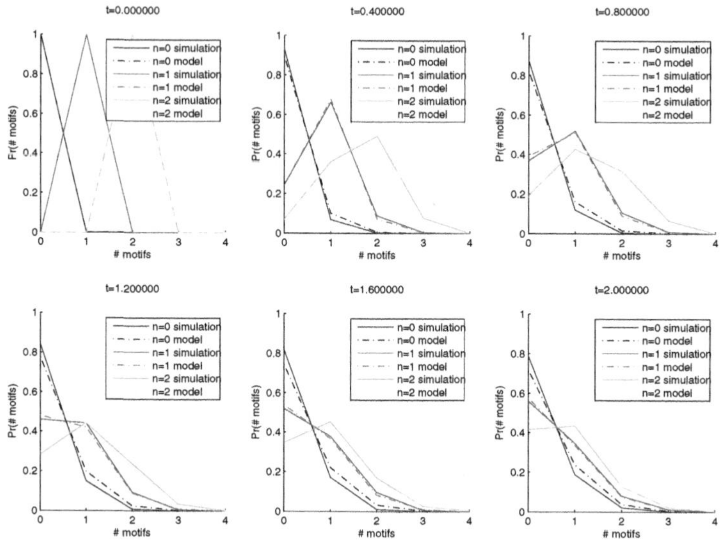

Fig. 2. Comparison of empirical and predicted distributions of motif counts in simulated data. Conditional probabilities of motif counts at 6 time points are shown. Solid lines indicate empirical distributions of 3 motifs with 0 (left bump), 1 (middle bump) and 2 (right bump) instances in the initial sequence respectively. Dashed lines are the distributions derived from equation (12).

3.3 Known Transcription Factor Binding Motifs Have Higher Selection Coefficients Than Random Sequences with High Occurrence Frequencies

388 transcription factor binding motifs were extracted from the TRANSFAC database [13]. Motif lengths ranged from 5 to 15 nucleotides and the mean length was 10.66 nucleotides. We applied the algorithm in Figure 1 to evaluate the selection coefficient of each motif on the 5kb promoters of 27667 orthologous families in 34 mammals. In addition to selection coefficients, we also evaluated the magnitudes of conservation by counting the fractions of species containing the motifs among 34 mammals and averaging the scores over all segments in all the gene families. As a negative control we generated 10000 random motif sequences (in IUPAC format) of 5 and 10 nucleotides and selected the top 500 sequences according to their occurrence frequencies on mammalian promoters. The left diagram of Figure 4 shows the distributions of selection coefficients in TRANSFAC motifs and two negative control sets. Intriguingly, there are many more high-scoring TRANSFAC motifs than the negative controls. About one quarter (96 of 388) of known motifs have selection coefficients ≥ 4.0. In contrast, only 11 and 24 of 500 5-mer and 10-mer control motifs pass the same threshold. The fraction of high-scoring motifs may be over-estimated as some transcription factors possess multiple similar motifs. By grouping motifs by their transcription factors, the same conclusion was reached (results not shown).

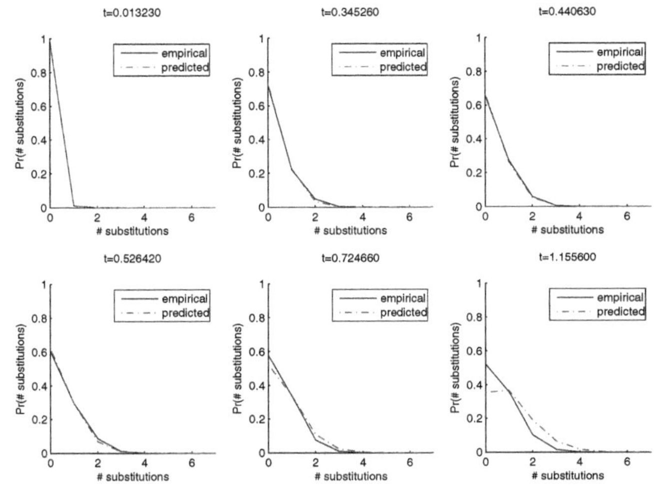

Fig. 3. Sequence substitutions on 5kb promoters of mammalian genes. Empirical distributions of sequence substitutions (solid lines) are obtained by the numbers of sequence changes along the paths connecting each pair of species in each position. Predicted distributions (dashed lines) are calculated by a Poisson process with $\lambda = 0.8937$. The distributions at 6 time intervals are shown.

The right diagram of Figure 4 shows the distributions of conservation magnitudes in TRANSFAC motifs and two negative control sets. The conservation mangitude of a motif on a promoter is the fraction of the species containing the motif. We report the average of the conservation magnitudes over the genes where the motif appears at least in one species. Clearly, natural selection is not revealed by conservation alone, as the conservation magnitudes of most TRANSFAC motifs are smaller than those of the control motifs. Moreover, unlike selection coefficients conservation magnitudes of control motifs are sensitive to their lengths. The results are sensible in two aspects. First, the "random motifs" in Figure 4 are the sequences with high occurrence frequencies. They often contain multiple degenerate sequences and are thus expected to appear in more species by chance. The birth-death model can eliminate these spurious motifs as the volumes of motif sequences are taken into account. Second, conservation of motifs is sensitive to sequence length as short sequences are likely to appear in more species by chance. The birth-death model also takes sequence length into account. Therefore, the selection coefficient distributions of 5-mer and 10-mer control motifs are similar.

These sequences in the negative control set occur more frequently on mammalian promoters thus are more likely to be conserved by chance. Alternatively, we also selected random sequences without ranking them by occurrence frequencies as the negative controls. The distributions of selection coefficients and conservation scores are similar to Figure 4. Thus the superiority of the selection coefficients in detecting TRANSFAC motifs sustains.

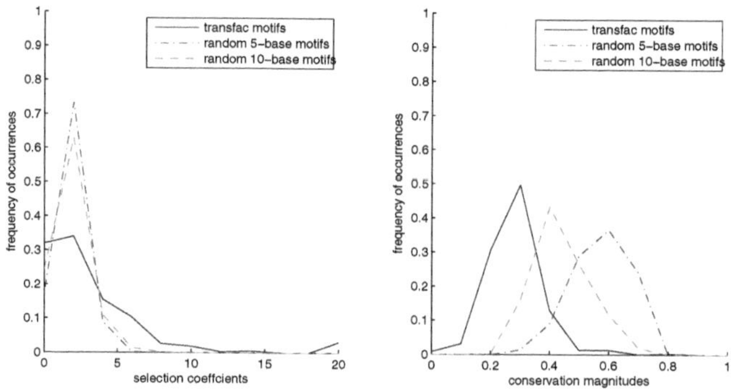

Fig. 4. Left: Distributions of selection coefficients of 388 TRANSFAC motifs (solid), 500 frequent 5-mer random motifs (dashed), and 500 frequent 10-mer random motifs (broken). Right: Distributions of conservation magnitudes of 388 TRANSFAC motifs (solid), 500 frequent 5-mer random motifs (dashed), and 500 frequent 10-mer random motifs (broken).

4 Discussions

In this work we propose a model and an algorithm to evaluate the strength of natural selection of a sequence motif from aligned sequences across gene families and species. The neutral model of motif occurrence distributions is based on a simple assumption that each position undergoes an independent Poisson process of sequence substitution. We consequently derive a birth-death model of motif occurrences according to the sequence substitution rate, motif sequence degeneracy and total sequence length. The selection coefficient is the penalty on the rate of motif deletion in the birth-death model. Predictions derived from the neutral model fit both simulated data and the statistics of random motifs on the aligned promoters of 34 mammals. In addition, many more known transcription factor binding motifs have high selection coefficients relative to negative controls, suggesting many of them are under purifying selection.

Despite the success in the preliminary study the current model and algorithm have several limitations. First, the model (and many other models of natural selection) focuses on sequence substitution and does not take other types of mutations – insertions/deletions, recombinations – into account. This simplification yields false negatives when there are gaps in the motifs. Second, the model considers a ubiquitous selection along each branch of phylogeny and discards lineage specific selection. Third, the model also discards the gene-specific selection on promoters and only considers the overall effects on all gene families. Fourth, numerical simulations and binary search of the algorithm are time-consuming. Analytic approximations to the transient responses of the birth-death model should be developed. Fifth, the inverse problem of this work – identify the motif sequences with high selection coefficients – is yet to be tackled. In spite of these limitations our model serves as a reasonable tool to validate the computationally or experimentally discovered candidate motifs.

References

1. Kellis, M., Patterson, N., Endrizzi, M., Birren, B., Lander, E.S.: Sequencing and comparison of yeast species to identify genes and regulatory motifs. Nature 423, 241–254 (2003)
2. Yang, Z., Bielawski, J.P.: Statistical methods for detecting molecular adaptation. Trends of Ecology and Evolution 15, 496–503 (2000)
3. Siepel, A., Haussler, D.: Combining phylogenetic and hidden Markov models in biosequence analysis. Journal of Computational Biology 11(2-3), 413–428 (2004)
4. McDonald, J.H., Kreitman, M.: Adaptive evolution at the Adh locus in Drosophila. Nature 351, 652–654 (1991)
5. Tajima, F.: Statistical method for testing the neutral mutation hypothesis by DNA polymorphism. Genetics 123, 585–595 (1989)
6. Atwal, G.S., Bond, G.L., Metsuyanim, S., et al.: Haplotype structure and selection of the MDM2 oncogene in humans. Proceedings of National Academy of Science USA 104, 4525–4529 (2007)
7. Raijman, D., Shamir, R., Tanay, A.: Evolution and selection in yeast promoters: analyzing the combined effect of diverse transcription factor binding sites. PLoS Computational Biology 4, 77–87 (2008)
8. Zuckerandl, E., Pauling, L.: Evolutionary divergence and convergence in proteins. In: Bryson, V., Vogel, H.J. (eds.) Evolving genes and proteins, pp. 97–166. Academic Press, New York (1965)
9. Felsenstein, J.: Evolutionary trees from DNA sequences: a maximum likelihood approach. Journal of Molecular Evoution 17, 368–376 (1981)
10. Kendall, D.G.: On the generalized birth-death process. The Annals of Mathematical Statistics 19(1), 1–15 (1948)
11. Bird, A.P.: CpG islands as gene markers in the vertebrate nucleus. Trends in Genetics 3, 342–347 (1987)
12. Kuhn, R.M., Karolchik, D., Zweig, A.S., et al.: The UCSC Genome Browser Database: update 2009. Nucleic Acids Research, D755–D761 (2009)
13. Matys, V., Fricke, E., Geffers, R., Gossling, E., Haubrock, M., et al.: TRANSFAC: transcriptional regulation, from patterns to profiles. Nucleic Acids Research 31(1), 374–378 (2003)

Author Index

GPSR Compliance

The European Union's (EU) General Product Safety Regulation (GPSR) is a set of rules that requires consumer products to be safe and our obligations to ensure this.

If you have any concerns about our products, you can contact us on ProductSafety@springernature.com

In case Publisher is established outside the EU, the EU authorized representative is:

Springer Nature Customer Service Center GmbH
Europaplatz 3
69115 Heidelberg, Germany

Batch number: 10584053

Printed by Printforce, the Netherlands